Textbook of Microbiology

Textbook of Microbiology

Editor: Ralph Becker

www.callistoreference.com

Callisto Reference,
118-35 Queens Blvd., Suite 400,
Forest Hills, NY 11375, USA

Visit us on the World Wide Web at:
www.callistoreference.com

ISBN: 978-1-63239-918-2 (Hardback)

Cataloging-in-Publication Data

Textbook of microbiology / edited by Ralph Becker.
     p. cm.
Includes bibliographical references and index.
ISBN 978-1-63239-918-2
1. Microbiology. 2. Molecular biology. I. Becker, Ralph.
QR41.2 .T49 2018
579--dc23

Table of Contents

Preface

The study of microorganisms, their interaction with each other, their environment and their hosts is known as microbiology. Research in this field deals with unicellular, multicellular and acellular organisms, such as bacteria, fungi, protozoa and viruses. It further branches out into sub-fields like parasitology, virology, etc. Microbiology, as a field, contributes to the advancements in the fields of biochemistry, genetic engineering and medical microbiology. It also plays a significant role in the food industry. This book is a compilation of chapters that discuss the most vital concepts and emerging trends in the field of microbiology. Coherent flow of topics, student-friendly language and extensive use of examples make this book an invaluable source of knowledge.

Various studies have approached the subject by analyzing it with a single perspective, but the present book provides diverse methodologies and techniques to address this field. This book contains theories and applications needed for understanding the subject from different perspectives. The aim is to keep the readers informed about the progresses in the field; therefore, the contributions were carefully examined to compile novel researches by specialists from across the globe.

Indeed, the job of the editor is the most crucial and challenging in compiling all chapters into a single book. In the end, I would extend my sincere thanks to the chapter authors for their profound work. I am also thankful for the support provided by my family and colleagues during the compilation of this book.

Editor

Cell-Free Culture Supernatant of *Bifidobacterium breve* CNCM I-4035 Decreases Pro-Inflammatory Cytokines in Human Dendritic Cells Challenged with *Salmonella typhi* through TLR Activation

Miriam Bermudez-Brito[1], **Sergio Muñoz-Quezada**[1], **Carolina Gomez-Llorente**[1], **Esther Matencio**[2], **Maria J. Bernal**[2], **Fernando Romero**[2], **Angel Gil**[1]*

1 Department of Biochemistry and Molecular Biology II, Institute of Nutrition and Food Technology "José Mataix", Biomedical Research Center, University of Granada, Granada, Spain, **2** Global Centre for Child Nutrition Technology, Hero Group, Alcantarilla, Murcia, Spain

Abstract

Dendritic cells (DCs) constitute the first point of contact between gut commensals and our immune system. Despite growing evidence of the immunomodulatory effects of probiotics, the interactions between the cells of the intestinal immune system and bacteria remain largely unknown. Indeed,, the aim of this work was to determine whether the probiotic *Bifidobacterium breve* CNCM I-4035 and its cell-free culture supernatant (CFS) have immunomodulatory effects in human intestinal-like dendritic cells (DCs) and how they respond to the pathogenic bacterium *Salmonella enterica* serovar *Typhi*, and also to elucidate the molecular mechanisms involved in these interactions. Human DCs were directly challenged with *B. breve*/CFS, *S. typhi* or a combination of these stimuli for 4 h. The expression pattern of genes involved in Toll-like receptor (TLR) signaling pathway and cytokine secretion was analyzed. CFS decreased pro-inflammatory cytokines and chemokines in human intestinal DCs challenged with *S. typhi*. In contrast, the *B. breve* CNCM I-4035 probiotic strain was a potent inducer of the pro-inflammatory cytokines and chemokines tested, i.e., TNF-α, IL-8 and RANTES, as well as anti-inflammatory cytokines including IL-10. CFS restored TGF-β levels in the presence of *Salmonella*. Live *B.breve* and its supernatant enhanced innate immune responses by the activation of TLR signaling pathway. These treatments upregulated *TLR9* gene transcription. In addition, CFS was a more potent inducer of *TLR9* expression than the probiotic bacteria in the presence of *S. typhi*. Expression levels of *CASP8* and *IRAK4* were also increased by CFS, and both treatments induced *TOLLIP* gene expression. Our results indicate that the probiotic strain *B. breve* CNCM I-4035 affects the intestinal immune response, whereas its supernatant exerts anti-inflammatory effects mediated by DCs. This supernatant may protect immune system from highly infectious agents such as *Salmonella typhi* and can down-regulate pro-inflammatory pathways.

Editor: Fabrizio Mattei, Istituto Superiore di Sanità, Italy

Funding: This study was supported by Hero Spain S. A. through a number 3143 contract signed with the Fundación General Universidad de Granada Empresa and co-sponsored by a CDTI project, a public entity of the Ministry of Economy and Competitiveness of the Spanish Government. CGLL is a recipient of a postdoctoral fellowship of Plan Propio of University of Granada. The funders had no role in study design, data collection and analysis, decision to publish, or preparation of the manuscript.

Competing Interests: Esther Matencio, Maria J. Bernal, and Fernando Romero are members of the Hero Institute for Infant Nutrition, Hero Spain S. A. The sponsor had no role in the biological sample analysis, statistical analysis or data interpretation.

* E-mail: agil@ugr.es

Introduction

Probiotic bacteria including lactobacilli and bifidobacteria are part of a normal intestinal microbiota in humans and generally considered as potentially beneficial to various aspects of host metabolism [1]. *Bifidobacterium* sp. are among the most relevant probiotic microorganisms because they colonize the intestinal tract soon after birth, are present at high levels in the guts of infants and adults and promote beneficial effects on intestinal ecology and immune responses [2,3]. Several mechanisms for the favorable influence of probiotic bacteria on the intestinal mucosa have been suggested including the secretion of antimicrobial products, resistance to pathogen colonization, barrier function enhancement and maintenance, modulation of epithelial cell signal transduction and innate and adaptive immunomodulation [3,4]. The beneficial effects of specific probiotic strains have been established for the treatment and prevention of many diseases [5], including diarrhea [6], the alleviation of lactose intolerance [7] and postoperative complications [8], antimicrobial [9] and anticolorectal cancer activity [10,11] and for reducing irritable bowel symptoms [12] and increasing the relapse time for some inflammatory bowel diseases [13].

The probiotic properties of commensal bacteria including lactobacilli and bifidobacteria are likely to be determined at least in part by their effects on dendritic cells (DCs) [1], a complex, heterogeneous group of multifunctional antigen-presenting cells (APCs) that comprise a critical arm of the immune system [14,15].

These cells play a critical role in the orchestration of the adaptive immune response by inducing tolerance and adaptive immunity. Understanding the direct interaction between commensal bacteria and DCs is particularly important to know how the immune system of the gut is locally able to distinguish these bacteria from pathogens and to elicit a tolerogenic response [16]. The primary response to these bacteria is triggered by the innate pattern recognition receptors (PRRs), which bind pathogen-associated molecular patterns (PAMPs). PRRs comprise Toll-like receptors (TLRs), nucleotide-binding oligomerization domain (NOD)-like receptors (NLRs), adhesion molecules and lectins [4,17]. The binding of microbe-associated molecules to these receptors can activate APCs and initiate a signaling transduction cascade that leads to the release of cytokines and initiation of the acquired immune response [18].

Mucosal DCs appear to have unique properties that distinguish them from peripheral DCs [19]. However, to date, probiotic activity has often been tested in monocyte-derived DCs (MoDCs) or murine DCs, which are quite different from human gut DCs [20]. For this reason, in this study, we used intestinal-like human DCs that were developed from umbilical cord blood CD34+ progenitor cells. These human DCs are Langerhans-like cells that extend dendritic processes and sample antigens similarly to the lamina propria DCs in the gut that sample luminal antigens [21].

We have previously reported some of the probiotic properties of *Bifidobacterium breve* CNCM I-4035, a novel bifidobacteria strain isolated from the feces of newborns that were exclusively breast-fed [22,23]. In the present work, we studied the immunomodulatory effects of *B.breve* CNCM I-4035 and its cell-free culture supernatant (CFS) on human intestinal-like DCs and how the treated DCs interact and respond to the pathogenic bacteria *Salmonella enterica* serovar *Typhi* at molecular level.

Materials and Methods

Ethic Statement

The ethical Committee of Granadas University approved this study.

Bifidobacterium breve was obtained from feces of breast-fed newborns, in a previous work [24]. Briefly, 12 healthy, exclusively breast-fed infants, aged 1 month, were selected for the study at the Clinic Hospital of the University of Granada. This study was conducted according to the guidelines laid down in the Declaration of Helsinki and all procedures involving human subjects were approved by the Ethical Committee of the University of Granada. Written informed consent was obtained from the parents after a careful explanation of the nature of the study.

Preparation of bacteria and cell-free culture supernatant

B. breve CNCM I-4035was isolated from the feces of breast-fed newborns and previously selected for its *in vitro* probiotic characteristics [22,23]. *B. breve* CNCM I-4035 was routinely anaerobically cultured for 24 hours at 37°C in de Man-Rogosa-Sharpe (MRS) broth medium (Oxoid, Basingstoke, United Kingdom) supplemented with 0.05% (wt/vol) cysteine (Sigma-Aldrich, St. Louis, MO) to promote the growth of *B. breve*. The supernatant of the culture medium was collected by centrifugation at 12,000×g for 10 min, neutralized to pH 7.0 by the addition of 1 N NaOH and concentrated ten-fold by lyophilization. The supernatants were passed through a 0.22-μm pore size filter unit (Minisart hydrophilic syringe filter; Sartorius Stedim Biotech GmbH, Goettingen, Germany) and stored at −20°C until use. The supernatant was added to the DC culture medium at a concentration of 7% v/v.

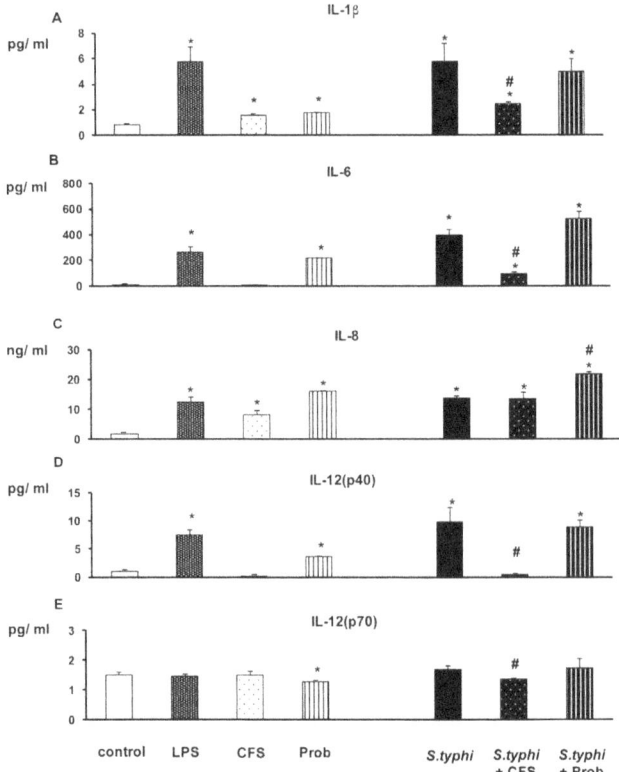

Figure 1. Effects of *B. breve* CNCM I-4035 and its cell-free culture supernatant on the secretion of pro-inflammatory cytokines by intestinal-like human dendritic cells. Dendritic cells (DCs) were incubated for 4 h with the *B. breve* CNCM I-4035 probiotic (Prob) or its cell-free supernatant (CFS), *Salmonella* (Sal) or both and further incubated for 20 h in medium containing antibiotics. *E. coli* lipopolysaccharide (LPS; 20 ng/ml) was used as a positive control. Negative-control cultures contained unstimulated DCs. Culture supernatants were collected, and the cytokine levels were assessed using an immunoassay. The production of IL-1β, IL-6, IL-8, IL-12(p40) and IL-12(p70) was measured. The data shown are the mean value ± SEM for three independent experiments. *, p<0.05 compared with the negative control; #, p<0.05 compared with *S. typhi*; N.D. indicates not detected.

Salmonella enterica serovar *Typhi* CECT 725 was provided by the Spanish Type Culture Collection (CECT; Burjassot, Spain) and aerobically cultured in tryptone soy broth (Panreac Química, Barcelona, Spain).

For experiments, *S. typhi* was cultured for 8 h at 37°C in tryptone soy broth and then subcultured 1:500 in RPMI 1640 medium (Sigma-Aldrich) containing 10% fetal bovine serum (FBS; Gibco Invitrogen, Paisley, United Kingdom) at 37°C overnight.

Cell preparation

DCs generated from umbilical cord blood CD34+ progenitor cells (hematopoietic stem cells) were supplied by MatTek Corporation (Ashland, MA). These cells were seeded in 24-well plates in DC maintenance medium (DC-MM; MatTek) containing cytokines and antibiotics.

Bacterial co-culture and DC stimulation

Cell cultures were seeded in 24-well plates at a density of 2×10^5 DCs/well. For incubations, DC-MM was replaced with RPMI-1640 medium. DCs were co-incubated with *B. breve* CNCM I-4035 bacteria (10^7 CFU/ml) or CFS as well as *S. typhi* (10^6 CFU/ml) or

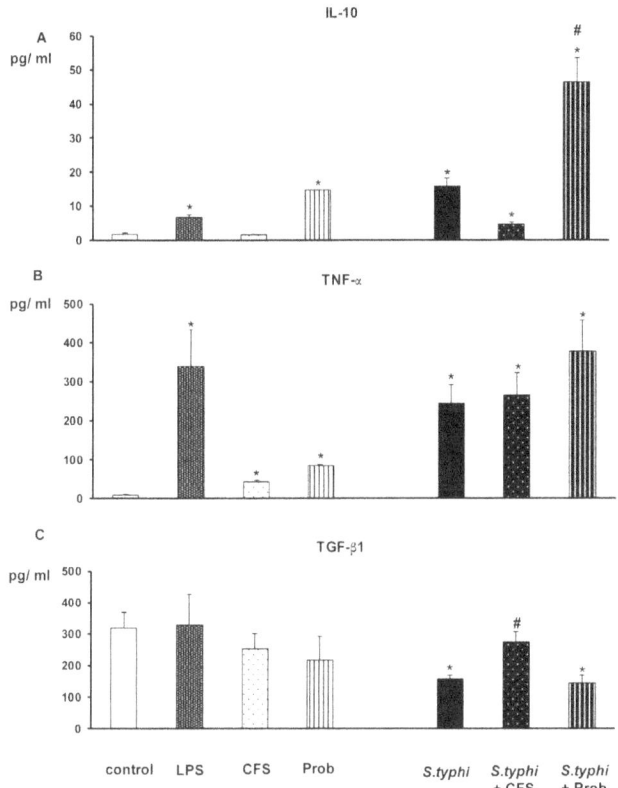

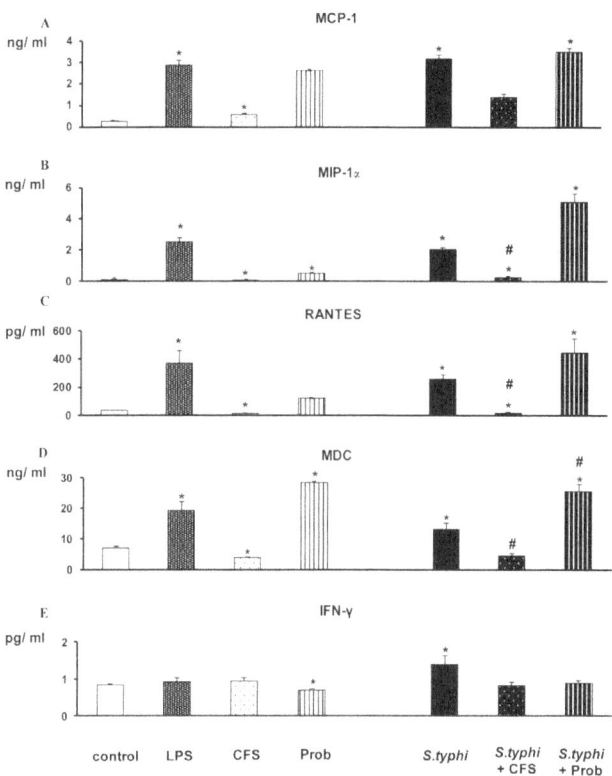

Figure 2. Measurement of anti-inflammatory cytokines and TNF-α in DCs after exposure to *B. breve*, *Salmonella* **or a combination of the two.** Dendritic cells (DCs) were incubated for 4 h with the *B. breve* CNCM I-4035 (Prob) probiotic or its cell-free supernatant (CFS), *Salmonella* (Sal) or both and then incubated for 20 h in medium containing antibiotics. *E. coli* lipopolysaccharide (LPS; 20 ng/ml) was used as a positive control. Negative-control cultures contained unstimulated DCs. Culture supernatants were collected, and the cytokine levels were assessed by an immunoassay in which the production of IL-10, TNF-α, TGF-β1 and TGF-β2 was measured. The data shown are the mean value ± SEM of three independent experiments. *, p<0.05 compared with controls; #, p<0.05 compared with *S. typhi*; N.D. indicates not detected.

Figure 3. Measurement of chemokines and IFNγ in DCs after exposure to *B. breve*, *Salmonella* **or a combination of the two.** Dendritic cells (DCs) were incubated for 4 h with the *B. breve* CNCM I-4035 (Prob) probiotic or its cell-free supernatant (CFS), *Salmonella* (Sal) or a combination of the two and further incubated for 20 h in medium containing antibiotics. *E. coli* lipopolysaccharide (LPS; 20 ng/ml) was used as a positive control. Negative control cultures contained unstimulated DCs. Culture supernatants were collected, and the cytokine and chemokine levels were assessed by an immunoassay. The production of IFNγ and the chemokines MCP-1/CCL2, MIP-1α/CCL3, RANTES/CCL5 and MDC/CCL22 was measured. The data shown are the mean values ± SEM of three independent experiments. *, p<0.05 compared with controls; #, p<0.05 compared with *S.typhi*; N.D. indicates not detected.

a combination of these treatments for 4 h at 37°C in a 5% CO₂/ 95% air atmosphere. After incubation, the DCs were washed with PBS, and DC-MM (containing cytokines and antibiotics) was added to the wells and incubated for an additional 20 h. Cell supernatants and cells were collected for cytokine analysis and RNA extraction, respectively. *Escherichia coli* lipopolysaccharide (LPS; Sigma-Aldrich) was applied at a concentration of 20 ng/ml as a positive control. Negative-control cultures contained unstimulated DCs.

Cytokine and chemokine quantification in culture supernatants

Cytokine production was measured by immunoassay with the MILLIplex^{TM} kit (Linco Research Inc., MO) using the Luminex 200 system according to the manufacturer's instructions. IL-1β, IL-6, IL-8, IL-10, IL-12(p40), IL-12(p70), TNF-α, IFN-γ, MCP-1/ CCL2, MIP-1α/CCL3, RANTES/CCL5, MDC/CCL22 and TGF-β were analyzed. We performed three independent experiments.

Reverse transcriptase (RT) reaction and polymerase chain reaction (PCR)

Total RNA was isolated from cells using the RNAqueous Kit (Ambion, Paisley, United Kingdom) and additional Turbo DNase treatment (Ambion) according to the manufacturer's recommendations. The RNA quality was verified using a Model 2100 Bioanalyzer (Agilent, Santa Clara, USA), and the RNA concentration was determined using a Rediplate 96 Ribogreen RNA Quantitation Kit (Gibco, Invitrogen). The total RNA was reversed-transcribed using an RT^2 First-strand Kit (SABiosciences Corporation, Frederick, MD). Real-time PCR was performed using an RT^2 Real-time PCR SYBR Green/ROX Kit (SABiosciences) on an ABI Prism 7500 sequence detector (Applied Biosystems, Foster City, CA). Real-time RT-PCR analysis of the samples was performed using a Human TLR Signaling Pathway PCR Array (SABiosciences), which includes primer pairs specific for the following 20 genes related to TLR-mediated signaling pathways: *TLR1, TLR2, TLR3, TLR4, TLR5, TLR9, MYD88, TNF-α, IRAK-1, IRAK-4, TOLLIP, CASP8, IL-10, TAK-1, JNK, NFKB1A, NFKB-1, TBK-1, MAPK14* and *IRF-3*. The housekeeping gene *GAPDH* was used as a control. The thermal profile for all reactions was: 1 cycle of 95°C for 10 min and 40 cycles of 95°C for 15 s and 60°C for 1 min. The expression levels of the target genes were normalized to those of untreated DCs (control).

Statistical analysis

The results shown are the mean ± SEM of three independent experiments.

The differences in cytokine levels and gene expression between treatments were compared using the Mann-Whitney U-test. Analyses were performed using NCSS 2007 software (Kaysville, UT). P<0.05 was considered significant and is indicated with an asterisk in the figures.

Differences between DCs treated with *Salmonella* and *Salmonella* plus *B. breve* CNCM I-4035 or its CFS were also evaluated. P<0.05 was considered to significant and is indicated in the figures with a pound sign (#).

Results

Supernatant of *B. breve* CNCM I-4035 decreases cytokine release in human DCs co-cultured with *S. typhi*

The addition of pathogenic bacteria (*S. typhi* CECT 725) or LPS to DCs markedly affected the expression of pro-inflammatory cytokines (Figures 1 and 2). These treatments lead to a strong secretion of IL-1β, IL-6, IL-8, IL-12p40 and TNF-α compared to the controls. Accordingly, as illustrated in figure 3, the release of MCP-1, MIP-1α, RANTES and MDC to the culture medium was significantly elevated by either *Salmonella* or LPS stimulation.

In DCs, *B.breve* CNCM I-4035 (live bacteria) and its CFS exerted different behaviors with regard to cytokine induction in response to *B. breve* CNCM I-4035 stimulation. The CFS decreased the release of pro-inflammatory cytokines (e.g., IL-6 and IL-12p40) and chemokines (e.g., RANTES/CCL5 and MIP-1α/CCL3) in human intestinal DCs challenged with *S. typhi* (Figures 1 to 3). In contrast, the *B. breve* CNCM I-4035 strain (live bacteria) was a potent inducer of pro-inflammatory cytokines (e.g., IL-8 and IL-6; Figures 1 and 2), chemokines (e.g., MDC/CCL22; Figure 3) and some anti-inflammatory cytokines (e.g., IL-10; Figure 2). Moreover, DCs interacting with the CFS, in absence of pathogenic bacteria, released low amounts of pro-inflammatory cytokines (e.g., IL-6 and IL-12p40) and chemokines (e.g., MDC and RANTES). In contrast, *B.breve* (live bacteria) stimulation increased overall cytokine and chemokine production, namely IL-6, IL-8 and MDC (Figure 1 to 3). In addition that treatment also produced high levels of IL-10 (Figure 2).

As shown in Figure 2, the *Bifidobacterium* strain was a potent TGF-β1 inducer. Interestingly, CFS restored TGF-β levels in the presence of *Salmonella*. In contrast, live *B.breve* was unable to increase TGF-β1 production. Finally, we did not detect TGF-β2 and TGF-β3 expression.

The effects of live *B.breve* CNCM I-4035 or its CFS, *S.typhi*, or a combination of the two on the production of pro-inflammatory cytokines, anti-inflammatory cytokines and chemokines by intestinal-like human DCs are summarized in tables 1, 2 and 3, respectively. The data shown are the mean value ± SEM of three independent experiments.

Differences between *B. breve* CNCM I-4035 and its supernatant were observed in the induction of TLR signaling pathway components in human DCs, particularly TLR9 expression

S. typhi induced the expression of other TLR genes including *TLR1*, *TLR2* and *TLR5* (Figure 4) and upregulated *TLR9* gene

Table 1. Effects of live *B.breve* CNCM I-4035 (Prob) or its cell-free culture supernatant (CFS), *S.typhi*, or a combination of the two on the secretion of IL-1β, IL-6, IL-8, IL-12p40 and IL-12p70 by intestinal-like human dendritic cells.

Treatment	IL-1β	IL-6	IL-8	IL-12p40	IL-12p70
Control	0,83± 0,05	1,46± 0,08	1629± 177,0	0,9± 0,36	1,49± 0,09
DCs + CFS	1,58± 0,08	10,2± 1,89	8214± 1405	0,14± 0,00	1,50± 0,12
DCs + Prob	1,81± 0,00	220,5± 0,10	16193± 125,6	3,65± 0,00	1,27± 0,05
DCs+ S.typhi	5,76± 1,43	396,8± 42,49	13794± 553,3	9,82± 2,54	1,69± 0,11
DCs+ S.typhi+ CFS	2,49± 0,11	96,1± 10,11	13573± 2131	0,43± 0,29	1,37± 0,02
DCs+ S.typhi+Prob	5,00± 0,97	526,4± 50,8	21936± 650,5	8,89± 1,23	1,74± 0,30

The data shown are the mean value ± SEM of three independent experiments.

expression in human DCs (Figure 5) and. A similar effect was observed for *IRAK4*, *TAK1*, *JNK* (Figures 5 and 6) and *IL-10* (Figure 7).

Differences between the probiotic bacteria *B. breve* CNCM I-4035 and its CFS were observed with regard to TLR expression in DCs (Figures 4 and 5). Both stimuli induced strong *TLR9* transcription. In addition, CFS was a more potent inducer of *TLR9* expression than live *B. breve* CNCM I-4035 in the presence of *S. typhi* (Figure 5). CFS and live *B. breve* CNCM I-4035 both induced strong and sustained *TLR2* transcription (Figure 4). The live bacteria upregulated *TLR4*, whereas CFS upregulated *TLR1* and *TLR5* (Figure 4).

Interestingly, in response to stimulation with strain *B. breve* CNCM I-4035 and *Salmonella*, *TLR1*, *TLR2* and *TLR5* expression was decreased, whereas exposure of the DCs to the probiotic and *Salmonella* upregulated *TLR3* gene expression (Figures 4 and 5). Upon stimulation with CFS plus *Salmonella*, the expression of the TLR genes increased (Figures 4 and 5). Both treatments induced the expression of *TOLLIP* (Figure 5), *JNK* (Figure 6), *TBK1* and *TNF-α* (Figure 7). We also observed differences between the treatments. CFS induced the expression of *IRAK4*, *MYD88*

Table 2. Effects of live *B.breve* CNCM I-4035 (Prob) or its cell-free culture supernatant (CFS), *S.typhi*, or a combination of the two on the secretion of IL-10, TNF-α and TGF-β1 by intestinal-like human dendritic cells.

Treatment	IL-10	TNF-α	TGF-β1
Control	1,75±0,02	9,82±1,28	319,2±48,4
DCs + CFS	1,66±0,05	42,6±3,31	252,1±50,6
DCs + Prob	14,9±0,00	85,1±1,11	216,2±77,2
DCs+S.typhi	15,8±2,46	244,3±47,9	157,9±11,5
DCs+S.typhi+CFS	4,70±0,67	264,9±57,5	273,9±33,3
DCs+S.typhi+Prob	46,5±7,10	377,7±80,8	143,5±25,4

The data shown are the mean value ± SEM of three independent experiments.

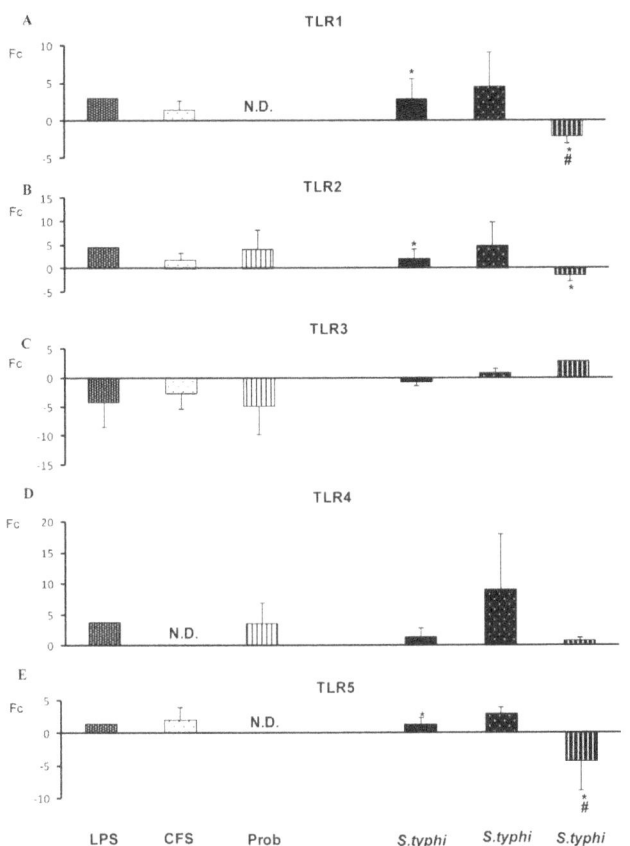

Figure 4. Expression of *TLR* genes in DCs in the presence of *B. breve*, *Salmonella* or a combination of the two. Comparison of the expression of *TLR1, TLR2, TLR3, TLR4* and *TLR5* in dendritic cells (DCs) in the presence of the probiotic (Prob), its supernatant (CFS), *Salmonella* (Sal) or a combination of these stimuli. *E. coli* lipopolysaccharide (LPS; 20 ng/ml) was used as a positive control. The fold change (Fc) represents the ratio of the expression in treated DCs to that in control cells. *, p<0.05 compared with controls; #, p<0.05 compared with *S. typhi*. N.D. indicates not detected.

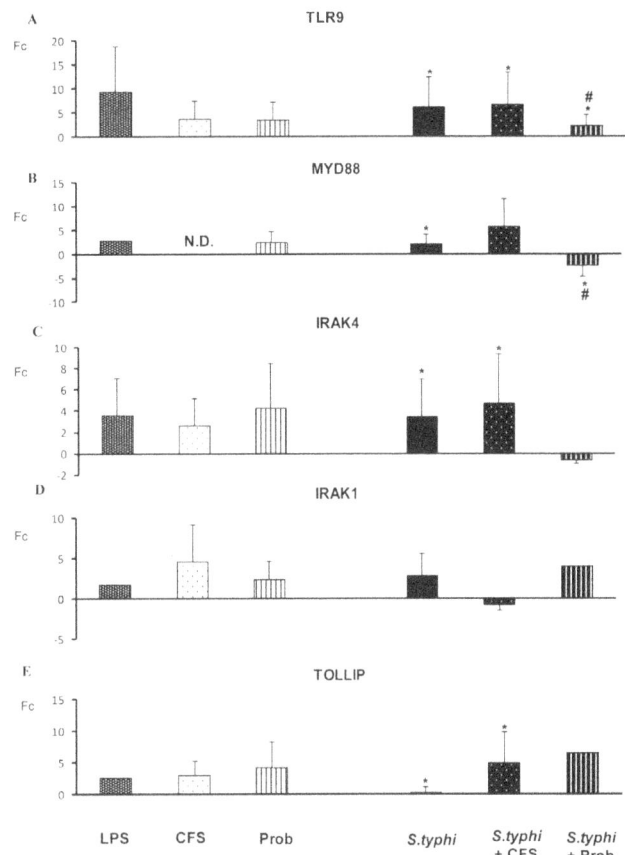

Figure 5. Expression of TLR signaling pathway components in DCs treated with *B. breve*, *Salmonella* or a combination of the two. Comparison of the expression of *TLR9, MYD88, IRAK-1, IRAK-4* and *TOLLIP* in dendritic cells (DCs) in the presence of the probiotic (Prob) *B. breve*, its supernatant (CFS), *Salmonella* (Sal) or a combination of these stimuli. *E. coli* lipopolysaccharide (LPS; 20 ng/ml) was used as a positive control. The fold change (Fc) represents the ratio of the expression in treated DCs to that in control cells. *, p<0.05 compared with controls; #, p<0.05 compared with *S. typhi*. N.D. indicates not detected.

Table 3. Effects of live *B.breve* CNCM I-4035 (Prob) or its cell-free culture supernatant (CFS), *S.typhi*, or a combination of the two on the secretion of chemokines MCP-1, MIP-1α, RANTES, MDC and IFN-γ by intestinal-like human dendritic cells.

Treatment	MCP-1	MIP-1α	RANTES	MDC	IFN-γ
Control	245,5± 8,5	131,8± 15,7	37,3± 2,4	6953± 589,6	0.84± 0,02
DCs + CFS	556,4± 94,0	50,3± 1,4	13,7± 1,5	3936± 105,7	0,95± 0,08
DCs + Prob	2624± 82,2	511,8± 30,8	120,3± 2,4	28472± 89,3	0,70± 0,00
DCs+ S.typhi	3201± 164,4	2043,1± 86,2	258,5± 28,8	13136± 2041	1,40± 0,22
DCs+ S.typhi+CFS	1372± 179,0	263,3± 29,4	19,2± 3,7	4535±8 11,0	0,83± 0,08
DCs+ S.typhi+Prob	3495± 185,0	5078± 535,1	442,4± 107,3	25523± 2433	0,90± 0,07

The data shown are the mean value ± SEM of three independent experiments.

(Figure 5) and *CASP8* (Figure 6), whereas these genes were downregulated by *B. breve* CNCM I-4035.

Discussion

Intestinal DCs, are known to sample microbes that continuously bombard the intestinal mucosa via PRRs such as TLRs and NLRs [25]. However, the underlying mechanisms are poorly understood. One main difficulty is the assessment of the interaction with intestinal immune system components, particularly DCs, which are key players in mucosal immunity [26]. A few studies have addressed the effects of bifidobacteria on human immunocompetent cells [27–29]; however, to the best of our knowledge, this is the first study to analyze the immune response to human intestinal-like DCs developed from CD34+ progenitor cells isolated from the umbilical cord blood.

The main finding of this study was that *B. breve* CNCM I-4035 and its supernatant could modify the release of cytokines by DCs in specific and differing manners. The CFS exhibited an anti-inflammatory behavior by decreasing pro-inflammatory cytokines (e.g., IL-6 and IL-12p40) and chemokines (e.g., RANTES/CCL5 and MIP-1α/CCL3) in DCs challenged with *S. typhi*. However, CFS did not increase IL-10 production. This observation is in

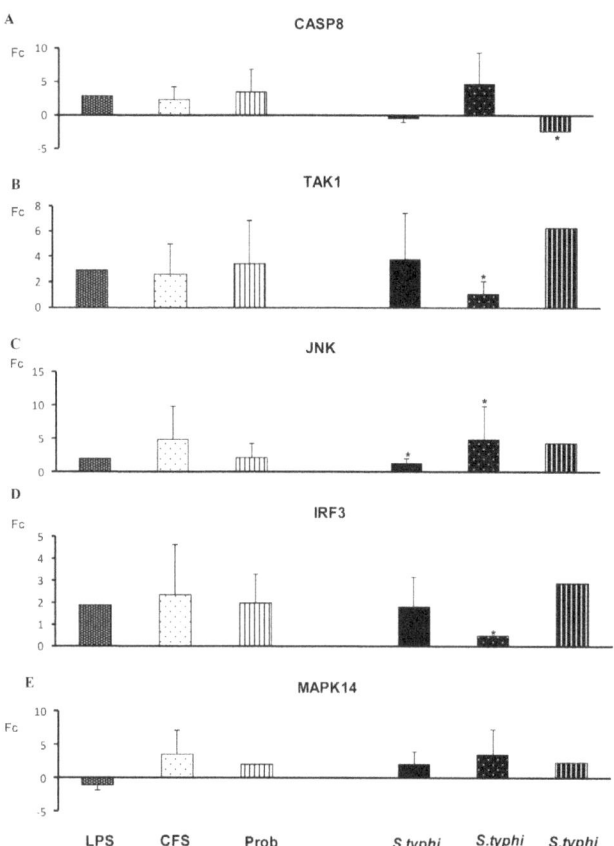

Figure 6. Expression of TLR signaling pathway components in DCs treated with _B. breve_, _Salmonella_ or a combination of the two. Comparison of the expression of _CASP8, TAK-1, JNK, IRF-3_ and _MAPK14_ in dendritic cells (DCs) in the presence of the probiotic (Prob) _B. breve_, its supernatant (CFS), _Salmonella_ (Sal) or a combination of these stimuli. _E. coli_ lipopolysaccharide (LPS; 20 ng/ml) was used as a positive control. The fold change (Fc) represents the ratio of the expression in treated DCs to that in control cells. *, p<0.05 compared with controls; #, p<0.05 compared with _S. typhi_. N.D. indicates not detected.

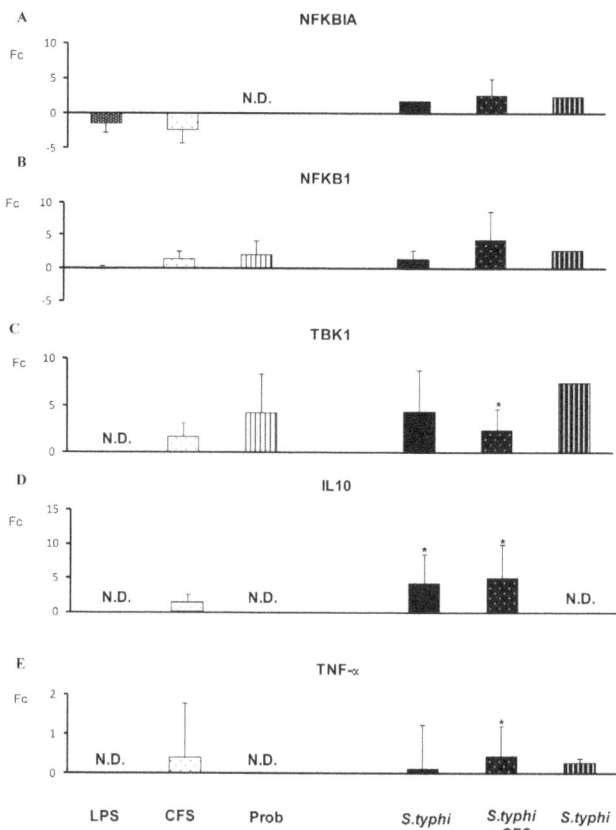

Figure 7. Expression of TLR signaling pathway components in DCs treated with _B. breve_, _Salmonella_ or a combination of the two. Comparison of the expression of _NFKBIA, NFKB-1, TBK-1, IL-10_ and _TNF-α_ in dendritic cells (DCs) in the presence of the probiotic (Prob) _B. breve_, its supernatant (CFS), _Salmonella_ (Sal) or a combination of these stimuli. _E. coli_ lipopolysaccharide (LPS; 20 ng/ml) was used as a positive control. The fold change (Fc) represents the ratio of the expression in the treated DCs to that in the control cells. *, p<0.05 compared with controls; #, p<0.05 compared with _S. typhi_. N.D. indicates not detected.

contrast with that of Hoarau _et al._, who reported that the _B. breve_ C50 supernatant induces high IL-10 levels in DCs [30]. It is important to note that the _B. breve_ supernatant by itself was a poor cytokine inducer (both inflammatory and non-inflammatory) in our study but had an important impact on the ability of human DCs to secrete lower levels of inflammatory cytokines in response to _Salmonella_, which suggests that this supernatant may have immunomodulatory properties and may be used to dampen inflammatory responses. These results are consistent with a recent study indicating similar effects for CFS from a novel probiotic strain isolated from the feces of newborns that were exclusively breast-fed (_Lactobacillus paracasei_ CNCM I-4034), i.e., decreased pro-inflammatory cytokines and chemokines in human DCs challenged with _S. typhi_ [23,31]. Altogether, these data indicate that soluble bacteria product(s) released by _B.breve_ CNCM I-4035 possess anti-inflammatory activity and should be identified. Work in progress in our lab using a proteomic view indicates that this bacterium secretes a number of proteins able to interact with the gut associated immune system (unpublished data).

In contrast to CFS, _B. breve_ CNCM I-4035 (live bacteria) was a potent inducer of the pro-inflammatory cytokines (e.g., IL-8) and chemokines tested (e.g., MDC/CCL22). Our results are in

agreement with several studies that reported that some members of the _Bifidobacterium_ genus are inducers of IL-6 [32] and, to a lesser degree, IL-12 [33]. Similarly to most _Bifidobacterium_ strains, live _B. breve_ CNCM I-4035 stimulated the production of high levels of IL-10 [34]. This increase may have an anti-inflammatory effect [35]. The co-incubation of DCs with live _B. breve_ and _S. typhi_ strongly increased the release of pro-inflammatory cytokines, particularly IL-6, IL-8 and TNF-α, as well as IL-10. IL-10 and TNF-α are pleiotropic cytokines that are produced by immune cells. These cytokines are mutually regulated and play opposing roles in inflammatory responses; therefore, their relative balance is of central relevance for controlling immune deviation [36,37]. We observed that IL-10 production was higher than TNF-α release; therefore, it appears that _B. breve_ CNCM I-4035 (live bacteria) promotes anti-inflammatory effects to restore homeostasis and prevents _Salmonella_-induced inflammation. In addition, it has been suggested that the high genomic cytosine and guanosine (CG) content of the _Bifidobacterium_ species (approximately 60%) increases IL-10 production [38]. Moreover, in line with other studies, live _B. breve_ CNCM I-4035 did not stimulate IL-12 p70, which is a characteristic effect of bifidobacteria. However, IL-12p40 expression was increased in the presence of _S. typhi_. This result was predictable because IL-12 induction is strongly correlated with TNF-α production [32].

TGF-β expression analysis demonstrated that the *Bifidobacterium* strain is a potent TGF-β1 inducer. CFS restored TGF-β levels in the presence of *Salmonella*. TGF-β1 is a pleiotropic cytokine known to inhibit immune responses at several levels including inhibition of T cell proliferation and differentiation [39,40] and inhibition of DC maturation [41]. Therefore, the elevation in TGF-β1 production could be responsible for the observed anti-inflammatory effects upon probiotic stimulation.

TLRs are pattern recognition receptors that recognize microbial components and initiate an innate immune response (4). *B. breve* and its supernatant possess different abilities to regulate the TLR signaling pathway. Our results demonstrate that live *B. breve* CNCM I-4035 and its CFS stimulated *TLR9* expression in the presence and absence of *Salmonella*. Plantinga *et al.* reported that cytokine induction by *B. breve* and lactobacilli is strongly dependent on TLR9 [42]. Their genomic DNA was identified as one of the anti-inflammatory components [43]. Ghadimi *et al.* reported that TLR9 signaling may at least in part mediate the anti-inflammatory effects of natural-commensal origin DNA [44]. Our results, consistent with several authors, indicate that TLR9 activation is one of the major pathways responsible for the anti-inflammatory effects of probiotics [43–45]. In our study, *TLR9* gene expression in response to the bacterial supernatant was significantly higher than that in response to *B. breve* CNCM I-4035 in DCs challenged with *Salmonella*. In consequence, this could explain the anti-inflammatory effects of the CFS compared with *B.breve* (live bacteria). Moreover, it has been proposed that the high frequency of CpG motifs in the DNA of the *Bifidobacterium* genus may play an important role in the immunostimulatory properties of commensal or probiotic bifidobacterial strains.

The strong upregulation of the *TLR2* gene in the presence of *B. breve* is not surprising because peptidoglycan and lipoteichoic acid, components of the cell wall of Gram-positive bacteria are TLR-2 ligands. A recent study reported that TLR2 recognition had the opposite effect of TLR9 recognition as it induced the expression of TNF-α, IL-1β and IFNγ [42]. This effect may explain the cytokine profile induced by the bacteria in the absence of *Salmonella*, which

is characterized by IL-10 and TNF-α secretion. Moreover, TLR2 has also been implicated in the induction of regulatory T-cell responses, which further emphasizes the immunosuppressive potential of TLR2 signaling. However, the involvement of TLR2 in this process remains unclear, although an immunoregulatory role of TLR2 in the recognition of probiotic strains has been described [30,46,47].

In line with several previous studies, our results suggest that probiotic live bacteria increase the expression of *TLR2* and *TLR9* to activate an innate immune response [48–51] and provide immunostimulation, whereas its CFS increases the expression of *TLR9* and *TLR5* to exert anti-inflammatory effects. In addition, Uematsu *et al.* [52] proposes that microbiota induce Ig A production through a mechanism mediated by TLR5. The major role of Ig A is to maintain a balance between the host and microbiota [53,54].

In contrast, in the presence of *Salmonella*, the CFS increased the expression of *CASP8* and *IRAK4*, whereas these genes were downregulated by the live bacteria. Both treatments induced *TOLLIP* gene expression. In this context, signal propagation is necessary to amplify the initiating signal to trigger the nuclear mobilization of transcription factors and induce gene expression. TOLLIP is an adaptor molecule that can bind to TLR2 and TLR4 to inhibit MyD88 binding and activation [55,56]. In addition, TOLLIP binds to and is phosphorylated by IRAK, which suppresses its ability to function in the TLR pathway [57].

Finally, our results coincide with those of another study indicating that live probiotic bacteria affect the intestinal immune response, whereas secreted components exert anti-inflammatory effects in the gastrointestinal tract [58]. This supernatant may protect immune system from highly infectious agents such as *Salmonella typhi* and can down-regulate pro-inflammatory pathways.

Author Contributions

Conceived and designed the experiments: AG CGL. Performed the experiments: MBB SMQ. Analyzed the data: MBB. Contributed reagents/materials/analysis tools: EM MJB FR. Wrote the paper: MBB AG.

References

1. Verbeek R, Bsibsi M, Plomp A, van Neerven RJ, te Biesebeke R, et al. (2010) Late rather than early responses of human dendritic cells highlight selective induction of cytokines, chemokines and growth factors by probiotic bacteria. Benef. Microbes. 1:109–19.
2. Guarner F, Malagueleda JR (2003) Gut flora in health and disease. Lancet. 361: 512–519.
3. Collado MC, Isolauri E, Salminen S, Sanz Y (2009) The impact of probiotic on gut health. Curr. Drug. Metab. 10: 68–78.
4. Bermudez-Brito M, Plaza-Díaz J, Muñoz-Quezada S, Gomez-Llorente C, Gil A (2012) Mechanisms of action of probiotics. Ann. Nutr. Metab. 61: 160–174.
5. Yan F, Polk DB (2011) Probiotics and immune health. Curr. Opin. Gastroenterol. 27: 496–501.
6. Huang JS, Bousvaros A, Lee JW, Diaz A, Davidson EJ (2002) Efficacy of probiotic use in acute diarrhea in children: a meta-analysis. Dig. Dis. Sci. 47: 2625–34.
7. Pelletier X, Laure-Boussuge S, Donazzolo Y (2001) Hydrogen excretion upon ingestion of dairy products in lactose-intolerant male subjects: importance of the live flora. Eur. J. Clin. Nutr. 55: 509–12.
8. Woodard GA, Encarnacion B, Downey JR, Peraza J, Chong K, et al. (2009) Probiotics improve outcomes after Roux-en-Y gastric bypass surgery: a prospective randomized trial. J. Gastrointest. Surg. 13: 1198–204.
9. Karska-Wysocki B, Bazo M, Smoragiewicz W (2010) Antibacterial activity of *Lactobacillus acidophilus* and *Lactobacillus casei* against methicillin-resistant *Staphylococcus aureus* (MRSA). Microbiol. Res. 165:674–86.
10. Liong MT (2008) Safety of probiotics: translocation and infection. Nutr. Rev. 66: 192–202.
11. Rafter J, Bennett M, Caderni G, Clune Y, Hughes R, et al. (2007) Dietary synbiotics reduce cancer risk factors in polypectomized and colon cancer patients. Am. J. Clin. Nutr. 85: 488–96.
12. Moayyedi P, Ford AC, Talley NJ, Cremonini F, Foxx-Orenstein AE, et al. (2010) The efficacy of probiotics in the treatment of irritable bowel syndrome: a systematic review. Gut. 59: 325–32.
13. Golowczyc MA, Mobili P, Garrote GL, Abraham AG, De Antoni GL (2007) Protective action of *Lactobacillus kefir* carrying S-layer protein against *Salmonella enterica* serovar *enteritidis*. Int. J. Food. Microbiol. 118: 264–73.
14. Ohnmacht C, Pullner A, King SB, Drexler I, Meier S, et al. (2009) Constitutive ablation of dendritic cells breaks self-tolerance of CD4 T cells and results in spontaneous fatal autoimmunity. J. Exp. Med. 206: 549–59.
15. Kushwah R, Hu J (2011) Role of dendritic cells in the induction of regulatory T cells. Cell. Biosci. 1: 20.
16. Rizzello V, Bonaccorsi I, Dongarrà ML, Fink LN, Ferlazzo G (2011) Role of natural killer and dendritic cell crosstalk in immunomodulation by commensal bacteria probiotics. J. Biomed. Biotechnol. 2011: 473097.
17. Gómez-Llorente C, Muñoz S, Gil A (2010) Role of Toll-like receptors in the development of immunotolerance mediated by probiotics. Proc. Nutr. Soc. 69: 381–9.
18. Hua MC, Lin TY, Lai MW, Kong MS, Chang HJ, et al. (2010) Probiotic Bio-Three induces Th1 and anti-inflammatory effects in PBMC and dendritic cells. World. J. Gastroenterol. 16: 3529–40.
19. Kelsall BL, Rescigno M (2004) Mucosal Dendritic Cells in Immunity and Inflammation. Nat. Immunol. 5: 1091–5.
20. Tsilingiri K, Barbosa T, Penna G, Caprioli F, Sonzogni A, et al. (2012) Probiotic and postbiotic activity in health and disease: comparison on a novel polarised ex-vivo organ culture model. Gut. 61: 1007–15.
21. Ayehunie S, Snell M, Child M, Klausner M (2009) A plasmacytoid dendritic cell (CD123+/CD11c−) based assay system to predict contact allergenicity of chemicals. Toxicology. 264: 1–9.

22. Vieites Fernández JM, Muñoz Quezada S, Llamas Company I, Maldonado Lozano J, Romero Braquehais R, et al. (2010) PCT AX090006WO.

23. Muñoz-Quezada S, Chenoll E, Vieites Fernández JM, Genovés S, Maldonado J, et al. (2013) Isolation, identification and characterization of three novel probiotic strains (*Lactobacillus paracasei* CNCM I-4034, *Bifidobacterium breve* CNCM I-4035 and *Lactobacillus rhamnosus* CNCM I-4036) from faeces of exclusively breast milk fed infants. Br. J. Nutr. 109: S51–62.

24. Muñoz-Quezada S, Bermudez-Brito M, Chenoll E, Genovés S, Gómez-Llorente C, et al (2013) Competitive inhibition of three novel bacteria isolated from faeces of breast milk-fed infants against selected enteropathogens. Br. J. Nutr. 109: S63–9.

25. Stagg AJ, Hart AL, Knight SC, Kamm MA (2003) The dendritic cell: its role in intestinal inflammation and relationship with gut bacteria. Gut. 52: 1522–9.

26. Evrard B, Coudeyras S, Dosgilbert A, Charbonnel N, Alamé J, et al. (2011) Dose-dependent immunomodulation of human dendritic cells by the probiotic *Lactobacillus rhamnosus* Lcr35. PLoS One 6: e18735.

27. Boyle RJ, Robins-Browne RM, Tang ML (2006) Probiotic use in clinical practice: what are the risks? Am. J. Clin. Nutr. 83: 1256–1264.

28. Young SL, Simon MA, Baird MA, Tannock GW, Bibiloni R, et al. (2004) Bifidobacterial species differentially affect expression of cell surface markers and cytokines of dendritic cells harvested from cord blood. Clin. Diagn. Lab. Immunol. 11: 686–690.

29. López P, Gueimonde M, Margolles A, Suárez A (2010) Distinct *Bifidobacterium* strains drive different immune responses in vitro. Int. J. Food. Microbiol. 138: 157–165.

30. Hoarau C, Lagaraine C, Martin L, Velge-Roussel F, Lebranchu Y (2006) Supernatant of *Bifidobacterium breve* induces dendritic cell maturation, activation, and survival through a Toll-like receptor 2 pathway. J. Allergy. Clin. Immunol. 117: 696–702.

31. Bermudez-Brito M, Muñoz-Quezada S, Gomez-Llorente C, Matencio E, Bernal MJ, et al. (2012) Human intestinal dendritic cells decrease cytokine release against Salmonella infection in the presence of *Lactobacillus paracasei* upon TLR activation. PloS One 7: e43197.

32. Weiss G, Christensen HR, Zeuthen LH, Vogensen FK, Jacobsen M, et al. (2011) Lactobacilli and bifidobacteria induce differential interferon-β profiles in dendritic cells. Cytokine. 56: 520–30.

33. Turroni F, Van Sinderen D, Ventura M (2011) Genomics and ecological overview of the genus *Bifidobacterium*. Int. J. Food. Microbiol. 149; 37–44.

34. Borchers AT, Selmi C, Meyers FJ, Keen CL, Gershwin ME (2009) Probiotics and immunity. J. Gastroenterol. 44: 26–46.

35. Madsen K (2006) Probiotics and the immune response. J. Clin. Gastroenterol. 40: 232–4.

36. Aujla SJ, Dubin PJ, Kolls JK (2007) Th17 cells and mucosal host defense. Semin Immunol. 19: 377–82.

37. Dubin PJ, Kolls JK (2008) Th17 cytokines and mucosal immunity. Immunol Rev. 226: 160–71.

38. Medina M, Izquierdo E, Ennahar S, Sanz Y (2007) Differential immunomodulatory properties of *Bifidobacterium longum* strains: relevance to probiotic selection and clinical applications. Clin. Exp. Immunol. 150: 531–8.

39. Marie JC, Letterio JJ, Gavin M, Rudensky AY (2005) TGF-beta 1 maintains suppressor function and Foxp3 expression in CD4$^+$ CD25$^+$ regulatory T cells. J Exp Med. 201:1061–7.

40. Nakamura K, Kitani A, Fuss I, Pedersen A, Harada N, et al. (2004) TGF-beta 1 plays an important role in the mechanism of CD4$^+$ CD25$^+$ regulatory T cell activity in both humans and mice. J Immunol. 172:834–4240.

41. Strobl H, Knapp W (1999) TGF-beta 1 regulation of dendritic cells. Microbes Infect.1:1283–90.

42. Plantinga TS, van Maren WW, van Bergenhenegouwen J, Hameetman M, Nierkens S, et al. (2011) Differential Toll-like receptor recognition and induction of cytokine profile by Bifidobacterium breve and Lactobacillus strains of probiotics. Clin. Vaccine. Immunol. 18: 621–8.

43. Hiramatsu Y, Satho T, Irie K, Shiimura S, Okuno T, et al. (2012) Differences in TLR9-dependent inhibitory effects of H(2)O(2)-induced IL-8 secretion and NF-kappa B/I kappa B-alpha system activation by genomic DNA from five Lactobacillus species. Microbes Infect. 2012.pii: S1286–4579(12)00274–2.

44. Ghadimi D, Vrese Md, Heller KJ, Schrezenmeir J (2010) Effect of natural commensal-origin DNA on toll-like receptor 9 (TLR9) signaling cascade, chemokine IL-8 expression, and barrier integritiy of polarized intestinal epithelial cells. Inflamm Bowel Dis. 16:410–27.

45. Lavelle EC, Murphy C, O'Neill LA, Creagh M (2010) The role of TLRs, NLRs, and RLRs in mucosal innate immunity and homeostasis. Mucosal. Immunol. 3: 17–28.

46. Kaji R, Kiyoshima-Shibata J, Nagaoka M, Nanno M, Shida K (2010) Bacterial teichoic acids reverse predominant IL-12 production induced by certain *Lactobacillus* strains into predominant IL-10 production via TLR2-dependent ERK activation in macrophages. J Immunol. 184: 3505–13.

47. Zeuthen LH, Fink LN, Frøkiaer H (2008) Toll-like receptor 2 and nucleotide-binding oligomerization domain-2 play divergent roles in the recognition of gut-derived lactobacilli and bifidobacteria in dendritic cells. Immunology. 124: 489–502.

48. Vizoso Pinto MG, Rodriguez Gomez M, Seifert S, Watzl B, Holzapfel WH, et al. (2009) Lactobacilli stimulate the innate immune response and modulate the TLR expression of HT29 intestinal epithelial cells in vitro. Int. J. Food. Microbiol. 109: 86–93.

49. Tao Y, Drabik KA, Waypa TS, Musch MW, Alverdy JC, et al. (2006) Soluble factors from *Lactobacillus* GG activate MAPKs and induce cytoprotective heat shock proteins in intestinal epithelial cells. Am. J. Physiol. 290: 1018–30.

50. Kim YG, Ohta T, Takahashi T, Kushiro A, Nomoto K, et al. (2006) Probiotic *Lactobacillus casei* activates innate immunity via NF-kappaB and p38 MAP kinase signaling pathways. Microbes. Infect. 8: 994–1005.

51. Voltan S, Castagliuolo I, Elli E, Longo S, Brun P, et al. (2007) Aggregating phenotype in *Lactobacillus crispatus* determines intestinal colonization and TLR2 and TLR4 modulation in murine colonic mucosa. Clin. Vaccine. Immunol. 14: 1138–48.

52. Uematsu S, Fujimoto K, Jang MH, Yang BG, Jung YJ, et al. (2008) Regulation of humoral and cellular gut immunity by lamina propria dendritic cells expressing Toll-like receptor 5. Nat. Immunol. 9: 769–76.

53. Feng T, Elson CO, Cong Y (2011) Treg cell-Ig A axis in maintenance of host immune homeostasis with microbiota. Int. Immunopharmacol. 11: 589–592.

54. Hansen J, Gulati A, Sartor RB (2010) The role of mucosal immunity and host genetics in defining intestinal commensal bacteria. Curr. Opin. Gastroenterol. 26: 564–71.

55. Bulut Y, Faure E, Thomas L, Equils O, Arditi M (2001) Cooperation of Toll-like receptor 2 and 6 for cellular activation by soluble tuberculosis factor and *Borrelia burgdorferi* outer surface protein A lipoprotein: role of Toll-interacting protein and IL-1 receptor signaling molecules in Toll-like receptor 2 signaling. J. Immunol. 167: 987–94.

56. Zhang G, Ghosh S (2002) Negative regulation of toll-like receptor-mediated signalling by Tollip. J. Biol. Chem. 277: 7059–65.

57. Burns K, Clatworthy J, Martin L, Martinon F, Plumpton C, et al. (2000) Tollip, a new component of the IL-1RI pathway, links IRAK to the IL-1 receptor. Nat Cell Biol. 2: 346–51.

58. Adams CA (2010) The probiotic paradox: live and dead cells are biological response modifiers. Nutr. Res. Rev. 23: 37–46.

2

Development of a Strain-Specific Real-Time PCR Assay for Enumeration of a Probiotic *Lactobacillus reuteri* in Chicken Feed and Intestine

Verity Ann Sattler[1]*, **Michaela Mohnl**[2], **Viviana Klose**[1]

1 University of Natural Resources and Applied Life Sciences, Department for Agrobiotechnology, Tulln, Austria, **2** BIOMIN Research Center, Tulln, Austria

Abstract

A strain-specific real-time PCR assay was developed for quantification of a probiotic *Lactobacillus reuteri* (DSM 16350) in poultry feed and intestine. The specific primers were designed based on a genomic sequence of the strain derived from suppression subtractive hybridization with the type strain *L. reuteri* DSM 20016. Specificity was tested using a set of non-target strains from several sources. Applicability of the real-time PCR assay was evaluated in a controlled broiler feeding trial by using standard curves specific for feed and intestinal matrices. The amount of the probiotic *L. reuteri* was determined in feed from three feeding phases and in intestinal samples of the jejunum, ileum, and caecum of three, 14, and 39 day old birds. *L. reuteri* DSM 16350 cells were enumerated in all feeds supplemented with the probiotic close to the inclusion rate of 7.0×10^3 cfu/g, however, were not detected in *L. reuteri* DSM 16350 free feed. In three day old birds *L. reuteri* DSM 16350 was only detected in intestinal samples from probiotic fed animals ranging from $8.2 \pm 7.8 \times 10^5$ cfu/g in the jejunum, $1.0 \pm 1.1 \times 10^7$ cfu/g in the ileum, and $2.5 \pm 5.7 \times 10^5$ cfu/g in the caecum. Similar results were obtained for intestinal samples of older birds (14 and 39 days). With increasing age of the animals the amount of *L. reuteri* signals in the control animals, however, also increased, indicating the appearance of highly similar bacterial genomes in the gut microbiota. The *L. reuteri* DSM 16350 qPCR assay could be used in future for feeding trials to assure the accurate inclusion of the supplement to the feed and to monitor it's uptake into the GIT of young chicken.

Editor: Jose Luis Balcazar, Catalan Institute for Water Research (ICRA), Spain

Funding: This work was part of a FFG (Austrian Research Promotion Agency) funded project (Gut function Project No. 820117). The funders had no role in study design, data collection and analysis, decision to publish, or preparation of the manuscript.

Competing Interests: This work was done in cooperation with the company BIOMIN, who provided the probiotic strain and housing facility for the feeding trial.

* E-mail: verityann.sattler@boku.ac.at

Introduction

Lactobacilli are widely used as probiotics in animals [1]. For application as feed additive in poultry production, a probiotic bacterium is commonly isolated from the intestine of healthy chickens and further selected for specific beneficial properties [2]. Identification of the probiotic is important to discriminate it from related strains with different properties. Differentiation can be achieved with molecular typing methods such as pulsed-field gel electrophoresis (PFGE) [3] or random amplified polymorphic DNA (RAPD)-PCR [4]. However, these techniques rely on isolation and cultivation capabilities of organisms and are of limited use in complex microenvironments such as feed or the gastro-intestinal tract (GIT). For evaluating the efficacy and persistence of a probiotic strain, it is important to assure correct inclusion rates in the feed and to trace the introduced strain through the GIT. Methods to monitor the probiotic should be strain-specific, quantitative and applicable for analysis of feed and GIT samples. Quantitative PCR (qPCR) is a technique that has been used to detect several bacterial species in food [5], rumen [6] or faeces [7]. It's high sensitivity enables quantification of microorganisms with low abundance within an environmental sample. The challenge in qPCR development is designing primers that specifically target species or strains of interest, despite the presence of closely related bacteria. Efficiency and accuracy of the qPCR depend on DNA quality. The main obstacles for good quality DNA are co-extraction of PCR-inhibitory substances from the environmental matrix and inefficient recovery of total genomic DNA from the bacterial community [8]. To create qPCR standard curves for absolute quantification of microorganisms, environmental samples are often spiked with a known amount of target cells before DNA is extracted. This allows a more precise determination of microorganisms of interest in a complex sample. Animal feed contain a diverse bacterial population originating from soil, water, or dust where the feed plant was grown, processed and/or stored [9]. *Enterobacteriacae* were mainly found in commercial poultry diets [10], while lactobacilli proliferate best under moist and anaerobic conditions and are predominantly present in grass silage feed [11]. Thus, lactobacilli are not expected to be prevalent in rather dry mashed or pelleted poultry feed, which makes it easier to specifically detect these, when added to grain feed. In contrast, lactobacilli are a prominent group of the autochthonous microbial community prevalent in the upper GIT part of chicken [12,13]. The genus *Lactobacillus* is taxonomically very complex and known for its extreme phenotypic and ecological diversity [14]. To differentiate closely related *Lactobacillus* strains based on 16S rRNA gene sequence is difficult due to high sequence homology in variable regions of the gene [15]. As an alternative to this gene, several reports describe the development of strain-specific primers using RAPD [16,17,18,19,20] or by identifying phage-related sequences [21]. Suppression subtractive hybridization (SSH) is a method to identify genomic DNA fragments that are present in one but not another closely related strain [22,23,24]. This method is especially useful, when genome sequence information is lacking for the strain of interest. In combination with qPCR, these unique genomic markers might be used to track probiotics from the feed through the animal's GIT. So far, only few studies have reported the use of a genome-hybridization based method combined with qPCR for identification and quantification of probiotics *in vivo* [25,26]. This study describes the development of a strain-specific qPCR assay for detection of the chicken derived strain *Lactobacillus reuteri* LR (DSM 16350), which was isolated and evaluated as a probiotic strain within the European Union project "C-EX" (QLK-CT-2002-71662) for the use as feed additive in young chicken [2]. To our knowledge, SSH in combination with qPCR was used for the first time for specific quantification of a *Lactobacillus* strain in environmental samples. Primers were tested for specificity with a set of non-target *L. reuteri* strains. Applicability of the qPCR assay was evaluated in a feeding trial with broiler chickens by using standard curves, specific for feed and intestinal matrix. Presence and amount of the probiotic were determined in feed from three different feeding phases, and additionally the probiotic was monitored in three compartments of the GIT (jejunum, ileum, caecum) of three, 14, and 39 day old birds.

Materials and Methods

Bacterial strains and growth conditions

Suppression subtraction hybridization (SSH) was performed with the probiotic strain *Lactobacillus reuteri* (DSM 16350) as tester, which is mentioned hereafter by the code LR, and with the type strain *Lactobacillus reuteri* (DSM 20016) as the driver. Other *L. reuteri* strains used for specificity testing were of distinct sources, either purchased from strain collections or previously isolated from animal intestinal samples (Table 1). All strains were grown in de Man Rogosa Sharpe medium (MRS; Oxoid, Hampshire, UK) under semi-anaerobic conditions at 37°C for 24 h. Electro competent cells *Escherichia coli* ElectroMAX DH10B (Invitrogen, Carlsbad, CA, USA), used for cloning, were grown at 37°C on Luria Bertani agar (LB; Oxoid, Hampshire, UK) supplemented with ampicillin 100 mg/ml.

Ethics statement

Feeding trial protocol and animal experiments were approved by the local authority for agriculture 'Amt der Niederösterreichischen Landesregierung für Agrarrecht' in accordance with the Austrian act on animal experimentation (1988, BGBL 501/1989). Newly hatched broiler chickens (Ross) of mixed sex were kept in an environmentally controlled poultry house at the Centre for Animal Nutrition (Mank, Austria).

Feeding trial and sample collection

Newly hatched broiler chickens (Ross) were randomly allocated to eight pens per group with 20 birds per pen (in total 320 birds). Birds from the control group received a standard formulated broiler feed without supplements and birds from the probiotic group received the standard feed supplemented with *Lactobacillus reuteri* LR with a final concentration of 7.0 x 10^3 cfu/g feed. Birds were fed manually once a day with diet and water available *ad libitum*. Starter feed was given from day 0–14, followed by grower diet from day 15–28, and ending with the finisher feed from day

29–41. Intestinal samples were taken from animals (n = 8 per group and day) at day three, 14 and 39 of the trial, which was equivalent to the age of birds. Contents of jejunum, ileum and caeca were collected in sterile tubes and immediately frozen.

DNA extraction

DNA from bacterial cultures was extracted following a protocol for Gram- positive bacteria [27]. Cells were lysed with lysozyme (2.5 mg/ml) and Proteinase K (250 µg/ml), cell suspension purified with phenol:chloroform:iso-amylalcohol (25:24:1), and DNA precipitated in two volumes of ethanol at −20°C for at least 2 h. The DNA pellet was washed with 70% ethanol and dissolved in 50 µl nuclease-free water.

Genomic DNA from about 250 mg intestinal digesta was extracted using the QIAamp DNA Stool Mini Kit (Qiagen GmbH, Hilden, Germany) according to the manufacturer's instructions for pathogen detection. Prior to the kit protocol, samples were incubated with lysozyme (50 mg/ml) for 45 min at 37°C and then homogenized for 10 s at 6000 rpm using Precellys® SK38 bead beating tubes and the Precellys® 24-Dual homogenizer (Peqlab Biotechnology GmbH, Erlangen, Germany).

Microbial DNA was extracted from 20 g feed sample of every feeding phase. To wash off bacterial cells from feed particles, the feed was mixed with 100 ml peptone water containing 1% Triton X-100 and shaken in a flask for 30 min. The mixture was then smashed and filtered through a stomacher bag. Bacterial cells of the filtrate were pelleted and further used for DNA extraction following the protocol for intestinal samples. Isolated DNA was visualized by agarose gel electrophoresis and concentration was determined by NanoDrop ND-1000 spectrophotometer (Peqlab Biotechnology GmbH, Erlangen, Germany).

Construction of suppression subtractive hybridization (SSH) clone library

For SSH, the PCR-select bacterial genome subtraction kit (Clonetech Laboratories, Mountain View, CA, USA) was used to subtract unique genomic DNA of the probiotic strain *L. reuteri* LR (tester DNA) from the type strain *L. reuteri* 20016^T (driver DNA) following the manufacturer's protocol with modifications. The first hybridization step was performed at 55°C for 90 min and the second hybridization at 55°C for 16 h. Primary and secondary (nested) PCR were conducted using the Advantage 2 Polymerase Mix (Clonetech Laboratories, Mountain View, CA, USA), 10 mM dNTP mix, 10 µM of each primer, and 1 µl template DNA. Subtracted PCR products were cloned into pJet1.2/blunt vectors using the CloneJet PCR Cloning Kit (Fermentas, Burlington, CA, USA) and transformed into ElectroMAX DH10B cells by electroporation for 5 sec at 1.8 kV using the GenePulser (Bio-Rad Laboratories, Hercules, CA, USA). Transformants were recovered in liquid LB medium and grown over night on LB agar plates with 100 µg/ml ampicillin. From the SSH clone library, 57 clones were picked and a rapid plasmid preparation was performed as previously described [28]. Briefly, each colony was picked with a tooth pick and suspended in 20 µl 0.2 M NaOH. RNA was removed using RNase (10 µg/µl) by incubation for 7 min at 37°C. The plasmid extract was neutralized by adding 40 µl of 0.1 M HCl. Plasmid inserts were amplified by PCR with primers flanking the insertion site of the cloning vector. Size of the PCR products were analyzed by agarose gel electrophoresis.

Differential screening of strain-specific SSH products

For differential screening of subtracted PCR products from the SSH clone library, a DNA dot blot hybridization was performed.

Table 1. *L. reuteri* (LR) qPCR specificity test (C_T values and melting curve analysis) using non-target strains from various sources, purchased from culture collections and isolated from intestinal samples.

L. reuteri strains	Source of origin	C_T mean	Tm (°C)	MC properties +/−
DSMZ[a] 16350[b](LR)	chicken intestine	10	78.7	+
DSMZ 8533	lab strain [41]	>31	b.t.	−
DSMZ 12246	Chr. Hansen strain	>31	78.4[*]	−
DSMZ 17509	rat gut	>36	78.9[*]	−
LMG[c] 18238	chicken	>40	b.t.	−
DSMZ 20015	manure	>40	81.8	−
DSMZ 20016[T]	human intestine	n.d.	-	−
DSMZ 20053	human faeces	>40	b.t.	−
DSMZ 20056	rat faeces	n.a.	-	−
LMG 22879	laying hen, cloacae	>31	b.t.	−
CA2	pig intestine	>36	78.9[*]	−
LRS	pig intestine	>36	78.8	+
F2	pig intestine	>36	79.1	−
R8A	pig intestine	>31	78.6[*]	−
R20	pig intestine	>31	80.5	−
R22	pig intestine	>36	80.6	−
R31	pig intestine	n.a.	-	−
R36	pig intestine	>31	79.5[*]	−
S2A	pig intestine	>31	79.1[*]	−
S4A	pig intestine	>40	80.0	−
S6A	pig intestine	>36	80.8	−
S8A	pig intestine	>36	83.5	−
S11A	pig intestine	>31	80.1	−
S14A	pig intestine	>40	b.t.	−
S21A	pig intestine	>31	b.t.	−
S21C	pig intestine	>31	79.7	−
S24B	pig intestine	>31	76.2	−

C_T cycle threshold.
Tm melting temperature.
MC melting curve.
[a]DSMZ - Deutsche Sammlung von Mikroorganismen und Zellkultur.
[b]target strain and SSH tester strain.
[c]LMG – Belgian coordinated collections of microorganisms.
[T]type strain and SSH driver strain.
n.a. no amplification.
b.t. below threshold (33%).
"−" MC was either below threshold, showed the formation of one or more products, or showed a shift in Tm>1°C compared to Tm of LR (78.7°C).
"+" MC properties were identical to LR.
*melting curves not reproducible.

To generate single stranded DNA (ssDNA), all DNA or PCR samples were melted at 95°C and immediately chilled on ice. One microliter of ssPCR product was spotted onto a neutral BioBond nylon membrane (Sigma-Aldrich, St. Louis, MO, USA) and fixed by UV cross-linker (Stratagene, La Jolla, CA, USA) using the auto-crosslinking mode (1200 mJ×100/cm^2). Spotted membranes were hybridized with either SSH tester (*L. reuteri* LR) or driver (*L. reuteri* 20016[T]) DNA. Before hybridization, genomic tester and driver DNA were digested with RsaI and then labelled with biotin using the Biotin Decalable DNA labelling Kit (Fermentas, Burlington, CA, USA). Membranes were pre-hybridized with salmon sperm DNA to block unspecific binding sites. Then the membranes were incubated over night at 60°C in hybridization buffer (5×Denhardt's solution, 5 X SSPE buffer, 1% SDS) with 100 ng/ml labelled ssDNA under moderate shaking. To remove unbound and unspecific tester or driver DNA, membranes were washed with a non-stringent buffer (2×SSC, 0.1% SDS) and a stringent (0.1 X SSC, 0.1% SDS) buffer. DNA dots were visualized using the Biotin Chromogenic Detection Kit (Fermentas, Burlington, CA, USA) following the manufacturer's instructions.

Strain-specific primer design and qPCR assay

Potential strain-specific SSH PCR products were sequenced and checked for sequence similarity by the BLAST web tool of the National Center of Biotechnology Information (NCBI) [29]. Sequences were used for primer design with the Primer Premier 5 software (Premier Biosoft, Palo Alto, CA, USA). A set of primers was tested for specificity and efficiency in an annealing temperature gradient-qPCR with genomic tester DNA and genomic driver DNA as targets. The following conditions were applied: initial denaturation at 95°C for 5 min followed by 45 cycles at 95°C for 15 sec, annealing at 55°C–65°C for 15 sec, and elongation at 72°C for 20 min. To assure that the correct PCR product was amplified, a melting curve analysis was added at the end of the PCR program using the default settings of the realplex^2 Mastercycler ep-gradient S instrument (Eppendorf, Hamburg, Germany). Amplification was carried out in a 15 µl final volume containing 1 × Mesa Green qPCR MasterMix Plus for SYBR (Eurogentec S.A., Seraing, Belgium), 300 nM of each primer, and 3 µl target DNA. The best performing primer set was chosen for further specificity testing with several *L. reuteri* non-target strains (Table 1) in the qPCR assay at optimal annealing temperature (results not shown).

Standard curves for LR quantification in environmental samples

For quantification of *L. reuteri* LR in feed and intestinal samples two matrix based standard curves were created. Therefore, *L. reuteri* LR free broiler feed was spiked with 8.0×10^8 cfu/g and gut digesta with 8.5×10^6 cfu/g of lyophilized LR cells. DNA was extracted according to the matrix dependent DNA extraction protocol and was serially diluted in nuclease-free water. Standard DNA was amplified by qPCR in triplicates applying the same conditions as for primer specificity testing. Standard curves were generated by plotting cycle threshold values (C_T) versus equivalent log cell numbers. The amplification efficiencies for feed and gut samples, determined by the slope of the standard curves, were calculated based on the equation $E = (10^{-1/\text{slope}} - 1) \times 100$. To test accuracy and application of the qPCR assays *in vivo*, feed and gut lumen samples were collected and analyzed from a feeding trial with broiler chickens as described above. The number of *L. reuteri* LR cells was assessed in triplicates by qPCR and expressed as cfu/g.

Results

Differential screening and sequencing of SSH subtracted products

Subtracted PCR products from clone 4 and 19 of the SSH clone library were chosen as potential *L. reuteri* LR specific DNA markers as they showed intense colour after hybridization with labelled tester DNA, but not with labelled driver DNA (Figure S1). Insert sequences of clone 4 (accession KJ152779) and 19 (accession KJ152780) had no significant sequence similarity to any known gene in a BLAST search in the NCBI nucleotide database. The highest similarity hit for clone 4 sequence was given with *L. salivarius* CECT 5713 plasmid pHN1 (47% query cover, E = 3e−43, 81% identity) and for clone 19 sequence with *L. reuteri* I5007 plasmid pLRI04 (15% query cover, E = 9e−47, 93% identity).

L. reuteri LR specific primer

Primers that were designed based on sequences of DNA inserts in clone 4 and 19 were tested for efficiency and specificity using tester and driver DNA in a temperature gradient qPCR assay. The primer pair for clone 4, 21f (5′-CAGGATCGGTAATTGATG-3′) and 190r (5′-TGGATATGGAAGTTCGTC-3′), was specific for LR. The best PCR efficiency of this primer pair was at 56°C annealing temperature, because with this temperature highest fluorescent signal occurred at an early cycle threshold (C_T) of 10. Specificity was tested under the same conditions and showed that

all non-targets (n = 26) had C_T values above 31 compared to the specific signal at C_T 10. Seven strains had C_T values above 36, and five strains C_T values above 40. The SSH driver strain *L. reuteri* 20016 and two isolates, one from rat (DSM 20056) and one from pig (R31) showed no PCR amplification product (Table 1). Cycle difference between target LR and non-target strains was at least 21, which corresponds to a cell number of about log 6 per gram sample. Unspecific PCR product formation could be detected for non-target strains by melting curve analysis. Melting temperatures of unspecific products were, however, higher (up to 83.5°C) or lower (down to 76.2°C) compared to the LR specific product melting temperature (78.7°C). Six non-target strains revealed melting curves below threshold (b.t.) and several other strains showed formation of two or more products, displayed by multiple peaks (Figure S2). For some samples melting curves were not reproducible between duplicates (Table 1). The pig isolate LRS was the only strain that could not be distinguished by melting curve analysis from the LR curve, however, by its C_T value of 31, compared to the LR C_T value of 10. Primers for clone 19 were not as specific as the clone 4 primer pair 21f/190r. The non-target strain LRS was efficiently amplified with clone 19 primers and the melting curve of the resulting product could not be distinguished from the LR melting curve. For this reason, the clone 19 primers were rejected for further analysis.

Standard curves and limit of detection for environmental samples

qPCR standard curves for LR strain-specific quantification in feed ranged from 8.0×10^8 cfu/g–8 cfu/g, whereas those for LR quantification in gut lumen samples ranged from 8.5×10^6 cfu/g–48 cfu/g. Amplification efficiency E was 96% for feed samples and 99% for gut samples. Based on qPCR results with non-target strains, the limit for specific and reliable detection of LR in any sample was set at C_T 31, although the standard curves showed a linear quantification range up to a cycle number of 32 (Figure 1). The detection limit at C_T 31 corresponded to 3.5×10^1 cfu/g, for feed samples, and to 4×10^2 cfu/g, for gut digesta samples.

In vivo evaluation of the *L. reuteri* LR-specific qPCR assay

The strain-specific qPCR was evaluated with samples from an *in vivo* feeding trial, where LR cell numbers were quantified in feed from three different feeding phases and monitored in gut lumen samples of three days, 14 day and 39 day old control and LR supplemented chicken (LR 7.0×10^3 cfu/g feed).

Feed samples. The starter feed contained $5.4 \pm 0.3 \times 10^3$ cfu/g *L. reuteri* LR, the grower feed $2.0 \pm 0.1 \times 10^3$ cfu/g *L. reuteri* LR, and the finisher feed $0.8 \pm 1.1 \times 10^3$ cfu/g *L. reuteri* LR. Control feeds from each phase were below the detection limit of 3.5×10^1 cfu/g *L. reuteri* LR.

Gut samples (day 3). *L. reuteri* LR was detected in intestines of three day old birds of the probiotic group, however, not in the control group (Table 2). The average amount of LR in the probiotic group was $8.2 \pm 7.8 \times 10^5$ cfu/g in the jejunum, $1.0 \pm 1.1 \times 10^7$ cfu/g in the ileum, and $2.5 \pm 5.7 \times 10^5$ cfu/g in the caecum. Enumeration of *L. reuteri* LR was possible in each animal and intestinal location (n = 8) from probiotic fed animals. In control animals, the LR cell numbers were below the detection limit in every sample.

Gut samples (day 14). In 14 day old birds, the average amount of LR in the probiotic group was $3.7 \pm 9.0 \times 10^5$ cfu/g in the jejunum, $1.5 \pm 3.9 \times 10^6$ cfu/g in the ileum, and $3.5 \pm 9.0 \times 10^6$ cfu/g in the caecum (Table 2). LR could be detected in all intestinal samples from the probiotic group except in two jejunum samples. LR cell numbers were 1–2 logs higher in

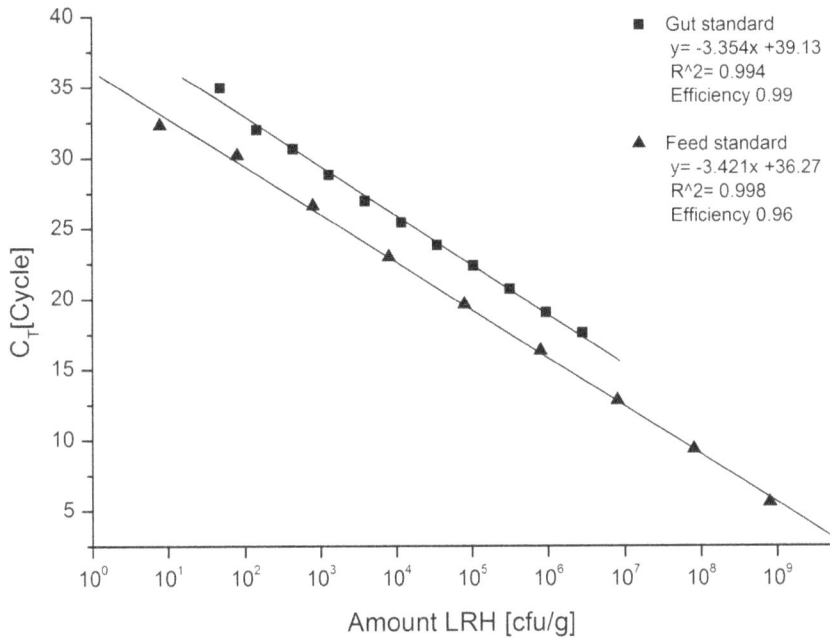

Figure 1. qPCR standard curve for quantification of *L. reuteri* **(LR) in feed and gut samples.** The standard curves were generated by amplification of serially diluted DNA from feed spiked with 8.0×10^8 cfu/g and from gut lumen content spiked with 8.5×10^6 cfu/g.

the probiotic group compared to the control group, where the average amount of putative LR detected was similar in all three intestinal locations with 1.5×10^4 cfu/g. The number of samples with LR signals above detection limit in the control group was 3 out of 8 in the jejunum, 5 out of 8 in the ileum, and 7 out of 8 in the caecum.

Gut samples (day 39). In 39 day old birds, the average amount of LR in the probiotic group was $2.0 \pm 1.9 \times 10^5$ cfu/g in the jejunum, $4.4 \pm 4.5 \times 10^5$ cfu/g in the ileum, and $1.4 \pm 2.3 \times 10^4$ cfu/g in the caecum (Table 2). Quantification of putative LR in control animals showed similar high results compared to probiotic fed animals, with an average LR cell count of $2.3 \pm 3.2 \times 10^5$ cfu/g in the jejunum, $1.6 \pm 2.5 \times 10^6$ cfu/g in the ileum, and $0.6 \pm 1.0 \times 10^4$ cfu/g in the caecum. In the probiotic feeding group, LR could be detected in all samples except in three caecum samples. In the control group, LR signals were also

detected in nearly every sample except in one jejunum and two caecum samples.

Discussion

Guidelines of the Food and Agriculture Organization and the World Health Organization for the evaluation of probiotics in food report that strain identification is important to link the claimed health effect to the probiotic and to enable correct surveillance during efficacy studies [30]. This emphasizes the need for strain-specific identification assays. In this study, DNA sequences that are unique for the genome of the probiotic *L. reuteri* LR were identified with suppression subtractive hybridization (SSH). Two out of 57 SSH clones harboured genomic sequences that were likely specific for LR and that could be used for primer design. Specificity testing with non-target strains showed that only one primer pair (21f/190r) was LR specific. A limitation of the SSH method is the occurrence of a certain

Table 2. Mean numbers of *L. reuteri* LR in chicken gut samples from animals fed without (control) or with LR (probiotic) over 39 days as determined by strain-specific qPCR.

	Jejunum		Ileum		Caecum	
	Control	Probiotic	Control	Probiotic	Control	Probiotic
n	0/8	8/8	0/8	8/8	0/8	8/8
Day 3	n.d.	$8.2 \pm 7.8 \times 10^5$	n.d.	$1.0 \pm 1.1 \times 10^7$	n.d.	$2.5 \pm 5.7 \times 10^5$
n	3/8	6/8	5/8	8/8	7/8	8/8
Day 14	$1.2 \pm 1.9 \times 10^4$	$3.7 \pm 9.0 \times 10^5$	$1.5 \pm 2.8 \times 10^4$	$1.5 \pm 3.9 \times 10^6$	$1.5 \pm 2.0 \times 10^4$	$3.5 \pm 9.0 \times 10^6$
n	7/8	8/8	8/8	8/8	6/8	5/8
Day 39	$2.3 \pm 3.2 \times 10^5$	$2.0 \pm 1.9 \times 10^5$	$1.6 \pm 2.5 \times 10^6$	$4.4 \pm 4.5 \times 10^5$	$0.6 \pm 1.0 \times 10^4$	$1.4 \pm 2.3 \times 10^4$

n.d. not detected above detection limit for gut sample 4×10^2 cfu/g.
n number of animals with cell counts above detection limit from a total number of eight animals.

portion of false positives [24]. A method called mirror orientated selection (MOS) has been reported to reduce the number of false positive clones from the SSH library [31] and should be considered for future SSH applications. The diversity of different *L. reuteri* strains in the gut is unknown. As it is impossible to isolate and screen all environmental strains, we chose to use a broad array of *L. reuteri* strains of distinct sources for primer specificity testing. Melting curve properties of non-targets were different from those of the LR strain, except for one pig isolate, probably because of homologous primer binding sites in its genome. Melting curves below threshold or with multiple peaks found for several non-target strains are likely due to primer binding to unspecific, partially complementary sequences resulting in inefficient amplification. The cycle threshold (C_T) difference between LR and non-target samples was equivalent to at least 6 logs of cell numbers. Thus, signals of unspecific targets in environmental samples were considered not to compromise the specific enumeration of *L. reuteri* LR. In order to avoid quantification of false-positive signals, the detection limit was not set according to the linear range of the standard curves but at the lowest cycle threshold that was detected for non-target strains. Detection limits reported in other studies for strain-specific real-time PCR assays varied; e.g. 10^4 cfu/U for spiked rumen feed and fluid [26] and 10^5 copies/g for spiked human faeces [20,32]. In these studies the standard curves were created by inoculating the environmental sample with decreasing concentrations of the target bacterium. This was different from our approach in which the environmental sample was first spiked and then DNA was diluted. This may have led to a relatively low detection limit allowing LR quantification at concentrations below 10^3 cfu/g, which may be of interest, when the strain is applied as a multi-strain probiotic, where single strains are mixed together [33].

Accuracy and applicability of the qPCR assay were examined in a controlled feeding trial, in which broiler chickens were administered feed supplemented with the probiotic *L. reuteri* LR and control feed. The LR cell number in probiotic-supplemented feeds was close to the theoretical inclusion rate (7.0×10^3 cfu/g), whereas that in the control feed was below the detection limit. This confirmed that accurate enumeration of LR in feed can be achieved. qPCR analysis revealed that the indigenous feed flora and potentially co-extracted PCR inhibiting substances did not interfere within the assay. In contrast to feed, the species *L. reuteri* is a common member of the gut microbiota in chicken [34]. In three day old birds, enumeration of LR was possible in each bird and intestinal location, indicating that the probiotic was taken up and did spread along the entire GIT. In young chickens, the highest LR cell count was obtained in the ileum. At the age of 14 days, the LR numbers were similar to those from three-day old birds. However, LR signals at a level of 10^4 cfu/g were also detected in control birds which indicated that two weeks after hatching the gut microbiota of the chickens harboured bacteria with a similar sequence in their genomes. Contamination of analysed control

birds was excluded after selective culturing of the LR strain by the use of strain typing methods (data not shown). Microbial colonization of the GI tract begins shortly after hatching [35,36], and the gut microbiota rapidly develops and becomes more complex as chickens age [34,37]. When diversity and abundance of microorganisms in the gut increase, at the same time the possibility of unspecific PCR amplification also raises. This is also indicated by the results of 39 day old birds, where the amount of LR found in the control animals was similar high as in the probiotic fed animals. To develop more specific primers, more isolates needed to be screened. However, this approach is limited, as isolation of all *L. reuteri* strains from complex intestinal samples is impossible. Some lactobacilli strains have been genetically labelled e.g. by introducing a fluorescent marker gene [38,39], or a silent mutation into a chromosomal gene [40]. A genetic label in the bacterial genome provides a way to monitor the specific strain in various environments, but introducing foreign DNA is generally a scientific approach not applicable for feeding farm animals, where this strain would then be released into the free environment. Aside from that, probiotics need to be alive to exert all their beneficial effects in the GIT. In this regard detecting live but not dead cells is of interest and may be achieved by pre-treating the intestinal content with propidium monoacid (PMA), a DNA intercalating chemical that inhibits PCR amplification of non-viable cells [19].

In conclusion, we were able to develop strain-specific qPCR primers using SSH to detect and enumerate *L. reuteri* LR in poultry feed and in the gut lumen of chickens in early days of life. It could be confirmed that qPCR is a sensitive and rapid tool to determine low abundant bacterial strains in samples with mixed microbial population. This technique replaces other genotypic identification techniques that include time consuming isolation of the target strain using selective culturing. In future, the LR specific qPCR assay might be used for feeding trials in order to assure the accurate inclusion of the supplement to the feed and to monitor it's uptake into the GIT of chicken.

Supporting Information

Figure S1 Dot blot hybridization of SSH clone inserts with A) driver DNA of *L. reuteri* 20016, and B) tester DNA of the target strain *L. reuteri* LR.

Figure S2 Representative melting curves from LR-qPCR specificity test with non-target *L. reuteri* strains showing the formation of unspecific PCR products.

Author Contributions

Conceived and designed the experiments: VAS VK MM. Performed the experiments: VAS. Analyzed the data: VAS. Contributed reagents/materials/analysis tools: MM. Wrote the paper: VAS.

References

1. Gaggia F, Mattarelli P, Biavati B (2010) Probiotics and prebiotics in animal feeding for safe food production. Int J Food Microbiol 141: 15–28.
2. Klose V, Mohnl M, Plail R, Schatzmayr G, Loibner AP (2006) Development of a competitive exclusion product for poultry meeting the regulatory requirements for registration in the European Union. Mol Nutr Food Res 50: 563–571.
3. Yeung PS, Kitts CL, Cano R, Tong PS, Sanders ME (2004) Application of genotypic and phenotypic analyses to commercial probiotic strain identity and relatedness. J Appl Microbiol 97: 1095–1104.
4. Welsh J, McClelland M (1990) Fingerprinting genomes using PCR with arbitrary primers. Nucleic Acids Res 18: 7213–7218.

5. De Martinis EC, Duvall RE, Hitchins AD (2007) Real-time PCR detection of 16S rRNA genes speeds most-probable-number enumeration of foodborne Listeria monocytogenes. J Food Prot 70: 1650–1655.
6. Tajima K, Aminov RI, Nagamine T, Matsui H, Nakamura M, et al. (2001) Diet-dependent shifts in the bacterial population of the rumen revealed with real-time PCR. Appl Environ Microbiol 67: 2766–2774.
7. Rinttila T, Kassinen A, Malinen E, Krogius L, Palva A (2004) Development of an extensive set of 16S rDNA-targeted primers for quantification of pathogenic and indigenous bacteria in faecal samples by real-time PCR. J Appl Microbiol 97: 1166–1177.

8. Zoetendal EG, Ben-Amor K, Akkermans AD, Abee T, de Vos WM (2001) DNA isolation protocols affect the detection limit of PCR approaches of bacteria in samples from the human gastrointestinal tract. Syst Appl Microbiol 24: 405–410.

9. Maciorowski KG, Herrera P, Jones FT, Pillai SD, Ricke SC (2007) Effects on poultry and livestock of feed contamination with bacteria and fungi. Anim Feed Sci Tech 133: 109–136.

10. Cox NA, Bailey JS, Thomson JE, Juven BJ (1983) Salmonella and other Enterobacteriaceae found in commercial poultry feed. Poult Sci 62: 2169–2175.

11. Maciorowski K, Herrera P, Jones F, Pillai S, Ricke S (2007) Effects on poultry and livestock of feed contamination with bacteria and fungi. Anim Feed Sci Tech 133: 109–136.

12. Lu J, Idris U, Harmon B, Hofacre C, Maurer JJ, et al. (2003) Diversity and succession of the intestinal bacterial community of the maturing broiler chicken. Appl Environ Microbiol 69: 6816–6824.

13. Hilmi HTA, Surakka A, Apajalahti J, Saris PEJ (2007) Identification of the most abundant Lactobacillus species in the crop of 1-and 5-week-old broiler chickens. Appl Environ Microbiol 73: 7867–7873.

14. Makarova KS, Koonin EV (2007) Evolutionary genomics of lactic acid bacteria. J Bacteriol 189: 1199–1208.

15. Yeung PS, Sanders ME, Kitts CL, Cano R, Tong PS (2002) Species-specific identification of commercial probiotic strains. J Dairy Sci 85: 1039–1051.

16. Tilsala-Timisjarvi A, Alatossava T (1998) Strain-specific identification of probiotic Lactobacillus rhamnosus with randomly amplified polymorphic DNA-derived PCR primers. Appl Environ Microbiol 64: 4816–4819.

17. Ahlroos T, Tynkkynen S (2009) Quantitative strain-specific detection of Lactobacillus rhamnosus GG in human faecal samples by real-time PCR. J Appl Microbiol 106: 506–514.

18. Fujimoto J, Matsuki T, Sasamoto M, Tomii Y, Watanabe K (2008) Identification and quantification of Lactobacillus casei strain Shirota in human feces with strain-specific primers derived from randomly amplified polymorphic DNA. Int J Food Microbiol 126: 210–215.

19. Fujimoto J, Tanigawa K, Kudo Y, Makino H, Watanabe K (2011) Identification and quantification of viable Bifidobacterium breve strain Yakult in human faeces by using strain-specific primers and propidium monoazide. J Appl Microbiol 110: 209–217.

20. Maruo T, Sakamoto M, Toda T, Benno Y (2006) Monitoring the cell number of Lactococcus lactis subsp. cremoris FC in human feces by real-time PCR with strain-specific primers designed using the RAPD technique. Int J Food Microbiol 110: 69–76.

21. Brandt K, Alatossava T (2003) Specific identification of certain probiotic Lactobacillus rhamnosus strains with PCR primers based on phage-related sequences. Int J Food Microbiol 84: 189–196.

22. Huang X, Li Y, Niu Q, Zhang K (2007) Suppression Subtractive Hybridization (SSH) and its modifications in microbiological research. Appl Microbiol Biot 76: 753–760.

23. Saxena D, Li Y, Caufield PW (2005) Identification of unique bacterial gene segments from Streptococcus mutans with potential relevance to dental caries by subtraction DNA hybridization. J Clin Microbiol 43: 3508–3511.

24. Agron PG, Macht M, Radnedge L, Skowronski EW, Miller W, et al. (2002) Use of subtractive hybridization for comprehensive surveys of prokaryotic genome differences. FEMS Microbiol Lett 211: 175–182.

25. Konstantinov SR, Smidt H, de Vos WM (2005) Representational difference analysis and real-time PCR for strain-specific quantification of Lactobacillus sobrius sp. nov. Appl Environ Microbiol 71: 7578–7581.

26. Peng M, Smith AH, Rehberger TG (2011) Quantification of Propionibacterium acidipropionici P169 bacteria in environmental samples by use of strain-specific primers derived by suppressive subtractive hybridization. Appl Environ Microbiol 77: 3898–3902.

27. Chan RK, Wortman CR, Smiley BK, Hendrick CA (2003) Construction and use of a computerized DNA fingerprint database for lactic acid bacteria from silage. J Microbiol Methods 55: 565–574.

28. Dong B, Ouyang LAN, Qin L, Li G (2007) A rapid and simple method for screening large numbers of recombinant DNA clones J Rapid Meth Aut Mic 15: 244–252.

29. Altschul SF, Gish W, Miller W, Myers EW, Lipman DJ (1990) Basic local alignment search tool. J Mol Biol 215: 403–410.

30. Guidelines for the evaluation of probiotics in food. Report of a joint FAO/WHO working group on drafting guidelines for the evaluation of probiotics in food. London, Ontario, Canada, April 30 and May 1, 2002.

31. Rebrikov DV, Britanova OV, Gurskaya NG, Lukyanov KA, Tarabykin VS, et al. (2000) Mirror orientation selection (MOS): a method for eliminating false positive clones from libraries generated by suppression subtractive hybridization. Nucleic Acids Res 28: E90.

32. Karjalainen H, Ahlroos T, Myllyluoma E, Tynkkynen S (2012) Real-time PCR assays for strain-specific quantification of probiotic strains in human faecal samples. Int Dairy J 27: 58–64.

33. Mountzouris KC, Tsitrsikos P, Palamidi I, Arvaniti A, Mohnl M, et al. (2010) Effects of probiotic inclusion levels in broiler nutrition on growth performance, nutrient digestibility, plasma immunoglobulins, and cecal microflora composition. Poult Sci 89: 58–67.

34. Guan LL, Hagen KE, Tannock GW, Korver DR, Fasenko GM, et al. (2003) Detection and identification of Lactobacillus species in crops of broilers of different ages by using PCR-denaturing gradient gel electrophoresis and amplified ribosomal DNA restriction analysis. Appl Environ Microbiol 69: 6750–6757.

35. Apajalahti J, Kettunen A, Graham H (2004) Characteristics of the gastrointestinal microbial communities, with special reference to the chicken. Worlds Poultry Sci J 60: 223–232.

36. Cason JA, Cox NA, Bailey JS (1994) Transmission of Salmonella typhimurium during hatching of broiler chicks. Avian Dis 38: 583–588.

37. Amit-Romach E, Sklan D, Uni Z (2004) Microflora ecology of the chicken intestine using 16S ribosomal DNA primers. Poult Sci 83: 1093–1098.

38. Drouault S, Corthier G, Ehrlich SD, Renault P (1999) Survival, physiology, and lysis of Lactococcus lactis in the digestive tract. Appl Environ Microbiol 65: 4881–4886.

39. Geoffroy MC, Guyard C, Quatannens B, Pavan S, Lange M, et al. (2000) Use of green fluorescent protein to tag lactic acid bacterium strains under development as live vaccine vectors. Appl Environ Microbiol 66: 383–391.

40. Malinen E, Laitinen R, Palva A (2001) Genetic labeling of lactobacilli in a food grade manner for strain-specific detection of industrial starters and probiotic strains. Food Microbiol 18: 309–317.

41. Rodtong S, Tannock GW (1993) Differentiation of Lactobacillus strains by ribotyping. Appl Environ Microbiol 59: 3480–3484.

3

Lactobacillus brevis Strains from Fermented *Aloe vera* Survive Gastroduodenal Environment and Suppress Common Food Borne Enteropathogens

Young-Wook Kim[1], Young-Ju Jeong[1], Ah-Young Kim[2], Hyun-Hee Son[1], Jong-Am Lee[1], Cheong-Hwan Jung[1], Chae-Hyun Kim[1], Jaeman Kim[3]*

1 KBNP Technology Institute, KBNP Inc., Yesan, Korea, **2** School of Biological Sciences and Technology, Chonnam National University, Gwangju, Korea, **3** Department of Biology, Mokpo National University, Muan, Korea

Abstract

Five novel *Lactobacillus brevis* strains were isolated from naturally fermented *Aloe vera* leaf flesh. Each strain was identified by Random Amplified Polymorphic DNA (RAPD) analysis and 16S rRNA sequence comparison. These strains were highly tolerant to acid, surviving in pH2.5 for up to 4 hours, and resistant to 5% bile salts at 37°C for 18 hours. Due to its tolerance to acid and bile salts, one strain passed through the gastric barrier and colonised the intestine after oral administration. All five strains inhibited the growth of many harmful enteropathogens without restraining most of normal commensals in the gut and hence named POAL (Probiotics Originating from Aloe Leaf) strains. Additionally, each strain exhibited discriminative resistance to a wide range of antibiotics. The *L. brevis* POAL strains, moreover, expressed high levels of the glutamate decarboxylase (GAD) gene which produces a beneficial neurotransmitter, γ-aminobutyric acid (GABA). These characteristics in all suggest that the novel *L. brevis* strains should be considered as potential food additives and resources for pharmaceutical research.

Editor: Shamala Devi Sekaran, University of Malaya, Malaysia

Funding: This study was supported by KBNP Inc., a manufacturing company specialising in animal health products for providing funding in this study. The funders had no role in study design, data collection and analysis, decision to publish, or preparation of the manuscript.

Competing Interests: Some authors have an affiliation to the commercial funders of this research (KBNP).

* E-mail: jkim@mokpo.ac.kr

Introduction

The *Aloe vera* L. plant has been used widely in herbal medicines for millennia [1] and it is a significant resource for cosmetic and pharmacological industries. It is one of a few edible species among approximately 500 species of *A. vera*. Its fleshy leaves have phenolic compounds such as aloe emodin, aloin, and aloesin, organic acids such as saponins and terpenoids [8]. With these ingredients, *Aloe vera* extracts have antibacterial and antifungal activities, which may be able to treat minor skin infections, such as boils and benign skin cysts [33 34]. *Aloe vera*, inner-leaf gel inhibits growth of *Streptococcus* and *Shigella* species *in vitro* [15]. In spite of these antimicrobial activities, the leaf gel is prone to oxidation at various temperatures, which leads to fermentation. This phenomenon is due to degradation of phenolic compounds in the aloe, which might be related a plant defense mechanism [11]. Oxidation of these compounds ablates the antimicrobial activity and results in bacterial growth. Preliminary investigation of aloe flesh fermentation revealed the presence of lactic acid bacteria (LAB). Therefore, we hypothesised that LAB produce compounds with antimicrobial activities in the aloe flesh and that these bacteria proliferate after aloe's antimicrobial compounds deteriorate.

Plant-derived LABs are generally found in fermenting fruits, vegetables or traditional foods and pickles in Asia. Plant-derived LABs were first investigated in Japan [40]. These studies focused on the isolation of acid and bile tolerant LAB because plant-derived LAB are much more resistant to artificial gastric juices and bile than animal-derived LAB [7,36]. In Korea, various acid-tolerant LAB strains have been isolated from kimchi, a traditional Korean food [4,19,21], fermented soybean paste [20] and pickles [17].

LABs are widely used in various industries as food and cosmetic additives [24,31]. These bacteria are "*probiotics*", which are viable bacilli of a single strain or mixture of strains that are beneficial to human and animals; they improve the composition of intestinal microflora when consumed as dried cells or in fermented products [2,3,6,12]. "*Probiotic*" bacteria must not only survive in the distal ileum and colon but also adhere and colonise there to effectively confer benefits on the host [5,9,32]. However, the low-pH and antimicrobial environment of the stomach forms a natural barrier to bacterial entry into the intestinal tract. Therefore, acid tolerance is one of the main criteria for selecting of "probiotics" [14].

The aims of this study were to isolate LAB from naturally fermented *A. vera* L. leaf flesh and to prove their probiotic properties, i.e., antimicrobial activity and tolerance to acid and bile salt. These putative probiotic bacteria and their metabolites may be useful resources for the development of antibiotics and antimicrobial substances. Moreover, probiotic bacteria may enhance the therapeutic value of *A. vera* L products.

Table 1. Screening of LAB from fermented aloe flesh.

Selection step	Number of colonies selected
1. Acid tolerance (pH 2.5 for 4 h at 37°C)	113
2. Bromocresol Purple (Lactic acid positive)	65
3. Haemolysis test (non-haemolytic)	30
4. RAPD Analysis	5 strains*

*5 strains were identified by RAPD analysis.

Materials and Methods

Ethics Statement

No specific ethics permits were required for the described studies. The collection of *A. vera* L. plants were purchased from the private farm owners of Jayeon aloe (Jindo Island, Jeonnam, Korea), under Licence Number 415/06/88966 of Korea Food and Drug Administration (KFDA). Collections were not performed in national parks or other protected area of land, and did not involve endangered or protected species. All animal experiments complied with the current laws of Korea. Animal care and treatment were conducted in accordance with guidelines established by the Institute Institutional Animal Care and Use Committee of Chonnam National University. The protocol was approved by the committee on the Ethics of Animal Experiments of the Chonnam National University.

Isolation of Acid-tolerant LAB

Leaves were collected from *A. vera* L. plants that had been grown for more than five years and allowed to ferment naturally for 20–30 days at RT. Fermented substances(pH 3.5–3.8) were centrifuged, and the supernatant was transferred to a sterile tube. MRS broth adjusted to pH 2.5 was inoculated with the fermentation supernatants (1% v/v final concentration). After 5 h incubation at 37°C, acid-resistant colonies were selected by the agar dilution method. Briefly, ten-fold serial dilutions of the MRS broth culture were spread onto MRS agar (Difco, USA), and plates were incubated at 37°C for 24 h in aerobic conditions. Isolated colonies were randomly picked for further selection.

A total of 113 colonies were subjected to a bromocresol purple (BCP) test. After 24 h incubation at 37°C, yellow colonies were selected as the positive clones of LAB [26]. From a total of 113 colonies, 65 colonies were subjected to a haemolysis test modified from [38]. This procedure ruled out 35 haemolytic colonies, and the other 30 colonies were investigated further to identify individual strains. The isolated colonies were stored at −80°C in MRS broth containing 10% glycerol.

Identification of Individual Strains

Each colony was streaked on an MRS agar plate and incubated at 37°C for 24 h in aerobic conditions. A single colony from the plate was transferred to MRS broth and incubated for 18 h at 37°C. To prepare genomic DNA and rRNA, bacteria were lysed with lysozyme and mutanolysin as described previously [39], and genomic DNA was extracted using an Accuprep Genomic DNA Extraction Kit (Bioneer, Korea).

RAPD analysis was carried out with 20 OPA and 20 OPC random primer sets (Operon, USA). PCR was performed with HiPi 5×PCR premix according to the manufacturer's instructions(Elpisbio, Korea). DNA templates (10 ng) and primers (5 pmol)

were added to the PCR premix and DNase-free distilled water was added to a final volume of 20 µl. Amplification was performed in a thermal cycler (Kyratech, Australia) programmed as follows: 94°C for 5 min; 40 cycles of 94°C for 1 min, 38°C for 1 min, and 72°C for 1.5 min; and 72°C for 5 min. PCR products were analysed on a 2% agarose gel. All procedures were repeated at least three times to ensure reproducibility. RAPD data were analysed with MVSP-32 software (Kovach Computing Services, UK) to calculate the genetic similarity between isolates and to draw a phylogenetic dendrogram by the UPGMA method.

The 16S rRNA gene was amplified using the following universal primers: (forward), 5′-AGAGTTTGATCCTGGCTCAG-3′and (reverse) 5′-GGTTACCTTTGTTA CGACTT-3′ (Bioneer, Korea). The thermal cycling parameters are as follows: denaturation at 94°C for 5 min, 30 cycles of 94°C for 1 min, 55°C for 1 min, and 72°C for 40 s and a final extension at 72°C for 5 min. Amplified products were separated on an agarose gel and purified for sequencing with a Gel Extraction Kit (Intron, Korea). The 16S rRNA gene products were subcloned into the pGEM-T easy vector (Promega, USA) and sequenced by Bioneer company (Korea). Sequence homologies were determined by BLASTn searches of the NCBI database (http://www.ncbi.nlm.nih.gov/BLAST) and by Molecular Evolutionary Genetics Analysis with maximum likelihood and neighbor-joining methods (Mega 4.0).

Characterisation of LAB Strains

The following reference strains were obtained from the Korean Collection for Type Cultures (KCTC, Daejeon, Korea), the Korean Culture Center of Microorganisms(KCCM, Seoul, Korea) and American Type Culture Collection(ATCC, Rockville, MD, USA): *B. cereus* KCTC1012[T], *C. perfringens* KCTC3269[T], *C. jejuni* ATCC33560[T], *E. coli* ATCC33694[T], ATCC8739[T] and ATCC43888[T], *E. aerogenes* KCTC2190[T], *E. faecalis* KCTC2011[T], *K. pneumonia* KCTC2208[T], *L. monocytogenes* KCTC3567[T], *P. acidilactici* KCTC1626[T], *P. aeroginosa* KCTC1750[T], *S. aureus* KCTC1621[T], *S. typhi* KCTC2424[T], *S. enteroritidis* KCCM12400[T], and *L. brevis* KCTC 3498[T], KCTC 3102[T].

Carbohydrate utilisation was investigated with an API kit (API 50 CHL) from Biomerieux (Marcy-l'Etoile, France) according to the manufacturer's instructions, and the results were analysed according to Biomerieux's guide.

For acid tolerance tests, the selected strains were subjected to a more accurate screening for acid tolerance in growth media, adjusted to pH 2.5, or 3.0, with 0.5 M HCl and SGF buffer [simulated gastric acid fluid,; 2 g/L NaCl, 3.2 g/L pepsin (Sigma, USA)]. Bacteria were grown on each type of media for the indicated time at 37°C. Afterwards, cell viability was determined by the agar dilution method on BCP agar plate. The original cell density (OCD) was calculated using the formula; OCD = [colonies on plate]/[volume of sample plated × dilution factor]. The survival rate was calculated as the percentage of OCD on MRS. Each experiment was performed in triplicates.

POAL strain bile salt tolerance was examined in 10 ml of MRS broth supplemented with 0.3, 0.5, 1.0, 2.0, and 5.0% (w/v) oxgall bile salt (Difco, USA) or a negative control (0% bile salt); cultures were incubated for 18 h at 37°C. Bacterial cultures were serially diluted in PBS (pH 7.4) and plated on BCP agar. After incubation for 24 h at 37°C, colonies were counted. The survival rate at each concentration of bile salt was calculated as the ratio of colonies to those on the negative control plate. The experiment was performed in triplicates.

Antibiotic susceptibility was tested by the agar dilution method published by the National Committee for Clinical Laboratory Standards [25]. To determine the minimum inhibitory concen-

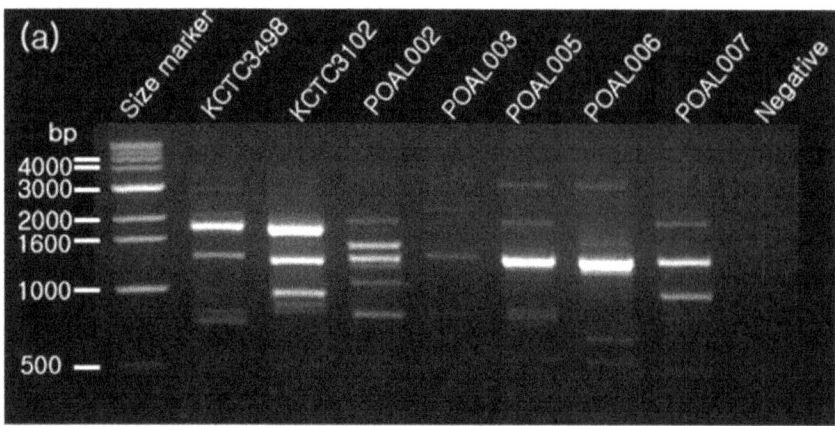

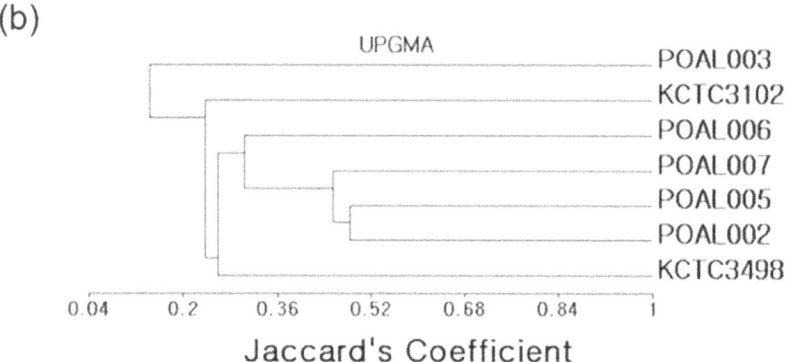

Figure 1. Phylogenetic analysis of isolated *L. brevis* strains by RAPD. (a) A representative RAPD analysis with primer OPA-08 (b) UPGMA phylogenetic tree of isolated strains.

tration (MIC) of antibiotics, concentrations recommended by the Scientific Committee on Animal Nutrition (SCAN) were used. The antibiotic concentrations (per ml of MRS agar) were as follows: neomycin (30 μg), streptomycin (10 μg), Amoxicillin/Clavulanic Acid (20/10 μg), Ampicillin (10 μg), Cefotaxime (30 μg), Chloramphenicol (30 μg), Ciprofloxacin (5 μg), Colistin (10 μg), Doxycycline (30 μg), Erythromycin (15 μg), Gentamicin (10 μg), Tetracycline (30 μg), and Trimethoprim/Sulfamethoxazole (1.25/ 23.75 μg). Antibiotic discs were obtained from Difco (USA). The

POAL strains were subcultured twice before susceptibility tests. MRS agar plates were inoculated with approximately 10^9– 10^{10} CFU/ml (final concentration; 100 μl final volume per well) and incubated for 24 h at 37°C before MICs were visually determined. Inhibition zones were measured and strains were categorised as resistant (R), intermediate (I), or susceptible (S).

Antibacterial activities. Antibacterial activities of the POAL strains against several Gram-positive and Gram-negative strains were tested using the penicylinder method, originally

site # Stains	4	19	60	143	204	293	294	582	747	778	1106	1154	1198	1203	1214	1308
KCTC3498	T	C	A	A	A	C	A	A	C	G	T	A	C	A	T	C
POAL 006	T	C	A	A	A	C	A	G	C	G	T	A	C	A	T	C
POAL 002	C	C	G	G	A	C	A	A	C	G	T	G	T	A	T	C
POAL 003	T	C	A	A	A	C	A	A	C	G	T	A	T	A	T	C
KCTC3102	T	C	A	A	A	C	A	A	C	G	T	A	T	A	T	C
POAL 005	T	T	A	A	G	A	A	A	–	A	T	A	T	A	A	C
POAL 007	T	T	A	A	G	C	T	A	C	G	C	A	T	G	T	T

Figure 2. Base changes in the 16S rRNA sequences of *L. brevis* strains.

Table 2. Biochemical and physiological characteristics of lactic acid bacteria*.

	POAL 002	POAL 003	POAL 005	POAL 006	POAL 007	KCTC 3498	KCTC 3102
Glycerol	−	−	−	−	−	−	−
Erythritol	−	−	−	−	−	−	−
D-Arabinose	−	−	−	−	−	−	−
L-Arabinose	+	+	+	+	+	+	?
Ribose	+	+	+	+	+	+	?
D-Xylose	+	+	+	+	+	+	+
L-Xylose	−	−	−	−	−	−	−
Adonitol	−	−	−	−	−	−	−
β-Methyl-xyloside	−	−	−	−	−	−	−
Galactose	−	−	−	−	−	−	−
D-Glucose	?	?	?	?	?	?	−
D-Fructose	?	?	?	?	?	+	?
D-Mannose	−	−	−	−	−	−	−
L-Sorbose	−	−	−	−	−	−	−
Rhamnose	−	−	−	−	−	−	−
Dulicitol	−	−	−	−	−	−	−
Inositol	−	−	−	−	−	−	−
Mannitol	−	−	−	−	−	−	−
Sorbitol	−	−	−	−	−	−	−
α Methyl-D-mannoside	−	−	−	−	−	−	−
α Methyl-D-glucoside	−	−	−	−	−	−	−
N acetyl glucosamine	?	?	?	−	?	?	?
Amygdaline	−	−	−	−	−	−	−
Arbutine	−	−	−	−	−	−	−
Esculin	−	−	−	−	−	−	−
Salicine	−	−	−	−	−	−	−
Cellobiose	−	−	−	−	−	−	−
Maltose	+	+	?	?	?	+	+
Lactose	−	−	−	−	−	−	−
Melibiose	−	−	−	−	−	−	−
Saccharose	−	−	−	−	−	−	−
Trehalose	−	−	−	−	−	−	−
Inuline	−	−	−	−	−	−	−
Melezitose	−	−	−	−	−	− −	−
D-Raffinose	−	−	−	−	−	−	−
Amidon	−	−	−	−	−	−	−
Glycogen	−	−	−	−	−	−	−
Xylitol	−	−	−	−	−	−	−
β Gentiobiose	−	−	−	−	−	−	−
D-Turanose	−	−	−	−	−	−	−
D-Lyxose	−	−	−	−	−	−	−
D-Tagatose	−	−	−	−	−	−	−
D-Fucose	−	−	−	−	−	−	−
L-Fucose	−	−	−	−	−	−	−
D-Arabitol	−	−	−	−	−	−	−
L-Arabitol	−	−	−	−	−	−	−
Gluconate	?	?	−	?	?	?	−
2 ceto-gluconate	−	−	−	−	−	−	−

Table 2. Cont.

	POAL 002	POAL 003	POAL 005	POAL 006	POAL 007	KCTC 3498	KCTC 3102
5 ceto-gluconate	?	?	?	?	?	?	–

*(−) = negative result, (+) = positive result, (?) = doubtful result, as described in Materials and Methods.

described by [13]. Overnight cultures of the indicator strains were diluted to a concentration of 10^5–10^6 CFU/ml, transferred into nutrient agar and brain heart infusion agar (Difco, USA) at $37°C$, and poured into plates. Two stainless steel penicylinders (Fisher Scientific, USA) were evenly spaced onto each plate. The penicylinder slots were filled with 150 µl POAL strains, and the plates were incubated at $37°C$ for 24 h in aerobic conditions. The antibacterial activity of each strain was determined by the size of the inhibition (clear) zone.

Intestinal survival of POAL strains. To evaluate the survival of POAL strains in vivo, bacteria were cultured in MRS broth for 24 h at $37°C$ and pelleted at $3,000×g$ for 10 min. The bacterial pellets were resuspended in PBS buffer (pH 7.4) and administered orally to each group of female BALB/c mice (n = 3/ group) for 1 week. Female BALB/c mice (5–6 weeks; $24±1.5$ g) were purchased from the Yang-Sung Experimental Animal Center (Korea) and housed in standard environmental conditions (12 h light–dark cycle, 50–70% humidity and 20–25°C). Food and water were provided ad libitum. The mice were sacrificed by the cervical dislocation. Abdominal cavities of mice were cut open by a midline abdominal incision to collect large intestines. Large intestines were obtained from test mice and flushed with PBS buffer (pH 7.4). The lavage fluids were diluted 10-fold, incubated in SGF buffer (pH 2.5) at $37°C$ for 4 h, plated on MRS agar (Difco, USA) containing 30 µg of tetracycline at $37°C$ for 24 h. The strain identities of selected colonies were determined by RAPD-PCR. Each experiment was performed in triplicate.

Quantification of GAD expression. Total RNA was isolated from POAL strains and two reference strains (KCTC3498[T], KCTC3102[T]) with a Qiagen RNeasy kit (Qiagen, USA). The isolated RNA was treated with RNase-free DNase (Roche, Switzerland) to remove contaminant DNA. Reverse transcription was performed with 1 µg total RNA and 20 µl reaction mix from an iNtRON Power cDNA synthesis Kit (Intron, Korea) and AMV reverse transcriptase (Intron, Korea), according to the manufacturer's directions. Real-time PCR amplification was performed with gene-specific primers and SYBR Green Master Mix (Qiagen, USA) using a Rotor-Gene 6000 real-time amplification operator (Corbett Research, Australia). The gene specific primers were *GAD128_F* (5′-GATGAAGTTTGCTTGGCG-TAAG-3′) and *GAD128_R* (5′-CGATGTCCC AATAGACACA-GAA-3′). *gapB* was used as the reference gene and was amplified with *gapB_F* (5′-ACGGAATTAGTTGCAATCTTAGAC-3′) and *gapB_R* (5′-GAAAGTAGTACC GATAACATCAGA-3′) primers. The PCR conditions were as follows: 1 cycle of $95°C$ for 15 min and 45 cycles of $95°C$ for 25 s, $56°C$ for 25 s, and $72°C$ for 30 s. Samples were run three times in triplicate, and the relative amount of target RNA for each sample was calculated by statistical analysis [30].

Statistical Analyses

The results are mean values and standard deviations of 3 measurements from 2 independent assays. Statistical analysis was performed using Sigma Plot 2000 software (SPSS, USA). The results were compared by an Analysis of Variance (ANOVA) general linear model followed by Tukey's post-hoc test. Statistical significance was determined as $p < 0.05$.

Results

Isolation of LAB

From the fermented *Aloe vera* flesh, 113 colonies that survived 4 h incubation at pH 2.5 were isolated (Table 1). The LABs were identified on an MRS agar plate containing bromocresol purple, and yellow colonies were scored as positive. Of the 113 colonies, 65 lactic acid bacteria colonies were screened for haemolytic activity. Thirty colonies were non-haemolytic.

Identification of Five LAB StrainsRAPD analysis. RAPD analysis of the 30 non-haemolytic colonies revealed five lactic acid bacterial strains (Figure 1a). Analysis of patterns produced with 40 RAPD primers resulted in a UPGMA phylogenetic tree (Figure 1b), which showed the relative genetic similarity between the 5 strains. All 5 strains were identified as *Lactobacillus brevis* in the API 50 CHL test (Biomereiux, France), and strains POAL002, 007, 005, and 006 were similar to the reference *L. brevis* strain, KCTC3498[T]. POAL006 was most closely related to KCTC3498[T]. POAL003 strain was least genetically similar to KCTC3498[T] but was related to another *L. brevis* reference strain, KCTC3102[T].

16S rRNA sequence comparisons. The genetic identities and phylogenetic relationships of the five strains were confirmed by 16S rRNA gene sequence analysis (Figure 2). The 16S rRNA sequences from all 5 strains were 97–99% similar to *L. brevis* and were most similar to the *L. brevis* ATCC 14869[T] strain. Thus, these strains were named *L. brevis* POAL 002, 003, 005, 006, and 007 deposited in Genebank with the accession numbers JX185493, JX185494, JX185495, JX185496, and JX185497, respectively). 16S rRNA sequence analysis confirmed the phylogenetic relationships determined by RAPD analysis (Figure 1b).

Characterisation of the Strains

Carbohydrate utilization. Carbohydrate utilisation by the POAL strains was consistent with that of *L. brevis* species (Table 2). The metabolic characteristics of POAL005 were most similar (97.4%) to *L. brevis*, followed by POAL006 (89.1%), POAL007 (88.1%), POAL002 and POAL003 (88%) strains. The reference strains KCTC 3498[T] and KCTC 3102[T] were also analysed to validate the accuracy of the assay.

Acid tolerance. All the POAL strains were strongly tolerant to acidic conditions (Figure 3). The POAL strains could withstand exposure to pH 2.5 and pH 3.0 conditions better than KCTC 3498[T], and KCTC 3102[T]. All the POAL strains survived at pH 3.0 for 4 hours. The POAL005, 006, 007 strains were especially acid-tolerant. While the other strains did not survive longer than 2 hours, these 3 strains survived for 4 hours at pH 2.5. The POAL003 strain was superior in particular at pH 3.0 but not at pH 2.5.

Bile salt tolerance. The POAL strains exhibited greater tolerance to bile salt than the reference strains KCTC 3498[T], and

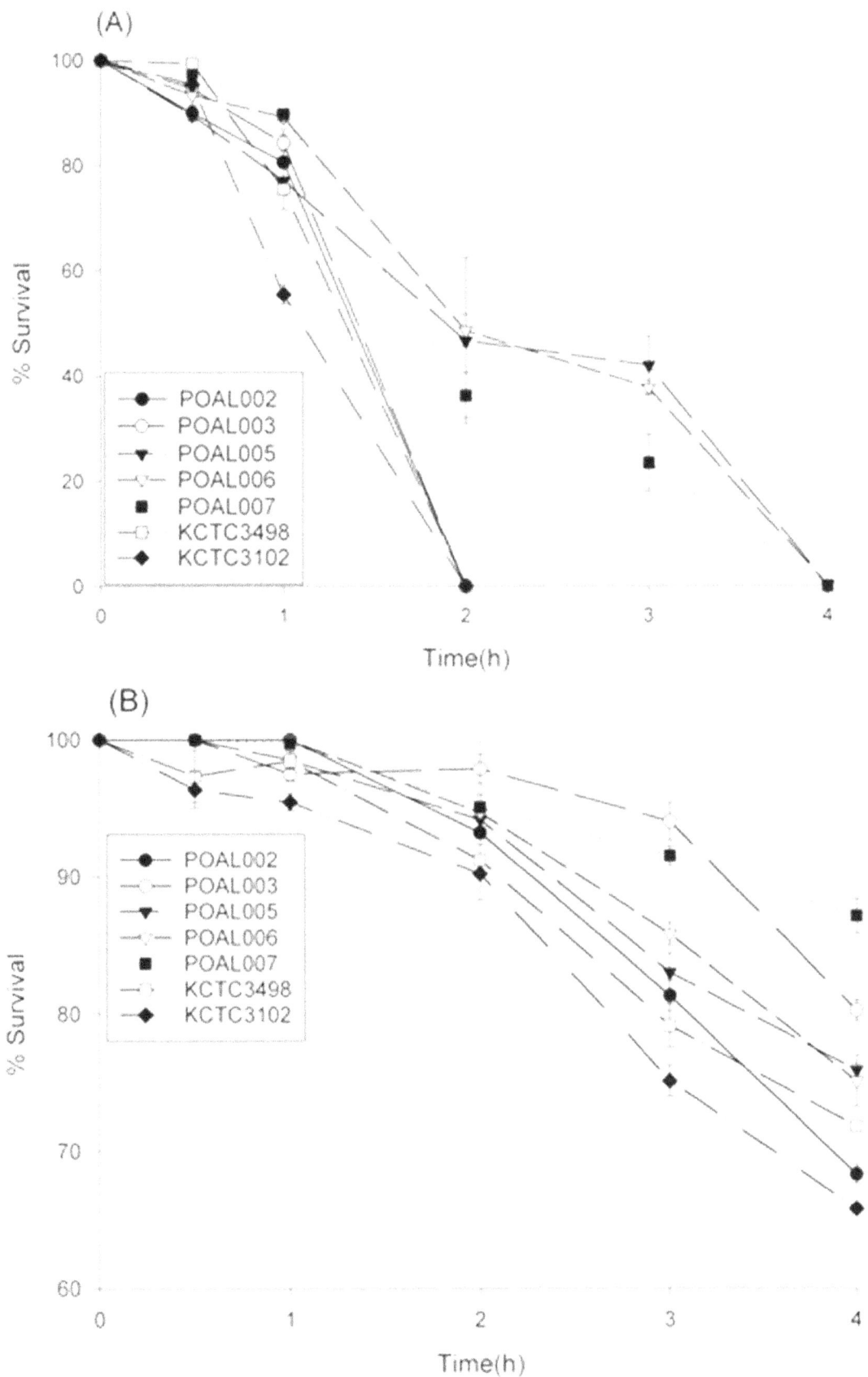

Figure 3. Survival of *L. brevis* POAL strains at pH 2.5 (A) and pH 3.0 (B), respectively.

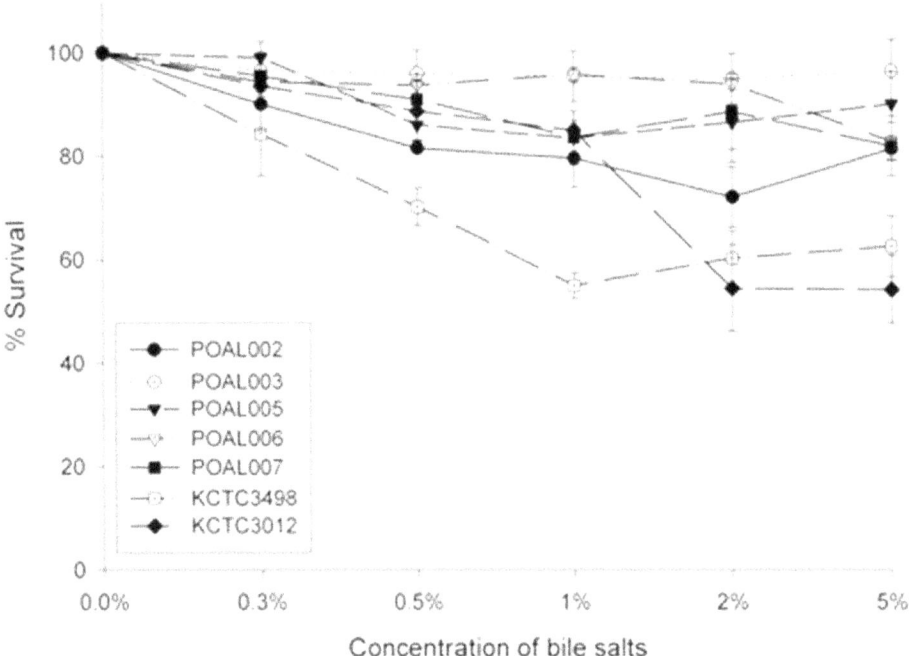

Figure 4. Effect of bile salt concentration on the viability of *L. brevis* **POAL strains.** Survival rates were estimated after 18-hours at 37°C in the indicated bile salt concentration.

KCTC 3102^T (Figure 4). All the POAL strains maintained more than 80% of plain control solutions below 1.0% after 18 h at 37°C. The POAL003, 005 and 006 strains were especially resistant, with more than 90% survival at >2% bile salt. The reference strains were more sensitive to bile salt than POAL strains. Survival rates were estimated after 18 hours at 37°C in the indicated bile salt concentration.

Antibiotic resistance. The *L. brevis* POAL strains and reference strains KCTC 3498^T, and 3012^T had specific resistance to various antibiotics (Table 3). All of the *L. brevis* strains were resistant to tetracycline, norfloxacin, sulfatrimethoxazole, colistin, and the common Gram-negative antibiotics. These strains were, on the contrary, susceptible to amoxicillin/clavulanic acid, chloramphenicol and ampicillin without discrepancy. However, each POAL strain also exhibited an individual antibiotic sensitivity

Table 3. Antibiotic susceptibility profiles of the POAL strains*.

Antibiotics	POAL 002	POAL 003	POAL 005	POAL 006	POAL 007	KCTC 3498	KCTC 3102
Apramycin	I	+	+	S	S	S	+
Gentamicin	+	S	I	S	I	S	+
Cefotaxime	I	I	+	+	I	I	S
Tetracycline	+	+	+	+	+	+	+
Amox./Clav Acid†	S	S	S	S	S	S	S
Chloramphenicol	S	S	S	S	S	S	S
Neomycin	+	I	I	I	I	I	I
Norfloxacin	+	+	+	+	+	+	+
Colistin	+	+	+	+	+	+	+
Ampicillin	S	S	S	S	S	S	S
Sulfa-trimetho	+	+	+	+	+	+	+
Erythromycin	I	I	I	I	I	S	I
Oxyteracycline	+	I	+	+	+	+	+
Doxycycline	I	I	+	I	I	I	S
Ciprofloxacin	+	+	+	+	+	+	+
Streptomycin	+	+	+	I	+	+	+

*+ = resistant; I = intermediate; S = susceptible †Amox/Clav acid; Amoxicillin/Clavulanic Acid.

Table 4. Antimicrobial spectrum of the selected *L. brevis* POAL strains*.

Indicator strains	*L. brevis* strains						
	POAL002	POAL003	POAL005	POAL006	POAL007	KCTC3498	KCTC3102
Bacilluis cereus KCTC 1012[T]	−	−	−	−	−	−	−
Campylobacter jejuni ATCC33560 [T]	++	++	+++	+++	++	++	++
Clostridium perfringens KCTC 3269 [T]	+++	+++	+++	+++	+++	+++	+++
Escherichia coli ATCC33694 [T]	−	−	−	−	−	−	−
Escherichia coli ATCC 8739 [T]	++	++	++	+++	+++	+++	+++
Escherichia coli O157:H7 ATCC43888 [T]	++	++	++	+++	+++	+++	+++
Enterobacter aerogenes KCTC 2190 [T]	+	+	+	++	++	+	+
Enterococcus faecalis KCTC 2011 [T]	−	−	−	−	−	−	−
Klebsiella pneumoniae KCTC 2208[T]	+	+	+	+	+	+	+
Listeria monocytogenes KCTC3567[T]	−	−	−	−	−	−	−
Pediococcus acidilactici KCTC 1626[T]	+	+	+	+	+	+	+
Pseudomonas aeruginosa KCTC 1750[T]	+	+	+	+	+	++	++
Staphylococcus aureus KCTC 1621[T]	++	++	++	++	++	++	++
Salmonella typhi KCTC 2424[T]	+	++	++	++	+	+	+
Salmonella enteritidis KCCM12400[T]	++	++	++	++	++	+	++

*Symbols for diameter of inhibition zones: +++ = >30 mm, ++ = >20~25 mm, + = >15 mm; − = 0.

profile. POAL002 strain was resistant to neomycin while others were not. POAL003 was moderately susceptible to oxytetracycline, while POAL006 was somewhat sensitive to streptomycin. In general, unique antibiotic resistance profile distinguished each of the *L. brevis* strains tested (Table 3).

Probiotic Properties

Antibacterial activities. The five POAL strains were tested for their antimicrobial activities against various foodborne bacteria using the penicylinder method (Table 4). All the POAL strains demonstrated a broad spectrum of antimicrobial activity against some harmful enteropathogens. *C. jejuni* and *C. perfringens* were most strongly suppressed and *S. aureus*, *Salmonella* were second most. On thecontrary, the normal commensals such as *E. faecalis*, *B. cereus* were not affected at all. In case of *E. coli*, a normal strain, ATCC33694 [T] avoid antimicrobial activities of the *L. brevis* strains while another non-pathogenic strain, ATCC 8739 [T] was severely inhibited. Beneficially, however, the enteropathogenic O157:H7 (ACTC43888[T]) strain also was prevailed by the POALs. Similar to the reference strains, the antimicrobial properties of POAL strains did not coincide with Gram satiability.

Intestinal survival of POAL strains. Another criterion for valid probiotic bacteria is viability in the intestines of live mammals. We investigated the viability of POAL strains in mice intestines, after 1 week of oral administration. After screening the bacterial colonies with pH 2.5 and tetracycline, we found that one strain, POAL006, survived in the intestine. The KCTC3498[T]

strain also passed the same selection, while another reference strain, KCTC3102[T], failed. The identities of POAL006 and KCTC3498[T] colonies were confirmed by RAPD analysis (Table 5).

Elevated expression of Glutamate decarboxylase (GAD). All POAL strains expressed GAD mRNA at high level compared with the KCTC3498[T] and KCTC3102[T] reference strains. GAD mRNA levels in POAL006 were about three fold that in KCTC3102[T]. POAL003, 006 and 007 expressed two times more GAD mRNA than the reference strains (Figure 5). Higher expression of GAD mRNA suggested that the POAL strains are potential probiotics.

Discussion

Despite its strong antibacterial properties, fermented of *Aloe vera* L. has been scarcely examined. It may due to the fact that effective ingredients in the leaf gel may be degraded after oxidation, and consequently, their antibiotic activity may be weakened to be hardly detectable. Because the fermentation reached at about pH 3.8, it is assumed that acid-tolerant bacteria will be found in the fermented stuff. Previous studies of the acidic fermentation of foods, plant-derived LAB such as *L. plantarum*, *L. acidophilus*, and *L. pentosus* were discovered, these observations led us to investigate LAB in fermented *Aloe vera*.

We identified five LAB strains in naturally fermented *Aloe vera* leaf flesh. API 50CHL analyses revealed that all LAB strains were

Table 5. Viability of *L. brevis* strains in the mouse intestine after administration for 1 week.

Strains	POAL 002	POAL 003	POAL 005	POAL 006	POAL 007	KCTC3498	KCTC3102
Identified	−	−	−	+	−	+	−
Matched	0	0	0	100%	0	100%	0

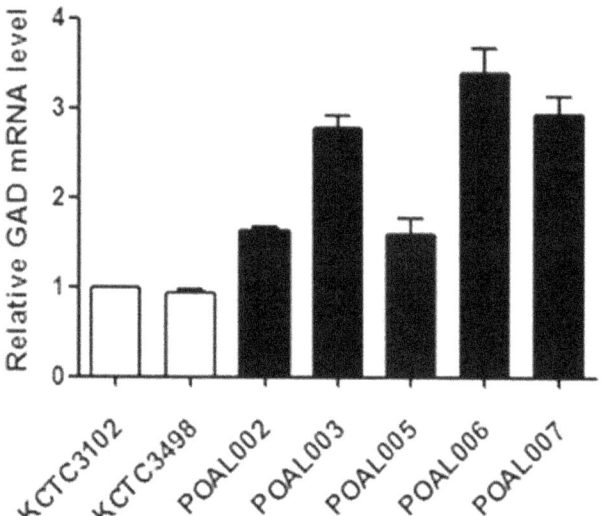

Figure 5. qRT- PCR analysis of GAD expression. Total RNA was isolated from *L. brevis,* converted to cDNA and analysed by qRT-PCR. KCTC3498[T] and KCTC3102[T] strains were used as references. Bars represent relative mean intensity. All values are expressed as the means ± SE, and the statistical significance was set at P<0.05.

L. brevis, and their identities were confirmed by RAPD pattern analysis and 16S rRNA sequence comparisons. These strains were unique in their acid and bile tolerance, resistance to antibiotics and GAD expression. We named these POAL (Probiotics Originating from Aloe Leaf) strains because they possess probiotic characteristics similar to those of the *L. brevis* reference strains, KCTC 3498[T] (isolated from animal faeces) and KCTC 3102[T] (isolated from pickles).

The five novel POAL strains identified in this study are all highly acid tolerant, surviving at pH 2.5 for as long as 4 hours and even longer(up to 6 hours) at pH 3.0. Generally, *L. brevis* strains do not survive at pH 2.5 [29,37], and the reference strains used in this study, KCTC3498[T] and KCTC3102[T], did not persist for 2 hours at pH 2.5 (Figure 3). High acid tolerance may confer sustainability on these strains to survive in the gastric and intestinal environments. Another essential characteristic of probiotics is tolerance to bile salts, which are another barrier to LAB colonisation of the intestine; the probiotics should withstand the bile salts in the environment. All of the five POAL strains had above 80% survival rate after incubation for 18 hours in 5% bile salts. However, the reference strains' survival rate was below 70%. Interestingly, strains POAL003 and 006 exhibited higher tolerance to both bile salts and strong acid, suggesting that these strains might survive in vivo passage to the intestine. In fact, viable POAL006 was

recovered from the intestines of mice one week after ingestion (Table 5). Although no more strain detected in this assay, it may refer to the selection in, pH 2.5 for 4 hours, too stringent conditions. In the acid tolerance test, POAL003 survived at pH 3.0 but not pH 2.5.

Along with acid and bile tolerance, antibacterial activity is another valuable property of probiotics. The five POAL strains identified in this study inhibited the growth of many harmful enteropathogens while did not restrain most of normal commensals in the gut. Contradictory results against *E. coli* strains await further intensive investigation. Antimicrobial properties of POAL strains had no relevance to Gram stainability of enteropathogenic strains [22].

The existence of GAD may also validate the POAL strains. GAD is a pyridoxal 50-phosphate (PLP)-dependent enzyme, that catalyses the irreversible GABA [34,35]. GABA is a well-characterised inhibitory neurotransmitter with hypotensive and analgesic properties. GABA also has tranquilising effect, particularly with regards to insomnia, depression and an autonomic disorder observed during the menopausal or presenium periods [16,18,23]. It induces insulin secretion from the pancreas and effectively relieves diabetic troubles [10]. Therefore, GABA attracts special interest as a potential bioactive component in foods and pharmaceuticals. In this study, GAD gene expression in the POAL strains was approximately two fold that in the reference strains (Figure 5). Several investigators have previously described GAD expression in LAB [27,28]. Although direct analysis of GABA synthesis should be performed, these results suggest that our POAL strains may be resources for GABA production.

In conclusion, the five novel *L. brevis* POAL strains have many desirable probiotic characteristics such as antimicrobial properties against enteropathogens, a little inhibition to normal fauna, acid resistance, bile tolerance and intestinal viability. Their higher levels of GAD expression implied some potential validity. Antibiotic resistance also suggested their practical usage as supplements for patients with intestinal disorders or undergoing antibiotic treatment. Further studies of POAL strains will help us utilise these bacteria in pharmaceutical and industrial settings.

Acknowledgments

We are grateful to Professor Je Chang, Woo at the Department of Biological Sciences, Mokpo National University, for helpful advice.

Author Contributions

Conceived and designed the experiments: YWK JK. Performed the experiments: YWK YJJ AYK HHS JAL. Analyzed the data: YWK YJJ AYK JK. Contributed reagents/materials/analysis tools: CHK CHJ JK. Wrote the paper: YWK JK. Acid abd bile tolerant test: YWK HHS JAL. DNA/RNA work and sequencing: YWK HHS JAL.

References

1. Barcroft M (2003) Aloe Vera: Nature's Silent Healer. BAAM, USA. ISBN 0-9545071-0-X.
2. Betoret N, Puente L, Diaz MJ, Pagan MJ, Garcia MJ, et al. (2003) Development of probiotic enriched dried fruits by vacuum impregnation. J Food Eng 56: 273-277.
3. Chandramouli V, Kailasapathy K, Peiris P, Jones M (2004) An improved method of microencapsulation and its evaluation to protect *Lactobacillus spp.* in simulated gastric conditions. J Microbiol Methods 56: 27-35.
4. Cho JK, Li GH, Cho SJ, Yoon YC, Hwang SG, et al. (2007) The identification and physiological properties of *Lactobacillus plantarum* JK-01 isolated from kimchi. Korean J Food Sci Ani Resour 27: 363-370.
5. Chou LS, Weimer B (1999) Isolation and characterization of acid- and bile-tolerant isolates from strains of *Lactobacillus acidophilus*. J Dairy Sci 82: 23-31.
6. Collado MC, Meriluoto J, Salminen S (2007) Role of commercial probiotic strains against human pathogen adhesion to intestinal mucus. Lett Appl Microbiol 45: 454-460.
7. Cotter PD, Hill C (2003) Surviving the acid test: Responses of Gram-positive bacteria to low pH. Microbiol Mol Biol Rev 67: 429-453.
8. Deng XH, Chen WX, He QM, Zhu LF (1999) Utilization and resources protection of *Aloe vera L. var. chinensis* (Haw.) Berger. J Plant Resour Environ 8: 26-30.
9. Ding WK, Shah NP (2009) An improved method of microencapsulation of probiotic bacteria for their stability in acidic and bile conditions during storage. J Food Sci 74: M53-M61.
10. Erlander MJ, Tobin AJ (1991) The structural and functional heterogeneity of glutamic acid decarboxylase: a review. Neurochem Rese 16: 215-226.

11. Esteban-Carrasco A, López-Serrano M, Zapata JM, Sabater B, Martín M (2001) Oxidation of phenolic compounds from *Aloe barbadensis* by peroxidase activity: Possible involvement in defence reactions. Plant Physiol Biochem 39: 521–527.

12. Ezendam J van, Loveren H (2006) Probiotics: immunomodulation and evaluation of safety and efficacy. Nutr Rev 64: 1–14.

13. Farnell MB, Donoghue AM, Cole K, Reyes-Herrera I, Blore PJ, et al. (2005) Campylobacter susceptibility to ciprofloxacin and corresponding fluoroquinolone concentrations within the gastrointestinal tracts of chickens. J Appl Microbiol 99: 1043–1050.

14. Favaro-Trindade CS, Grosso CR (2002) Microencapsulation of *L. acidophilus* La-05 and *B. lactis* Bb-12 and evaluation of their survival at the pH values of the stomach and in bile. J Microencapsul 19: 485–494.

15. Ferro VA, Bradbury F, Cameron P, Shakir E, Rahman SR, et al. (2003) In vitro susceptibilities of *Shigella flexneri* and *Streptococcus pyogenes* to inner gel of *Aloe barbadensis* Miller. Antimicrob Agents Chemother 47: 1137–9.

16. Ge S, Pradhan DA, Ming GL, Song H (2007) GABA sets the tempo for activity-dependent adult neurogenesis. Trends Neurosci 30: 1–8.

17. Kang TK, Kim WJ (2010) Characterization of an amylase-sensitive bacteriocin DF01 produced by *Lactobacillus brevis* DF01 isolated from dongchimi, Korean fermented vegetable. Korean J Food Sci Ani Resour 5: 795–803.

18. Kaufman DL, Houser CR, Tobin AJ (1991) Two forms of the gamma-aminobutyric acid synthetic enzyme glutamate decarboxylase have distinct intraneural distributions and cofactor interactions. J Neurochem 56: 720–723.

19. Kang HJ, Kim BC, Park W (2004) Isolation of tetracycline-resistant lactic acid bacteria from Kimchi. Korean J Microbiol 40: 1–6.

20. Kim TW, Lee JH, Kim SE, Park MH, Chang HCh, et al. (2009) Analysis of microbial communities in doenjang, a Korean fermented soybean paste, using nested PCR-denaturing gradient gel electrophoresis. Int J Food Microbiol 131: 265–271.

21. Lee YM, Kwon MJ, Kim JK, Suh HS, Choi JS, et al. (2004) Isolation and identification of active principle in Chinese cabbage Kimchi responsible for antioxidant activity. Korean J Food Sci Technol 36: 129–133.

22. Makras L, Vuyst De L (2006) The in vitro inhibition of Gram-negative pathogenic bacteria by bifidobacteria is caused by the production of organic acids. Int Dairy J 16: 1049–1057.

23. Mody I, Dekoninck Y, Otis TS, Soltesz I (1994) Bringing the cleft at GABA synapses in the brain. Trends Neurosci 17: 517–525.

24. Morelli L (2000) In Vitro Selection of Probiotic Lactobacilli: A Critical Appraisal. Curr Issues Intest Microbiol 1: 59–67.

25. NCCLS (2004) Methods for Antimicrobial Susceptibility Testing of Anaerobic Bacteria. Approved Standard, 6th ed. NCCLS document M11-A6. Clinical and Laboratory Standards Institute, Wayne, PA.

26. Noritoshi T, Jin-Zhong X, Kazuhiro M, Tomoko Y, Akinori H, et al. (2004) Selection of acid tolerant Bifidobacteria and evidence for a low-pH-inducible acid tolerance response in *Bifidobacterium longum*. J Dairy Res 71: 340–345.

27. Park KB, Oh SH (2004) Cloning and expression of a full-length glutamate decarboxylase gene from *Lactobacillus plantarum*. J Food Sci Nutri 9: 324–329.

28. Park KB, Oh SH (2007) Cloning, sequencing and expression of a novel glutamate decarboxylase gene from a newly isolated lactic acid bacterium, *Lactobacillus brevis* OPK-3. Biores Technol 98: 312–319.

29. Park SY, Ko YT, Jeong HK, Yang JO, Chung HS, et al. (1996) Effect of various lactic acid bacteria on the serum cholesterol levels in rats and resistance to acid, bile and antibiotics. Korean J Appl Microbiol Biotechnol 24: 304–310.

30. Pfaffl MW, Horgan GW, Dempfle L (2002) Relative expression software tool (REST©) for group-wise comparison and statistical analysis of relative expression results in real-time PCR. Nucleic Acids Res 9: e36.

31. Saxelin M, Tynkkynen S, Mattila-Sandholm T, M de Vos W (2005) Probiotic and other functional microbes: from markets to mechanisms. Curr Opin Biotechnol 16: 204–211.

32. Salminen S, Deighton M, Gorbach S (1993) Lactic acid bacteria in health and disease. ISBN 0-8247-8907-5.

33. Shahzad K, Ahmad R, Nawaz L, Saeed S, Iqbal Z (2009) Comparative antimicrobial activity of Aloe Vera gel on microorganisms of public health significance. Pharmacol online 1: 416–423.

34. Sumbul S, Ahmed SW, Azhar I (2004) Antifungal activity of *Allium*, *Aloe*, and *Solanum* species. Pharmaceut Biol 42: 491–498.

35. Ueno H (2000) Enzymatic and structural aspects on glutamate decarboxylase. J Mol Catal B: Enzym 10: 67–79.

36. Van de, Guchte M, Serror P, Chervaux C, Smokvina T, et al. (2002) Stress responses in lactic acid bacteria. Antonie Van Leeuwenhoek 82: 187–216.

37. Vernazza CL, Gibson GR, Rastall RA (2006) Carbohydrate preference, acid tolerance and bile tolerance in five strains of *Bifidobacterium*. J Appl Microbiol 100: 846–853.

38. Vesterlund S, Vankerckhoven V, Saxelin M, Goossens H, Salminen S, et al. (2007) Safety assessment of Lactobacillus strains: Presence of putative risk factors in faecal, blood and probiotic isolates. Int J Food Microbiol 116: 325–331.

39. Vidgrén G, Palva I, Pakkanen R, Lounatmaa K, Palva A (1992) S-layer protein gene of *Lactobacillus brevis*: cloning by polymerase chain reaction and determination of the nucleotide sequence. J Bacteriol 174: 7419–7427.

40. Yamada K (1977) Recent advances in industrial fermentation in Japan. Biotechnol Bioeng 19: 1563–1621.

4

Are There Any Different Effects of *Bifidobacterium, Lactobacillus* and *Streptococcus* on Intestinal Sensation, Barrier Function and Intestinal Immunity in PI-IBS Mouse Model?

Huan Wang[1], Jing Gong[1], Wenfeng Wang[2], Yanqin Long[3], Xiaochao Fu[4], Yu Fu[1], Wei Qian[1], Xiaohua Hou[1]*

1 Division of Gastroenterology, Union Hospital of Tongji Medical College, Huazhong University of Science and Technology, Wuhan, Hubei, China, **2** Division of Gastroenterology, Hospital of Dongfeng Motor Company of Hubei University of Medicine, Shiyan, Hubei, China, **3** Division of Gastroenterology, Sir Run Run Shaw Hospital, School of Medicine, Zhejiang University, Hangzhou, Zhejiang, China, **4** Division of Culture Collection, Hubei Center of Industrial Culture Collection and Research, Wuhan, Hubei, China

Abstract

Background and Aims: Research has increasingly suggested that gut flora plays an important role in the development of post-infectious irritable bowel syndrome (PI-IBS). Studies of the curative effect of probiotics for IBS have usually been positive but not always. However, the differences of treatment effects and mechanisms among probiotic stains, or mixture of them, are not clear. In this study, we compared the effects of different probiotics (*Befidobacterium, Lactobacillus, Streptococcus* or mixture of the three) on intestinal sensation, barrier function and intestinal immunity in PI-IBS mouse model.

Methods: PI-IBS model was induced by *Trichinella spiralis* infection in mice. Different probiotics were administered to mice after 8 weeks infection. Visceral sensitivity was measured by scores of abdominal withdrawal reflex (AWR) and the threshold intensity of colorectal distention. Colonic smooth muscle contractile response was assessed by contraction of the longitudinal muscle strips. Plasma diamine oxidase (DAO) and d-lactate were determined by an enzymatic spectrophotometry. Expression of tight junction proteins and cytokines in ileum were measured by Western blotting.

Results: Compared to control mice, PI-IBS mice treated either alone with *Befidobacterium* or *Lactobacillus* (but not *Streptococcus*), or the mixture of the three exhibited not only decreased AWR score and contractile response, but also reduced plasma DAO and D-lactate. These probiotic treatments also suppressed the expression of proinflammatory cytokine IL-6 and IL-17 and promoted the expression of major tight junction proteins claudin-1 and occludin. The mixture of the three probiotic strains performed better than the individual in up-regulating these tight junction proteins and suppressing IL-17 expression.

Conclusions: *Bifidobacterium* and *Lactobacillus*, but not *Streptococcus*, alleviated visceral hypersensitivity and recovered intestinal barrier function as well as inflammation in PI-IBS mouse model, which correlated with an increase of major tight junction proteins. In addition, Mixture of three species was indicated to be superior to a single one.

Editor: Michael Koval, Emory University School of Medicine, United States of America

Funding: Our sources of funding are from the National Natural Science Fundation of China.Their granted numbers are 81070300 and 81000159.Its website is http://www.nsfc.gov.cn/Portal0/default152.htm. The funders had no role in study design, data collection and analysis, decision to publish, or preparation of the manuscript.

Competing Interests: The authors have declared that no competing interests exist.

* E-mail: houxh@public.wh.hb.cn

Introduction

Irritable bowel syndrome (IBS) is a common functional gastrointestinal disorder usually originated from gut dysfunction with an estimated worldwide prevalence of 10–20% [1,2,3]. 7–30% of IBS patients have a history of acute gastrointestinal infection developed post-infectious IBS (PI-IBS)[4,5]. There are two major pathophysiological findings, including visceral hypersensitivity and dysmotility in the PI-IBS patients. The mechanisms underlying the development of PI-IBS are not fully understood, but are believed to be associate with changes in intestinal permeability and persistent low-grade inflammation[6,7]. Recently, researches have increasingly suggested that gut flora interacts with the bowel in a complex and dynamic relationship. Therefore, gut flora plays an important role in the induction and progression of PI-IBS[8,9,10].

Based on this latter possibility, therapeutic approaches designed to manipulate gut flora with the replenishment of probiotics have been tested in both of patients and animal models with IBS. Several systematic reviews and meta-analyses report that probiotics have a statistically significant effect in improving overall and individual symptoms of IBS patients[11,12,13,14,15,16]. Unfortunately, studies of the curative effect of probiotics in IBS have usually, but not always, been positive. Four trails involved with four kinds of *Lactobacillus* fail to show reduction in global symptom score over placebo[17,18,19,20]. Thus, we reasonably conclude that the benefit of probiotics for IBS is likely to be strains specific; nevertheless, comparable studies among species are rare. What's worse, the study design, probiotic strains and dose are different among studies, which make it difficult to compare the existing results. Which is more effective, single or mixture of species for IBS? Are there any differences among them in the mechanism?

The mechanisms influenced by probiotics that are of potential relevance to the treatment of IBS include enhancement of mucosal permeability, restraining immune activation as well as changes of visceral sensation and motility[21]. Other modes of action, involving enteric neuromuscular modulation and brain-gut axis regulation, are also plausible. In a study of mouse model of PI-IBS, *Lactobacillus paracasei* normalized muscle hypercontractility resulted from a modulation of gut immunologic response to infection[22]. Early life administration of VSL#3, a combination of probiotic strains of various *Bifidobacterium*, *Lactobacillus* and *Streptococcus*, reduces visceral pain perception in a model of IBS[23]. Knowledge of their mechanisms of action is still relatively incomplete. Besides, the results among studies are hard to compare. Can probiotics only be responsible for the changes of the pathophysiologic aspects linked with physiological function of different species in PI-IBS?

In a previous study from our laboratory, NIH mice infected with *Trichinella spiralis* produced alteration in visceral sensitivity, intestinal motility and T helper lymphocytes in lamina propria[24,25]. These abnormalities persisted after recovery from infection, thus well establishing a model of PI-IBS. To compare the different effects of probiotics, we chose three popular species: *Bifidobacterium longum*, *Lactobacillus acidophilus* and *Streptococcus faecalis*, which were contained by a common probiotic called Bifid Lriple Viable in China and VSL#3 for IBS treatment. We want to investigate the hypothesis that these three stains or their mixture separately change visceral hypersensitivity, contractile hyperresponsiveness, intestinal permeability and inflammation in PI-IBS mouse model.

Materials and Methods

Animals

Male NIH mice (6–8 weeks old) were purchased from Medical Animal Laboratory center of Guangdong and kept under specific pathogen-free conditions at Animal Laboratory Center of Tongji Medical College, Huazhong University of Science and Technology. All experiments were approved by the Ethics Committee of Tongji Medical College. (No. 2010–72).

Trichinella spiralis Infection

T.spiralis parasites were obtained from the department of Parasitology at Huazhong University of Science and Technology, Wuhan, China. The colony was maintained through infection among Sprague-Dawley rats. The larvae were obtained from the infected rodents by using a modified technique described by Castro and Fairbairn[26]. Each mouse was infected by gavaging of 350 *T.spiralis* larvae in 0.2 ml of phosphate-buffered saline (PBS).

Probiotics Preparation and Administration

Live bacterial strains of *Bifidobacterium longum HB55020* (1.6×10^{12} CFU/g), *Lactobacillus acidophilus HB56003* (2.15×10^{11} CFU/g) and *Streptococcus faecalis HB62001* (3.81×10^{12} CFU/g) were obtained from Hubei Center of Industrial Culture Collection and Research, HBCC. Each strain was mixed with glucose and was converted to freeze dried powder. The mixed powder was packed in sealed bags of 2 g and stored at $-20°C$ for further use.

T. spiralis-infected mice after 8 weeks were divided into 5 groups. Each group had 8 mice. Controls were daily gavaged with 0.2 ml PBS for 7 days. The other four groups were separately treated with B*ifidobacterium longum HB55020* (2×10^9 CFU/d), *Lactobacillus acidophilus HB56003* (1×10^9 CFU/d), *Streptococcus faecalis HB62001* (0.67×10^9 CFU/d) and all three probiotics mixture (3.67×10^9 CFU/d, *Bifidobacterium*: *Lactobacillus*: *Streptococcus* = 3:2:1) for one week.

Study Design

Visceral sensitivity of each mouse was assessed by behavioral responses to colorectal distention (CRD), which was measured by a semiquantitative score abdominal withdrawal reflex (AWR) and the threshold intensity of CRD that elicits an express contraction in the abdominal wall musculature[25]. Colonic smooth muscle contractile response was studied by measuring the contraction of the longitudinal muscle strips in the organ bath. Plasma diamine oxidase (DAO) activity has been reported to be significantly correlated with lesions and integrity of the intestinal mucosa[27]. D-lactate cumulation in plasma reflects membrane permeability and barrier function of the intestinal mucosa[28,29]. So plasma DAO activity and D-lactate concentration was used to indirectly evaluate intestinal permeability. The tight junction (TJ) forms a barrier which keeps the apical fluid compartments on opposite sides of the epithelial cell layer and contributes to epithelial paracellular permeability[30]. To explore whether TJ takes effect on intestinal permeability after infection, we analyzed the content of TJ structure proteins in ileum including transmembrane components (claudin-1 and occludin) and cytosolic components (ZO-1) in ileum. Intestinal inflammation was assessed by proinflammatory cytokine profiles of IFN-γ, IL-6 and IL-17. The temporary infection caused by *Trichinella spiralis* mainly occur in the small intestine. Furthermore, gut flora, especially probiotics, becomes more and more from proximal intestine to distal intestine. Based on these facts, we choose terminal ileum to analyse expession of cytokines and tight junction proteins. We studied the parameters that mentioned above in the *T. spiralis*-infected mice after one week treatment or without treatment of *Bifidobacterium longum*, *Lactobacillus acidphilus* and *Streptococcus faecalis*, or mixture of three strains.

AWR recording to CRD

CRD was performed as described previously [31]. AWR and thresholds were recorded during plastic balloon inflation to 20, 40, 60 and 80 mmHg. AWR score scale was as previously described. The stimulus intensity that evokes a visually identifiable contraction of the abdominal wall was recorded as the threshold intensity of CRD. During the measurements, mice were given CRD for 20 seconds every 4 minutes. To achieve an accurate results, balloon inflation was done 5 times for each value and was observed by two persons.

Measurement of Contractile Response of Colonic Smooth Muscle to Ach

A piece of mid colon was pinned flat (mucosal side up) in a paraffin-bottomed dissecting dish filled with Krebs solution.

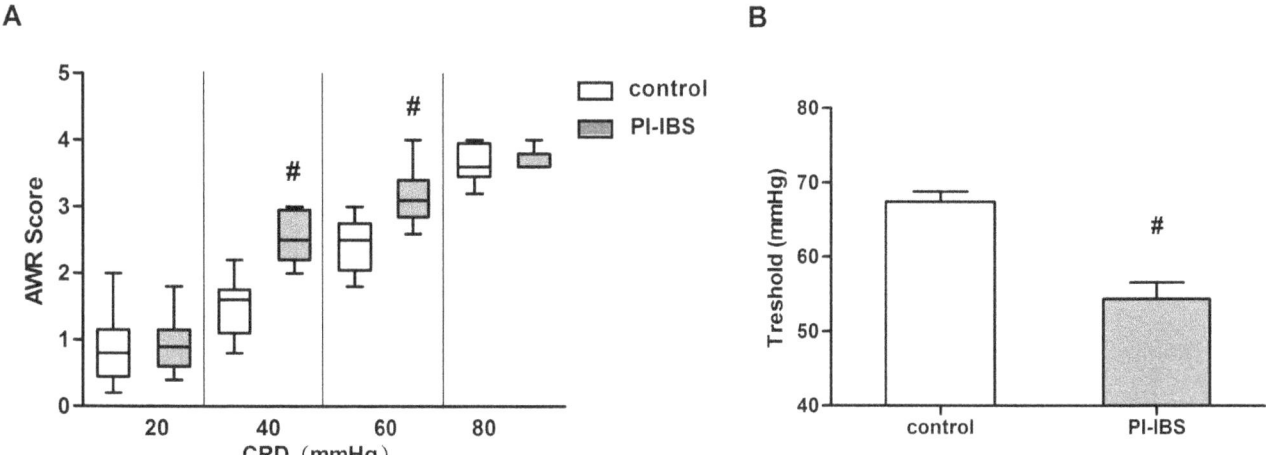

Figure 1. AWR scores and thresholds of control and PI-IBS mouse model. (a) Box plot of AWR scores. Lines represent the median within the box, the 25th and 75th centiles at the ends of the box, and the error bars define the 5th and 95th centiles; n≥8 mice per group. (b) Thresholds of the CRD intensities that evoke abdominal contraction of the mice. Mean±SEM values were plotted; n≥8 mice per group. #p<0.05 PI-IBS group versus control.

Longitudinal muscle strips were taken from each mouse and cut into 3 mm×10 mm pieces and then placed in 25 ml organ bath containing warm (37°C) oxygenated (95%O, 5%CO2) Krebs solution. One end of each strip was attached to an isometric force transducer (Fort-10, WPI, USA) and the other to the armature of the bath. The digitized data were collected by a computer equipped with Acknowledge 3.7.1 software (BIOPAC system, USA). Strips were preloaded with the weight of 1.0 g and allowed to equilibrate in the baths for 60 min with flushing every 20 min. After a stable baseline was attained for 5 min, 10^{-5} mol/L Acetylcholine chloride (Sigma, USA) were added cumulatively to the bath every 5 min. The area under curve (AUC, g·s) was measured at time intervals of 5 min after Ach addition. The response in different groups was quantified by calculating the AUC.

Measurement of DAO and D-lactate levels

Mouse blood was collected in special centrifugal tube and centrifuged at 3000 bpm for 20 min at 4°C and stored at −80°C. DAO and D-lactate levels were measured by spectrophotometry at 436 nm[32]and enzyme-linked ultraviolet spectrophotometry at 340 nm[33], respectively. O-dianisidine, Gadaverine Dihydrochloride and DAO were purchased from Sigma Chemical Company.

Western Blot

Ileum tissues were homogenized by mechanical disruption in RIPA buffer with a protease inhibitor cocktail and incubated on ice for 30 min. Lysates were centrifuged at 12000 rpm for 10 min and the protein content of the supernatant was determined by using the BCA protein assay kit. After being diluted with loading buffer and heated to 95°C for 10 min. Depending on the molecular weight, a total of 60 μg of protein lysates derived from ileum tissue samples were loaded onto 8–12% SDS-PAGE gel. Membranes were probed overnight at 4°C with antibodies against tight junction proteins of ZO-1, occludin and claudin-1 (Invitrogen, USA) and cytokines of IFN-γ, IL-17 (R&D, USA) and IL-6 (abcam, USA) or β-actin/GAPDH (Pierce, USA) antibodies, followed by using the appropriate species-specific HRP conjugate (Pierce, USA) and developing in the SuperSignal West Pico Substrate (Pierce, USA). Band intensities were quantified by the Quantity One 4.6.2 software (BioRad, USA).

Statistical Analysis

AWR scores at each pressure of CRD among the 6 groups were compared using the Kruskal-Wallis one-way analysis of variance on ranks, if the result was significant (P<0.05), a Wilcoxon rank sum test with a Bonferroni correction at 0.05/3 to correct for multiple comparisons. Other data were expressed as mean±SEM, and one-way ANOVA was performed among six groups, followed by LSD or DunnettT3 multiple range analysis. A value of P<0.05 was considered significant. Statistical analyses were performed with SPSS version 17.

Result

Animal model

Morphology: Consistent with previous findings, there were no overt damages of the ileum and colon seen under the microscope after 8 weeks infection. Likewise, the histological scores compared with controls indicated resolution of inflammation.

Visceral sensation: After infection, mice presented increased visceral sensation contrasted to control (Figure 1). Even when the intestinal inflammation subsided, 8 weeks PI group showed a significant increase of AWR scores for intensities 40, 60 mmHg of CRD, coinciding with lower nociceptive threshold. It suggested that 8week PI group as a good model of PI-IBS with visceral hypersensitivity was managed.

Effect of Probiotic Strains on Visceral Sensation

Except *Streptococcus*, the decrease of visceral hypersensitivity was noted in the *Bifidobacterium*, *Lactobacillus* and Mixture (Figure 2). In the comparison of PI-IBS group, AWR scores of the groups of *Bifidobacterium*, *Lactobacillus* and Mixture obviously decreased at 40 mmHg and 60 mmHg, whereas *Streptococcus* remained higher. Correspondingly, the nociceptive threshold was up regulation in groups of *Bifidobacterium*, *Lactobacillus* and Mixture.

Effect of Probiotic Strains on Contractile Response of Colonic Smooth Muscle to Ach

Decreasing contractile responses to Ach was observed in longitudinal muscle strips in *Bifidobacterium* and Mixture (Figure 3). Compared with control, PI-IBS mice presented

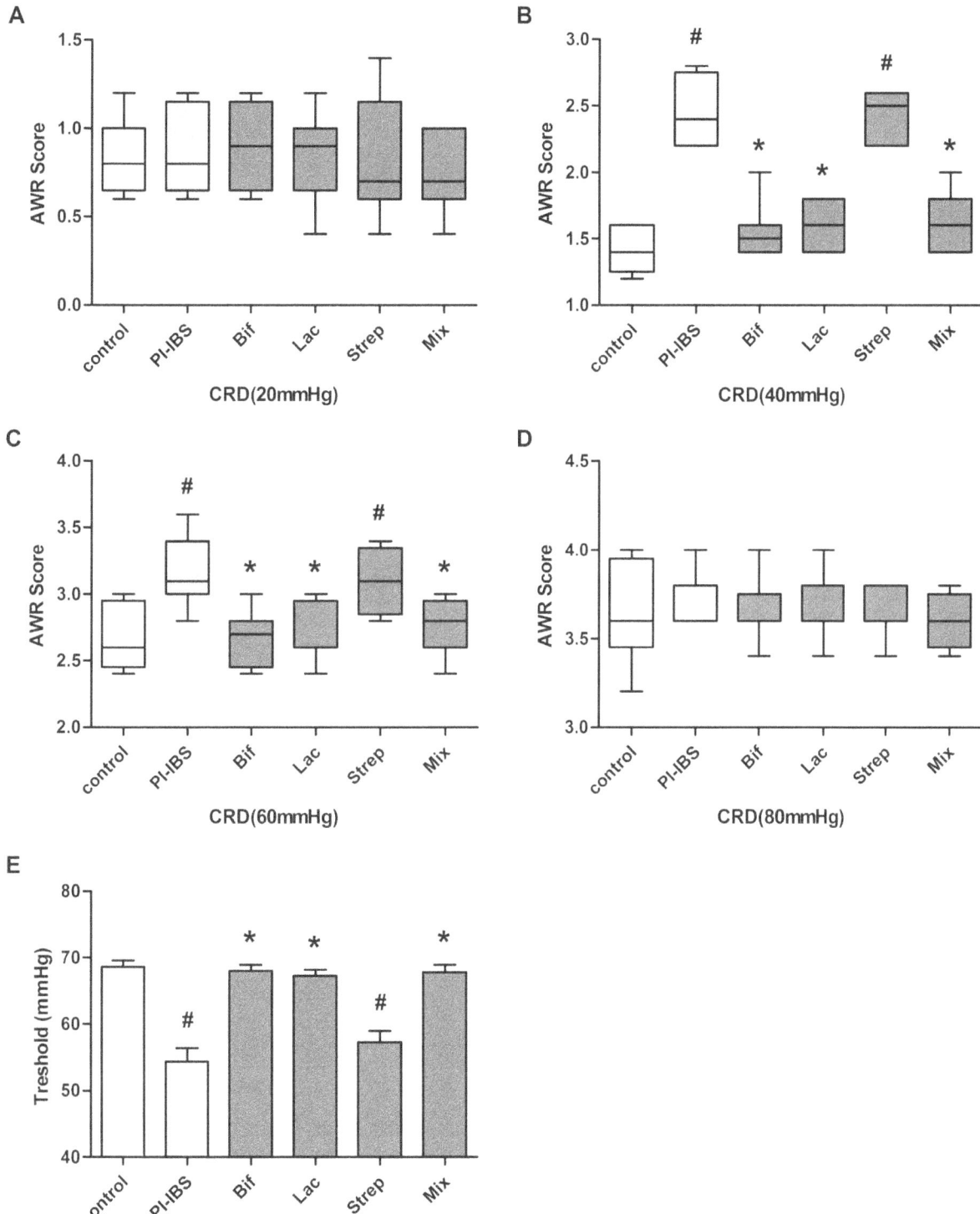

Figure 2. Effect of Probiotic Strains on Visceral Sensation. PI-IBS mice were administered daily with *Bifidobacterium longum*(Bif), *Lactobacillus acidophilus*(Lac), *Streptococcus faecalis* (Strep) or Mixture(Mix) of them for one weeks. Box plot of AWR scores were shown at 20 mmHg(a), 40 mmHg(b), 60 mmHg(c) and 80 mmHg(d). Lines represent the median within the box, the 25th and 75th centiles at the ends of the box, and the error bars define the 5th and 95th centiles; n = 8 mice per group. (d) Thresholds of the CRD intensities. Mean±SEM values were plotted; n = 8 mice per group. #p<0.05 PI-IBS group versus control; *p<0.05 probiotic group versus PI-IBS group.

hyperresponsive tendency(p = 0.08) when the response was generated by the longitudinal muscle response to Ach. The change rate of AUC in *Bifidobacterium* and Mixture was significantly decreased over PI-IBS (31.66±7.21vs 62.6±14.25, p = 0.024; 33.31±5.68 vs.62.6±14.25, p = 0.026). However, no difference was found in groups of *Lactobacillus* and *Streptococcus* compared to PI-IBS.

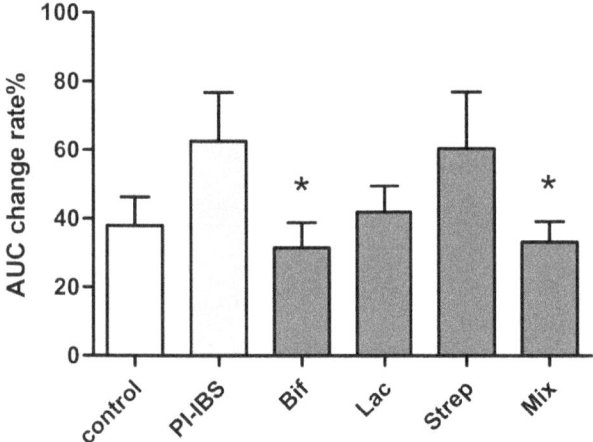

Figure 3. Effect of Probiotic Strains on Contractile Response of Colonic Smooth Muscle to Ach. The AUC change rate of each group was shown by Mean±SEM, n = 8 mice per group. *p<0.05 probiotic group versus PI-IBS group.

Effect of Probiotic Strains on intestinal permeability

Probiotics could relieve raised intestinal permeability of PI-IBS (Figure 4).

The difference of plasma DAO activity among the groups was statistically significant (p<0.05) (Figure 4a). The mean DAO remained higher in PI-IBS than control (p = 0.004), suggesting that intestinal mucosa injury was persistent even though the histological changes were not obvious. Except *Streptococcus* (22.31±1.80), *Bifidobacterium*, *Lactobacillus* and Mixture showed an evident reduction of plasma DAO activity compared to the group without probiotics given (18.25±1.52, 19.60±1.57 and 19.19±1.85vs 25.30±2.90 U/ml). Three effective groups approached control but the *Streptococcus* was still higher. In addition, these three effective groups showed no statistical difference among them.

Similarly, the difference of plasma D-lactate concentration among the groups was statistically significant (p = 0.001) (Figure 4b). The mean D-lactate concentration of PI-IBS and *Streptococcus* group was still higher than control. However, concentration in the groups of *Bifidobacterium*, *Lactobacillus* and Mixture were obviously decreased compared to PI-IBS (151.12±9.51, 163.78±7.92 and 147.88± 17.53vs 232.55± 19.84 µg/ml). In contrast with control, no differences were found among the three effective groups. Likewise, there was no statistical difference among the three effective groups.

The results showed that the expression of claudin-1 in terminal ileum was significantly down regulated in PI-IBS group compared with control (0.76±0.04vs1.06±0.07, p = 0.031)(Figure 4c). After administration of probiotics, the groups of *Bifidobacterium*, *Lactobacillus* and Mixture, but not *Streptococcus* group showed an obvious up-regulation in contrast with PI-IBS group. Moreover, three effective groups were in proximity to control and presented no difference among them. Likewise, the difference of expression of occludin among the groups was statistically significant (p = 0.015)(Figure 4d). A lower level of expression of occludin was not noted in PI-IBS group compared with control. However the level of occludin in the *Bifidobacterium* as well as Mixture was higher than PI-IBS (1.35±0.19, 1.74±0.23vs0.85±0.11). Nonetheless, neither *Streptococcus* nor *Lactobacillus* showed statistically difference compared with PI-IBS. Interestingly, the Mixture group was even higher than control. The statistical difference of ZO-1 expression among the groups was not found.

To estimate whether the increased intestinal permeability could have effect on down-regulation of the expression of TJ proteins, the correlation of plasma D-lactate concentration and expression of claudin-1in ileum was analyzed. We found that plasma D-lactate concentration was negatively correlated with the expression of claudin-1 in ileum(r = −0.421, p = 0.004), indicating that up-regulation of the TJ protein by probiotics contributes to recovery of intestinal permeability.

Effect of Probiotic Strains on cytokine profiles

In terminal ileum, IL-6 and IL-17 expression reduced after *Bifidobacterium*, *Lactobacillus* and Mixture administration. But the difference of IFN-γ expression among the groups was not statistically significant. (Figure 5).

In our study, the expression of IL-6 was significantly elevated in PI-IBS group compared with control. Except *Streptococcus* group, the level of IL-6 in the groups of *Bifidobacterium*, *Lactobacillus* and Mixture shared an obvious down-regulation compared to PI-IBS group. Moreover, the level of IL-6 in Mixture was even lower than control (1.08±0.09vs 0.88±0.1).

In comparison with control, IL-17 expression was increased in PI-IBS as well as *Streptococcus* group (p<0.05).The groups of *Bifidobacterium*, *Lactobacillus* and Mixture, but not *Streptococcus* group cut down the expression of IL-17 (0.75±0.07,0.66±0.06, 0.47±0.10vs1.19±0.13). In addition, Mixture was more effective than *Bifidobacterium* (p = 0.02).

Correlation between intestinal permeability and Visceral Sensation

To explore whether visceral hypersensitivity could be affected by the increased intestinal permeability, the relationship of plasma D-lactate concentration and threshold was analyzed. We found that threshold was negatively correlated with plasma D-lactate concentration (r = −0.508, p = 0.001).

Discussion

In this study, after treatment with *Bifidobacterium*, *Lactobacillus* and Mixture, PI-IBS mouse model presented not only lower AWR scores and contractile response, but also reduction of plasma DAO and D-lactate and cytokines in ileum, suggesting improvement of intestinal hypersensitivity as well as recovery of intestinal barrier function and inflammation. Moreover, our results suggested that probiotic-induced protection of epithelial barrier function may be due to prevention of down-regulation in tight junction proteins expression. However, *Streptococcus* failed to show any favorable effects. What's more notable was that the Mixture of three stains was supposed to be a bit superior to single one.

As described in the results, *Bifidobacterium longum* presented favorable effects, equally with *Lactobacillus*, on sensation, intestinal barrier and inflammation. Nevertheless, *Bifidobacterium* but not *Lactobacillus* reduced contractile hyperresponsiveness to Ach of longitudinal muscle strips. Therefore, *Bifidobacterium longum* was partly superior to other species for treatment of PI-IBS. *Bifidobacterium* is reported to have a great ability to colonize at the intestine, which modify the gut microbiota by producing organic acids such as butyrate acid and competitively adhering to the mucosa and epithelium[34]. Not only does strengthen the gut epithelial barrier, it also modulates the immune system to convey an advantage to the host[35,36]. As the most commonly used probiotics, *Bifidobacterium* have been extensively studied in IBS[18,37,38,39]. The majority of studies of the therapeutic effect of it in IBS has been positive, indicating mainly beneficial impact on bloating, abdominal pain and flatulence[11]. In particular, a

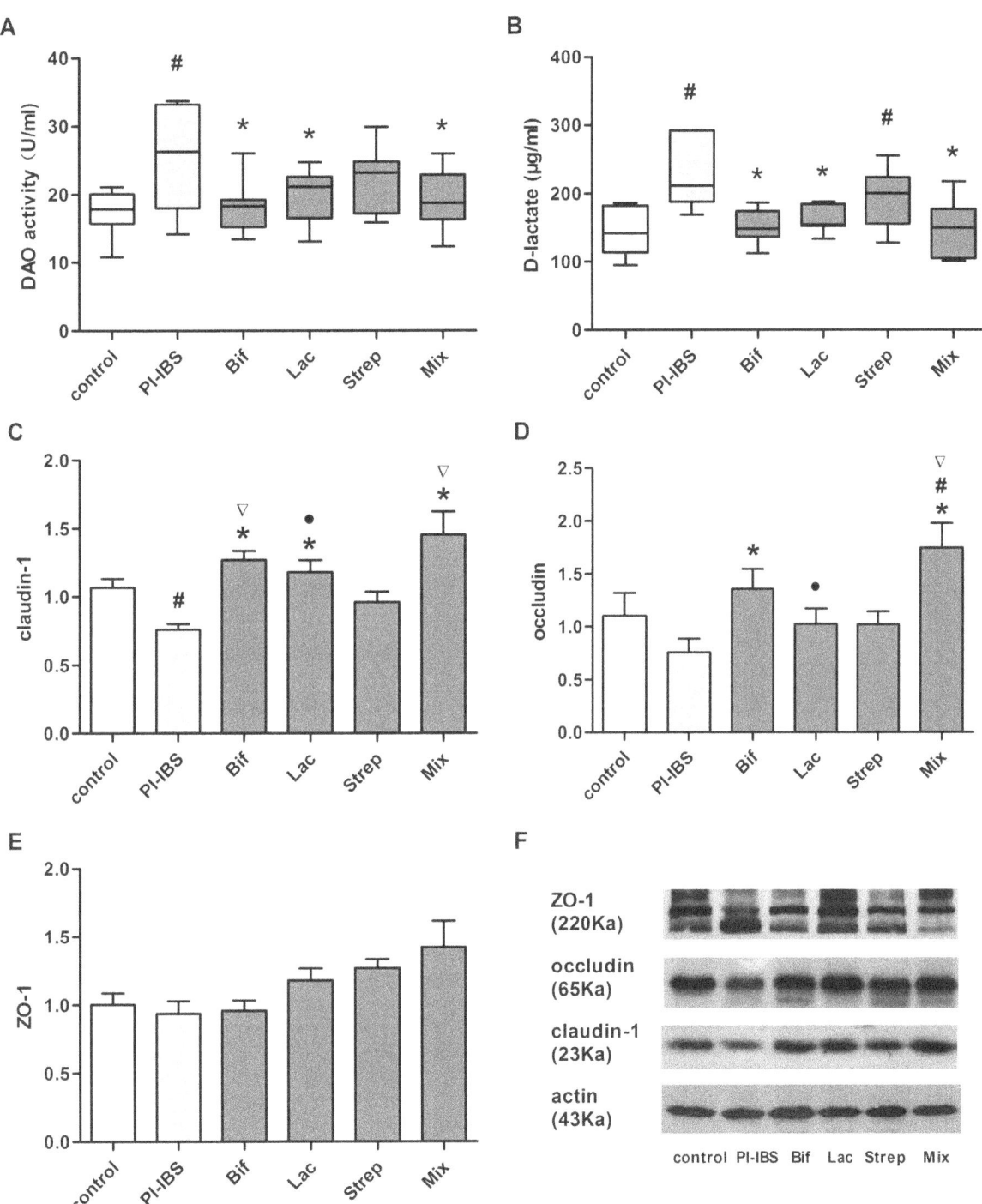

Figure 4. Effect of Probiotic Strains on intestinal permeability. (a) Plasma DAO activity. (b) Pasma D-lactate concentration.Expression of tight junction proteins(c) in ileum: claudin-1, occludin(d) and ZO-1(e). All data were presented by Mean±SEM, n = 5–8 mice per group. #p<0.05 PI-IBS group versus control; *p<0.05 probiotic group versus PI-IBS group; ▽p<0.05 Strep group versus other probiotic groups; ●p<0.05 Mix group versus other probiotic groups.

well-designed and frequently quoted trail reveals that *Bifidobacterium infantis 35624*, not *Lactobacillus salivarius UCC4331* significantly improves in abdominal pain/discomfort, bloating/distension and bowel movements compared with placebo[18]. Our result, cioinciding with previous study, showed the possible superiority of *Bifidobacterium* for treatment in IBS.

Lactobacillus acidophilus, in our study, revealed the improvement of barrier function and reduction of cytokines secretion, thus extending for visceral sensitivity. A lot of studies highlighted the properties of different strains of *Lactobacillus*, mentioning their ability to product the intracolonic short chain fat acid (SCFA) with a consequent improvement in colonic propulsion[40]. However, some of clinical studies are negative and show either no effect or a favorable effect.

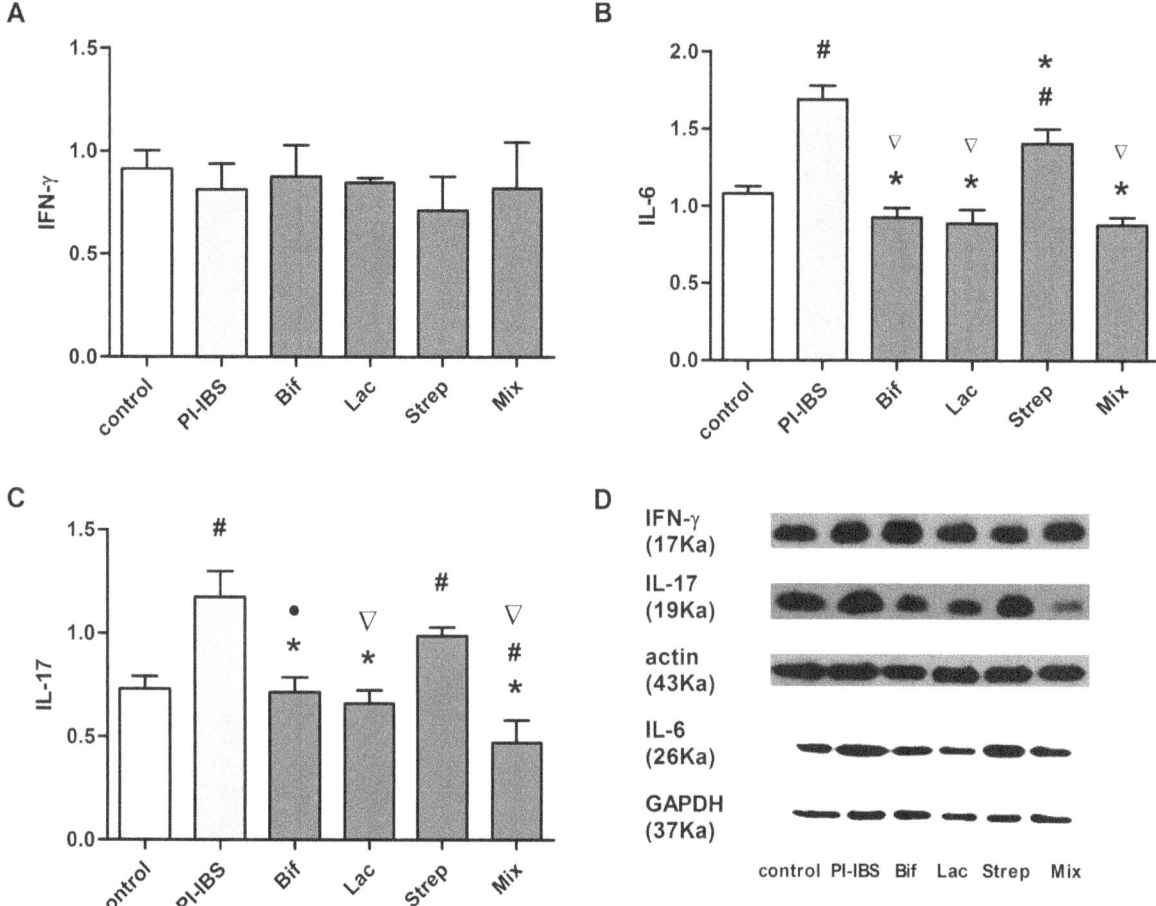

Figure 5. Effect of Probiotic Strains on cytokine profiles. Expression of cytokines(d) in ileum: INF-γ(a), IL-10(d) and IL-17(c). All data were presented by Mean±SEM, n = 5–8 mice per group. #p<0.05 PI-IBS group versus control; *p<0.05 probiotic group versus PI-IBS group; ▽p<0.05 Strep group versus other probiotic groups; ●p<0.05 Mix group versus other probiotic groups.

The divergent results of the efficacy of the *Lactobacillus* used in IBS could be related to different species and doses, suggesting that the effects of *Lactobacillus* may be stains-specific.

Beyond *Bifidobacterium* and *Lactobacillus*, *Streptococcus* has less frequently been used alone in IBS. *Streptococcus faecalis* in this study proved to be ineffective in visceral hypersensitivity, gut permeability and immunomodulatory effects. Although the outcomes of an inactive *Escherichia coli* and *Enterococcus faecalis* bacterial preparation for therapy of IBS have been favorable[41,42], the overall rationality for their use in IBS has been doubted, because a lack of specific mechanism of action has been confirmed. However, almost all probiotic combinations contained *Streptococcus*, it is therefore possible that *Streptococcus* cooperated with other species of probiotics are synergistic in promoting a therapeutic effect in IBS.

In this study, PI-IBS mouse after gavaged with mixture of three species, ameliorated visceral sense, intestinal permeability and cytokine profiles. Compared with single species, the Mixture has, to some extent, evident advantages. According to the expression of occludin, the Mixture group was higher than *Lactobacillus*. In addition, the Mixture group showed decreased expression of IL-17 compared to *Bifidobacterium*. Based on these results, we could conclude that mixture of three stains was superior to single species. VSL#3, probiotic 'cocktail', was reported to be a novel probiotic for the treatment of IBS[43,44,45,46]. In IBS patients with

predominant bloating, VSL#3 significantly reduced flatulence scores and retarded colonic transit in contrast to placebo [43]. The comparison between single probiotic and combination probiotic was not reported before, but it turned out that combination was superior to single species in this study. Thus, we have demonstrated the superiority of mixture of three species in barrier protection as well as immunoregulation.

In summary, the literature confirms the benefit of *Bifidobacterium* and *Lactobacillus* alone or the combination of the three species on the gut sensation, intestinal permeability in PI-IBS mouse model; the mechanisms supporting these beneficial effects may be up-regulation of tight junction proteins and restriction inflammation. Nonetheless, *Streptococcus* shows either no effect or a favorable effect. Most importantly, we have demonstrated the superiority of mixture of three species over a single one. This study may aid our understanding of the mechanisms underlying probiotic treatments for PI-IBS, which might offer referrences to select appropriate probitic species for IBS patients with different symptoms.

Author Contributions

Conceived and designed the experiments: HW XHH. Performed the experiments: HW JG WFW. Analyzed the data: HW. Contributed reagents/materials/analysis tools: YQL XCF YF WQ. Wrote the paper: HW. Revised my manuscript: YF.

References

1. Wilson S, Roberts L, Roalfe A, Bridge P, Singh S (2004) Prevalence of irritable bowel syndrome: a community survey. Br J Gen Pract 54: 495–502.
2. Gwee KA, Wee S, Wong ML, Png DJ (2004) The prevalence, symptom characteristics, and impact of irritable bowel syndrome in an asian urban community. Am J Gastroenterol 99: 924–931.
3. Drossman DA, Camilleri M, Mayer EA, Whitehead WE (2002) AGA technical review on irritable bowel syndrome. Gastroenterology 123: 2108–2131.
4. Gwee KA, Graham JC, McKendrick MW, Collins SM, Marshall JS, et al. (1996) Psychometric scores and persistence of irritable bowel after infectious diarrhoea. Lancet 347: 150–153.
5. Rodriguez LA, Ruigomez A (1999) Increased risk of irritable bowel syndrome after bacterial gastroenteritis: cohort study. BMJ 318: 565–566.
6. Dunlop SP, Jenkins D, Spiller RC (2003) Distinctive clinical, psychological, and histological features of postinfective irritable bowel syndrome. Am J Gastroenterol 98: 1578–1583.
7. Spiller R, Garsed K (2009) Postinfectious irritable bowel syndrome. Gastroenterology 136: 1979–1988.
8. Saulnier DM, Riehle K, Mistretta TA, Diaz MA, Mandal D, et al. (2011) Gastrointestinal microbiome signatures of pediatric patients with irritable bowel syndrome. Gastroenterology 141: 1782–1791.
9. Jeffery IB, O'Toole PW, Ohman L, Claesson MJ, Deane J, et al. (2012) An irritable bowel syndrome subtype defined by species-specific alterations in faecal microbiota. Gut 61: 997–1006.
10. Rajilic-Stojanovic M, Biagi E, Heilig HG, Kajander K, Kekkonen RA, et al. (2011) Global and deep molecular analysis of microbiota signatures in fecal samples from patients with irritable bowel syndrome. Gastroenterology 141: 1792–1801.
11. Moayyedi P, Ford AC, Talley NJ, Cremonini F, Foxx-Orenstein AE, et al. (2010) The efficacy of probiotics in the treatment of irritable bowel syndrome: a systematic review. Gut 59: 325–332.
12. Whelan K (2011) Probiotics and prebiotics in the management of irritable bowel syndrome: a review of recent clinical trials and systematic reviews. Curr Opin Clin Nutr Metab Care 14: 581-587.
13. Clarke G, Cryan JF, Dinan TG, Quigley EM (2012) Review article: probiotics for the treatment of irritable bowel syndrome-focus on lactic acid bacteria. Aliment Pharmacol Ther 35: 403–413.
14. Hoveyda N, Heneghan C, Mahtani KR, Perera R, Roberts N, et al. (2009) A systematic review and meta-analysis: probiotics in the treatment of irritable bowel syndrome. BMC Gastroenterol 9: 15.
15. McFarland LV, Dublin S (2008) Meta-analysis of probiotics for the treatment of irritable bowel syndrome. World J Gastroenterol 14: 2650–2661.
16. Ortiz-Lucas M, Tobias A, Saz P, Sebastian JJ (2013) Effect of probiotic species on irritable bowel syndrome symptoms: A bring up to date meta-analysis. Rev Esp Enferm Dig 105: 19–36.
17. Niedzielin K, Kordecki H, Birkenfeld B (2001) A controlled, double-blind, randomized study on the efficacy of Lactobacillus plantarum 299V in patients with irritable bowel syndrome. Eur J Gastroenterol Hepatol 13: 1143–1147.
18. O'Mahony L, McCarthy J, Kelly P, Hurley G, Luo F, et al. (2005) Lactobacillus and bifidobacterium in irritable bowel syndrome: symptom responses and relationship to cytokine profiles. Gastroenterology 128: 541–551.
19. Nobaek S, Johansson ML, Molin G, Ahrne S, Jeppsson B (2000) Alteration of intestinal microflora is associated with reduction in abdominal bloating and pain in patients with irritable bowel syndrome. Am J Gastroenterol 95: 1231–1238.
20. Niv E, Naftali T, Hallak R, Vaisman N (2005) The efficacy of Lactobacillus reuteri ATCC 55730 in the treatment of patients with irritable bowel syndrome a double blind, placebo-controlled, randomized study. Clin Nutr 24: 925–931.
21. Bixquert Jimenez M (2009) Treatment of irritable bowel syndrome with probiotics. An etiopathogenic approach at last? Rev Esp Enferm Dig 101: 553–564.
22. Verdú EF, Bercik P, Bergonzelli GE, Huang X-X, Blennerhasset P, et al. (2004) Lactobacillus paracasei normalizes muscle hypercontractility in a murine model of postinfective gut dysfunction. Gastroenterology 127: 826–837.
23. Distrutti E, Cipriani S, Mencarelli A, Renga B, Fiorucci S (2013) Probiotics VSL#3 protect against development of visceral pain in murine model of irritable bowel syndrome. PLoS ONE 8: e63893.
24. Fu Y, Wang W, Tong J, Pan Q, Long Y, et al. (2009) Th17: A New Participant in Gut Dysfunction in Mice Infected with Trichinella spiralis. Mediators of Inflammation 2009: 1–7.
25. Long Y, Wang W, Wang H, Hao L, Qian W, et al. (2012) Characteristics of intestinal lamina propria dendritic cells in a mouse model of postinfectious irritable bowel syndrome. J Gastroenterol Hepatol 27: 935–944.
26. Castro GA, Fairbairn D (1969) Carbohydrates and lipids in Trichinella spiralis larvae and their utilization in vitro. J Parasitol 55: 51–58.
27. Wolvekamp MC, de Bruin RW (1994) Diamine oxidase: an overview of historical, biochemical and functional aspects. Dig Dis 12: 2–14.
28. Nakayama M, Yajima M, Hatano S, Yajima T, Kuwata T (2003) Intestinal adherent bacteria and bacterial translocation in breast-fed and formula-fed rats in relation to susceptibility to infection. Pediatr Res 54: 364–371.
29. Ewaschuk JB, Naylor JM, Palmer R, Whiting SJ, Zello GA (2004) D-lactate production and excretion in diarrheic calves. J Vet Intern Med 18: 744–747.
30. Anderson JM, Van Itallie CM (1995) Tight junctions and the molecular basis for regulation of paracellular permeability. Am J Physiol 269: G467–475.
31. Jones RC 3rd, Otsuka E, Wagstrom E, Jensen CS, Price MP, et al. (2007) Short-term sensitization of colon mechanoreceptors is associated with long-term hypersensitivity to colon distention in the mouse. Gastroenterology 133: 184–194.
32. Li JY, Lu Y, Hu S, Sun D, Yao YM (2002) Preventive effect of glutamine on intestinal barrier dysfunction induced by severe trauma. World J Gastroenterol 8: 168–171.
33. Brandt RB, Siegel SA, Waters MG, Bloch MH (1980) Spectrophotometric assay for D-(-)-lactate in plasma. Anal Biochem 102: 39–46.
34. Singh N, Arioli S, Wang A, Villa CR, Jahani R, et al. (2013) Impact of Bifidobacterium bifidum MIMBb75 on mouse intestinal microorganisms. FEMS Microbiol Ecol 85: 369–375.
35. Matsuki T, Pedron T, Regnault B, Mulet C, Hara T, et al. (2013) Epithelial cell proliferation arrest induced by lactate and acetate from Lactobacillus casei and Bifidobacterium breve. PLoS ONE 8: e63053.
36. Fanning S, Hall LJ, Cronin M, Zomer A, MacSharry J, et al. (2012) Bifidobacterial surface-exopolysaccharide facilitates commensal-host interaction through immune modulation and pathogen protection. Proc Natl Acad Sci U S A 109: 2108–2113.
37. Brenner DM, Chey WD (2009) Bifidobacterium infantis 35624: a novel probiotic for the treatment of irritable bowel syndrome. Rev Gastroenterol Disord 9: 7–15.
38. Guyonnet D, Chassany O, Ducrotte P, Picard C, Mouret M, et al. (2007) Effect of a fermented milk containing Bifidobacterium animalis DN-173 010 on the health-related quality of life and symptoms in irritable bowel syndrome in adults in primary care: a multicentre, randomized, double-blind, controlled trial. Aliment Pharmacol Ther 26: 475–486.
39. Whorwell PJ, Altringer L, Morel J, Bond Y, Charbonneau D, et al. (2006) Efficacy of an encapsulated probiotic Bifidobacterium infantis 35624 in women with irritable bowel syndrome. Am J Gastroenterol 101: 1581–1590.
40. Vanderhoof JA (2001) Probiotics: future directions. Am J Clin Nutr 73: 1152S–1155S.
41. Martens U, Enck P, Zieseniss E (2010) Probiotic treatment of irritable bowel syndrome in children. Ger Med Sci 8: Doc07.
42. Enck P, Zimmermann K, Menke G, Muller-Lissner S, Martens U, et al. (2008) A mixture of Escherichia coli (DSM 17252) and Enterococcus faecalis (DSM 16440) for treatment of the irritable bowel syndrome a randomized controlled trial with primary care physicians. Neurogastroenterol Motil 20: 1103–1109.
43. Kim HJ, Vazquez Roque MI, Camilleri M, Stephens D, Burton DD, et al. (2005) A randomized controlled trial of a probiotic combination VSL# 3 and placebo in irritable bowel syndrome with bloating. Neurogastroenterol Motil 17: 687–696.
44. Guandalini S, Magazzu G, Chiaro A, La Balestra V, Di Nardo G, et al. (2010) VSL#3 improves symptoms in children with irritable bowel syndrome: a multicenter, randomized, placebo-controlled, double-blind, crossover study. J Pediatr Gastroenterol Nutr 51: 24–30.
45. Michail S, Kenche H (2011) Gut microbiota is not modified by Randomized, Double-blind, Placebo-controlled Trial of VSL#3 in Diarrhea-predominant Irritable Bowel Syndrome. Probiotics Antimicrob Proteins 3: 1–7.
46. Kim HJ, Camilleri M, McKinzie S, Lempke MB, Burton DD, et al. (2003) A randomized controlled trial of a probiotic, VSL#3, on gut transit and symptoms in diarrhoea-predominant irritable bowel syndrome. Aliment Pharmacol Ther 17: 895–904.

5

Lactobacillus amylovorus Inhibits the TLR4 Inflammatory Signaling Triggered by Enterotoxigenic Escherichia coli via Modulation of the Negative Regulators and Involvement of TLR2 in Intestinal Caco-2 Cells and Pig Explants

Alberto Finamore[1], Marianna Roselli[1], Ambra Imbinto[1], Julie Seeboth[2,3], Isabelle P. Oswald[2,3], Elena Mengheri[1]*

1 Consiglio per la Ricerca e la Sperimentazione in Agricoltura (CRA), Centro di Ricerca per gli Alimenti e la Nutrizione (Research Center on Food and Nutrition, CRA-NUT), Rome, Italy, **2** INRA, UMR 1331 Toxalim, Research Center in Food Toxicology, Toulouse, France, **3** University of Toulouse, National Polytechnic Institute of Toulouse (INP), UMR 1331 Toxalim, Toulouse, France

Abstract

Inflammation derived from pathogen infection involves the activation of toll-like receptor (TLR) signaling. Despite the established immunomodulatory activities of probiotics, studies relating the ability of such bacteria to inhibit the TLR signaling pathways are limited or controversial. In a previous study we showed that *Lactobacillus amylovorus* DSM 16698[T], a novel lactobacillus isolated from unweaned pigs, protects the intestinal cells from enterotoxigenic *Escherichia coli* (ETEC) K88 infection through cytokine regulation. In the present study we investigated whether the ability of *L. amylovorus* to counteract the inflammatory status triggered by ETEC in intestine is elicited through inhibition of the TLR4 signaling pathway. We used the human intestinal Caco-2/TC7 cells and intestinal explants isolated from 5 week-old crossbreed Pietrain/Duroc/Large-White piglets, treated with ETEC, *L. amylovorus* or *L. amylovorus* cell free supernatant, either alone or simultaneously with ETEC. Western blot analysis showed that *L. amylovorus* and its cell free supernatant suppress the activation of the different steps of TLR4 signaling in Caco-2/TC7 cells and pig explants, by inhibiting the ETEC induced increase in the level of TLR4 and MyD88, the phosphorylation of the IKKα, IKKβ, IκBα and NF-κB subunit p65, as well as the over-production of inflammatory cytokines IL-8 and IL-1β. The immunofluorescence analysis confirms the lack of phospho-p65 translocation into the nucleus. These anti-inflammatory effects are achieved through modulation of the negative regulators Tollip and IRAK-M. We also found that *L. amylovorus* blocks the up-regulation of the extracellular heat shock protein (Hsp)72 and Hsp90, that are critical for TLR4 function. By using anti-TLR2 antibody, we demonstrate that TLR2 is required for the suppression of TLR4 signaling activation. These results may contribute to develop therapeutic interventions using *L. amylovorus* in intestinal disorders of piglets and humans.

Editor: Markus M. Heimesaat, Charité, Campus Benjamin Franklin, Germany

Funding: The research leading to these results has received funding from the European Community's Seventh Framework Programme (FP7/2007-2013) under the grant agreement n° 227549, and partly from the Italian MiPAAF funds, Qualità Alimentare e Funzionale-QUALIFU programme, (D.M.2087/7303/09-29/1/2009). The funders had no role in study design, data collection and analysis, decision to publish, or preparation of the manuscript.

Competing Interests: The authors have declared that no competing interests exist.

* E-mail: elena.mengheri@entecra.it

Introduction

The intestinal mucosa is colonized by a vast community of bacteria and should be able to defend against pathogen infections. The Toll-like receptor (TLR) family plays a critical role in the host defense or in the development of inflammation by recognizing microbe-associated molecular patterns. Among these receptors, TLR4 has been associated with pathogenesis of several diseases [1–4]. Indeed, binding of lipopolysaccharide (LPS) to TLR4 caused intestinal inflammation through production of pro-inflammatory cytokines [5,6], and elimination of TLR4 increased the susceptibility to dextran sodium sulfate-induced disease [7]. In addition, the expression of TLR4 was increased in intestinal epithelial cells and dendritic cells of patients suffering of ulcerative colitis and Crohn's disease and in macrophages of inflamed tissues [8–10], while mice knockout for TLR4 showed reduced myocardial ischemic injury [11]. TLR4 was found to be the most strongly expressed TLR in porcine intestinal cells derived from neonatal pigs [6], that can be related to the high incidence of inflammation associated with pig weaning. TLR4 detects Gram-negative bacteria, but recent studies identified other molecules able to bind to and activate this receptor, namely the extracellular heat shock proteins (Hsps), such as the extracellular Hsp72 and Hsp90 [12–14]. When released from cells, these Hsps may induce inflammation in a TLR4- and NF-κB-dependent mechanism [15,16], and circulating Hsp72 has been found increased in pathological conditions including renal disease, hypertension, atherosclerosis and sickle cell disease [17].

Induction of TLR4 may lead to inflammatory cytokine overproduction through activation of two signaling pathways, the early myeloid differentiation primary response gene 88 (MyD88)-dependent and delayed MyD88-independent response [18]. The MyD88-dependent cascade includes activation of the NF-κB pathway, involving recruitment of the IL-1R-associated kinases (IRAKs), phosphorylation of IκB kinase (IKK) and subsequent phosphorylation and degradation of the family of IκB proteins, which allow phosphorylation of NF-κB followed by its translocation into the nucleus and transcription of pro-inflammatory cytokines such as TNF-α, IL-1β, IL-6 and IL-8 [19–22].

Targeting the TLR4-mediated inflammatory signaling may represent a way to counteract the pathogen induced damages. Probiotic bacteria are microorganisms that may confer health benefits to the host, including prevention of inflammatory intestinal diseases [23–25]. There is some evidence that probiotic bacteria can inhibit the activation of TLR4 signaling pathway, although the studies are limited and the results sometimes contradictory. For instance, a down-regulation of TLR4 expression by L. paracasei associated with a decreased cytokine and chemokine release against Salmonella typhi infection was found in dendritic cells [26]. L. jensenii reduced the mRNA level of pro-inflammatory cytokines by inhibiting the pathogen induced TLR4 activation in porcine intestinal epithelial cells [27]. However, it was also shown that L. rhamnosus and L. plantarum did not change the TLR4 expression neither the secretion of IL-8 in cells infected with Salmonella [28].

In a previous study, we showed that treatment of porcine intestinal cells with L. amylovorus strain 16698[T] (formerly called L. sobrius), a new lactobacillus species isolated from intestine of unweaned piglets, protects against enterotoxigenic E. coli (ETEC) K88 infection by inhibiting pathogen adhesion and membrane damages through cytokine modulation [29]. This lactobacillus is able to reduce the diarrhea caused by ETEC, decrease colonization of ETEC, and improve weight gain of infected piglets [30]. Other than in piglets, the probiotic characteristics of L. amylovorus were shown in an in vitro model that simulates the human upper gastrointestinal tract, suggesting a potential use of L. amylovorus for animal as well as for human health [31]. In the present study we aimed to examine the ability of L. amylovorus to counteract the inflammatory stimulus triggered by ETEC in intestine through inhibition of the TLR4 signaling pathways and modulation of the negative regulators. We used an in vitro model of intestinal cells and ex vivo model of piglet explants that more closely mimics the gut mucosal environment [32]. We found in both human Caco-2/TC7 cells and pig intestinal explants, that L. amylovorus and its cell free supernatant are able to inhibit the different steps of TLR4 signaling activated by ETEC K88 and the production of inflammatory cytokines through modulation of the negative regulators Toll-interacting protein (Tollip) and IRAK-M, as well as down-regulation of the extracellular Hsp72 and Hsp90. We also show that TLR2 activation is required for these anti-inflammatory activities.

Materials and Methods

Ethic statement

All animal experiments were carried out in strict accordance with the recommendation of the European Guidelines for the Care and Use of Animals for Research Purposes. The protocol was approved by the Committee of the Ethics of Animal Experiments of Pharmacology-Toxicology of France Midi-Pyrénées, Toco-méthique, agreement number TOXCOM/0017. This Committee is affiliated to INP and INRA. The two authors, Isabelle Oswald and Julie Sebooth are from INRA. All efforts were made to minimize suffering and immediately after electrical stunting, animals were killed by exanguination prior to samples collection, as already described [33].

Epithelial cell culture

The human intestinal Caco-2/TC7 cell line (developed by Chantret et al. [34] and kindly provided by Monique Rousset, Institut National de la Santé et de la Recherche Médicale, INSERM, France) was used. These cells derive from parental Caco-2 cells at late passage, and have been reported to express higher metabolic and transport activities than the original line [35]. Cells were used between passages 100 and 105 and were routinely grown on plastic tissue culture flasks (75 cm^2 growth area, Becton Dickinson, Milan, Italy) in Dulbecco's modified minimum essential medium (DMEM; 3.7 g/L NaHCO$_3$, 4 mM glutamine, 10% heat inactivated fetal calf serum, 1% non essential amino acids, 10^5 U/L penicillin and 100 mg/L streptomycin). All cell culture reagents were from Euroclone (Milan). The cells were routinely maintained at 37°C in an atmosphere of 5% CO$_2$/95% air at 90% relative humidity. For the experiments, cells were seeded on Transwell filters (polyethylene terephtalate filter inserts for cell culture; Becton Dickinson) of 12 mm diameter, 0.45 μm pore diameter, as described below. After confluency, cells were left for 17–21 days to allow differentiation. Medium was changed three times a week.

Pig jejunal explants

Jejunal tissues were obtained from five Pietrain/Duroc/Large-White piglets that were 5 week-old and weaned at 4 weeks (8.3±0.15 Kg), housed in groups in normal conditions, and fed a standard commercial diet (Solignac, Bessière, France), as previously described [36]. Briefly, the external tunica muscularis was removed from middle jejunum segment, then explants were excised with punch trocards (Centravet, Lapalisse, France) and placed in Williams culture medium (Sigma-Aldrich, St Louis, MO) supplemented with 1% penicillin/streptomycin, 0.5% gentamycin (Eurobio, Les Ulis, France), 4.5 g/L glucose (Sigma-Aldrich), 10% fetal bovine serum (Eurobio) and 30 mM amino acids (Ala/Glu, Eurobio).

Bacterial growth

ETEC strain K88 (kindly provided by The Lombardy and Emilia Romagna Experimental Zootechnic Institute, Reggio Emilia, Italy) was grown in Luria-Bertani (LB) broth containing 1% tryptone and 0.5% yeast extract (both from OXOID, Basingstoke, England), plus 1% NaCl, pH 7.0. After overnight incubation at 37°C with vigorous shaking, bacteria were diluted 1:200 in fresh LB and grown until mid-log phase. Bacterial cells were then harvested by centrifugation at 3000×g for 10 min at 4°C and resuspended in antibiotic- and serum-free DMEM.

L. amylovorus strain DSM 16698[T], isolated from piglet small intestine [37], was grown in DeMan Rogosa Sharp (MRS) medium (DIFCO, Milan) at 37°C under anaerobic conditions. After overnight incubation, bacteria were diluted 1:15 in fresh MRS, grown until mid-log phase and processed as described above for ETEC. Bacterial concentrations were determined in preliminary experiments by densitometry and confirmed by serial dilutions followed by CFU counts of ETEC on LB agar after 16 hrs incubation, and of L. amylovorus on MRS agar after 48 hrs incubation at 37°C, under anaerobic conditions.

The viability of ETEC and *L. amylovorus* grown on DMEM did not differ from that of bacteria grown on LB or MRS media, as tested in preliminary experiments by CFU counts after agar plating of bacterial inocula from the different media.

Preparation of *L. amylovorus* cell free culture supernatant

Cell free culture supernatant was prepared from overnight cultures of *L. amylovorus* after centrifugation at $4000 \times g$ for 10 min at $4°C$ followed by filtration of the supernatant fractions through a 0.22 μm pore-size filter, to remove any remaining bacteria. The cell free supernatant equivalent to 5×10^7 CFU/mL was added to 1 mL antibiotic- and serum-free DMEM for the experiments described below. Since the addition of bacterial medium to DMEM can reduce the pH and previous studies reported that a pH value below 5.8 affects Caco-2 cell viability [38], the pH value of supernatants in DMEM was checked and resulted not lower than 6.

Analysis of TLR4 signaling in Caco-2/TC7 cells and pig jejunal explants by Western blot

Caco-2/TC7 cells differentiated on Transwell filters (1×10^6 cells/filter), were untreated (control) or apically treated with 1 mL of medium containing ETEC (5×10^6 CFU/mL), *L. amylovorus* (5×10^7 CFU/mL) or *L. amylovorus* supernatant (equivalent to 5×10^7 CFU/mL), either alone or simultaneously with ETEC, for 2.5 hrs. We chose the pathogen concentration and time of incubation based on preliminary experiments to allow triggering of the inflammation pathway without disruption of the cell monolayer. The 1:10 ratio of ETEC to *L. amylovorus* was that used in our previous study [29]. In some experiments, neutralizing anti-TLR2 antibody was apically added to Caco-2/TC7 cells in antibiotic- and serum-free DMEM (150 μg/L; R&D System Inc., Minneapolis, MN), for 1 hr before the addition of bacteria as above described.

Pig jejunal explants (6 mm diameters) were untreated (control) or treated with 2.5 mL of medium containing ETEC (5×10^6 CFU/mL) or *L. amylovorus* (1.25×10^8 CFU/mL), either alone or simultaneously, for 1.5 hrs at $39°C$ in a humidified atmosphere of 5% CO_2. After treatment, tissues were collected, immediately snap-frozen and maintained at $-80°C$ before analysis.

For analysis of the TLR4 cascade proteins, Caco-2/TC7 cells and intestinal explants were lysed or homogenized respectively, in cold radioimmunoprotein assay (RIPA: 20 mM Tris-HCl pH 7.5, 150 mM NaCl, 0.1% SDS, 1% Na deoxycholate, 1% Triton X-100) buffer supplemented with 1 mM phenylmethylsulphonyl fluoride, protease inhibitor cocktail (Complete Mini, Roche, Milan) and phosphatase inhibitor cocktail (PhosSTOP, Roche). For analysis of the extracellular Hsp72 and Hsp90, 30 μL of medium collected from basolateral compartment were added to 30 μL of $2 \times$ RIPA. The cells, explants and basolateral media were centrifuged at $15000 \times g$ for 20 min and supernatants were recovered. Cell lysates (50 μg total proteins) intestinal homogenates (50 μg total proteins) and 50 μL of basolateral media were dissolved in sample buffer (50 mM Tris-HCl, pH 6.8, 2% SDS, 10% glycerol, 100 g/L bromophenol blue, 10 mM β-mercaptoethanol), heated for 5 min, fractionated by SDS-polyacrylamide gel (4–20% gradient) electrophoresis and transferred to 0.2 μm nitrocellulose filters (Trans-Blot Turbo, Biorad, Milan). Membranes were incubated with the following primary antibodies: rabbit polyclonal anti-human TLR4, MyD88, IRAK-4, IKKα, IKKβ, phospho(P)-IKKα/β, IκBα, P-IκBα, NF-κB p65, P-p65, IRAK-M, Tollip, or mouse monoclonal anti-human Hsp90, Hsp72, α-tubulin antibodies. All primary antibodies were from

Cell Signaling Technology (Danvers, MA), except for TLR4, that was from Santa Cruz Biotechnology Inc. (Santa Cruz, CA). Preliminary experiments verified the complete cross reactivity of the human antibodies with the analyzed pig proteins (data not shown). Proteins were detected with horseradish peroxidase-conjugated secondary antibodies (Cell Signaling Technology) and enhanced chemiluminescence reagent (ECL kit LiteAblot Extend, Euroclone), followed by analysis of chemiluminescence with the charge-coupled device camera detection system Las4000 Image Quant (GE Health Care Europe GmbH, Milan).

Relative levels of TLR4, MyD88, IRAK-4, Tollip and IRAK-M were normalized to α-tubulin, whereas the phosphorylated proteins were normalized to their corresponding unphosphorylated forms.

TLR4 and P-p65 immunolocalization in intestinal cells

Caco-2/TC7 cells differentiated on Transwell filters (1×10^6 cells/filter), were untreated (control) or apically treated with 1 mL of medium containing ETEC (5×10^6 CFU/mL) or *L. amylovorus* (5×10^7 CFU/mL), either alone or simultaneously, for 2.5 hrs. At the end of treatment, cells were washed with PBS and fixed in ice-cold methanol for 3 min. Localizations of TLR4, P-p65, and occludin were analyzed as follows. Briefly, the cells were treated with rabbit polyclonal anti-TLR4 or P-p65 antibodies (Cell Signaling Technology), or mouse monoclonal anti-occludin antibody (Zymed Laboratories, Milan). For secondary detection, the cells were incubated with FITC-conjugated goat anti-rabbit IgG for TLR4, TRITC-conjugated goat anti-mouse IgG (Jackson Immunoresearch, Milan) for occludin or TRITC-conjugated goat anti-rabbit IgG for P-p65. For P-p65 analysis, the cell nuclei were stained with 300 nM DAPI, added directly to the mounting medium. Stained monolayers were mounted on glass slides by using Prolong Gold antifade Reagent (Molecular Probes, Invitrogen, Milan) and analyzed under a confocal microscope (LSM 700, Zeiss, Jena, Germany).

Measurement of inflammatory cytokine production by intestinal cells

Caco-2/TC7 cells differentiated on Transwell filters (1×10^6 cells/filter), were untreated (control) or apically treated with 1 mL of medium containing ETEC (5×10^6 CFU/mL), *L. amylovorus* (5×10^7 CFU/mL) or *L. amylovorus* supernatant (equivalent to 5×10^7 CFU/mL), either alone or simultaneously with ETEC, for 4 hrs. At the end of bacterial incubation time, antibiotics (10^5 U/L penicillin and 100 mg/L streptomycin) were added to the apical media for 20 hrs. IL-8 and IL-1β levels were analyzed in basolateral media by a cytometric bead array inflammatory kit (Becton Dickinson Biosciences, Milan), according to the manufacturer's specifications. Briefly, microbeads with distinct fluorescence intensities, coated with capture antibodies specific for each cytokine, were incubated with supernatant samples (50 μL) and PE-conjugated detection antibodies for 2 hrs. The samples were then washed, resuspended in 300 μL wash buffer, and analyzed by flow cytometry using FCAP array software (BD Biosciences).

Statistical analysis

The significance of the differences was evaluated by one-way ANOVA followed by Fisher's test. Significance was set at P values <0.05. All statistical analyses were performed with "Statistica" software program (version 4.5; StatSoftInc, Tulsa, OK).

Results

Inhibition of ETEC induced TLR4 signaling pathways by L. amylovorus in Caco-2/TC7 cells and pig intestinal explants

Infection of Caco-2/TC7 cells with ETEC resulted in a stimulation of TLR4 inflammatory cascade. ETEC induced a significant increase in the levels of TLR4 and MyD88, and caused a strong phosphorylation of the IKK family proteins IKKα and IKKβ, IκBα and NF-κB subunit, p65 (Figure 1). The immuno-fluorescence analysis of TLR4 in Caco-2/TC7 cells confirmed the increase of the receptor caused by ETEC, that was abolished by co-treatment with *L. amylovorus* (Figure 2). In addition, the negative regulators Tollip and IRAK-M were reduced by ETEC, as compared with untreated cells. All these inflammatory responses to ETEC were inhibited by *L. amylovorus* and its cell free supernatant. Of relevance, *L. amylovorus* when added simultaneously with ETEC, induced an up-regulation of Tollip, as compared with control cells. Treatment of uninfected cells with *L. amylovorus* did not activate the TLR4 signaling, and induced an increased level of Tollip, as compared with control cells (Figure 1). Similar results were found in pig intestinal explants (Figure 3), where ETEC induced a higher level of TLR4, P-IKKα, P-IκBα, and P-p65, while *L. amylovorus* completely abolished all these alterations and upregulated Tollip and IRAK-M expression when co-treated with ETEC. In addition, *L. amylovorus* did not activate the TLR4 cascade and increased the levels of the two negative regulators when added alone to the tissue culture.

L. amylovorus abolishes P-p65 translocation into the nucleus induced by ETEC

Since phosphorylation of p65 is necessary for its translocation into the nucleus to activate transcription of inflammatory cytokines, we verified whether this final step of TLR4 cascade was inhibited by *L. amylovorus*. The immunofluorescence experiments showed nuclear localization of P-p65 upon ETEC stimulation that was not present in cells simultaneously treated with ETEC and *L. amylovorus* (Figure 2). The P-p65 translocation into the nucleus did not occur after the addition of *L. amylovorus* alone to the cells.

L. amylovorus inhibits the ETEC induced up-regulation of extracellular Hsp72 and Hsp90

Since extracellular Hsps can activate the TLR4 signaling, we further investigated whether these proteins were regulated by ETEC and *L. amylovorus*. The infection of Caco-2/TC7 cells with ETEC caused an increase in Hsp72 and Hsp90 levels, that was inhibited by co-treatment of the cells with ETEC and *L. amylovorus* (Figure 4). Similar effects were obtained when the cells were treated with ETEC and *L. amylovorus* supernatant. No change in the levels of Hsp72 and Hsp90 was induced by treatment with *L. amylovorus* alone.

L. amylovorus abolishes the ETEC induced increase of pro-inflammatory cytokine production

A strong secretion of IL-8 and IL-1β was induced by ETEC in Caco-2/TC7 cells, compared with control cells (Figure 5). *L. amylovorus* and its cell free supernatant when added simultaneously with ETEC, were able to inhibit the up-regulation of both these cytokines. The level of IL-8 and IL-1β was not increased by treatment of *L. amylovorus* or its cell free supernatant alone.

The regulation of TLR4 signaling is TLR2 dependent

In order to evaluate whether TLR2 could play a role in *L. amylovorus* anti-inflammatory activity, we performed immunoneutralization experiments using anti-TLR2 antibodies. The neutralization of TLR2 resulted in a lack of protection by *L. amylovorus*, since the levels of P-IκBα and P-p65 were up-regulated, while the levels of Tollip and IRAK-M decreased in cells co-treated with ETEC, *L. amylovorus* and anti-TLR2, as compared with control cells, reaching similar levels to those of ETEC infected cells (Figure 6). The addition of anti-TLR2 antibody to ETEC infected cells did not induce any change in the levels of P-IκBα, P-p65, Tollip and IRAK-M, as compared with ETEC infected-anti-TLR2 untreated cells. The levels of these proteins were not modified by the addition of anti-TLR2 antibody to control and *L. amylovorus* treated cells, as compared with control cells.

Discussion

In a previous study we showed that *L. amylovorus* DSM 16698[T] protects the intestinal cells against the inflammatory status and mucosal injury triggered by ETEC K88 infection in intestinal cells through repression of pro-inflammatory cytokine production, such as IL-1β and IL-8. Of relevance, IL-8 was responsible of the ETEC induced membrane barrier damages of the cells [29]. The results of the present study indicate a mechanism through which the protective activity of *L. amylovorus* DSM 16698[T] is exerted, since we found that *L. amylovorus* blocks the ETEC induced increase in IL-8 and IL-1β by inhibiting the various steps of TLR4 signaling and modulating the cascade negative regulators. These findings are consistent with the notion that TLR4 down-regulation is important for the resolution of inflammation and repair of membrane damage, as established by previous studies showing that TLR4 knockout mice had reduced inflammation in response to pathogen infection [39], and that TLR4 antagonist inhibited pro-inflammatory cytokine production and mucosal damages in dextran sodium sulfate-induced colitis mice and in spontaneous chronic colitis model [40]. In addition, downregulation of TLR4 expression in pigs challenged with LPS was associated with a decrease in pro-inflammatory cytokine levels and improvement of intestinal barrier integrity [41]. Our results contribute to the understanding of the mechanisms through which probiotics may inhibit the intestinal injury caused by pathogenic infection. In fact, while the role of probiotics in the maintenance of membrane barrier function is supported by several studies [29,42], only recently their protective activity through modulation of TLR signaling has been addressed, but the results are sometime controversial (26–28).

We show that the results obtained in intestinal cells are confirmed in porcine explants treated with ETEC K88 and *L. amylovorus* DSM 16698[T], supporting the ability of this lactobacillus to prevent or reduce the inflammatory response to ETEC infection in piglets through regulation of the TLR4 signaling. Notably, a previous study showed that *L. amylovorus* DSM 16698[T] (formerly called *L. sobrius*) was effective in reducing ETEC infection and in improving weight gain of piglets [43].

Pathogen binding to TLR4 triggers aggregation of TLR4 with its co-adaptor MyD88, which initiates the inflammatory cascade leading to the production of pro-inflammatory cytokines such as TNF-α, IL-1β, IL-6 and IL-8 [19,44]. Our results show an elevated level of TLR4 and MyD88 after ETEC infection, in agreement with the findings of an increased expression of TLR4 in Caco-2 cells after *E. coli* K4 infection [45]. Treatment of ETEC infected cells with *L. amylovorus* or its cell free supernatant inhibits the increase in both TLR4 and MyD88, indicating that *L.*

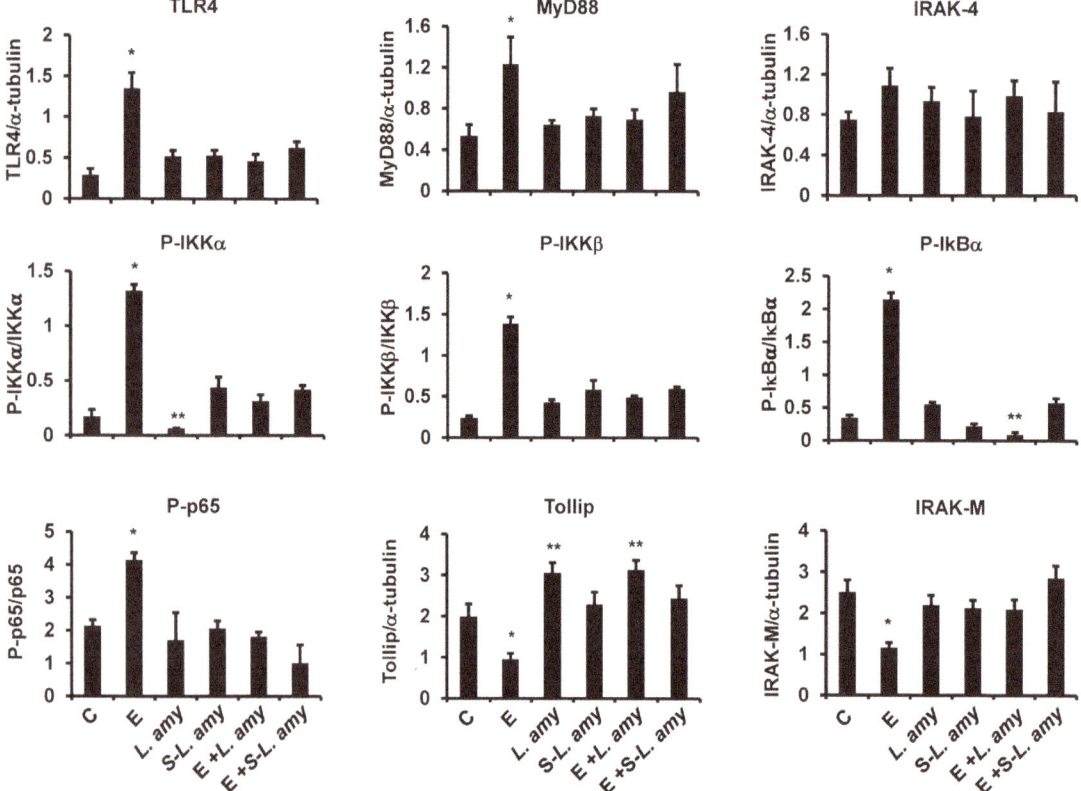

Figure 1. Inhibition of ETEC induced TLR-4 signaling pathways by *L. amylovorus* **in Caco-2/TC7 cells.** The cells were untreated (control, C), infected with ETEC (E), or treated with *L. amylovorus* (*L. amy*) or its supernatant (S-*L. amy*), either alone or simultaneously with ETEC. The figure shows the densitometric values of proteins involved in TLR4 signaling, analyzed by Western blot. The relative expression levels of TLR4, MyD88, IRAK-4, Tollip and IRAK-M were normalized to α-tubulin, whereas the phosphorylated IKKα, IKKβ, IκBα and p65 were normalized to their corresponding unphosphorylated forms. Values represent means $\pm$ SD of three independent experiments, carried out in triplicate. *$P<0.01$ compared with all. **$P<0.05$ compared with C.

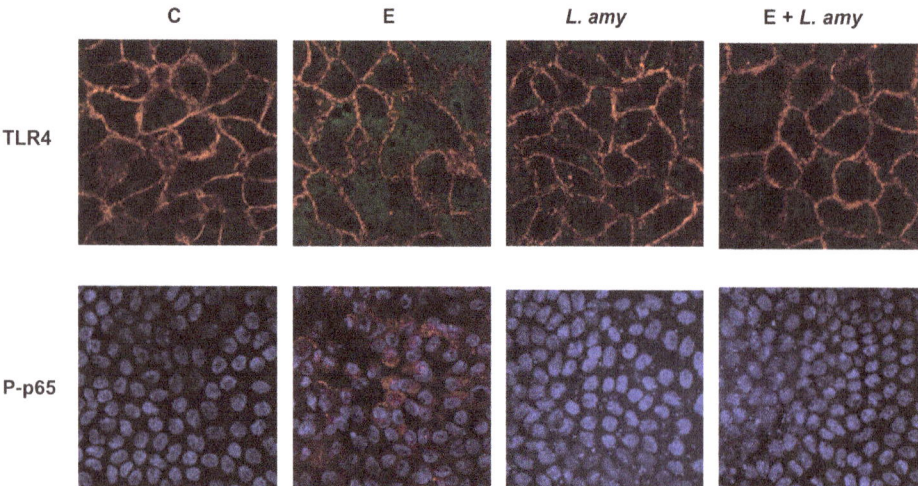

Figure 2. Inhibition of ETEC induced TLR-4 up-regulation and P-p65 translocation into the nucleus by *L. amylovorus.* Caco-2/TC7 cells were untreated (control, C), infected with ETEC (E), or treated with *L. amylovorus* (*L. amy*), either alone or simultaneously with ETEC, and analyzed by immunofluorescence. For TLR4 analysis, cells were stained in green for TLR4 and in red for occludin. For P-p65 analysis, cells were stained in red for P-p65, while nuclei were stained in blue. Each figure is representative of three independent assays ($63\times$ and $40\times$ magnification for TLR4 and P-p65, respectively).

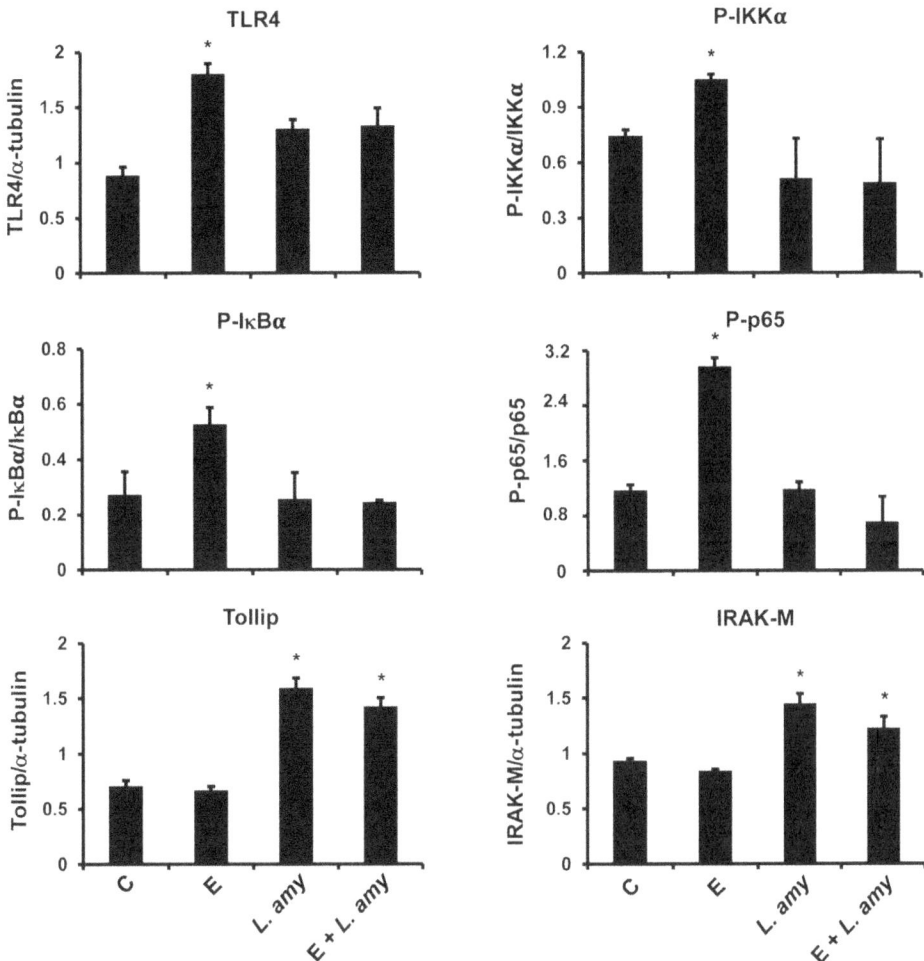

Figure 3. Inhibition of ETEC induced TLR-4 signaling pathways by *L. amylovorus* **in intestinal explants.** Pig jejunal explants were untreated (control, C), infected with ETEC (E), or treated with *L. amylovorus* (*L. amy*), either alone or simultaneously with ETEC. The figure shows the densitometric values of proteins involved in TLR4 signaling, analyzed by Western blot. The relative expression levels of TLR4, Tollip and IRAK-M were normalized to α-tubulin, whereas the phosphorylated IKKα, IκBα and p65 were normalized to their corresponding unphosphorylated forms. Values represent means ± SD of two independent experiments of five animals each.*P<0.01 compared with all.

amylovorus is able to block the first steps of the TLR4 inflammatory signaling. An ability to downregulate the expression of TLR4 has been recognized in other lactobacillus strains. For instance, *L. paracasei* inhibited the increase in TLR4 caused by *Salmonella* infection [26]. Treatment with *L. reuteri* induced a down-regulation of TLR4 associated with a reduction of inflammation in experimental enterocolitis [46]. By contrast, *L. rhamnosus* and *L. plantarum* did not modify the TLR4 expression in cells infected with *Salmonella* [28]. All together, these data indicate that the regulation of TLR4 expression may depend on the bacterial strain, and that TLR4 is a target of *L. amylovorus* anti-inflammatory activity. In uninfected cells, *L. amylovorus* does not modify the expression of TLR4 and MyD88, and this result was expected since TLR4 is the essential receptor for Gram-negative bacteria. Accordingly, previous studies showed no influence on TLR4 gene expression by lactic acid bacteria in Caco-2 cells [45,47]. Of relevance, our data underline that the immunomodulatory activity of *L. amylovorus* may change with the presence or absence of pathogens, suggesting that this property, likely extendable to other lactobacillus strains, should be considered for attributing anti-inflammatory properties to lactobacilli.

Another way by which *L. amylovorus* may counteract the ETEC induced inflammation is the inhibition of the extracellular secretion of Hsp72 and Hsp90. These proteins induce inflammation and are critical for the regulation of TLR4 complex formation and function [13–17,48]. In fact, the increase in IL-8 secretion induced by extracellular Hsp90 was suppressed by a dominant-negative form of TLR4 in vascular smooth muscle cells [16]. Furthermore, the induction of IL-8 production by extracellular Hsp72 in leukemic HL-60 cells was dependent upon activation of TLR4 and NF-κB [14]. In the present study we found strong increases in the secretion of both Hsp72 and Hsp90 after ETEC infection, which are associated with TLR4 up-regulation. These findings are consistent with previous studies showing extracellular release of Hsp72 in virally infected airway epithelial cells via activation of TLR4 [14]. Although further experiments are necessary to verify whether the reduction in Hsp72 and Hsp90 is associated with inhibition of Hsp72- and Hsp90-TLR4 binding, our results reasonably indicate that the down-regulation of the extracellular Hsp72 and Hsp90 by *L. amylovorus* contributes to the inhibition of TLR4 inflammatory signals. In addition, we demonstrate that the factor/factors secreted by *L. amyovorus* are

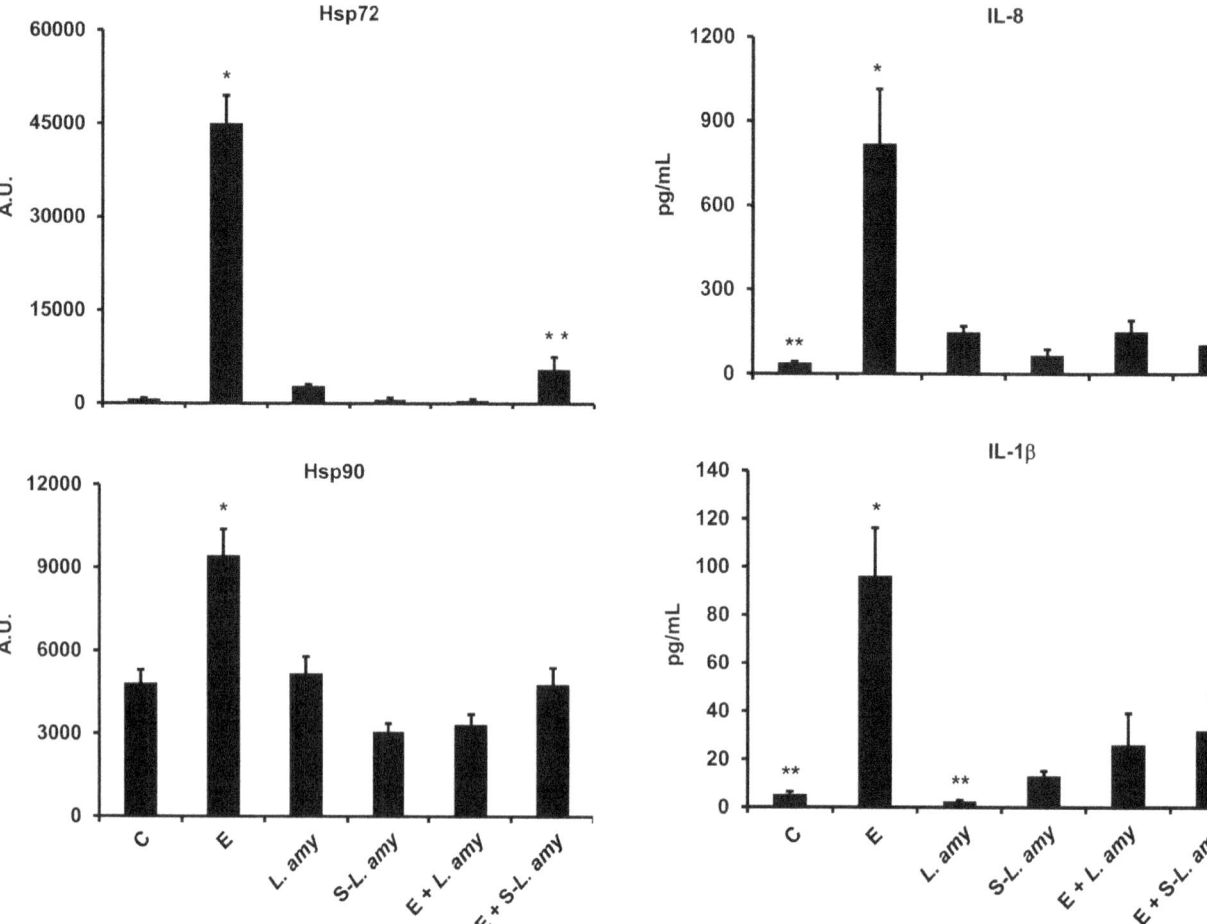

Figure 4. *L. amylovorus* **inhibits the ETEC induced up-regulation of extracellular Hsp72 and Hsp90.** Caco-2/TC7 cells were untreated (control, C), infected with ETEC (E), or treated with *L. amylovorus* (*L. amy*) or its supernatant (*S-L. amy*), either alone or simultaneously with ETEC. The Hsp72 and Hsp90 levels were analyzed by Western blot and expressed in arbitrary units (A.U.). Values represent means ± SD of three independent assays, carried out in triplicate. *P<0.001 compared with all. **P<0.05 compared with C.

Figure 5. *L. amylovorus* **abolishes the ETEC induced increase of pro-inflammatory cytokine production.** Caco-2/TC7 cells were untreated (control, C), infected with ETEC (E), or treated with *L. amylovorus* (*L. amy*) or its supernatant (*S-L. amy*), either alone or simultaneously with ETEC. IL-8 and IL-1β levels were analyzed in basolateral media by a cytometric bead array inflammatory kit. Values represent means ± SD of three independent experiments, carried out in triplicate. *P<0.001 compared with all. **P<0.05 compared with C.

able to inhibit the Hsp72 and Hsp90 release, as shown by the decreased levels of Hsp72 and Hsp90 in cells treated with *L. amylovorus* supernatant. To our best knowledge, this is the first study showing the ability of a lactobacillus strain and of its secreted products to regulate the expression of extracellular Hsps that are involved in TLR4 mediated inflammatory response to a pathogen.

The classical signaling pathway of NF-κB activation triggered by pathogens includes activation of the IKK complex followed by phosphorylation and degradation of IκB proteins, that allow phosphorylation of p65 subunit and NF-κB translocation into the nucleus to activate the transcription of inflammatory genes [19–22]. In the present study, ETEC activates all these steps leading to higher levels of IL-1β and IL-8. On the other hand, *L. amylovorus* is able to block the phosphorylation of the IKK complex, IκBα and p65, and consequently to inhibit the enhanced production of IL-1β and IL-8. Interestingly, similar inhibitory effects are triggered by *L. amylovorus* supernatant. These results together with those on Hsp72 and Hsp90, indicate that the released soluble factors from *L. amylovorus* possess anti-inflammatory activity and may act throughout the TLR4 cascade, by inhibiting either the microbial

receptor with the involvement of the extracellular Hsp72 and Hsp90, or the steps downstream TLR4, up to the inhibition of the pro-inflammatory cytokine production. Recently, an ability of *L. paracasei* and *B. breve* supernatants to modulate TLR signaling and reduce pro-inflammatory cytokine secretion in dendritic cells challenged with *S. typhi* was shown [26,49]. However, contrary to our results, these studies also found an up-regulation of TLR4 or TLR9, suggesting different anti-inflammatory properties of probiotic secreted products.

The inhibition of the NF-κB signaling may be achieved through the activity of negative regulators such as Tollip and IRAK-M [50,51]. The important role played by these regulators to control the TLR induced inflammatory responses was shown by several studies. Tollip-deficient mice mounted a decreased immune response to LPS stimulation [52]. The expression of Tollip was up-regulated in intestinal cells hyporesponsive to TLR activation, and overexpression of Tollip resulted in decreased pro-inflammatory response [53]. In addition, IRAK-M-deficient cells stimulated by TLR ligands or bacteria had increased NF-κB activation and pro-inflammatory cytokines production [50,54]. Less clear is whether the negative regulators are the targets of pathogen

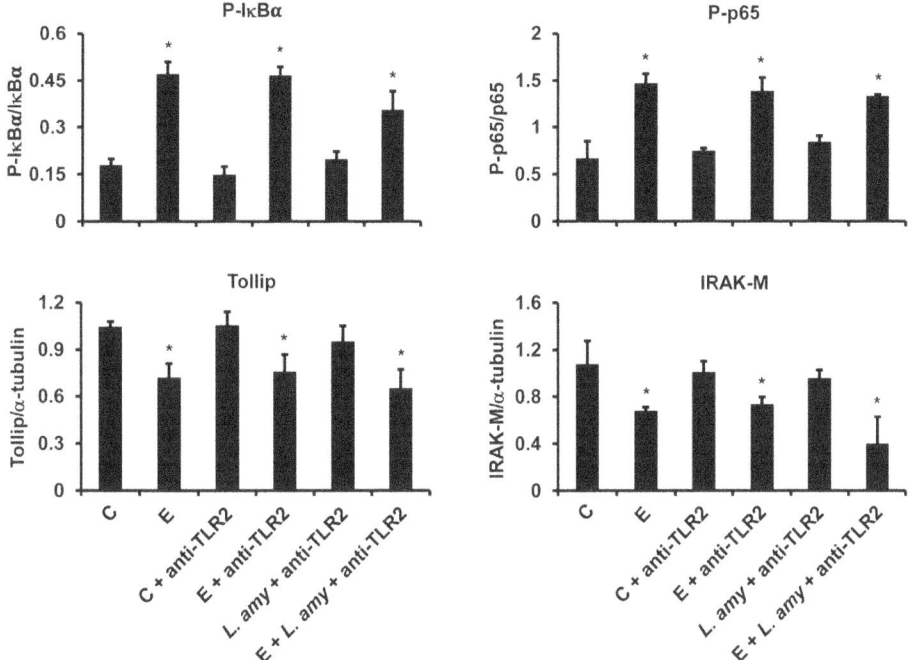

Figure 6. The regulation of TLR4 signaling is TLR2 dependent. Caco-2/TC7 cells were pretreated with anti-TLR2 antibodies and then treated with *L. amylovorus (L. amy)*, either alone or simultaneously with ETEC. Control (C) and ETEC (E) infected cells were not pretreated with anti-TLR2 antibodies. The figure shows the densitometric values of Western blots. The relative expression levels of Tollip and IRAK-M were normalized to α-tubulin, whereas the phosphorylated IκBα and p65 were normalized to their corresponding unphosphorylated forms. Values represent means $\pm$ SD of three independent assays, carried out in triplicate. *P<0.05 compared with all.

infection. For instance, IRAK-M was up-regulated in LPS treated macrophages or in lung epithelial cells in response to *S. pneumonia* [50,55], while ETEC infection did not modify IRAK-M and Tollip expression in bovine epithelial cells, and *S. typhi* caused a decrease in Tollip level in dendritic cells [26,56]. Our results show a decrease in Tollip and IRAK-M levels triggered by ETEC infection in Caco-2/TC7 cells, whereas these negative regulators are unaffected by ETEC in intestinal explants. Despite these different effects of ETEC, our findings in both intestinal cells and pig explants clearly indicate that the repression of the NF-κB signaling involves modulation of Tollip and IRAK-M. In addition, we report that *L. amylovorus* and its supernatant up-regulate Tollip either in the presence or absence of ETEC. These findings are in line with those of previous studies showing an enhanced Tollip gene expression in dendritic cells upon stimulation with *S. typhi* and *B. breve* [26], and in bovine intestinal cells treated with *L. casei* alone or simultaneously with ETEC [56]. However, Tollip gene expression was unaffected or increased by the addition of several strains of lactobacilli in uninfected porcine intestinal epithelial cells [27,57], suggesting a strain-specific effect on Tollip expression. Our study by investigating the regulation of Tollip protein expression by *L. amylovorus* in both uninfected and pathogen infected cells, provides advanced knowledge of Tollip modulation by lactobacilli. The inhibition of TLR4 signaling is triggered as well via modulation of IRAK-M, since *L. amylovorus* is able to counteract its decrease induced by ETEC in Caco-2/TC7 cells or even to up-regulate its level in pig explants. There is some evidence that lactobacilli can modulate the expression of IRAK-M, however the related studies were conducted in uninfected cells. In fact, Biswas *et al.* showed that IRAK-M activity was restored in germfree mice by colonization with *L. plantarum* [58]. Villena *et al.* reported that the immunomodulatory effect of *L. jensenii* in porcine

antigen-presenting cells was dependent on the increased expression of three negative regulators of TLRs, including IRAK-M [59]. Thus, our results show a novel ability of lactobacilli, specifically of *L. amylovorus* DSM 16698[T], to counteract the pathogen induced activation of TLR4 signaling by controlling the IRAK-M protein expression.

The intestinal epithelial cells are exposed to a myriad of commensal and pathogenic bacteria, and we cannot exclude that other cascades than TLR4 signaling may be stimulated by ETEC infection, all leading to NF-κB activation, and possibly *L. amylovorus* dampens all these signaling. Further studies are necessary to elucidate these aspects. We hypothesized that activation of TLR2 by *L. amylovorus* binding may be required for the anti-inflammatory activity. This hypothesis was supported by previous findings that treatment of epithelial cells with anti-TLR2 antibody abolished the *L. plantarum* induced blockade of IL-17 and IL-23 production triggered by *S. pyogenes* infection [60]. In addition, anti-TLR2 antibody blocked the up-regulation of IRAK-M in porcine antigen-presenting cells [59], and activation of TLR2 reduced the mucosal inflammation in mice [61]. Our results show for the first time that the binding of *L. amylovorus* to TLR2 is necessary for the inhibition of TLR4 signaling steps, as well as for the regulation of Tollip and IRAK-M, further supporting the relevance of probiotic mediated TLR2 activation to counteract the inflammatory signaling and restore intestinal immune balance.

In conclusion, the results reported in this manuscript provide advance in the knowledge of the mechanisms of probiotic anti-inflammatory activity by demonstrating that *L. amylovorus* DSM 16698[T] and its cell free supernatant inhibit the ETEC K88 induced activation of the TLR4 signaling pathway through modulation of the negative regulators Tollip and IRAK-M, as well as down-regulation of the extracellular Hsp72 and Hsp90, which are important for TLR4 functioning, leading to reduced pro-inflammatory cytokine production. In addition, we show that these anti-inflammatory activities are TLR2 dependent. Of relevance, the results obtained in the *in vitro* model of Caco-2/TC7 cells were confirmed in the *ex vivo* model of piglets mucosal explants. This study may provide helpful information for the development of potential therapeutic strategies using *L. amylovorus* to prevent or ameliorate intestinal disorders in piglets and humans. Notably, a recent study indicates that *L. amylovorus* may be considered a probiotic strain for animal as well as for human health [31].

Acknowledgments

We thank Hauke Smidt, Wageningen University, the Netherlands, for providing *L. amylovorus* strain DSM 16698[T].

Author Contributions

Conceived and designed the experiments: AF MR IPO EM. Performed the experiments: AF MR AI JS. Analyzed the data: AF MR IPO EM. Contributed reagents/materials/analysis tools: AF MR. Wrote the paper: EM.

References

1. Akira S, Uematsu S, Takeuchi O (2006) Pathogen recognition and innate immunity. Cell 124: 783–801.
2. Kawai T, Akira S (2009) The roles of TLRs, RLRs and NLRs in pathogen recognition. Int Immunol 21: 317–337.
3. O'Neill LA, Bryant CE, Doyle SL (2009) Therapeutic targeting of Toll-like receptors for infectious and inflammatory diseases and cancer. Pharm Rev 61: 177–197.
4. Lavelle EC, Murphy C, O'Neill LA, Creagh EM (2010) The role of TLRs, NLRs, and RLRs in mucosal innate immunity and homeostasis. Mucosal Immunol 3: 17–28.
5. Long KZ, Rosado JL, Santos JI, Haas M, Al Mamun A, et al. (2010) Associations between mucosal innate and adaptive immune responses and resolution of diarrheal pathogen infections. Infect Immun 78: 1221–1228.
6. Moue M, Tohno M, Shimazu T, Kido T, Aso H, et al. (2008) Toll-like receptor 4 and cytokine expression involved in functional immune response in an originally established porcine intestinal epitheliocyte cell line. Biochim Biophys Acta 1780: 134–144.
7. Rakoff-Nahoum S, Pagina J, Eslami-Varzaneh F, Edberg S, Medzhitov R (2004) Recognition of commensal microflora by toll-like receptors is required for intestinal homeostasis. Cell 118: 229–241.
8. Ogino T, Nishimura J, Barman S, Kayama H, Uematsu S, et al. (2013) Increased Th17-Inducing Activity of CD14(+) CD163(low) Myeloid Cells in Intestinal Lamina Propria of Patients With Crohn's Disease. Gastroenterology 145: 1380–1391.
9. Hausmann M, Kiessling S, Mestermann S, Webb G, Spöttl T, et al. (2002) Toll-like receptors 2 and 4 are up-regulated during intestinal inflammation. Gastroenterology 122: 1987–2000.
10. Hart AL, Al-Hassi HO, Rigby RJ, Bell SJ, Emmanuel AV, et al. (2005) Characteristics of intestinal dendritic cells in inflammatory bowel disease. Gastroenterology 129: 50–56.
11. Oyama J, Blais CJr, Liu X, Pu M, Kobzik L, et al. (2004) Reduced myocardial ischemia-reperfusion injury in toll-like receptor 4-deficient mice. Circulation 109: 784–789.
12. Asea A (2008) Heat shock proteins and toll-like receptors. Handb Exp Pharmacol 183: 111–127.
13. Triantafilou M, Triantafilou K (2004) Heat-shock protein 70 and heat-shock protein 90 associate with Toll-like receptor 4 in response to bacterial lipopolysaccharide. Biochem Soc Trans 32: 636–639.
14. Wheeler DS, Chase MA, Senft AP, Poynter SE, Wong HR, et al. (2009) Extracellular Hsp72, an endogenous DAMP, is released by virally infected airway epithelial cells and activates neutrophils via Toll-like receptor (TLR)-4. Respir Res 10: 31.
15. Chase MA, Wheeler DS, Lierl KM, Hughes VS, Wong HR, et al. (2007) Hsp72 induces inflammation and regulates cytokine production in airway epithelium through a TLR4- and NF-{kappa}B dependent mechanism. J Immunol 179: 6318–6324.
16. Chung SW, Lee JH, Choi KH, Park YC, Eo SK, et al. (2009) Extracellular heat shock protein 90 induces interleukin-8 in vascular smooth muscle cells. Biochem Biophys Res Commun 378: 444–449.
17. Asea A (2007) Hsp72 release: mechanisms and methodologies. Methods 43: 194–198.
18. Zughaier SM, Zimmer SM, Datta A, Carlson RW, Stephens DS (2005) Differential induction of the toll-like receptor 4-MyD88-dependent and -independent signaling pathways by endotoxins. Infect Immun 73: 2940–2950.
19. Zeytun A, Chaudhary A, Pardington P, Cary R, Gupta G (2010) Induction of cytokines and chemokines by Toll-like receptor signaling: strategies for control of inflammation. Crit Rev Immunol 30: 53–67.
20. Medvedev AE, Kopydlowski KM, Vogel SN (2000) Expression chemokine, and toll-like receptor 2 and 4 gene macrophages: Deregulation of cytokine, transduction in endotoxin-tolerated mouse inhibition of lipopolysaccharide-induced signal. J Immunol 164: 5564–5574.
21. Haddad JJ, Abdel-Karim NE (2011) NF-κB cellular and molecular regulatory mechanisms and pathways: Therapeutic pattern or pseudo-regulation. Cell Immunol 271: 5–14.
22. O'Neill LA, Bowie AG (2007) The family of five: TIR-domain-containing adaptors in Toll-like receptor signalling. Nat Rev Immunol 7: 353–364.
23. Haller D, Antoine JM, Bengmark S, Enck P, Rijkers GT, et al. (2010) Guidance for substantiating the evidence for beneficial effects of probiotics: probiotics in chronic inflammatory bowel disease and the functional disorder irritable bowel syndrome. J Nutr 140: 690S–697S.
24. Roselli M, Finamore A, Nuccitelli S, Carnevali P, Brigidi P, et al. (2009) Prevention of TNBS-induced colitis by different *Lactobacillus* and *Bifidobacterium* strains is associated with an expansion of γδT and regulatory T cells of intestinal intraepithelial lymphocytes. Inflamm Bow Dis 15: 1526–1536.
25. Finamore A, Roselli M, Britti MS, Merendino N, Mengheri E (2012) *Lactobacillus rhamnosus* GG and *Bifidobacterium animalis* MB5 induce intestinal but not systemic antigen-specific hyporesponsiveness in ovalbumin immunized rats. J. Nutr. 142: 375–381.
26. Bermudez-Brito M, Muñoz-Quezada S, Gomez-Llorente C, Matencio E, Bernal MJ, et al. (2013) Cell-free culture supernatant of *Bifidobacterium breve* CNCM I-4035 decreases pro-inflammatory cytokines in human dendritic cells challenged with *Salmonella typhi* through TLR activation. PLoS One 8: e59370.
27. Shimazu T, Villena J, Tohno M, Fujie H, Hosoya S, et al. (2012) Immunobiotic *Lactobacillus jensenii* elicits anti-inflammatory activity in porcine intestinal epithelial cells by modulating negative regulators of the Toll-like receptor signaling pathway. Infect Immun 80: 276–288.
28. Vizoso Pinto MG, Rodriguez Gómez M, Seifert S, Watzl B, Holzapfel WH, et al. (2009) Lactobacilli stimulate the innate immune response and modulate the TLR expression of HT29 intestinal epithelial cells *in vitro*. Int J Food Microbiol 133: 86–93.
29. Roselli M, Finamore A, Britti MS, Konstantinov SR, Smidt H, et al. (2007) The novel porcine *Lactobacillus sobrius* strain protects intestinal cells from enterotoxigenic *Escherichia coli* K88 infection and prevents membrane barrier damage. J Nutr 137: 2709–2716.
30. Konstantinov SR, Smidt H (2006) Commensal microbiota is required for the normal development and function of the porcine host immune system and physiology. In: Mengheri E, editor. Nutrition and immunity. Kerala (India): Research Signpost, p. 23–38.
31. Martinez RC, Aynaou AE, Albrecht S, Schols HA, De Martinis EC, et al. (2011) *In vitro* evaluation of gastrointestinal survival of *Lactobacillus amylovorus* DSM 16698 alone and combined with galactooligosaccharides, milk and/or *Bifidobacterium animalis* subsp. lactis Bb-12. Int J Food Microbiol 149: 152–158.
32. Pinton P, Tsybulskyy D, Lucioli J, Laffitte J, Callu P, et al (2012) Toxicity of deoxynivalenol and its acetylated derivatives on the intestine: differential effects on morphology, barrier function, tight junctions proteins and MAPKinases. Tox Sci 130: 180–190.
33. Grenier B, Loureiro-Bracarense AP, Lucioli J, Drociunas-Pacheco G, Cossalter AM, et al (2011) Individual and combined effects of subclinical doses of deoxynivalenol and fumonisins in piglets. Mol Nutr Food Res 55: 761–771.
34. Chantret I, Rodolosse A, Barbat A, Dussaulx E, Brot-Laroche E, et al. (1994) Differential expression of sucrase-isomaltase in clones isolated from early and late passages of the cell line Caco-2: evidence for glucose-dependent negative regulation. J Cell Sci 107: 213–225.
35. Caro I, Boulenc X, Rousset M, Meunier V, Bourrié M, et al. (1995) Characterisation of a newly isolated Caco-2 clone (TC-7), as a model of transport processes and biotransformation of drugs. Int J Pharm 116: 147–158.
36. Kolf-Clauw M, Castellote J, Joly B, Bourges-Abella N, Raymond-Letron I, et al. (2009) Development of a pig jejunal explant culture for studying the gastrointestinal toxicity of the mycotoxin deoxynivalenol: histopathological analysis. Toxicol In Vitro 23: 1580–1584.
37. Konstantinov SR, Poznanski E, Fuentes S, Akkermans AD, Smidt H, et al. (2006) *Lactobacillus sobrius* sp. nov., abundant in the intestine of weaning piglets. Int J Syst Evol Microbiol 56: 29–32.
38. Lehto EM, Salminen SJ (1997) Inhibition of *Salmonella typhimurium* adhesion to Caco-2 cell cultures by *Lactobacillus* strain GG spent culture supernate: only a pH effect? FEMS Immunol Med Microbiol 18: 125–132.
39. Takeuchi O, Hoshino K, Kawai T, Sanjo H, Takada H, et al. (1999) Differential roles of TLR2 and TLR4 in recognition of gram-negative and gram-positive bacterial cell wall components. Immunity 11: 443–451.

40. Fort MM, Mozaffarian A, Stöver AG, Correia Jda S, Johnson DA, et al. (2005) A synthetic TLR4 antagonist has anti-inflammatory effects in two murine models of inflammatory bowel disease. J Immunol 174: 6416–6423.

41. Liu Y, Chen F, Odle J, Lin X, Jacobi SK, et al. (2012) Fish oil enhances intestinal integrity and inhibits TLR4 and NOD2 signaling pathways in weaned pigs after LPS challenge. J Nutr 142: 2017–2024.

42. Ohland CL, Macnaughton WK (2010) Probiotic bacteria and intestinal epithelial barrier function. Am J Physiol Gastrointest Liver Physiol 8: G807–G819.

43. Konstantinov SR, Smidt H, Akkermans AD, Casini L, Trevisi P, et al. (2008) Feeding of *Lactobacillus sobrius* reduces *Escherichia coli* F4 levels in the gut and promotes growth of infected piglets. FEMS Microbiol Ecol 66: 599–607.

44. Newton K, Dixit VM (2012) Signaling in innate immunity and inflammation. Cold Spring Harb Perspect Biol 4: a006049.

45. Cammarota M, De Rosa M, Stellavato A, Lamberti M, Marzaioli I, et al. (2009) *In vitro* evaluation of *Lactobacillus plantarum* DSMZ 12028 as a probiotic: emphasis on innate immunity. Int J Food Microbiol 135: 90–98.

46. Liu Y, Fatheree NY, Mangalat N, Rhoads JM (2012) *Lactobacillus reuteri* strains reduce incidence and severity of experimental necrotizing enterocolitis via modulation of TLR4 and NF-κB signaling in the intestine. Am J Physiol Gastrointest Liver Physiol 302: G608–G617.

47. Wang S, Ng LH, Chow WL, Lee YK (2008) Infant intestinal *Enterococcus faecalis* down-regulates inflammatory responses in human intestinal cell lines. World J Gastroenterol 14: 1067–1076.

48. Triantafilou M, Miyake K, Golenbock DT, Triantafilou K (2002) Mediators of innate immune recognition of bacteria concentrate in lipid rafts and facilitate lipopolysaccharide-induced cell activation. J Cell Sci 115: 2603–2611.

49. Bermudez-Brito M, Plaza-Diaz J, Mun~oz-Quezada S, Gomez-Llorente C, Gil A (2012) Mechanisms of action of probiotics. Ann Nutr Metab 61: 160–174.

50. Kobayashi K, Hernandez LD, Galan JE, Janeway CA Jr, Medzhitov R, et al. (2002) IRAK-M is a negative regulator of Toll-like receptor signaling. Cell 110: 191–202.

51. Zhang G, Ghosh S (2002) Negative regulation of toll-like receptor-mediated signaling by Tollip. J Biol Chem 277: 7059–7065.

52. Didierlaurent A, Brissoni B, Velin D, Aebi N, Tardivel A, et al. (2006) Tollip regulates proinflammatory responses to interleukin-1 and lipopolysaccharide. Mol Cell Biol 26: 735–742.

53. Otte JM, Cario E, Podolsky DK (2004) Mechanisms of cross hyporesponsiveness to Toll-like receptor bacterial ligands in intestinal epithelial cells. Gastroenterology 126: 1054–1070.

54. Deng JC, Cheng G, Newstead MW, Zeng X, Kobayashi K, et al. (2006) Sepsis-induced suppression of lung innate immunity is mediated by IRAK-M. J Clin Invest 116: 2532–2542.

55. Lagler H, Sharif O, Haslinger I, Matt U, Stich K, et al. (2009) TREM-1 activation alters the dynamics of pulmonary IRAK-M expression *in vivo* and improves host defense during pneumococcal pneumonia. J Immunol 183: 2027–2036.

56. Takanashi N, Tomosada Y, Villena J, Murata K, Takahashi T, et al. (2013) Advanced application of bovine intestinal epithelial cell line for evaluating regulatory effect of lactobacilli against heat-killed enterotoxigenic *Escherichia coli*-mediated inflammation. BMC Microbiol 13: 54.

57. Hosoya S, Villena J, Chiba E, Shimazu T, Suda Y, et al. (2013) Advanced application of porcine intestinal epithelial cells for the selection of immunobiotics modulating toll-like receptor 3-mediated inflammation. J Microbiol Immunol Infect 46: 474–481.

58. Biswas A, Wilmanski J, Forsman H, Hrncir T, Hao LH (2011) Negative regulation of Toll-like receptor signaling plays an essential role in homeostasis of the intestine. Eur J Immunol 41: 182–194.

59. Villena J, Suzuki R, Fujie H, Chiba E, Takahashi T, et al. (2012) Immunobiotic *Lactobacillus jensenii* modulates the Toll-like receptor 4-induced inflammatory response via negative regulation in porcine antigen-presenting cells. Clin Vaccine Immunol 19: 1038–1053.

60. Rizzo A, Losacco A, Carratelli CR, Domenico MD, Bevilacqua N (2013) *Lactobacillus plantarum* reduces *Streptococcus pyogenes* virulence by modulating the IL-17, IL-23 and Toll-like receptor 2/4 expressions in human epithelial cells. Int Immunopharmacol 17: 453–461.

61. Cario E, Gerken G, Podolsky DK (2007) Toll-like receptor 2 controls mucosal inflammation by regulating epithelial barrier function. Gastroenterology 132: 1359–1374.

6

In Vitro Selection and Characterization of New Probiotic Candidates from Table Olive Microbiota

Cristian Botta[1], Tomaz Langerholc[2], Avrelija Cencič[2†], Luca Cocolin[1]*

1 Department of Forestry, Agriculture and Food Sciences, University of Torino, Torino, Italy, **2** Department of Microbiology, Biochemistry, Molecular Biology and Biotechnology, Faculty of Agriculture and Life Sciences, University of Maribor, Maribor, Slovenia

Abstract

To date, only a few studies have investigated the complex microbiota of table olives in order to identify new probiotic microorganisms, even though this food matrix has been shown to be a suitable source of beneficial lactic acid bacteria (LAB). Two hundred and thirty eight LAB, belonging to *Lactobacillus plantarum*, *Lactobacillus pentosus* and *Leuconostoc mesenteroides* species, and isolated from *Nocellara Etnea* table olives, have been screened in this survey through an *in vitro* approach. A simulation of transit tolerance in the upper human gastrointestinal tract, together with autoaggregation and hydrophobicity, have been decisive in reducing the number of LAB to 17 promising probiotics. None of the selected strains showed intrinsic resistances towards a broad spectrum of antibiotics and were therefore accurately characterized on an undifferentiated and 3D functional model of the human intestinal tract made up of H4-1 epithelial cells. As far as the potential colonization of the intestinal tract is concerned, a high adhesion ratio was observed for *Lb. plantarum* O2T60C (over 9%) when tested in the 3D functional model, which closely mimics real intestinal conditions. The stimulation properties towards the epithelial barrier integrity and the *in vitro* inhibition of *L. monocytogenes* adhesion and invasion have also been assessed. *Lb. plantarum* S1T10A and S11T3E enhanced trans-epithelial electrical resistance (TEER) and therefore the integrity of the polarized epithelium in the 3D model. Moreover, S11T3E showed the ability to inhibit *L. monocytogenes* invasion in the undifferentiated epithelial model. The reduction in *L. monocytogenes* infection, together with the potential enhancement of barrier integrity and an adhesion ratio that was above the average in the 3D functional model (6.9%) would seem to suggest the *Lb. plantarum* S11T3E strain as the most interesting candidate for possible *in vivo* animal and human trials.

Editor: Ivo G. Boneca, Institut Pasteur Paris, France

Funding: The conducted research received funding from the EU (FP7/2007–2013), under grant agreement no. 243471- PROBIOLIVES (www.probiolives.eu). The information in this document only reflects the authors' views and the Community is not liable for any use that may be made of the information contained herein. The funders had no role in study design, data collection and analysis, decision to publish, or preparation of the manuscript.

Competing Interests: The authors have declared that no competing interests exist.

* E-mail: lucasimone.cocolin@unito.it

† Deceased

Introduction

In the past, the gastro intestinal tract (GIT) was considered the main potential source of probiotic bacteria, but the scientific community has recently focused attention on fermented foods, recognizing them as valid and heterogeneous sources of probiotic microorganism. Although dairy products have been exploited extensively as both a source and a carrier of lactic acid bacteria (LAB) or bifidobacteria, few researches have been focused on fermented vegetable products. Their native microbiota offer a broad range of LAB species, such as *Lactobacillus* (*Lb.*) *plantarum*, *Lb. casei*, *Lb. paracasei*, *Lb. delbrueckii* and *Lb. brevis*, all of which present strains with probiotic features [1].

In this context, table olives are the most important fermented vegetables on the international food market, and their spontaneous fermentation, which occurs in different production processes, is usually the result of the competitive activities of the autochthonous microbiota, together with a variety of contaminating microorganisms from fermentation environments. This fermentation is mainly caused by the synergic metabolic activity of yeasts and LAB. It is generally recognized that LAB are the main inducers of brine

acidification and are therefore fundamental for the stability of the final product, whereas yeasts are mainly involved in the development of the organoleptic characteristics [2]. As far as LAB are concerned, *Lb. plantarum* and *Lb. pentosus* are the most representative species involved in fermentation. The LAB microbiota of table olives is also characterized by the presence of *Lb. casei* and heterofermentative cocci, such as *Leuconostoc* (*Ln.*) *mesenteroides* [3]. All of these species have shown probiotic potentiality in many studies. For example, *in vitro Lb. plantarum* tests have highlighted their ability to modulate the immune response and to potentially inhibit pathogens [4], [5], as well as strains belonging to *Lb. casei* and *Lb. paracasei* species, which have proved able to inhibit Gram negative pathogens [6], [7] and *Listeria* (*L.*) *monocytogenes* [8]. *In vitro* inhibition of *L. monocytogenes* infection was recently discovered for *Ln. mesenteroides* species [9]. Table olives could also be regarded as a promising probiotic food considering that, compared to dairy products, they do not pose problems for those people who are intolerant to milk and milk products or those who need low-cholesterol diets. Moreover, it should be pointed out that an edible portion of about 100 g of olives contains more than 10^9 live cells of selected *Lb. paracasei* or *Lb. plantarum* strains, which corresponds to the daily dose recommended to obtain beneficial effects [10].

The use of table olives as a probiotic source has already been explored in several studies [11], [12], [13], which, through *in vitro* methods, have evaluated the probiotic and technological characteristics of autochthonous LAB isolated from table olive fermentations. These studies have confirmed that table olives area suitable source of probiotic LAB [14]. They have also highlighted the importance of an *in vitro* approach as the first step towards a rational selection of new probiotics, which should take into account criteria such as antibiotic resistance and survival ability in simulated GIT conditions, and the interaction with epithelial human cells.

As established by the Food and Agriculture Organization and the World Health Organization (FAO/WHO), a potential probiotic LAB has to be Generally Recognized as Safe (GRAS) and a possible resistance towards antibiotics is the main undesirable feature [15]. Genetic resistance to antibiotics might be transferred from LAB to other commensal microorganisms through plasmids or conjugative transposons, thus increasing the danger of the pathogens that could be present in the gut environment [16]. As far as the survival ability in simulated GIT conditions is concerned, many different simulations of human digestion have been reported in literature [17] – [21]. These *in vitro* assessments of digestion resistance differ by the transit time, the modality of probiotic assimilation (alone or inserted in food matrices), and the complexity of the GIT model. These different simulations of the GIT conditions are extensively used as first discriminatory test for the selection of potential probiotic candidates [22].

Moreover, it is universally recognized that probiotics must be able to colonise the digestive tract [1], [23], On the other hand, it is also currently acknowledged that a strong persistence of probiotics in the GIT could generate dysbiosis via the excessive deconjugation of bile salts and/or degradation of intestinal mucus layer [24]. Anyhow, no serious adverse effects have been described in clinical trials, [25]. Accordingly, adhesion properties have been proposed as a crucial factor for the selection of new probiotic and they could easily be investigated using *in vitro* models of the intestine. These tools are also fundamental for the study of the interaction between probiotics, pathogens and human cells, such as the enhancing of the innate immune function and the inhibition of pathogen action [26].

Therefore, the objectives of the current paper were: (i) to identify new probiotic candidates in a collection of LAB isolated from table olives, through a comprehensive approach which would initially consider the safety features of the strains; and (ii) to establish their interaction with human epithelial cells. For this second purpose, the most suitable strains have been characterized with *in vitro* models of the gut, focusing attention on the inhibition of *L. monocytogenes* and stimulating the effect of intestinal barrier integrity.

Materials and Methods

2.1 Bacteria and Sources of Isolations

Initially, 238 LAB were collected separately from brines and drupe surfaces of table olive fermentations carried out with the *Nocellara Etnea* variety. Most of this collection (191 strains) was isolated from two industrial processes conducted in small enterprises located in Sicily, as previously described [27]. These strains were identified as *Lb. plantarum* (182 isolates) and *Lb. pentosus* (9 isolates) species by means of multiplex PCR analysis of the *recA* gene, with species-specific primers for *L. pentosus*, *L. plantarum* and *L. paraplantarum*, according to the protocol described by Torriani et al. [28]. The remaining 47 isolates were recovered from experimental table olive fermentations carried out at both industrial and laboratory scale and followed as previously described [27], with the aim of determining the technological performances of potential starter cultures. This second collection was composed by: 24 isolates of *Lb. plantarum*, 7 *Lb. pentosus*, and 16 belonging to *Ln. mesenteroides*.

All the strains collected were purified by streaking and checked through Gram staining and catalase activity. Isolates were grown in Man Rogosa Sharp broth (MRS, Oxoid, Milan, Italy) for 24 h at 30°C and stored at - 80°C with 20% (w/v) of glycerol (Sigma, Milan, Italy).

2.2 Phenotypic Tests

Initial screening of the 238 isolated strains was performed according to the *in vitro* phenotypic tests described by Bautista-Gallego et al. [13]. Briefly, production of antimicrobial compounds, hemolytic activity, bile salt hydrolysis (BSH activity), autoaggregation, bacterial surface hydrophobicity and survival in a simulated human digestion process were used for the discrimination. In all tests the probiotic strains *Lb. rhamnosus* GG (LGG) and *Lb. casei Shirota* were used as reference controls.

The detection of bacteriocins production was assessed using the agar-well diffusion assay (AWDA) as described by Toba et al. [29] with some minor modifications. Briefly lawn of BHI (Oxoid) soft agar (10 g L^{-1}) medium containing each indicator microorganism, namely *L. monocytogenes* FMCC B-128 and NCTC 10527, was poured onto Petri dishes. After solidification, 5 mm wells were made in the plates and filled with 50 μL of overnight BHI broth culture of each strain and left to diffuse for 30 min. After an overnight incubation at the optimum growth temperature for each indicator strains, the plates were examined for halos around the wells. For those strains, which showed the inhibition zones, the test was repeated adding a volume (4 μL) of proteinase K solution (25 mg mL^{-1}; Sigma) in each well. The proteolytic enzyme was added in order to confirm the proteinaceous nature of the inhibitor compound.

To test for haemolytic activity, overnight lactobacilli broth cultures were streaked onto Columbia agar plates containing 50 g L^{-1} of horse blood (Oxoid), and incubated for 48 h at 30°C. Then, the plates were examined for signs of α, β or γ-haemolysis.

The BSH activity was tested by using the plate assay described by Dashkevicz and Feighner [30]. Briefly, bile salt MRS agar plates containing 5 g L^{-1} of bovine bile (Sigma) were inoculated with an overnight MRS culture, incubated at 37°C for 72 h, and then observed for colonies with precipitated bile salts.

Autoaggregation assays were performed according to the methodology described by Kos et al. [31]. The autoaggregation percentage was expressed as a function of time until it was constant, using the formula $1 - (A_t/A_0) \times 100$, where A_t represents the absorbance measured at 600 nm (A600) at any time (1, 2, 3, 4 or 5 h), and A_0 the absorbance at time t = 0 h. Final value of autoaggregation, after 5 h (AA), was used as index of the bacterial cells capability to aggregate among them.

Bacterial cell surface hydrophobicity was assessed by measuring microbial adhesion to hydrocarbons using the procedure described by Crow et al. [32] with some modifications. Briefly, bacteria growth at 37°C for 48 h were centrifuged (10 000×g for 5 min). The resulting pellet was washed twice in PBS, re-suspended in 3 mL of KNO$_3$ 0.1 M and the A600 was measured (A_0). One mL of o-xylene (Sigma) was then added to the cell suspension to form a two-phase system. After a 10 min pre-incubation at room temperature, the two-phase system was mixed by vortexing for

2 min. Then, the water and xylene phases were separated by incubation for 20 min at room temperature. The aqueous phase was carefully removed and the A600 was measured (A_1). The percentage of the cell surface hydrophobicity (H) was calculated using the formula $H = (1 - A_1/A_0) \times 100$.

A simulated process of human digestion process was assessed performing consecutively a gastric and intestinal step of *in vitro* digestion. The gastric digestion step was simulated using the synthetic gastric juice described by Corcoran et al. [18]. The lactobacilli were grown for 24 h at 37°C, centrifuged and the pellet was washed with PBS. Initial count (T_0) of the overnight culture was performed pleating serial dilution on MRS agar. Plates were incubated for 3–5 days at 37°C in microaerophilic conditions. Then, the bacteria were re-suspended in the synthetic gastric juice and incubated for 2 h at 37°C in an orbital shaker (~ 200 rpm) to simulate peristaltic movements. Harvested cells from the gastric digestion step were washed in PBS and the enumeration of bacteria cells was performed as previously (T_g). The pellet was then re-suspended in the same volume of the simulated intestinal juice, which was formulated using bile (3 g L^{-1}, Oxoid) and pancreatin (0.1 g L^{-1}, Sigma) in a buffer at pH 8.0 consisting of 50.81 g L^{-1} of sodium phosphate dibasic heptahydrate, 8.5 g L^{-1} of NaCl and 1.27 g L^{-1} of KH_2PO_4 [13]. After shaking at 200 rpm in an orbital shaker for 4 h at 37°C, the pellet was washed and then re-suspended in a volume of PBS and the enumeration on MRS plates after the intestinal step (T_i) was performed as described above. Overall digestion survival (ODS) was obtained by comparison of the initial lactobacilli counts at the start of the simulated gastric digestion (T_0) and those remaining at the end of the simulated intestinal digestion (T_i). Data were expressed in percentage according to the formula: $T_i/T_0 \times 100$.

2.3 Chemometric Analysis and Selection

The final value of autoaggregation (AA), bacterial surface hydrophobicity (H) and ODS of the strains able to survive to the simulated digestion were used as input variables in hierarchical cluster analysis (HCA) based on Euclidian distances (single linkage distance). Considering these three variables, the probiotic strains LGG and *Lb. casei Shirota* were compared to the tested strains in the HCA. The HCA allowed the strains to be allocated into homogeneous groups according to their characteristics, identifying those deserving to be characterized subsequently by *in vitro* gut models [12].

Moreover, the physiological characteristics (ODS, AA, and H) of the selected strains, together with those of two reference probiotics, were subjected to ANOVA and Duncan's test, in order to highlight any significant differences between the phenotypes. The data were analyzed using Statistica, ver. 7.0, (StatSoft Inc., Tulsa, USA).

2.4 Antibiotic Resistance

The micro-dilution broth test described by Argyri et al. [12] was used with slight modifications to test the antibiotic resistance of each strain selected in the previous phenotypic tests. The analysis was performed with eight antibiotics (ampicillin, gentamicin, kanamycin, streptomycin, erythromycin, clindamycin, tetracycline, chloramphenicol; Sigma), which were initially added at the species-specific breakpoint concentrations proposed by the European Food Safety Agency (EFSA) [33]. If resistant, the strains were progressively tested with higher antibiotic concentrations until their Minimum Inhibitory Concentrations (MICs) were found.

The strains were grown in MRS broth for 18 h at 37°C and, after a centrifuge step, the media were removed and the cells were washed twice with an isotonic solution (Ringer, Oxoid). Concentrations of bacteria suspensions were quantified through optical density (OD) measurement at 630 nm in order to standardize the inoculum at $7.0 \pm 0.5 \log_{10}$ CFU mL^{-1}. The suspensions were then inoculated in MRS broth (1% v/v), supplemented with each antibiotic, and incubated for 24 hours at 37°C. The experiment was performed in 96 microtiter well plates (Corning, New York, USA) and reference controls were performed by inoculating the bacteria in MRS not supplemented with antibiotics. Bacterial growth was monitored using an ELx880 microtiter plate reader (Savatec, Turin, Italy) by measuring OD at 630 nm after the incubation period. Three independent experiments were carried out and each assay was performed in duplicate.

The data were expressed as percentages of the bacterial growth in the antibiotic supplemented MRS (A_{atb}) and compared with the growth in pure MRS (A_n), using an $A_{atb}/A_n \times 100$ formula. Strains with a percentage of growth ratio ≤5% were considered as not resistant.

2.5 *In vitro* Gut Models and Experimental Conditions

The experiments were performed using intestinal epithelial and monocyte/macrophage derived cell lines of human origin, named respectively H4 clone 1 (H4-1) and TLT. Both cell lines have been prepared and characterized at the Department of Microbiology, Biochemistry, Molecular Biology and Biotechnology at the Faculty of Agriculture and Life Sciences, University of Maribor (Maribor, Slovenia) [34]. H4 cell line was initially prepared from neonatal intestinal epithelia [35], [36] and further cloned to H4-1, which is characterized by developing high trans-epithelial electrical resistance. Macrophage cell line TLT was prepared from human blood-derived PBMC, isolated from a healthy donor [37]. Both cell lines were grown in an advanced Dulbecco Modified Eagle Medium (DMEM) (Gibco), supplemented with 5% foetal calf serum (Lonza, Basel, Switzerland), L-glutamine (2 mM, Sigma), penicillin (100 U mL^{-1}, Sigma) and streptomycin (1 $mgmL^{-1}$, Fluka, Buchs, Switzerland). The cell lines were routinely grown in 25 cm^2 culture flasks (Corning, New York, USA) at 37°C in a humidified atmosphere containing 5% CO_2 and 95% air, until confluent monolayers were obtained. The culture medium was changed routinely and once the cells reached confluence, after 3–4 days, they were subpassaged.

Different *in vitro* experimental set-ups of human gut epithelium were used: cell growing as an undifferentiated monolayer and in two functional 3D model, as shown schematically in Figure 1.

For the tests performed with the undifferentiated monolayer, 96-well flat bottom plates (Corning, New York, USA) were filled with a suspension of 100 000 cells/well (c/w) in DMEM supplemented with antibiotics and a serum, in order to obtain a complete and confluent monolayer after 24 hours of incubation at 37°C in a humidified atmosphere of 5% CO_2 and 95% air (Fig. 1 A).

For the functional 3D model, the cells were seeded on polyester Transwell filter inserts with microporous (0.4 μm pore size, 12 mm, Corning) and placed in 12 well plates (22.1 mm, Corning) at a density of 50 000 cells/cm^2. The filters were maintained with a volume of 0.5 mL in the apical compartment and 1.5 mL in the basolateral compartment. The cells were grown in the same medium and under the same conditions as those described above for 14–15 days with regular changes of the media, until functional polarization was reached. Polarization of the cells was established by measuring the trans-epithelial electrical resistance (TEER) with the Millicell Electrical Resistance System (Millipore, Bedford, MA), as described previously [37]. At the same time, TLT cells were seeded in 12 well plates (200 000 c/w) and further incubated until confluence (24–48 hours) (Fig. 1B). In order to establish the

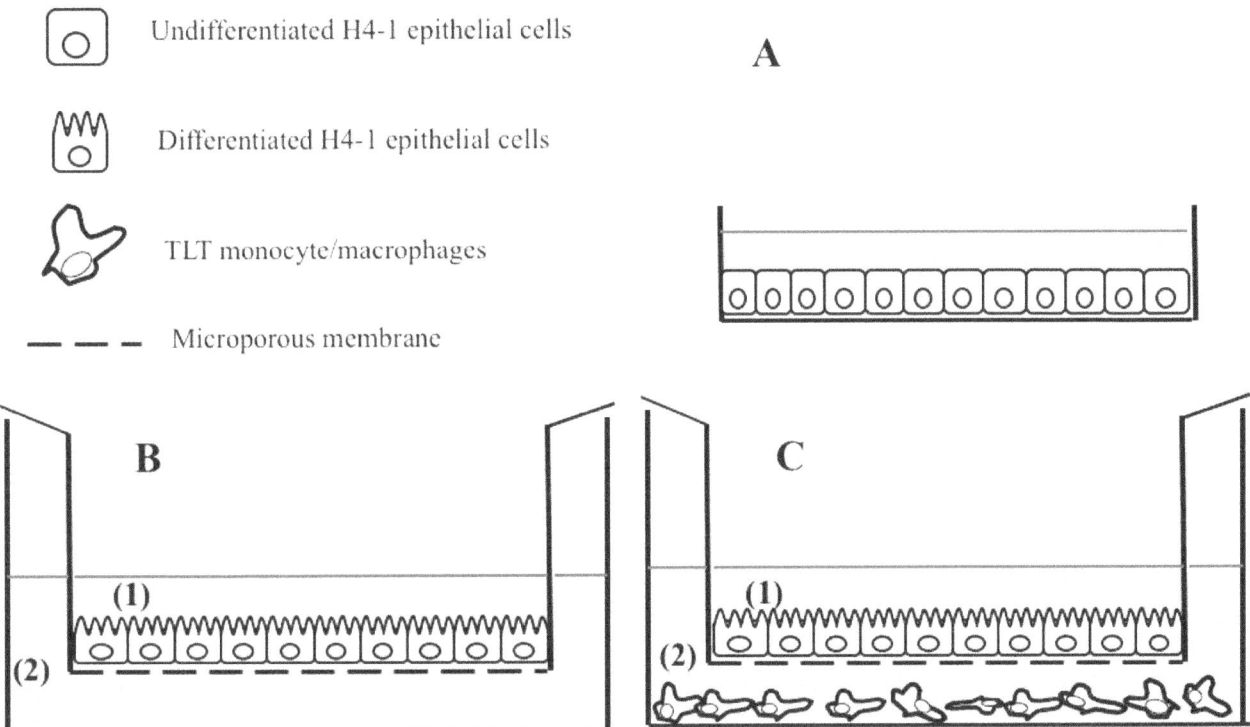

Figure 1. Experimental intestinal cell model settings used in this survey: (A) undifferentiated model with H4-1 epithelial cells grown on plastic surface; (B) functional 3D model where H4-1 were grown until the differentiation on microporous membrane; and (C) the complete 3D H4-1/TLT model with a differentiated layer of H4-1 in the apical compartment (1) and TLT monocyte/macrophages seeded in the basolateral side (2). In the models (B) and (C) the polarization of H4-1 cells was obtained after 14–15 days of incubation with regular changes of the media (cells initially seeded at a density of 50 000 cells per cm^2), and confirmed by measuring the TEER. In the model (C) inserts with polarized H4-1 cells were transferred into wells that had been underlain with TLT cells (200 000 cells each well and growth for 24–48 h).

complete 3D H4-1/TLT model, inserts with H4-1 cells were transferred into wells that had been underlain with TLT cells (Fig. 1C).

All the selected strains tested with the cell models were cultured for routine use in MRS broth and grown for 18 h at 37°C, in order to test them in an exponential phase. In all the experiments performed with human cells, the initial concentration of bacterial strains was determined by OD at 595 nm and all the suspensions were set to the same initial count using an internal standard curve. Before the experiment, the bacteria was washed twice in PBS and then resuspended in DMEM w/o antibiotics and serum. Moreover, in the human gut experimental set-ups, the H4-1 epithelial cells were washed three times with PBS before the addition of the bacteria, in order to eliminate any traces of the antibiotics.

The results of the experiments with human cells were expressed as the mean and standard error mean (SEM) of at least three experiments with duplicate or triplicate determinations. Statistical analyses were performed with Statistica, ver. 7.0, (StatSoft Inc., Tulsa, USA). In order to assess the overall variation and differences between the multiple groups, the numerical values were analyzed by ANOVA (One way-Analysis of Variance) and with Duncan's post-hoc test. The Student's t-test was also used to compare individual groups with the control or with other individual groups. In all experiments performed with *in vitro* gut models LGG strain was used as the probiotic reference strain (positive control).

2.6 Cellular Metabolic Activity

The cellular metabolic activity with 3-(4,5-dimethylthiazole-2-yl)-2,5-phenyl tetrazolium bromide (MTT; Sigma) was assessed in order to evaluate the potential cytotoxic effect of the selected strains towards H4-1 and TLT human cells. Undifferentiated monolayer and the complete 3D model (H4-1/TLT) were incubated with heat inactivated and live bacteria, respectively. In short, overnight bacterial cultures were set to 8.0 ± 0.5 log$_{10}$ CFU mL^{-1}, washed twice with PBS and heat inactivated at 95°C for 30 minutes [38]. This inactivation was made to avoid a misinterpretation of the results, since the MTT reduction may be lead by bacterial enzymes as well [39]. After the inactivation, the bacteria were resuspended in fresh DMEM and added to the H4-1 cells growing in the monolayer. The heat inactivated bacteria and H4-1 cells were co-incubated for 24 h at 37°C in a humidified atmosphere containing 5% CO_2 and 95% air. In order to evaluate cytotoxicity towards the TLT cells, an 8.0 ± 0.5 log$_{10}$ CFU mL^{-1} suspension of live bacteria, prepared in DMEM w/o antibiotics and serum, was seeded in the upper compartment of the 3D model (H4-1/TLT), but not in direct contact with the macrophages, in order to simulate the real conditions of the human gut. The live bacteria were incubated in the 3D model (H4-1/TLT) for 24 h at 37°C in a humidified atmosphere containing 5% CO_2 and 95% air.

After 24 h of incubation, the supernatant was removed and H4-1 and TLT cells were put in contact with DMEM w/o phenol red supplemented with 5 mg mL^{-1} of MTT (Sigma) for 3 h. At the end of MTT degradation, the supernatant was gently removed and the formazan on the bottom was dried and solubilized with a

0.04% HCl solution in isopropanol (Sigma).The cellular metabolic activity of the H4-1 and TLT cells was detected spectrophoto-metrically at 570 nm [40] and data were expressed by means of the following equation: 100- (OD of test sample/OD of blank×100). Cells not treated with bacteria were used as blanks.

2.7 Adhesion Assay

The adhesion ability of the selected strains was assessed on both the undifferentiated monolayer and on the functional 3D model of the H4-1 cells. The bacterial strains were seeded in an 8.0 ± 0.5 $\log_{10}$ CFU mL^{-1} concentration and incubated for 90 min in a modified atmosphere of 5% CO_2 and 95% air. The inoculum of each tested strain was quantified in each experiment using the standard CFU method. The H4-1 cells were washed five times with PBS and the cells were homogenized with a Triton-X solution (0.25% in PBS; Sigma). After 30 minutes of incubation, the solution with released bacteria was serially diluted and plated on MRS agar. The plates were incubated for 48 hours at 37°C in microaerophilic conditions. Adhesion ability was expressed as the percentage ratio between the counts initially seeded and the counts after the washing steps (CFU mL^{-1}). In parallel, assays were carried out on each bacterial strain to exclude any potential harmful effect on the survival of bacteria due to the treatment with the 0.25% Triton-X solution.

2.8 In vitro Adhesion and Invasion of Listeria monocytogenes

The capability of the LAB strains to inhibit in vitro the adhesion and invasion of L. monocytogenes WT (collected at the Faculty of Agriculture and Life Sciences, University of Maribor) was established on the H4-1 cells according to the method described by Corr et al. [4], with some minor modifications. Trials were performed with an undifferentiated monolayer treated with probiotic strains. The treatment was carried out by seeding 8.0 ± 0.5 $\log_{10}$ CFU mL^{-1} of each selected LAB onto the cell model. The human cells and LAB were then incubated for 90 min in a modified atmosphere containing 5% CO_2 and 95% air and washed five times with PBS, as described in the adhesion assay, in order to remove non adherent bacteria. L. monocytogenes was added in the next step. A multiplicity of infection (MOI) 10 was used for the ratio between the pathogen and the human cells, as suggested by Yamada et al. [41] for non tumorigenic cell lines of animal origin. This MOI was also established as being optimal for the H4-1 cells through a cytoxicity assay. In both assays, an 18 h pre-cultivated pathogen was inoculated in Nutrient Broth (Oxoid), incubated at 37°C for 13 h, washed twice with PBS, resuspended in DMEM w/o antibiotics, diluted in order to reach the appropriate bacterial count (7.0 ± 0.5 $\log_{10}$ CFU mL^{-1}) and then inoculated on the cell monolayer surface. After 2 h of co-incubation at 37°C, the monolayers were washed three times with PBS in order to remove any unattached pathogenic bacteria. The cells with attached L. monocytogenes were lysed in PBS containing 0.25% Triton-X, as described above for the adhesion assay. Counts were performed by means of the CFU method, plating the dilutions on Listeria Selective Agar (Fluka, St. Gallen, Switzerland) and incubating them at 30°C for 48 h. The result of the adhesion inhibition assay can be considered representative of the overall action of the strains, both against the L. monocytogenes attached to the cell membranes and against the pathogen present inside the cell cytosol.

As far as the invasion assay is concerned, after the final washing step to remove non-adhered L. monocytogenes cells, the DMEM media supplemented with 50 μg mL^{-1} gentamicin sulfate (Sigma) was added to the cell monolayers and further incubated for 2 h at 37°C to kill all the extracellular bacteria [4]. The monolayer was washed three times, the cells were lysed with Triton-X 0.25% and the bacterial counts were performed as described above. The action of the strains against pathogen invasion of the human cells was evaluated specifically in this second experiment.

The results of the adhesion and the invasion assays were expressed as percentage ratios of the L. monocytogenes recovered from the treated wells and the count from the wells not treated with LAB strains (CFU mL^{-1}).

2.9 Trans-epithelial Electrical Resistance (TEER) as a Measurement of the Epithelial Barrier Integrity

The bacterial effect on the epithelial barrier was evaluated by measuring the TEER using the Millicell Electrical Resistance System, as described by Jensen et al. [42], with some minor modifications. A functional 3D model of the H4-1 cells was constructed as previously reported. An optimal functional polarity was developed after 14 days of growth on the membrane, and an average TEER of 740 ± 100 Ω/cm^2 was reached. On the day before the experiment, the filters were washed with PBS to remove traces of the antibiotics and the cell media was changed to the original media without antibiotics. TEER was measured before the addition of bacteria (TEER$_{t0}$), and after 1, 3, 5, 7, 18 and 24 h of incubation (TEER$_{tn}$). The bacterial effect on TEER was tested with 10^7 CFU mL^{-1} of inoculated bacteria. L. monocytogenes WT and DMEM were used in the same concentration as the negative and positive controls, respectively.

The TEER (TEER$_{tx}$/TEER$_{t0}$×100) ratio was calculated and the bacterial effect on the epithelial barrier over time was compared for the investigated strains.

Results

3.1 Selection of the most Promising Strains

In the present study, 238 LAB strains collected from different Nocellara Etnea green table olive fermentation processes were screened in vitro for their phenotypic features related to probiotic traits. The results of the phenotypic screening of all the isolates are summarized in Table S1.

With regard to the autoaggregation indices, the strains exhibited a normal distribution, with the largest number of the observations located in the interval between 10 and 20%. Furthermore, Lb. plantarum strains showed a higher autoaggregation phenotype (23% on average) with respect to the other two species isolated, Lb. pentosus and Ln. mesenteroides (19% in average). Lb. plantarum O1T90E, S1T30B and O2T60D showed the highest autoaggregation values (more than 50%). However, most of the isolates showed hydrophobicity values of less than 40%, with a strongly skewed distribution of the observations tending towards the lower values. However, two strains of Lb. plantarum showed values of over 90% (S3T60C and S2T30B).

As far as the potential inhibition of the selected indicator pathogens is concerned, the first AWDA highlighted 22 strains of Lb. plantarum and the Ln. mesenteroides FO50O that were able to inhibit the growth of L. monocytogenes FMCC B-128 and NCTC 10527. However, the confirmation screening, with proteinase K, attributed this antagonist effect to the production of organic acids (Table S1). It therefore resulted that no strains were able to produce bacteriocins and none showed BSH or hemolytic activities (data not shown).

Regarding the effect of the digestion simulation over the strains (gastric and pancreatic transits), we considered as not resistant those which presented after the two passages undetectable viable counts (<10 CFU mL^{-1}), corresponding to an ODS value of

0.00001%. Overall, 55 strains (23.1% of the tested bacteria) were resistant to the simulated digestion process (Table S1), Using the ODS rates, together with the percentages of hydrophobicity and autoaggregation, hierarchical cluster analysis highlighted two well-defined groups (Fig. 2). The lower group (cluster II) was larger and contained strains with less resistance to simulated human digestion together with a weak auto-aggregating and hydrophobic phenotypes. The reference strain, *Lb. rhamnosus* GG, which showed a poor probiotic potential in the tests, was also allocated to this cluster, whereas *Lb. casei Shirota* was grouped in cluster I.

This group collected 17 strains with the most promising phenotypes, and these were therefore selected and further characterized for antibiotic susceptibility and tested with human cell models. The phenotypic features (ODS, H, AA) of these strains are shown in Figure 3. The majority of them belonged to the *Lb. plantarum* species (15 isolates), moreover, two strains of *Ln. mesenteroides* coded FS50Q and FO50E, and the *Lb. pentosus* S3T60C were included. Focusing attention on the ODS (Fig. 3 A), it was observed that three strains, O1T90E, S11T3E and O2T60C, showed values of over 0.003%, which were significantly higher than the positive reference strains *Lb. casei Shirota* and LGG ($P<0.05$), and the highest survival ratio was shown for the *Lb. plantarum* O1T90E, with an overall survival rate of 0.00677%. As far as the hydrophobicity (Fig. 3B) and autoaggregation (Fig. 3C), no significant differences were observed among the 17 tested candidates ($P>0.05$).

3.2 Antibiotic Resistance

The resistance levels detected for the selected strains are shown in Table 1. It can be observed that ampicillin, erythromycin, clindamycin and chloramphenicol completely inhibited the growth of all the strains at the breakpoint concentrations proposed by the EFSA. The same behavior was shown by streptomycin for the *Lb. pentosus* and *Ln. mesenteroides* species at the suggested MICs, whereas the EFSA does not require any specific limit for these antibiotics in the case of *Lb. plantarum* strains. Tetracycline inhibited all the strains, with the exception of *Lb. pentosus* S3T60C, which was only completely inhibited at 24 µg mL^{-1}, that is, three times the breakpoint proposed for facultative heterofermentative lactobacilli such as *Lb. pentosus*. The S3T60C strain was even resistant to the suggested kanamycin MIC, which, together with gentamicin, could not inhibit the majority of the *Lb. plantarum* isolates. In 15 strains of *Lb. plantarum*, only S2T10D was completely inhibited at 64 µg mL^{-1} of kanamycin, whereas 5 strains were able to grow to concentrations of 192 µg mL^{-1} and the remaining 8 strains showed an MIC of 296 µg mL^{-1}. Moreover, the O1T90E, S1T3B, O4T10E and S11T3E strains, resulted to be resistant to both gentamicin and kanamycin. Finally, the *Ln. mesenteroides* strains were susceptible to all the antibiotics tested at the MICs reported as breakpoints.

3.3 Modulation of Cellular Metabolic Activity

All the inactivated bacteria strains left the cellular metabolic activity of the H4-1 cells unchanged or even improved (Fig. 4A). The reference probiotic, LGG, did not enhance the metabolic activity of H4-1cells, and similar behavior was observed for *Lb. plantarum* O1T90C and S1T10A and for *Ln. mesenteroides* FS50Q and FO50E. All the other tested LAB determined an increasing in the metabolic activity of the H4-1 cells from 15 to 70%, thus showing significantly higher results than the control LGG ($P<0.05$).

Concerning the experiment performed incubating live bacteria with TLT cells in the 3D model (H4-1/TLT), it can be observed in Figure 4B that *Ln. mesenteroides* FS50Q caused a remarkable increase in the metabolic activity, with significantly higher values ($P<0.05$) than for the probiotic reference (LGG). On the other hand, *Lb. plantarum* strains O2T60C, O4T10E, and S4T30C significantly decreased the TLT metabolic index with respect to LGG ($P<0.05$).

3.4 Adhesion Capability

The adhesion capacity of the 17 most promising strains was variable and generally low (from 1% to 10%). Moreover, an enhancement of adhesion during the step from the undifferentiated to the 3D model of the H4-1 cells was observed for most of the strains (14) (Fig. 5). The overall average adhesion to the undifferentiated cells was 3.82%, whereas adhesion to the apical side of the cells reached 5.27% in the 3D functional model. This behavior was statistically significant for the *Lb. plantarum* O2T60C and *Ln. mesenteroides* F050E strains ($P<0.05$). However, *Lb. plantarum* S4T30C adhered significantly less in the 3D model ($P<0.05$).

Focusing attention on the experiment performed with the undifferentiated model, it was observed that 5 strains (O1T90C, O11T30D, O4T10E, S4T30C and O2T60D) showed a higher adhesion ratio than the reference probiotic LGG ($P<0.05$). These strains did not confirm their adhesive potential in the 3D model, in which the O1T90E and O2T60C strains showed the best adhesiveness, with 8.67% and 9.41%, and were significantly higher than LGG ($P<0.05$).

3.5 Strains Effect Towards Adhesion and Invasion of *L. monocytogenes*

In order to identify which strains could inhibit epithelial cell infection with *L. monocytogenes in vitro*, adhesion and invasion assays were carried out in parallel. Ten strains, including the reference probiotic LGG, showed potential inhibition (Fig. 6). The *Lb. plantarum* strains S1T30B, O3T15B, O2T60D, together with *Lb. pentosus*S3T60C, significantly reduced the adherence of the bacteria to the monolayer surface with respect to the untreated controls ($P<0.05$). *Lb. pentosus* S3T60C as well as the *Lb. plantarum* strains S11T3E and S2T10D were even able to significantly ($P<0.05$) reduce the intracellular invasion of the pathogen, by 57%, 61% and 66%, respectively.

3.6 Epithelial Barrier Integrity

The measure of TEER in polarized *in vitro* gut models is commonly used to verify the cell monolayer integrity after the incubation with probiotic microorganisms [42], [43]. Overall, after 3 and 5 hours of incubation, the time dependent TEER values of cells co-inoculated with bacteria were lower than the control, but not significantly. The O2T60C strain instead showed a significant reduction in the TEER from the 5th to the 7th hour of incubation, compared to the untreated control (Fig. 7). As expected, the negative control, *L. monocytogenes* WT, quickly degenerated the junctions between the cells, and this resulted in a significant reduction in the TEER values from the 5th hour of incubation until the end of the trial. Only the *Lb. plantarum* strains S1T10A and S11T3E showed a significant increase in the TEER values, compared to the control, after 18and 24 hours of incubation ($P<0.05$). The data of all the tested strains are shown in Table S2.

Discussion

In this paper, a collection of 238 LAB strains, isolated from different green table olive fermentation processes, has been screened progressively, using as selective phenotypic traits the

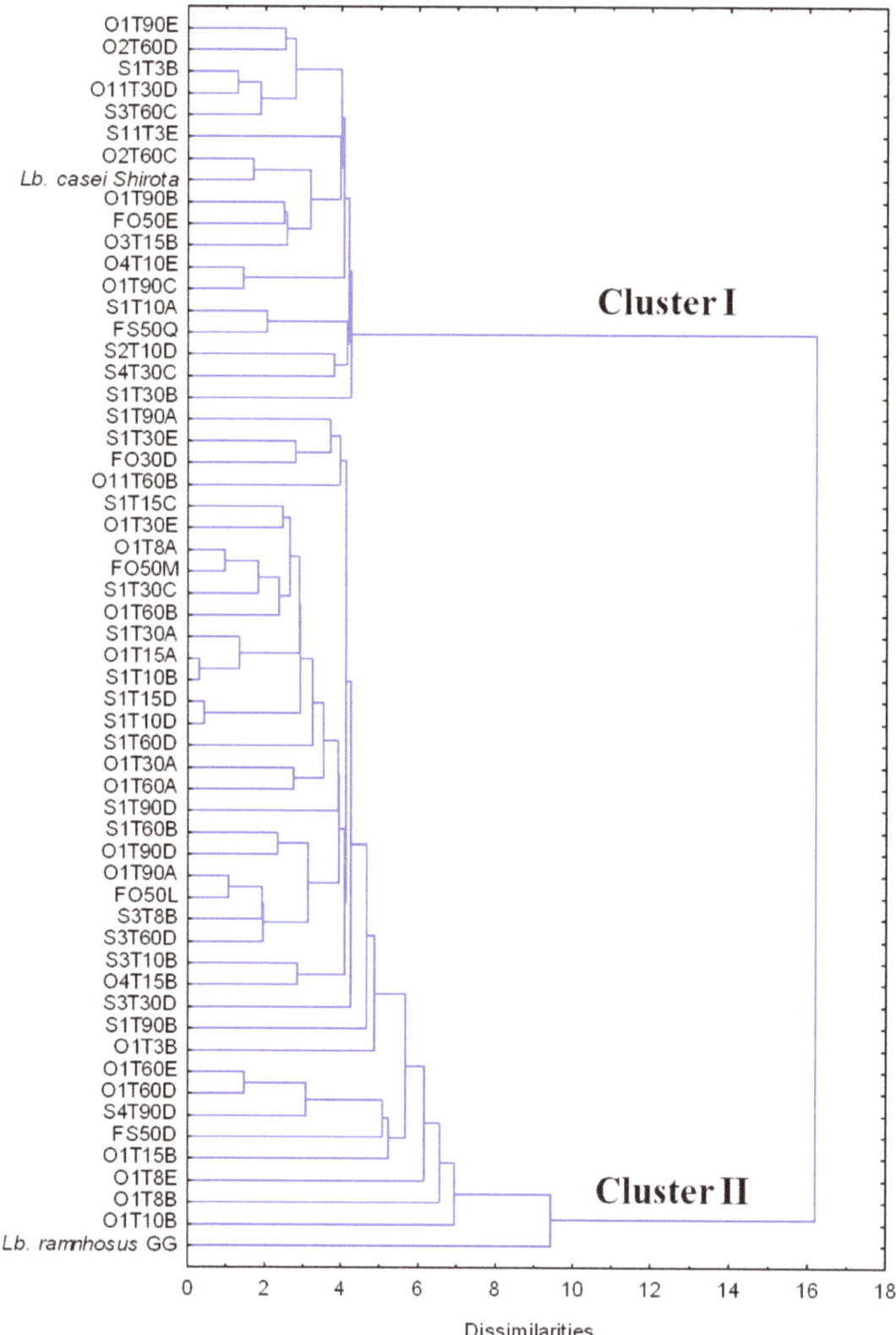

Figure 2. Dendrogram resulting from the hierarchical cluster analysis (HCA) of the 55 strains resistant to the simulation of digestion. The calculation of the dissimilarity between the cases is based on three independent indices expressed in percentage: ODS, autoaggregation (AA) and hydrophobicity (H). *Lb casei Shirota* and *Lb. rhamnosus* GG (LGG) were inserted in the HCA as reference probiotic controls.

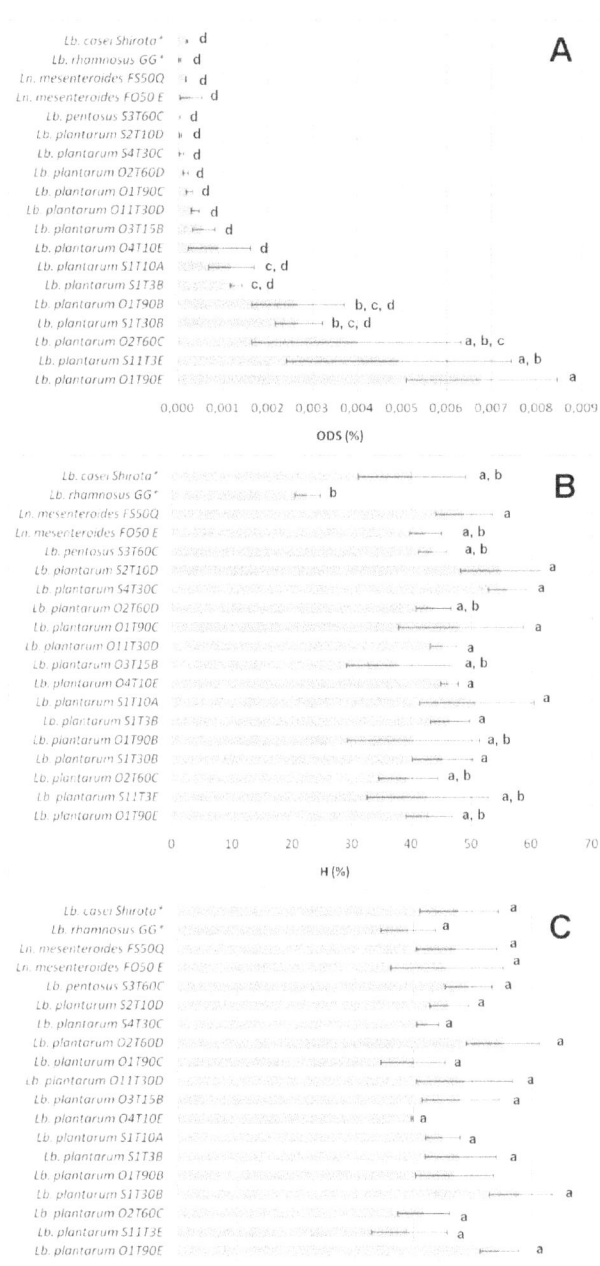

Figure 3. **Overall digestion survival (ODS %) index (A), hydrophobicity (H %) index (B), and autoaggregation (AA %) index (C) of the 17 selected strains.** The data are the means of three independent experiments ± SEM and the bars marked with different letters (a,b,c,d) indicate a significant difference at $P<0.05$ (ANOVA with Duncan's test as post hoc). (*): reference strains for the experiment.

autoaggregation, the adhesion to hydrocarbons (hydrophobicity), and the capability to survive in the simulated GIT conditions [13], [22], [44]–[46].

Other interesting phenotypic traits, such the production of bacteriocins and BSH activity were also assessed, but no strain showed these properties. Even though BSH activity is correlated to a decrease in cholesterol, it is not completely clear whether BSH

activity is a desirable property for probiotics, since large amounts of de-conjugated bile salts may have undesirable effects on the human host [12], [13]. As far as direct pathogen inhibition is concerned, many probiotic LAB have been observed to produce anti-microbial substances, and mainly organic acids, especially lactic and acetic acids [47]. Recently, the effective production of bacteriocins in the human intestinal conditions has been confirmed by *in vitro* and *in vivo* trials [48]. However, bacteriocins actively produced *in vitro* may not necessarily be sufficiently high in quantities, or at all, within the GIT [49]. Therefore, the production of bacteriocins might not be considered a fundamental tract for a new probiotic candidates.

The ability to autoaggregate and the cell surface hydrophobicity, are instead often suggested as suitable ways of identifying potential adherent bacteria, because related to the adhesiveness upon the intestinal epithelial layer [50], [51]. As far as the ODS is concerned, the *in vitro* resistance to the simulated GIT conditions is considered one of the most important feature for a probiotic candidate, since a microorganism without this prerequisite cannot develop its beneficial effect inside the gut [22]. For this reason, these two indices were used as variables, together with the ODS, in the HCA performed on the 55 strains able to survive to the simulated digestion. The dendrogram resulting from the HCA (Fig. 2) showed an opposite collocation of the two reference strains included in the analysis, *Lb. casei Shirota* and LGG, mainly due to the low ODS and AA rates of the LGG. A similar result was achieved in the study by Argyri et al. [12], in which the HCA was performed without autoaggregation and hydrophobicity variables and considering only strain resistance to simulated digestion. Anyhow, the chemometric analysis performed allowed the number of isolates to be reduced from the initial 55 strains to a final set of 17 probiotic candidates, which, considering together these three variables, highlighted the best phenotypic profiles.

As far as the phenotypic features of this set of probiotic candidates is concerned (Fig. 3), a limited difference in the autoaggregation and hydrophobicity values was observed among the strains, whereas a broad range of ODS values was detected with all values lower than 0.01%. However, as already reported by Bautista-Gallego et al. [13], the simulation of human digestion performed in the present study significantly reduced the viability of LAB, and a survival ratio of 0.008% could be considered as an appreciable result.

The 17 selected probiotic candidates were therefore tested for evaluating potential resistance to antibiotics or cytotoxic effect towards human cells. Concerning the antibiotic resistance, total inhibition for erythromycin, a Gram-positive spectrum antibiotic, which is known for its effective inhibition of LAB [52], was observed for the species-specific MICs suggested by the EFSA (Tab. 1). All the selected LAB were susceptible to β-Lactam antibiotic ampicillin and to the broad spectrum antibiotics clindamycin, chloramphenicol and tetracycline, with the exception of the latter against the *Lb. pentosus* strain S3T60C. Numerous strains were instead resistant to kanamycin and gentamicin, two aminoglycoside antibiotics. This resistance had already been observed for *Lb. plantarum* strains isolated from fermented vegetables [53] and as suggested by Hummel et al. [54], it could be considered as an intrinsic feature of LAB, that could be attributable to the absence of cytochrome-mediated electron transport, which mediates drug uptake. Therefore, membrane impermeability plays an important role in this intrinsic aminoglycoside resistance and it has been proved that the sensitivity of LAB is increased in the presence of bile, as in the intestine [55]. In light of these reports, the low susceptibility to kanamycin and gentamicin can be considered acceptable. Always in the context

Table 1. Minimum Inhibitory Concentrations (MICs) detected for the selected strains towards the antibiotics: ampicillin (A), gentamicin (G), kanamycin (K), tetracycline (T), streptomycin (S), erythromycin (E), clindamycin (C), chloramphenicol (Ch).

Strains	MICs (µg mL^{-1})							
	A	G	K	T	S	E	C	Ch
Lb. plantarum:								
O1T90C	<2	<16	296^R	<32	n.r.	<1	<1	<8
O11T30D	<2	<16	296^R	<32	n.r.	<1	<1	<8
O1T90E	<2	32^R	296^R	<32	n.r.	<1	<1	<8
S1T3B	<2	32^R	296^R	<32	n.r.	<1	<1	<8
O1T90B	<2	<16	296^R	<32	n.r.	<1	<1	<8
O2T60C	<2	<16	192^R	<32	n.r.	<1	<1	<8
S1T10A	<2	<16	296^R	<32	n.r.	<1	<1	<8
O4T10E	<2	32^R	192^R	<32	n.r.	<1	<1	<8
S4T30C	<2	<16	192^R	<32	n.r.	<1	<1	<8
S1T30B	<2	<16	192^R	<32	n.r.	<1	<1	<8
O3T15B	<2	<16	296^R	<32	n.r.	<1	<1	<8
O2T60D	<2	<16	192^R	<32	n.r.	<1	<1	<8
S11T3E	<2	32^R	296^R	<32	n.r.	<1	<1	<8
S2T10D	<2	<16	<64	<32	n.r.	<1	<1	<8
Lb. pentosus:								
S3T60C	<4	<16	192^R	24^R	<64	<1	<1	<4
Ln. mesenteroides:								
FS 50 Q	<2	<16	<16	<8	<64	<1	<1	<4
FO 50 E	<2	<16	<16	<8	<64	<1	<1	<4

(^{R}): resistance according to the EFSA breakpoints [33].
(n.r.): test not required [33].

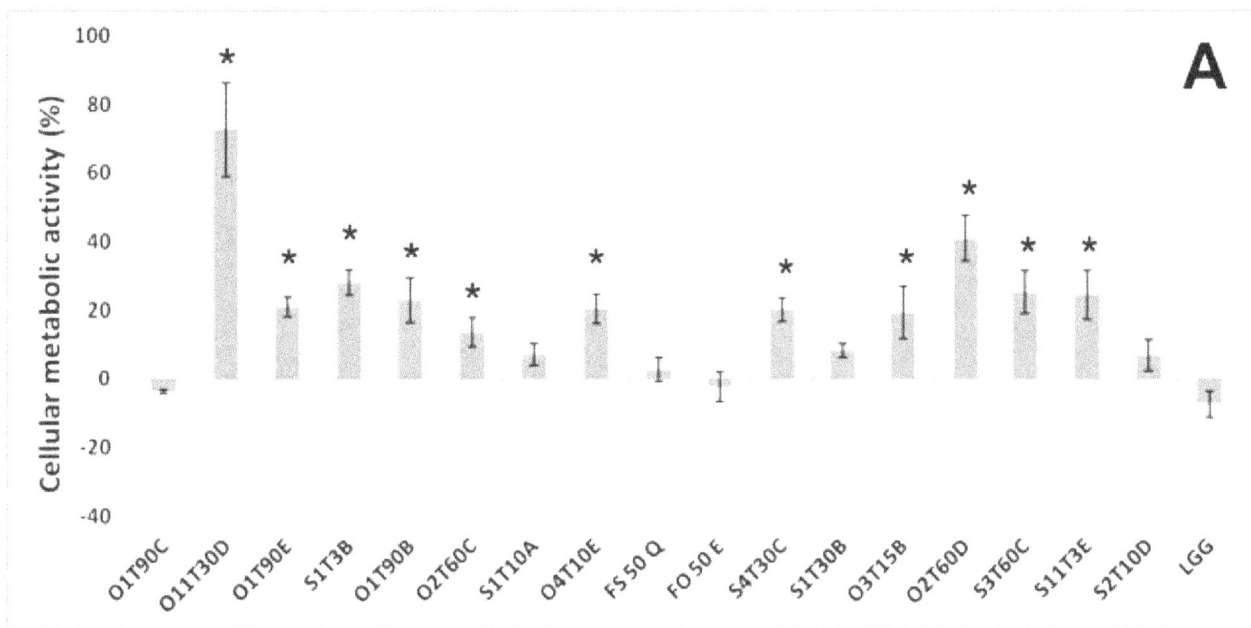

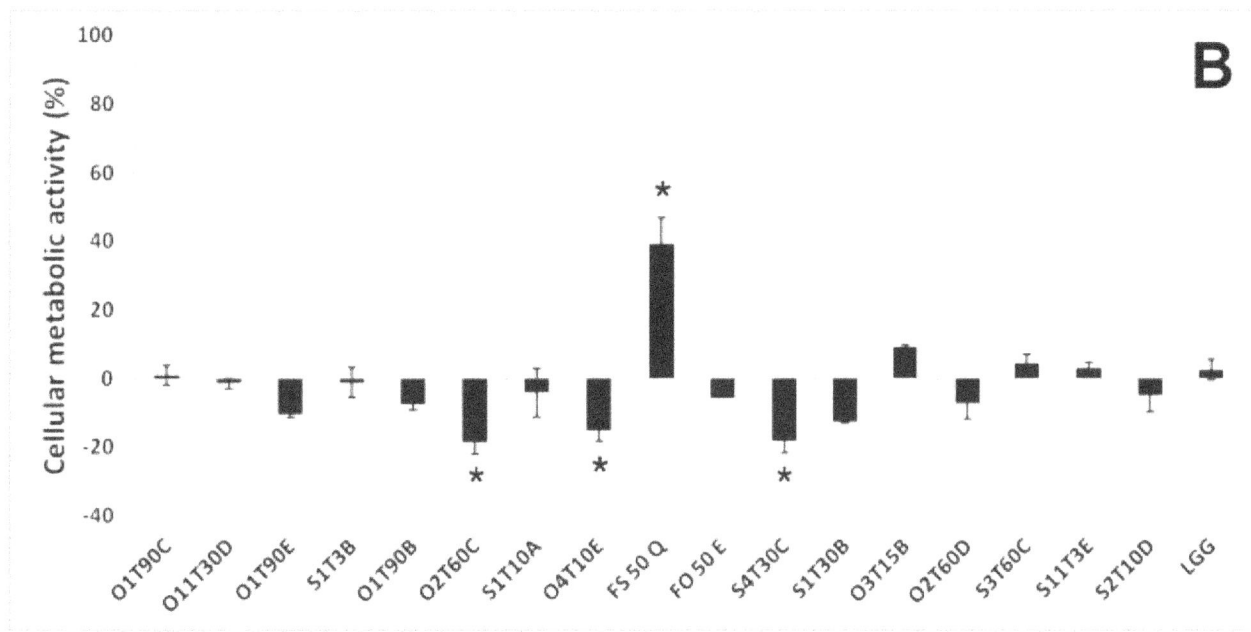

Figure 4. Variation of the cellular metabolic activity in H4-1 (A) and TLT (B) cells, respectively incubated with heat inactivated and live bacteria for 24 h. Undifferentiated model was used for the H4-1, whereas a complete 3D model (H4-1/TLT) was developed to assess the cellular metabolic activity in TLT. The data are expressed as mean ratios (% -100; ±SEM) of absorbance in the treated well (OD of test sample) and those not treated (OD of blank) with bacteria: 100- (OD of test sample/OD of blank×100). The asterisks (*) highlight significantly different values (Duncan's test; $P<0.05$) compared to the reference strain *Lb. rhamnosus* GG (LGG).

of strain safety, cytotoxicity has been evaluated towards human cells through an MTT assay. This is a well-known tool for the evaluation of cellular metabolic activity [56], since this reaction occurred in both the mitochondrial respiratory chain and outside the mitochondria, [57]. The heat-killed bacteria generally enhanced the viability of the H4-1 cells (Fig. 4A). A slow

decreasing effect in the cellular metabolic activity of the TLT macrophages was shown by three *Lb. plantarum* strains (Fig. 4B).

After the safety assessments, the 17 selected LAB were further tested in adhesion to intestinal cell line experiments, in order to indirectly measure their ability to colonize the intestinal tract. In this study, it was decided to only focus attention on the interaction between the selected strains and the H4-1 cells, without

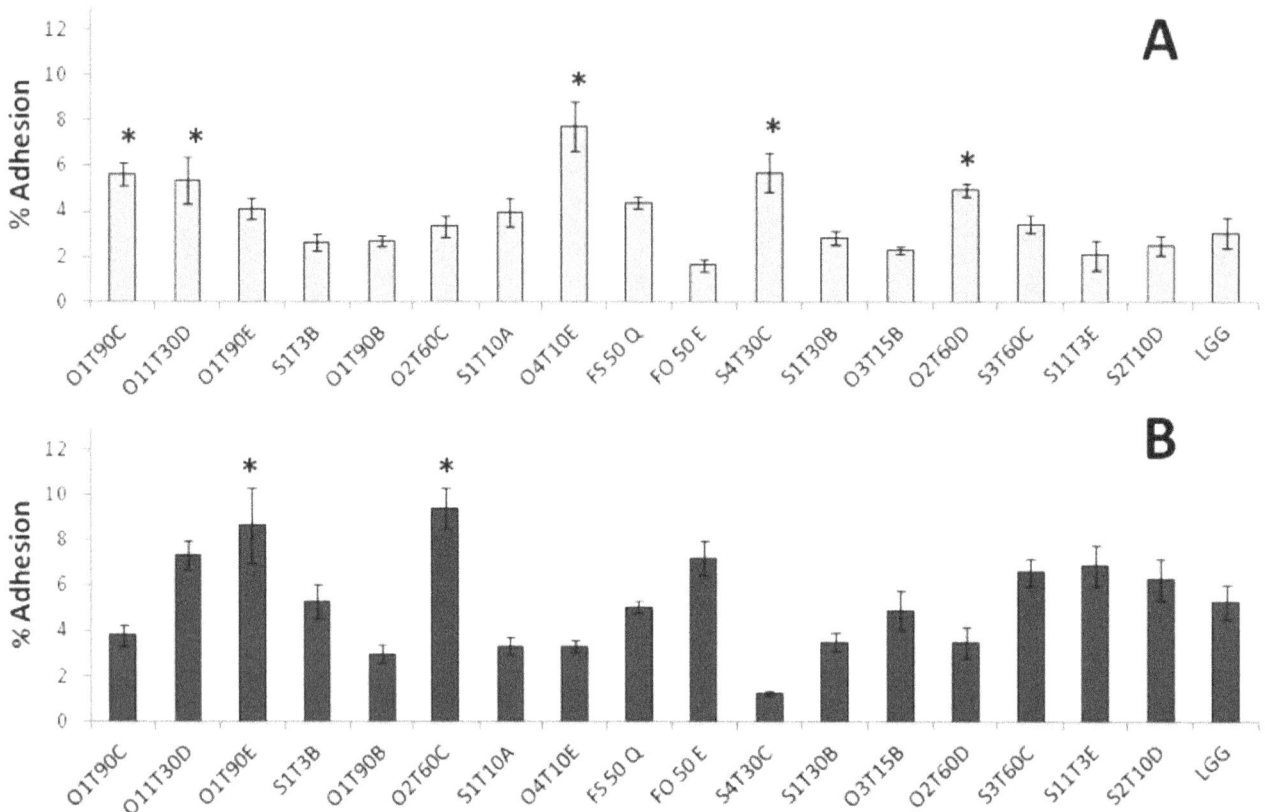

Figure 5. Adhesion profile of the LAB to the undifferentiated monolayer (A) and 3D functional model of H4-1 (B), expressed as the mean ratio (%) between the bacteria recovered from the human cells after incubation and the washing steps and the initial bacterial count ($8.0\pm0.5\log_{10}$ CFU mL^{-1}) of the inoculum (n = 3; ± SEM). The marked bars (*) indicate significantly ($P<0.05$) higher values (ANOVA with Duncan's test as post hoc) to the reference strain Lb. rhamnosus GG (LGG).

considering colon tumorigenic cells, for example, Caco-2 and HT-29, since they have a tumorigenic phenotype that distinguishes them from normal gut epithelia [34], [37]. As far as the results of our selected LAB are concerned, a better capability to adhere to the polarized monolayer (3D model) than to the normal monolayer was noted (Fig. 5). Both in the undifferentiated and 3D model, highly variable adhesion values were observed (from 1 to 10%), which confirmed that the adhesion was a strain-specific characteristic [51]. The adhesion values recorded for the control strain LGG are in agreement with the results obtained by others [58] on H4-1 undifferentiated cells. In a similar study, L. plantarum isolated from table olives showed the same adhesion range when tested on Caco-2 [11]. However, it should pointed out that the adhesiveness of a selected probiotic strain or in probiotic candidates may not guarantee its probiotic health promoting activity in the host [59]. Moreover it is more and more accepted that probiotics can provide beneficial effects even without true colonization of the GIT [60].

The modulation of gut microbiota, which causes a reduction in harmful microorganisms, is one of the main features of bacterial strains that can be defined as probiotic [61]. The characterization of the 17 probiotic candidates was therefore followed by testing their capability to inhibit both adhesion and invasion of L. monocytogenes in undifferentiated human gut model. This facultative intracellular bacterial pathogen has evolved a number of mechanisms to exploit host processes, growing and spreading from cell to cell without damaging the host cell [62]. As shown (Fig. 6), 9 strains, that could concurrently inhibit the adhesion and

the invasion of L. monocytogenes in the H4-1 undifferentiated cell model, were identified. These two potential capabilities did not seem related in the remaining 8 potential probiotics. For example, the O1T90C strain determined a significant reduction in L. monocytogenes adhesion compared to the untreated monolayer ($P<0.05$), but, at the same time, it caused an increase in the pathogen invasion to the H4-1 monolayer (data not shown). Moreover, the inhibition of L. monocytogenes adhesion did not depend on the capability of the LAB to adhere to human cells. As support of this finding, it was observed that strains S1T30B, O3T15B and S3T60C significantly reduced pathogen adhesion on the undifferentiated monolayer compared to the untreated control ($P<0.05$). However, these strains showed poor attachment to the H4-1 cells (Fig. 4). The non-correlation between the attachment capability of LAB and their activity against L. monocytogenes adhesion has already been highlighted by Koo et al. [63] and Botes et al. [64] using Caco-2 cells as a model. Accordingly, L. monocytogenes adhesion inhibition cannot be attributed solely to the simple competition between pathogens and LAB for the attachment sites on the cells.

As for the invasive action of L. monocytogenes, only three strains were able to reduce the pathogen invasion in a significant manner ($P<0.05$), compared to the control (Fig. 6), and of these, only Lb. pentosus S3T60C was able to reduce adhesion of the pathogen as well. Compared to the results obtained in other similar studies [4], [9], [63], the invasion inhibition shown by the S3T60C, S11T3E and S2T10D strains can be considered interesting, since they reduced the invasion of L. monocytogenes in the undifferentiated cells

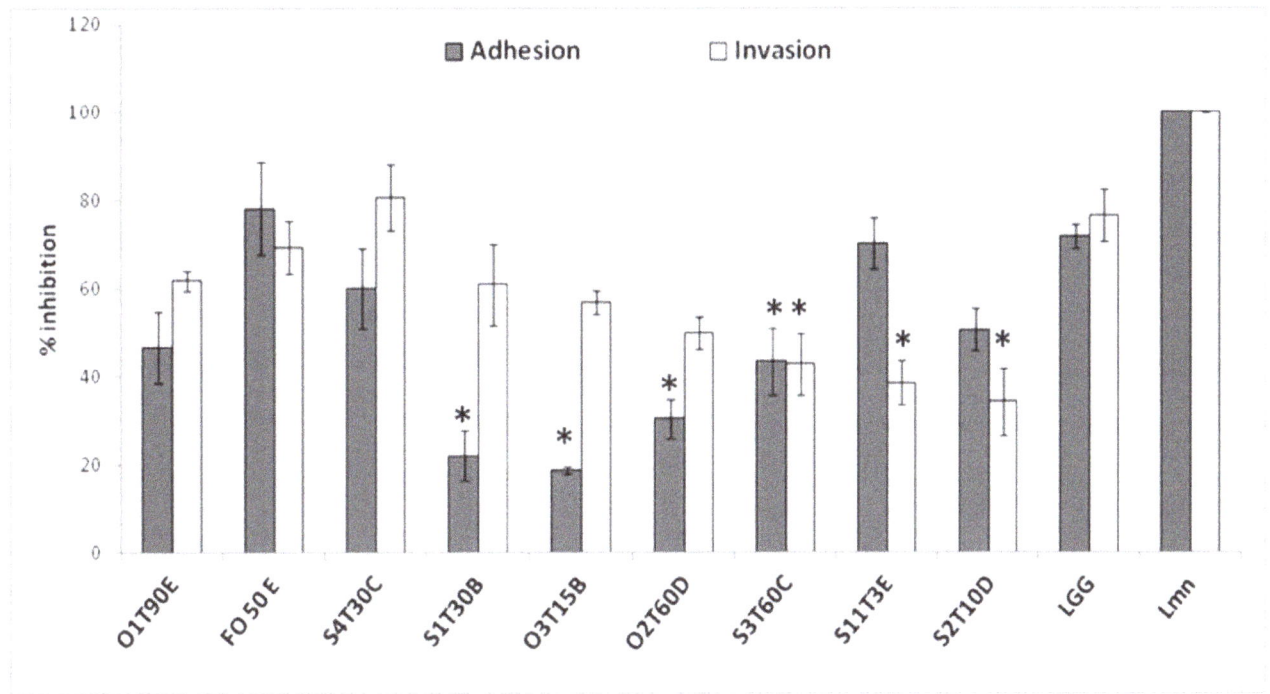

Figure 6. LAB strains that are able to inhibit the adhesion and invasion of *L. monocytogenes* (MOI:10). The data are the average (± SEM) of three independent experiments and are expressed as percentage ratios of the recovered pathogens in the treated wells with respect to those recovered from the non -treated well (Lmn). The marked bars (*) indicate significantly lower counts ($\log_{10}$ CFU mL^{-1}) than those of the cells infected with *L. monocytogenes* alone (Student's *t*-test, *P*<0.05).

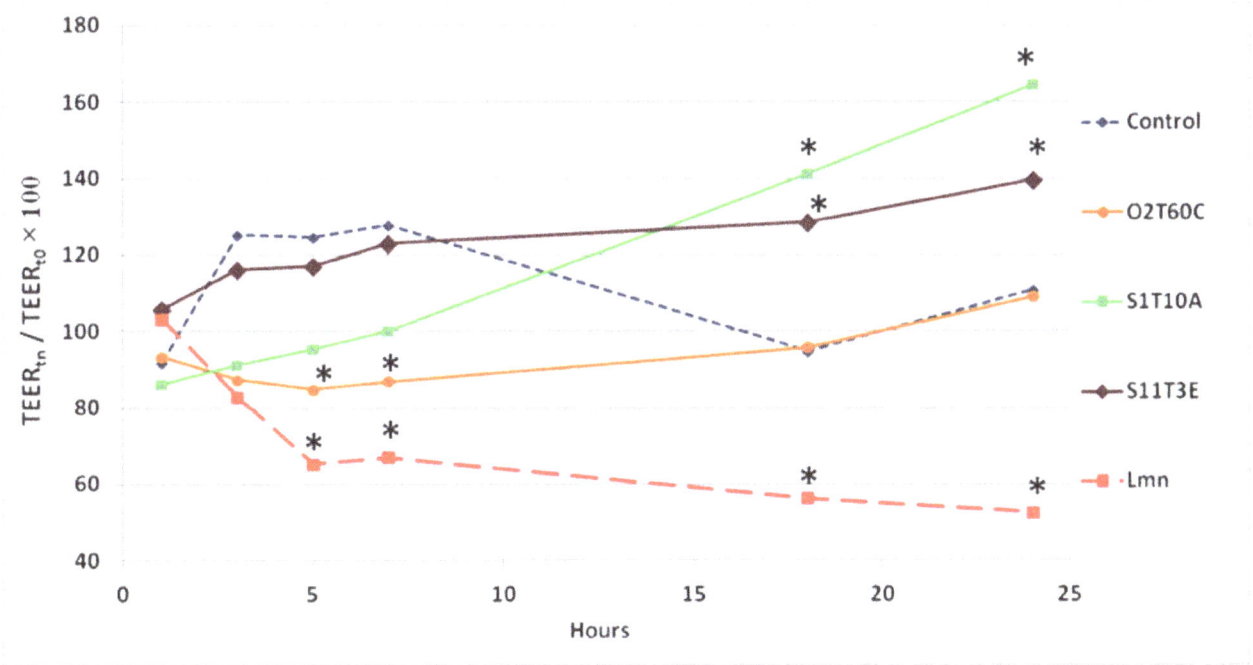

Figure 7. Time dependent TEER of the polarized H4 monolayer exposed to the strains: O2T60C, S1T10A, S11T3E and *L. monocytogenes* WT (Lmn). The untreated monolayer was used as a reference control. The bacterial effect on TEER dynamics was tested with 10^7 CFU mL^{-1} of bacterial inoculum. The data of three independent experiments are expressed as the ratio (%) of TEER at time n in relation to the initial value (t$_0$). The marked bars (*) indicate significantly different values (*P*<0.05) from the untreated control (ANOVA with Duncan's test as post hoc). The complete dataset of the experiment (mean ± SEM), with the results of the statistical analysis, is reported in Table S2.

by over 60%. Since the pathogenicity of *L. monocytogenes* is mainly determined by its capability to cross the cell plasma membrane and remain alive and active inside the cytosol [65], results obtained by using S3T60C, S11T3E and S2T10D strains can be considered interesting and worthy of further in-depth studies. *L. monocytogenes* invasion inhibition mechanisms might be caused by the modulation of human cells or by direct inhibition of the pathogen, by acidification or the production of antimicrobial compounds. The second suggestion can be excluded for S3T60C and S2T10D strains, since, in previous AWDA, they did not inhibit *L. monocytogenes* growth, either through bacteriocin production or through acidification, whereas *Lb. plantarum* S11T3E was able to inhibit the growth of the pathogen through the production of organic acids (Table S1).

As far as the cell modulation activity of the strains is concerned, the potential enhancement of epithelial integrity was investigated by measuring the TEER. Improvements in barrier integrity are associated with changes in tight junctions (TJ) via changes in TJ protein expression and distribution. Strains of *Lb. plantarum* have been shown to regulate human epithelial TJ proteins *in vivo* [66] and *in vitro* [67]. In the present study, two strains of *Lb. plantarum* (S11T3E and S1T10A) enhanced the values of TEER compared to the control (P<0.05) when applied in a differentiated 3D model (Fig. 7). As already demonstrated in another study [43], the pathogen *L. monocytogenes* significantly reduced the integrity of the epithelial layer from almost the very beginning of the incubation period. Like other pathogens, *L. monocytogenes* may alter TJ permeability to allow its own translocation through the epithelial barrier [68]. In light of this, *Lb. plantarum* S11T3E might be prevented from invading *L. monocytogenes* through an amelioration of the integrity of the epithelial barrier. However, *L. monocytogenes* is able to directly cross the cell membrane and hence the reinforcement of the paracellular spaces may not be the only reason for the inhibitory action of S11T3E, which then needs further studied.

Conclusions

In the current study, an extensive collection of autochthonous LAB from table olive fermentations has been screened, focusing on their tolerance to the hostile environment of the stomach and intestine, together with other suitable phenotypic features, such as high autoaggregation and hydrophobicity. A cluster analysis on their phenotypes has allowed the LAB set to be reduced to 17 strains belonging to the *Lb. plantarum*, *Lb. pentosus* and *Ln. mesenteroides* species. None of these selected strains showed resistance to broad spectrum antibiotics and they were confirmed to be safe and not cytotoxic when in contact with H4-1 human epithelial cells. Therefore, all the strains were characterized in relation to their probiotic potential using *in vitro* undifferentiated and 3D functional models of the H4-1 human intestinal epithelial cell line.

In conclusion, the S11T3E strain belonging to the *Lb. plantarum* species has overall shown the best probiotic performance, due to

its high resistance to simulated gastric digestion, an increased transepithelial resistance of polarized H4-1 cells and a significant reduction in *L. monocytogenes* invasion in undifferentiated gut model cells. The latter characteristic was also shown by the *Lb. pentosus* S3T60C and *Lb. plantarum* S2T10D strains, whereas the increase of transepithelial resistance was shown by *Lb. plantarum* S1T10A, as well.

In vivo assays in mice and clinical trials are now needed for a further characterization of these probiotic candidates before their incorporation into novel functional foods.

Supporting Information

Table S1   Overall digestion survival (ODS), hydrophobicity (H), and autoaggregation (AA) of the 238 LAB collected from green table olives. Data are the means of three independent experiment and are represented in percentage (%). ODS is obtained by comparison of the initial lactobacilli counts at the start of the simulated digestion (T_0) and those remaining at the end of the simulation (T_i), according to the formula: $T_i/T_0 \times 100$. The H values are calculated using the formula: $H = (1 - A_1/A_0) \times 100$; where A_0 represents the initial absorbance at 600 nm (A600) of the bacterial suspension in o-xylene/PBS solution, whereas the A_1 represents the A600 measured for the aqueous phase after 1 h of incubation. The AA is an index of the bacterial cells capability to aggregate among them. The data are expressed using the formula $1 - (A_5/A_0) \times 100$, where the A_0 represents the A600 of a well resuspended bacterial culture, and A_5 the A600 of the upper phase recovered from the same suspension after 5 h of incubation. Potential inhibition of *L. monocytogenes* growth due to acidification is reported in the last column.

Table S2   Time dependent TEER of polarized H4 monolayer exposed to the 17 probiotic candidates, *Lb. rhamnosus* GG (probiotic reference) and *L. monocytogenes* (negative control). Bacterial effect on TEER dynamics was tested with 10^7 CFU mL^{-1} of bacterial inoculum. Data (n = 3; ± SEM) are expressed as the ratio (%) of TEER at time t in relation to the initial value (t_0). Different letters (a, b, c) indicate significant differences among the values at P<0.05 (ANOVA with Duncan's test).

Acknowledgments

All the research activities performed with the H4-1 and TLT human cells were performed at the Faculty of Agriculture and Life Sciences, University of Maribor.

Author Contributions

Conceived and designed the experiments: LC AC. Performed the experiments: CB TL. Analyzed the data: CB TL. Contributed reagents/materials/analysis tools: LC AC TL. Wrote the paper: CB TL LC.

References

1. Rivera-Espinoza Y, Gallardo-Navarro Y (2010) Non-dairy probiotic products. Food Microbiol 27: 1–11.
2. Arroyo-López FN, Querol A, Bautista-Gallego J, Garrido-Fernández A (2008) Role of yeasts in table olive production. Int J Food Microbiol, 128: 189–196.
3. Botta C, Cocolin L (2012) Microbial dynamics and biodiversity in table olive fermentation: culture-dependent and -independent approaches. Front Microbiol, 3: 245–245.
4. Corr SC, Gahan CG, Hill C (2007) Impact of selected *Lactobacillus* and *Bifidobacterium* species on *Listeria monocytogenes* infection and the mucosal immune response. FEMS Immunol Med Mic, 50: 380–388.
5. Paolillo R, Carratelli CR, Sorrentino S, Mazzola N, Rizzo A (2009) Immunomodulatory effects of *Lactobacillus plantarum* on human colon cancer cells. Int Immunopharmacol, 9: 1265–1271.
6. Ogawa M, Shimizu K, Nomoto K, Takahashi M, Watanuki M, et al. (2001) Protective effect of *Lactobacillus casei* strain *Shirota* on Shiga toxin-producing *Escherichia coli* O157: H7 infection in infant rabbits. Infect Immun, 69: 1101–1108.
7. Tien MT, Girardin SE, Regnault B, Le Bourhis L, Dillies MA, Coppee JY, et al. (2006) Anti-inflammatory effect of *Lactobacillus casei* on Shigella-infected human intestinal epithelial cells. J Immunol, 176: 1228–1237.

8. Archambaud C, Nahori M-A, Soubigou G, Becavin C, Laval L, et al. (2012) Impact of lactobacilli on orally acquired listeriosis. Proc Natl Acad Sci U S A,109: 16684–16689.

9. Nakamura S, Kuda T, An C, Kanno T, Takahashi H, et al. (2012) Inhibitory effects of *Leuconostoc mesenteroides* 1RM3 isolated from *narezushi*, a fermented fish with rice, on *Listeria monocytogenes* infection to Caco-2 cells and A/J mice. Anaerobe, 18: 19–24.

10. Lavermicocca P (2006) Highlights on new food research. Digestive and liver disease: official journal of the Italian Society of Gastroenterology and the Italian Association for the Study of the Liver, 38 Suppl 2.

11. Bevilacqua A, Altieri C, Corbo MR, Sinigaglia M, Ouoba LI (2010) Characterization of lactic acid bacteria isolated from italian *Bella di Cerignola* table olives: selection of potential multifunctional starter cultures. J Food Sci, 75: M536–M544.

12. Argyri AA, Zoumpopoulou G, Karatzas K-AG, Tsakalidou E, Nychas G-JE, et al. (2013) Selection of potential probiotic lactic acid bacteria from fermented olives by in vitro tests. Food Microb, 33: 282–291.

13. Bautista-Gallego J, Arroyo-Lopez FN, Rantsiou K, Jimenez-Diaz R, Garrido-Fernandez A, Cocolin L (2013) Screening of lactic acid bacteria isolated from fermented table olives with probiotic potential. Food Res Int, 50: 135–142.

14. Peres C M, Peres C, Hernandez-Mendoza A, Malcata FX (2012) Review on fermented plant materials as carriers and sources of potentially probiotic lactic acid bacteria. With an emphasis on table olives. Trends Food Sci Tech, 26: 31–42.

15. FAO/WHO (2002) Guidelines for the Evaluation of Probiotics in Food Report of a Joint FAO/WHO group Available: ftp://ftpfaoorg/es/esn/food/wgreport2pdf Accessed 2013 Sept 1.

16. Teuber M, Meile L, Schwarz F (1999) Acquired antibiotic resistance in lactic acid bacteria from food. Anton Leeuw Int J, 76: 115–137.

17. Oomen AG, Rompelberg CJ, Bruil MA, Dobbe CJ, Pereboom DP, et al. (2003) Development of an in vitro digestion model for estimating the bioaccessibility of soil contaminants. Arch Environ Contam Toxicol, 44: 281–287.

18. Corcoran BM, Stanton C, Fitzgerald GF, Ross RP (2007) Growth of probiotic lactobacilli in the presence of oleic acid enhances subsequent survival in gastric juice. Microbiol-SGM, 153: 291–299.

19. de Vries MC, Vaughan EE, Kleerebezem M, de Vos WM (2006) *Lactobacillus plantarum*- survival, functional and potential probiotic properties in the human intestinal tract. Int Dairy J, 16: 1018–1028.

20. Ortakci F, Broadbent JR, McManus WR, McMahon DJ (2012) Survival of microencapsulated probiotic *Lactobacillus paracasei* LBC-1e during manufacture of Mozzarella cheese and simulated gastric digestion. J Dairy Sci, 95: 6274–6281.

21. Possemiers S, Pinheiro I, Verhelst A, Van den Abbeele P, Maignien L, et al. (2013) A dried yeast fermentate selectively modulates both the luminal and mucosal gut microbiota and protects against inflammation, as studied in an integrated *in vitro* approach. J of Agric Food Chem, 61: 9380–9392.

22. Dicks LM, Botes M (2010) Probiotic lactic acid bacteria in the gastro-intestinal tract: health benefits, safety and mode of action. Benef Microbes 1: 11–29.

23. Ouwehand AC, Salminen S, Isolauri E (2002) Probiotics: an overview of beneficial effects. Anton Leeuw Int J, 82: 279–289.

24. Marteau P, Shanahan F (2003) Basic aspects and pharmacology of probiotics: an overview of pharmacokinetics, mechanisms of action and side-effects. Best Pract Res Clin Gastroenterol, 17: 725–740.

25. Williams NT (2010) Probiotics. Am J Health Syst Pharm, 67: 449–458.

26. Lievin-Le Moal V, Servin AL (2013) Pathogenesis of human enterovirulent bacteria: lessons from cultured, fully differentiated human colon cancer cell lines. Microbiol Mol Biol Rev: MMBR, 77: 380–439.

27. Cocolin L, Alessandria V, Botta C, Gorra R, De Filippis F, et al. (2013) NaOH-debittering induces changes in bacterial ecology during table olives fermentation. PloS One, 8: e69074–e69074.

28. Torriani S, Felis GE, Dellaglio F (2001) Differentiation of *Lactobacillus plantarum*, *L pentosus*, and *L paraplantarum* by recA gene sequence analysis and multiplex PCR assay with recA gene-derived primers. Appl Environ Microbiol 67: 3450–3454.

29. Toba T, Samant SK, Itoh T (1991) Assay system for detecting bacteriocin in microdilution wells. Lett Appl Microbiol, 13: 102–104.

30. Dashkevicz MP, Feighner SD (1989) Development of a differential medium for bile-salt hydrolase-active *lactobacillus spp*. Appl Environ Microbiol 55: 11–16.

31. Kos B, Suskovic J, Vukovic S, Simpraga M, Frece J, et al. (2003) Adhesion and aggregation ability of probiotic strain *Lactobacillus acidophilus* M92. J Appl Microbiol, 94: 981–987.

32. Crow VL, Gopal PK, Wicken AJ (1995) Cell-surface differences of lactococcal strains. Int Dairy J 5: 45–68.

33. EFSA (2008) Update of the criteria used in the assessment of bacterial resistance to antibiotics of human or veterinary importance (Technical guidance). The EFSA Journal 732: 1–15.

34. Langerholc T, Maragkoudakis PA, Wollgast J, Gradisnik L, Cencič A (2011) Novel and established intestinal cell line models - An indispensable tool in food science and nutrition. Trends Food Sci Tech, 22: S11–S20.

35. Nanthakumar NN, Fusunyan RD, Sanderson I, Walker WA (2000) Inflammation in the developing human intestine: A possible pathophysiologic contribution to necrotizing enterocolitis. Proc Natl Acad Sci USA, 97: 6043–6048.

36. Sanderson IR, Ezzell RM, Kedinger M, Erlanger M, Xu ZX, et al. (1996) Human fetal enterocytes in vitro: modulation of the phenotype by extracellular matrix. Proc Natl Acad Sci USA, 93: 7717–7722.

37. Cencič A, Langerholc T (2010) Functional cell models of the gut and their applications in food microbiology – A review. Int J Food Microbiol, 141: S4–S14.

38. Puertollano MA, de Pablo MA, de Cienfuegos GA (2002) Relevance of dietary lipids as modulators of immune functions in cells infected with *Listeria monocytogenes*. Clin Diagn Lab Immun, 9.

39. Tsukatani T, Suenaga H, Higuchi T, Akao T, Ishiyama M, Ezoe K, et al. (2008) Colorimetric cell proliferation assay for microorganisms in microtiter plate using water-soluble tetrazolium salts. J Microbiol Meth, 75: 109–116.

40. Ivec M, Botic T, Koren S, Jakobsen M, Weingartl H, et al. (2007) Interactions of macrophages with probiotic bacteria lead to increased antiviral response against *vesicular stomatitis* virus. Antivir Res, 75: 266–274.

41. Yamada F, Ueda F, Ochiai Y, Mochizuki M, Shoji H, et al. (2006) Invasion assay of *Listeria monocytogenes* using Vero and Caco-2 cells. J Microbiol Methods, 66.

42. Jensen H, Grimmer S, Naterstad K, Axelsson L (2012) *In vitro* testing of commercial and potential probiotic lactic acid bacteria. Int J Food Microbiol, 153: 216–222.

43. Klingberg TD, Pedersen MH, Cencic A, Budde BB (2005) Application of measurements of transepithelial electrical resistance of intestinal epithelial cell monolayers to evaluate probiotic activity. Appl Environ Microbiol, 71: 7528–7530.

44. Dias FS, Duarte WF, Schwan RF (2013) Evaluation of adhesive properties of presumptive probiotic *Lactobacillus plantarum* strains. Biosci J, 29: 1678–1686.

45. Tuo Y, Yu H, Ai L, Wu Z, Guo B, et al. (2013) Aggregation and adhesion properties of 22 *Lactobacillus* strains. J Dairy Sci 96: 4252–4257.

46. Lin WH, Hwang CF, Chen LW, Tsen HY (2006) Viable counts, characteristic evaluation for commercial lactic acid bacteria products. Food Microbiol, 23: 74–81.

47. Collado MC, Isolauri E, Salminen S, Sanz Y (2009) The impact of probiotic on gut health. Curr Drug Metab, 10: 68–78.

48. Dobson A, Crispie F, Rea MC, O'Sullivan O, Casey PG, et al. (2011) Fate and efficacy of lacticin 3147-producing *Lactococcus lactis* in the mammalian gastrointestinal tract. FEMS Microbiol Ecol, 76: 602–614.

49. Dobson A, Cotter PD, Ross RP, Hill C (2012) Bacteriocin production: a probiotic trait? Appl Environ Microbiol, 78: 1–6.

50. Del Re B, Sgorbati B, Miglioli M, Palenzona D (2000) Adhesion, autoaggregation and hydrophobicity of 13 strains of *Bifidobacterium longum*. Lett Applied Microbiol, 31: 438–442.

51. Collado MC, Meriluoto J, Salminen S (2008) Adhesion and aggregation properties of probiotic and pathogen strains. Eur Food Res Technol, 226: 1065–1073.

52. Zhou JS, Pillidge CJ, Gopal PK, Gill HS (2005) Antibiotic susceptibility profiles of new probiotic *Lactobacillus* and *Bifidobacterium* strains. Int J Food Microbiol, 98: 211–217.

53. Petrovic T, Dimitrijevic S, Radulovic Z, Mirkovic N, Rajic J, et al. (2012) Comparative analysis of the potential probiotic abilities of lactobacilli of human origin and from fermented vegetables. Arch Biol Sci, 64: 1473–1480.

54. Hummel AS, Hertel C, Holzapfel WH, Franz CM (2007) Antibiotic resistances of starter and probiotic strains of lactic acid bacteria. Appl Environ Microbiol, 73: 730–739.

55. Elkins CA, Mullis LB (2004) Bile-mediated aminoglycoside sensitivity in *Lactobacillus* species likely results from increased membrane permeability attributable to cholic acid. Appl Environ Microbiol, 70: 7200–7209.

56. Bergamini A, Perno CF, Capozzi M, Mannella E, Salanitro A, et al. (1992) A tetrazolium-based colorimetric assay for quantification of HIV-1-induced cytopathogenicity in monocyte/macrophages exposed to macrophages-colony-stimulating factor. Virol Methods, 40: 275–286.

57. Berridge MV, Herst PM, Tan AS (2005) Tetrazolium dyes as tools in cell biology: new insights into their cellular reduction. Biotechn Annu Rev, 11: 127–52.

58. Maragkoudakis PA, Chingwaru W, Gradisnik L, Tsakalidou E, Cencič A (2010) Lactic acid bacteria efficiently protect human and animal intestinal epithelial and immune cells from enteric virus infection. Int J Food Microbiol, 141: S91–S97.

59. Nissen L, Chingwaru W, Sgorbati B, Biavati B, Cencic A (2009) Gut health promoting activity of new putative probiotic/protective *Lactobacillus spp* strains: A functional study in the small intestinal cell model Int J Food Microbiol, 135: 288–294.

60. Ohland CL, MacNaughton WK (2010) Probiotic bacteria and intestinal epithelial barrier function. Am J Physiol-Gastr L,298: G807–G819.

61. Saad N, Delattre C, Urdaci M, Schmitter JM, Bressollier P (2013) An overview of the last advances in probiotic and prebiotic field. Lwt-Food Sci Technol, 50: 1–164.

62. Rowan NJ, Candlish AA, Bubert A, Anderson JG, Kramer K, et al. (2000) Virulent rough filaments of *Listeria monocytogenes* from clinical and food samples secreting wild-type levels of cell-free p60 protein. J Clin Microbiol, 38: 2643.

63. Koo OK, Amalaradjou MA, Bhunia AK (2012) Recombinant probiotic expressing *Listeria* adhesion protein attenuates *Listeria monocytogenes* virulence in vitro. PloS One, doi:e29277101371/journalpone0029277.

64. Botes M, Loos B, van Reenen CA, Dicks LM (2008) Adhesion of the probiotic strains *Enterococcus mundtii* ST4SA and *Lactobacillus plantarum* 423 to Caco-2 cells under conditions simulating the intestinal tract, and in the presence of antibiotics and anti-inflammatory medicaments. Archiv Microbiol, 190: 573–584.

65. Portnoy DA, Auerbuch V, Glomski IJ (2002) The cell biology of *Listeria monocytogenes* infection: the intersection of bacterial pathogenesis and cell-mediated immunity. J Cell Biol, 158(3) doi:101083/jcb200205009.

66. Karczewski J, Troost FJ, Konings I, Dekker J, Kleerebezem M, et al. (2010) Regulation of human epithelial tight junction proteins by *Lactobacillus plantarum in vivo* and protective effects on the epithelial barrier. Am J Physiol-Gastr L, 298: G851–G859.

67. Anderson RC, Cookson AL, McNabb WC, Park Z, McCann MJ, et al. (2010) *Lactobacillus plantarum* MB452 enhances the function of the intestinal barrier by increasing the expression levels of genes involved in tight junction formation. BMC Microbiol, doi:101186/1471-2180-10-316.

68. Burkholder KM, Bhunia AK (2010) *Listeria monocytogenes* uses *Listeria* adhesion protein (LAP) to promote bacterial transepithelial translocation and induces expression of LAP Receptor *Hsp*60. Infect Immun, 78: 5062–5073.

7

Mechanistic Approach to Stability Studies as a Tool for the Optimization and Development of New Products Based on *L. rhamnosus* Lcr35® in Compliance with Current Regulations

Claudia Muller[1,2]*, **Virginie Busignies**[2], **Vincent Mazel**[2], **Christiane Forestier**[3], **Adrien Nivoliez**[1], **Pierre Tchoreloff**[2]

1 Département Recherche et Développement, Probionov, Aurillac, France, **2** Laboratoire Matériaux et Santé EA401, Univ. Paris Sud, Châtenay-Malabry, France, **3** Laboratoire Microorganismes : Genome Environnement (LMGE) UMR CNRS 6023 Univ. d'Auvergne-Clermont 1, Clermont-Ferrand, France

Abstract

Probiotics are of great current interest in the pharmaceutical industry because of their multiple effects on human health. To beneficially affect the host, an adequate dosage of the probiotic bacteria in the product must be guaranteed from the time of manufacturing to expiration date. Stability test guidelines as laid down by the ICH-Q1A stipulate a minimum testing period of 12 months. The challenge for producers is to reduce this time. In this paper, a mechanistic approach using the Arrhenius model is proposed to predict stability. Applied for the first time to laboratory and industrial probiotic powders, the model was able to provide a reliable mathematical representation of the effects of temperature on bacterial death ($R^2>0.9$). The destruction rate (k) was determined according to the manufacturing process, strain and storage conditions. The marketed product demonstrated a better stability (k = 0.08 months^{-1}) than the laboratory sample (k = 0.80 months^{-1}). With industrial batches, k obtained at 6 months of studies was comparable to that obtained at 12 months, evidence of the model's robustness. In addition, predicted values at 12 months were greatly similar ($\pm30\%$) to those obtained by real-time assessing the model's reliability. This method could be an interesting approach to predict the probiotic stability and could reduce to 6 months the length of stability studies as against 12 (ICH guideline) or 24 months (expiration date).

Editor: Hauke Smidt, Wageningen University, Netherlands

Funding: This work was supported by a grant from ANRT (Association Nationale Recherche Technologie) supporting the PhD thesis performed by CM. The funders had no role in study design, data collection and analysis, decision to publish, or preparation of the manuscript.

Competing Interests: The authors have read the journal's policy and have the following conflicts: Claudia Muller and Adrien Nivoliez have an institutional affiliation with the company (Probionov, Aurillac, France) which commercializes Bacilor (Lcr restituo® capsules).

* E-mail: c.muller@probionov.com

Introduction

Probiotics are defined by the World Health Organization (WHO) as viable live microorganisms, which, when administered in adequate amounts, confer a health benefit on the host [1]. Because of their large potential applications in human diseases such as gastrointestinal disorders, immunomodulation or cholesterol reduction, probiotic products are of great interest to the pharmaceutical industry [2–6]. The most commonly used strains for applications in both the food and pharmaceutical industries belong to the *Lactobacillus* and *Bifidobacterium* bacterial genus [4,7].

In addition to having beneficial health effects, probiotic bacteria must possess the properties required to be used as pharmaceutical products, such as safety, genetic stability, ability to be produced on a large scale, product stability (viability), and acceptable flavor or taste [5,7–9]. One of the most important criteria is the maintenance of viability during the product life cycle [5,10], which can be affected by the culture medium, manufacturing process and residual humidity [11–17].

Depending on the application (food or pharmaceutical) and the target (disease), the number of viable bacteria in the probiotic can vary, and has to be previously established by the producer [3,9,15]. The method used for determining the stability of pharmaceutical products is defined by the ICH Guideline Q1A [18] and requires a minimum of 12 months to test products under different storage conditions (Table 1). The stability studies can be performed until the expiration date (24 or 36 months) leading to a time-consuming method. Faced with the increasing cost of laboratory technology, producers would welcome any new strategy that could offer savings.

Previous studies have shown the interest of a mechanistic approach in predicting the stability of active pharmaceutical ingredients (API) in formulations, especially for chemical molecules, in vitamin denaturation [19,20]. Data predictions using an appropriate kinetic model could shorten the time required for a stability study. The Arrhenius equation has been widely applied for this purpose [19,21,22]. The ICH Q8 guideline outlines how time can be saved during product development with the aid of

Table 1. Storage conditions requested by guideline ICH Q1A for the development of new drug products.

Study	Storage condition	Minimum time period covered by data at submission
Long term*	25°C ±2°C/60% RH ±5% RH or 30°C ±2°C/65% RH ±5% RH	12 months
Intermediate**	30°C ±2°C/65% RH ±5% RH	6 months
Accelerated	40°C ±2°C/75% RH ±5% RH	6 months

*It is up to the applicant to decide whether long term stability studies are performed at 25°C ±2°C/60% RH ±5% RH or 30°C ±2°C/65% RH ±5% RH.
**If 30°C ±2°C/65% RH ±5% RH is the long-term condition, there is no intermediate condition.

Table 2. Stability testing for the products according to the ICH Guideline Q1A.

	20°C, 25°C, 30°C	40°C
Laboratory assay	0, 3, 6, 9, 12	/
Eleven batches	0, 3, 6, 9, 12, 18, 24	/
Recent batch	0, 1, 3, 6	0, 1, 2, 3, 4, 5, 6

Storage conditions and check points (months) realized on powders obtained with the laboratory assay and the industrial batches (Bacilor®).

Table 3. Kinetic parameters of the laboratory powders.

Temperature (°C)	k (months^{-1})	Decimal reduction time (months)		
		D_1	D_2	D_3
20	0.51	4.5	9.0	14
25	0.80	2.9	5.8	8.6
30	1.34	1.7	3.4	5.2

Destruction rates (k) and decimal reductions (D) obtained for the powder of the laboratory assay (values obtained with the mean of stability studies performed on three assays).

Design Space, which can be performed with a study design or with a mathematical equation [23].

Just like chemical compounds, bacteria are affected by increases in temperature. Accelerated storage conditions (40°C) lead to a greater loss of viability than long-term storage conditions or storage at room temperature [14,19,24,25]. The stability and the kinetic constant (k) depend on temperature but also on the formulation, manufacturing process and intrinsic resistance of the bacterial strain [8,14,26]. Each formulation should have its own equation. A new constant k obtained in real time can be compared with a reference kinetic to anticipate possible problems in the manufacturing of new industrial batches or during laboratory trials.

This paper analyses the stability of the probiotic strain *Lactobacillus rhamnosus* Lcr35® in manufactured powders stored at different temperatures. Isolated from human intestinal microbiota, this probiotic strain is used for intestinal and gynecological applications to restore flora after a disruption such as antibiotic treatment or stress [27,28]. Its biological and physical properties including stability during manufacture and storage have been described elsewhere [27,29–31]. We assessed the ability of the

Arrhenius model to determine the viability of *L. rhamnosus* Lcr35® at different storage temperatures using laboratory and industrial powders. We obtained a specific Arrhenius equation to predict stability irrespective of storage length and temperature. The repeatability and the robustness of this approach were verified by comparing predicted and observed stability values from industrial batches.

Materials and Methods

1. Microorganisms and Growth Production

For the laboratory assay, the probiotic bacteria *L. rhamnosus* Lcr35®, derived from reference cryotubes (Probionov, Aurillac, France) containing the native strain. *L. rhamnosus* Lcr35®, was grown in Man, Rogosa, Sharpe (MRS) broth (AES, Bruz, France) for 48 h at 37°C (three samples). Reconstituted milk was added (110 g/l) as cryoprotectant before lyophilization. This medium composed of MRS and reconstituted milk, will hereafter be designated "critical medium". The resulting bacterial suspensions were then lyophilized and the obtained cakes were ground in a Magimix® blender. The powders obtained from the laboratory assay were packaged in a glass bottle with a hermetic plug.

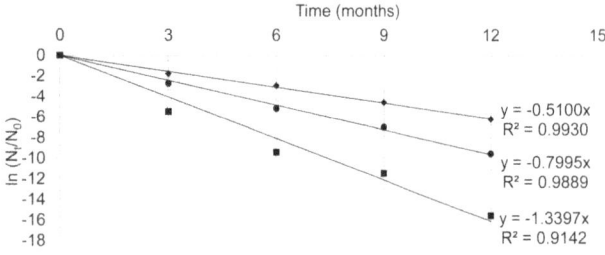

Figure 1. Effect of storage conditions on bacterial stability within the laboratory powder. Plot of $\ln(N_t/N_0)$ for powders obtained with the laboratory assay against time and temperature 20°C (♦), 25°C (●) and 30°C (■) (Mean of three values), according to the current regulation (ICH Guideline Q1A).

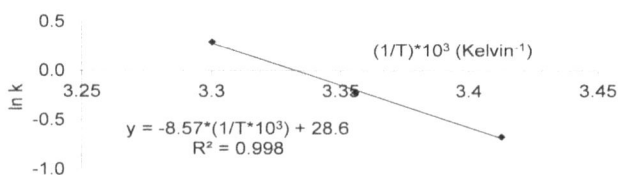

Figure 2. Arrhenius representation for the laboratory powder (Lcr35® bacteria). The observed k values were previously obtained with $\ln(N_t/N_0)$ against time for the laboratory powder (Fig. 1). These k values (y axis) were represented against 1/T (Kelvin^{-1}) (x axis) to obtain the Arrhenius equation of this assay (Equation 4).

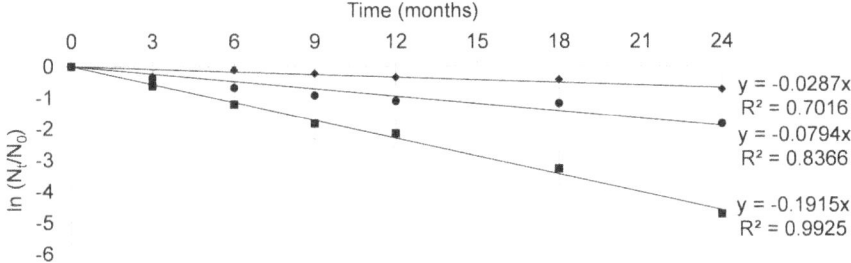

Figure 3. Effect of storage conditions on bacterial stability within Bacilor® (*Lcr restituo*® capsules). Plot of ln(N$_t$/N$_0$) for the batches of Bacilor® (*Lcr restituo*® capsules) against time and temperature 20°C (♦), 25°C (●) and 30°C (■) (Mean of eleven values), according to the current regulation (ICH Guideline Q1A).

An industrial product containing *L. rhamnosus* Lcr35®, manufactured and provided by Probionov (Aurillac, France), was also included in the study. Its commercial name is Bacilor® and it contains the active pharmaceutical ingredient (API) *Lcr restituo*®. Eleven batches of Bacilor® (*Lcr restituo*®) capsules manufactured between 2006 and 2011 and one manufactured in 2012 were taken to make comparisons between predicted and observed stability values.

2. Storage Conditions and Stability Testing (viability of *Lactobacillus*)

Powders obtained with the critical medium and the 12 industrial batches of Bacilor® (*Lcr restituo*® capsules) were stored and checked as recommended by the ICH Guideline Q1A (Table 2). At each checkpoint, the number of viable *Lactobacillus* was determined by a plate count method as follows. The powders were re-suspended in peptone water (Pastone 1 g/l, Biorad, Marnes-la-coquette, France) and a serial 10-fold dilution was performed. Each dilution was plated on MRS agar (Biomérieux, Marcy l'Etoile, France) plates and incubated at 37°C for 72 h. The results were then expressed as colony forming units (CFU.g^{-1}).

3. Linear-Arrhenius Model

A standard two-step method was used to obtain the Arrhenius model and to assess the influence of temperature on the stability of *Lactobacillus rhamnosus* Lcr35®.

Predictive microbiology describes the exponential loss of bacterial viability over time by the following equation (first-order low):

$$N_t = N_0 e^{-kt} \qquad (1)$$

t: time.
k: destruction rate.
N$_0$: number of viable microorganisms at t = 0.
N$_t$: number of viable microorganisms at t = t.

By plotting ln (N$_t$/N$_0$) over time t, k can be determined for each temperature. The effect of temperature on k can be represented by the Arrhenius equation:

$$k = Ae^{-Ea/RT} \qquad (2)$$

k: destruction rate (time^{-1}).
A: frequency factor (time^{-1}).
R: gas constant (8,314 J.mol^{-1}.K^{-1})
T: temperature (Kelvin).
Ea: activation energy (J.mol^{-1}).

By plotting ln(k) versus 1/T, a straight line is obtained, which leads to the parameters of equation (2). By extrapolation, k is accessible for any temperature. Using equation (1), the number of viable microorganisms can be predicted at any time and for any storage condition.

Finally, k is also related to the decimal reduction time D$_1$, the time for which the initial bacterial population is reduced by 90% [19]. D$_1$ is defined as:

$$D_1 = \ln_{10}/k \qquad (3)$$

By extension, it is possible to define the time for which the initial bacterial population is reduced by 99 or 99.9% (D$_2$ and D$_3$).

Results and Discussion

1. Model Validation on Laboratory Powders

The powders obtained from the laboratory assay were used to validate the use of the Arrhenius model in assessing bacterial stability over a 12-month period. The critical medium does not provide high bacterial stability during product process and storage (data not shown) and was therefore chosen as a model of damaged stability conditions.

The destruction rate k was determined at three storage temperatures (20, 25 and 30°C) using equation (1) (Fig. 1). When

Table 4. Kinetic parameters of Bacilor® (*Lcr restituo*® capsules).

Temperature (°C)	k (months^{-1})	Decimal reduction time (months)		
		D$_1$	D$_2$	D$_3$
20	0.03	80.2	161.0	241.0
25	0.08	29.0	58.0	87.0
30	0.19	12.0	24.1	36.1

Destruction rates (k) and decimal reductions (D) obtained with the commercial product powders (values obtained with the mean of stability studies performed on eleven batches of Bacilor® (*Lcr restituo*® capsules)).

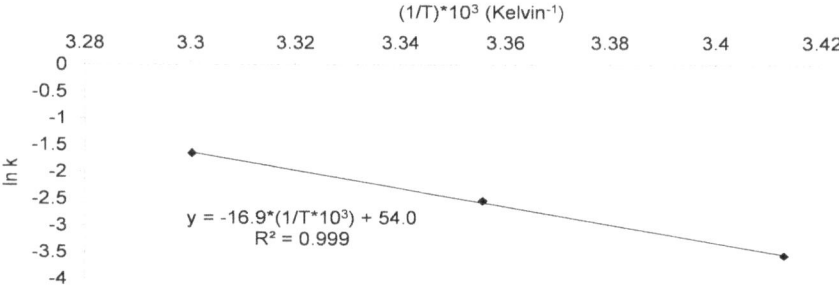

Figure 4. Arrhenius representation for Bacilor® (*Lcr restituo®* **capsules).** The observed k values were previously obtained with $\ln(N_t/N_0)$ against time for Bacilor® (*Lcr restituo®* capsules) (Fig. 3). These k values (y axis) were represented against $1/T$ (Kelvin^{-1}) (x axis) to obtain the Arrhenius equation of this assay (Equation 5).

the temperature increased so did the k values ($k_{20°C} < k_{25°C} < k_{30°C}$), which demonstrates, in agreement with Davey's study [19], that, due to the sensitivity of the microorganisms to heat, viability was lost (Table 3). The values of decimal reduction times D1, D2 and D3, unlike the destruction rate, decreased as the temperature increased. At 20°C, it took 4.5 months to reduce the initial bacterial population by 90% whereas at 30°C only 1.7 months was required. Thus, the loss of bacterial viability was dependent on temperature (k values) and the length of storage (D values).

The obtained Arrhenius model was plotted (Fig. 2). k was related to temperature by the following Arrhenius equation:

$$\ln(k) = -8.57^*(1/T^*10^3) + 28.6 \qquad (4)$$

With this mathematical relation, k could be extrapolated to a range of temperatures to obtain its variation under different storage conditions. Activation energy (Ea) extracted from this equation was 71 kJ.mol^{-1}. The Arrhenius equation obtained from experimental data was applied to the $\ln(N_t/N_0)$ data with a high correlation ($R^2 = 0.998$), which suggests that the model is suitable to determine the loss of bacterial viability in different storage conditions.

2. Application on Industrial Batches: Bacilor® (Lcr restituo®) Capsules

2.1 Determination of the arrhenius equation for the 11 batches (2006–2011). A mean of the stability values obtained

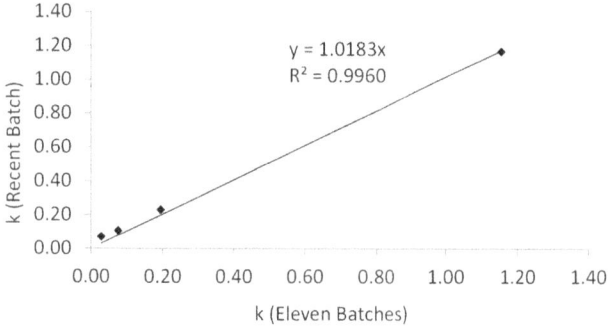

Figure 5. Comparison of the obtained and predicted k values. Observed k values obtained with stability studies of the recent batch during six months, against the k values obtained with the Arrhenius equation of the eleven batches of Bacilor® (*Lcr restituo®* capsules) (the k value at 40°C for the eleven batches was not available and was predicted with the Arrhenius equation).

for the 11 batches of Bacilor® (*Lcr restituo®*) capsules was calculated at each checkpoint. The Arrhenius model was then applied to these mean values. The guideline stipulates a minimum study time of 12 months, but stability can be tested as long as needed (i.e. until expiration date), sometimes up to 36 months. The industrial batches tested in this study were monitored over 24 months at three temperatures (20, 25, 30°C).

As shown in Figure 3, k increased with the temperature, but the values were about 10 times lower than those obtained previously with the laboratory powders (at 25°C, $k_{(Bacilor)} = 0.08/k_{(critical media)} = 0.80$ months^{-1}) (Table 4). The powders obtained with the critical medium led to low D values and a high destruction rate (k) whatever the temperature ($D_{1(20°C)} = 4.5$ months, $k_{20°C} = 0.51$ months^{-1}). The industrial batches were less affected by temperature ($D_{1(20°C)} = 80.2$ months, $k_{20°C} = 0.03$ months^{-1}), demonstrating their better stability over time. The failure of the critical medium to promote bacterial stability during manufacturing and storage was thus confirmed. In agreement with Oliveira et al. [15], our results illustrate the real influence of formulation on maintaining bacterial viability during these two phases. The use of lyophilization to dry the fermented culture media is one of the most useful processes to maintain optimal bacterial viability during product process and storage. However, as shown by the different results obtained with laboratory samples and industrial powders, the culture growth medium also has an important role in maintaining stability and must be carefully formulated during product development. Depending on the nature of the cryoprotectants and the culture components, the stability of any given bacterial strain can differ according to the stress encountered during the product's life cycle [13,14,16,17,32]. The impact of any modification (excipient, formulation or bacterial strain) on product stability could therefore be easily detected by determining the k value and comparing it with a reference kinetic.

The Arrhenius model was also applied to the results on the stability of the 11 batches (Fig. 4), and gave the following equation:

$$\ln(k) = -16.9^*(1/T^*10^3) + 54.0 \qquad (5)$$

The equation yielded an activation energy (Ea) of 140 kJ.mol^{-1} higher than that obtained with the laboratory powders (Ea = 71 kJ.mol^{-1}), evidence that bacterial stability and product formulation are closely interdependent.

The straight line of the equation showed a high correlation ($R^2 = 0.999$) and evidenced the relation between the loss of bacterial viability (k) and the storage temperature of industrial

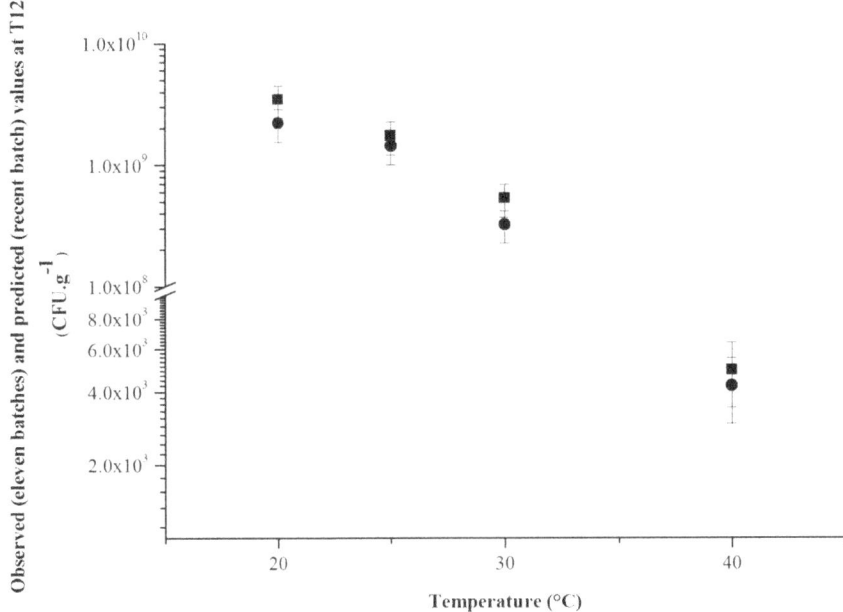

Figure 6. Comparison of observed and predicted stability values. Comparison between observed values (mean of 11 values) at 12 months on the eleven batches of Bacilor® (*Lcr restituo®* capsules) (■) and predicted values at 12 months for the recent one (●) against temperature.

batches. From equation (5), k was calculated for accelerated storage conditions (40°C) and had a value of 1.15 months^{-1}.

2.2 Application on a recent industrial batch (2012). To assess the accuracy of this model, k values obtained with stability studies of the 11 batches were compared with those obtained with the recent batch. Stability values were recorded at four temperatures (20, 25, 30 and 40°C) during a 6-month follow-up, which gave $k_{20°C} < k_{25°C} < k_{30°C} < k_{40°C}$, as previously shown. Concerning the loss of viability, k observed at 40°C for the recent batch ($k_{40°C} = 1.17$ months^{-1}) was closely similar to that predicted for the old batches of Bacilor® ($k_{40°C} = 1.15$ months^{-1}). Comparison of these values yielded a guiding coefficient equal to 1.0183 ($R^2 = 0.996$) (Fig. 5).

To assess this correlation, the number of viable lactobacilli in the recent batch was determined at 12 months (Equation 1) and compared with those of the 11 other batches (Fig. 6). A standard deviation of 30%, due to biological variability, sampling and dilution, is commonly accepted and was added to each value [33,34]. In spite of this constraint, it was clearly shown that whatever the temperature, the observed and predicted numbers of viable lactobacilli were closely similar after 12 months. A stability study performed during the 6 first months of storage gave very similar Arrhenius parameters to those in a stability study performed over 12 months in accordance with the ICH guideline Q1A. Thus, 6-month stability studies performed at three temperatures (Arrhenius representation) should be sufficient to predict the behavior of the product as successfully as a 12-month study (ICH Q1A) or even longer ones (up to expiration date).

As previously shown for chemical APIs [19,20], stability studies of probiotic products can be improved by data predictions using the Arrhenius equation [35]. It would therefore be possible to predict the behavior of the product under storage conditions before the end of the follow-up period. In the intermediate and long-term, problems of product quality or stability (i.e. problems encountered during the development process) could be detected and resolved earlier as the QbD approach (ICH Q8).

Conclusion

In the pharmaceutical industry, regulations are more restrictive than in the food industry, owing, in great part, to the ICH Q1A, which imposes check-ups over 12 months for each new product. The present work is the proof that the follow-up period needed to perform stability studies can be reduced and the results anticipated by using the Arrhenius model.

The stability measurements obtained in our study were similar for commercial batches of *Lactobacillus rhamnosus* Lcr35® produced at different times. This demonstrates the robustness of the model, which could be widely applied to probiotic products to predict the influence of different temperatures at any time.

Probiotics must be viable at sufficient dosage levels at the time of consumption and until their expiration date to have a health effect. As for sterilization, the manufacturers could determine the time at which viability meets the regulatory standards. The destruction rate obtained in real time for a new product could be compared with a reference kinetic to identify any change in formulation, bacterial strain or manufacturing process. Applied to probiotic products, this approach could be of great value in their development and market approval.

To conclude, reducing the time dedicated to research and development is one of the most important challenges for the pharmaceutical industry (ICH Q8). This aim could be achieved in compliance with ICH Q1A guidelines by using the Arrhenius model in stability studies.

Acknowledgments

We would like to thank Hadjer Bouaziz for her work on Arrhenius model and Gaëlle Castel, Caroline Dausset, Aurélie Lacalmontie, Amandine Pralus and Mathilde Roche for their technical support.

Author Contributions

Conceived and designed the experiments: CM VB VM AN PT. Performed the experiments: CM. Analyzed the data: CM VB VM AN PT.

Contributed reagents/materials/analysis tools: CM VB VM CF AN PT. Wrote the paper: CM VB VM.

References

1. FAO/WHO (2002) Guidelines for the evaluation of probiotics in food. Working group report. Rome and Geneva: Food and Agricultural Organization of the United Nation/World Health Organization.
2. Kaur IP, Chopra K, Saini A (2002) Probiotics: potential pharmaceutical applications. Eur. J. Pharm. Sci. 15: 1–9.
3. Ouwehand AC, Salminen SJ (1998) The health effects of cultured milk products with viable and non-viable bacteria. Int. Dairy J. 8: 749–758.
4. Quigley EMM (2010) Prebiotics and probiotics; modifying and mining the microbiota. Pharmacol. Res. 61: 213–218.
5. Saarela M, Mogensen G, Fondén R, Mättö J, Mattila-Sandholm T (2000) Probiotic bacteria : safety, functional and technological properties. J. Biotechnol. 84: 197–215.
6. Scheinbach S (1998) Probiotics: functionality and commercial status. Biotechnol. Adv. 16 (3): 581–608.
7. Kailasapathy K, Chin J (2000) Survival and therapeutic potential of probiotic organisms with reference to *Lactobacillus acidophilus* and *Bifidobacterium* spp. Immunol. Cell Biol. 78: 80–88.
8. Forssten SD, Sindelar CW, Ouwehand AC (2011) Probiotics from an industrial perspective. Anaerobe 17: 410–413.
9. Kosin B, Rakshit SK (2006) Criteria for production of probiotics. Food Technol. Biotech. 44 (3): 371–379.
10. Tuomola E, Crittenden R, Playne M, Isolauri E, Salminen S (2001) Quality assurance criteria for probiotic bacteria. Am. J. Cli. Nutr. 73 (suppl): 393S–398S.
11. Cerf O, Davey KR, Sadoudi AK (1996) Thermal inactivation of bacteria-a new predictive model for the combined effect of three environmental factors: temperature, pH and water activity. Food Res. Int. 29 (3–4): 219–226.
12. Chan ES, Zhang Z (2002) Encapsulation of probiotic bacteria Lactobacillus acidophilus by direct compression. Trans IChem E 80: Part C.
13. Grześkowiak Ł, Isolauri E, Salminen S, Gueimonde M (2010) Manufacturing process influences properties of probiotic bacteria. Brit. J. Nutr.: 1–8.
14. Makinen K, Berger B, Bel-Rhlid R, Ananta E (2012) Science and technology for the mastership of probiotic applications in food products. J. Biotechnol. 162: 356–365.
15. Oliveira MN, Sodini I, Remeuf F, Corrieu G (2001) Effect of milk supplementation and culture composition on acidification, textural properties and microbiological stability of fermented milks containing probiotic bacteria. Int. Dairy J. 11: 935–942.
16. Savini M, Cecchini C, Verdenelli MC, Silvi S, Orpianesi C, et al. (2010) Pilot-scale production and viability analysis of freeze-dried probiotic bacteria using different protective agents. Nutrients 2: 330–339.
17. Zarate G, Nader-Macias ME (2006) Viability and biological properties of probiotic vaginal lactobacilli after lyophilization and refrigerated storage into gelatin capsules. Process Biochem. 41: 1779–1785.
18. CPMP/ICH/2736/99, ICH Q1A (R2) (2003) Note for guidance on stability testing: Stability testing of new drug substances and products. European Medicines Agency.
19. Davey KR (1993) Linear-Arrhenius models for bacterial growth and death and vitamin denaturations. J. Ind. Microbiol. 12: 172–179.
20. Slater JG, Stone HA, Palermo BT, Duvall RN (1979) Reliability of Arrhenius Equation in Predicting Vitamin A Stability in Multivitamin Tablets. J. Pharm. Sci. 68: 49–52.
21. La Saponara, V (2011) Environmental and chemical degradation of carboxy/epoxy and structural adhesive for aerospace applications: Fickian and anomalous diffusion, Arrhenius kinetics. Compos. Struct. 93: 2180–2195.
22. Waterman KC (2011) The application of the Accelerated Assessment Program (ASAP) to Quality by Design (QbD) for Drug Product Stability. AAPS PharmSciTech 3 (12): 932–937.
23. CHMP/ICH/167068/04, ICH Q8(R2) (2009) Pharmaceutical development. European Medicines Agency.
24. Koutsoumanis K, Stamatiou A, Skandamis P, Nychas GJE (2006) Development of a microbial model for the combined effect of temperature and pH on spoilage of ground meat, and validation of the model under dynamic temperature conditions. Applied environ. microb. 72 (1): 124–134.
25. Madras G, Chattopadhyay S (2001) Optimum temperature for oxidative degradation of poly(vinyl acetate) in solution. Chem. Eng. Sci. 56: 5058–5089.
26. Carvalho AS, Silva J, Ho P, Teixeira P, Malcata FX, et al. (2004) Relevant factors for the preparation of freeze-dried lactic acid bacteria. Int. Dairy J. 14: 835–847.
27. Forestier C, De Champs C, Vatoux C, Joly B (2001) Probiotic activities of *Lactobacillus casei rhamnosus*: *in vitro* adherence to intestinal cells and antimicrobial properties. Res. Microbiol. 152: 167–173.
28. Petricevic L, Witt A (2008) The role of *Lactobacillus casei rhamnosus* Lcr35 in restoring the normal vaginal flora after antibiotic treatment of bacterial vaginosis. BJOG 115 (11): 1369–1374.
29. Coudeyras S, Jugie G, Vermerie M, Forestier C (2008) Adhesion of human probiotic *Lactobacillus rhamnosus* to cervical and vaginal cells and interaction with vaginosis-associated pathogens. Infect. Dis. Obstet. Gynecol.: 549640.
30. Coudeyras S, Marchandin H, Fajon C, Forestier C (2008) Taxonomic and strain-specific identification of the probiotics strain *Lactobacillus rhamnosus* 35 within the *Lactobacillus casei* group. Appl. Environ. Microb.: 2679–2689.
31. Nivoliez A, Camares O, Paquet-Gachinat M, Bornes S, Forestier C, et al. (2012) Influence of manufacturing processes on *in vitro* properties of the probiotic strain *Lactobacillus rhamnosus* Lcr35®. J. Biotechnol. 160: 236–241.
32. Berner D, Viernstein H (2006) Effect of protective agents on the viability of Lactococcus lactis to freeze-thawing and freeze-drying. Sci. Pharm. 74: 137–149.
33. Jennison MW, Wadsworth GP (1940) Evaluation of the errors involved in estimating bacterial numbers by the plating method. J. Bacteriol. 39(4): 389–397.
34. Sutton S (2011) Accuracy of plate counts. Journal of validation technology 3 (17): 42–46.
35. Waterman KC, Adami RC (2005) Accelerated aging: Prediction of chemical stability of pharmaceuticals. Int. J. Pharm. 293: 101–125.

8

Probiotics Protect Mice from Ovariectomy-Induced Cortical Bone Loss

Claes Ohlsson[1], Cecilia Engdahl[1,2], Frida Fåk[3], Annica Andersson[1,2], Sara H. Windahl[1], Helen H. Farman[1], Sofia Movérare-Skrtic[1], Ulrika Islander[1,2], Klara Sjögren[1]*

1 Centre for Bone and Arthritis Research, Institute of Medicine, Sahlgrenska Academy at University of Gothenburg, Gothenburg, Sweden, 2 Department of Rheumatology and Inflammation Research, Institute of Medicine, Sahlgrenska Academy at University of Gothenburg, Gothenburg, Sweden, 3 Applied Nutrition and Food Chemistry, Department of Food Technology, Engineering and Nutrition, Lund University, Lund, Sweden

Abstract

The gut microbiota (GM) modulates the hosts metabolism and immune system. Probiotic bacteria are defined as live microorganisms which when administered in adequate amounts confer a health benefit on the host and can alter the composition of the GM. Germ-free mice have increased bone mass associated with reduced bone resorption indicating that the GM also regulates bone mass. Ovariectomy (ovx) results in bone loss associated with altered immune status. The purpose of this study was to determine if probiotic treatment protects mice from ovx-induced bone loss. Mice were treated with either a single Lactobacillus (L) strain, L. paracasei DSM13434 (L. para) or a mixture of three strains, L. paracasei DSM13434, L. plantarum DSM 15312 and DSM 15313 (L. mix) given in the drinking water during 6 weeks, starting two weeks before ovx. Both the L. para and the L. mix treatment protected mice from ovx-induced cortical bone loss and bone resorption. Cortical bone mineral content was higher in both L. para and L. mix treated ovx mice compared to vehicle (veh) treated ovx mice. Serum levels of the resorption marker C-terminal telopeptides and the urinary fractional excretion of calcium were increased by ovx in the veh treated but not in the L. para or the L. mix treated mice. Probiotic treatment reduced the expression of the two inflammatory cytokines, TNFα and IL-1β, and increased the expression of OPG, a potent inhibitor of osteoclastogenesis, in cortical bone of ovx mice. In addition, ovx decreased the frequency of regulatory T cells in bone marrow of veh treated but not probiotic treated mice. In conclusion, treatment with L. para or the L. mix prevents ovx-induced cortical bone loss. Our findings indicate that these probiotic treatments alter the immune status in bone resulting in attenuated bone resorption in ovx mice.

Editor: Bernhard Ryffel, French National Centre for Scientific Research, France

Funding: This work was supported by the Swedish Research Council, Swedish Foundation for Strategic Research, COMBINE, Avtal om Läkarutbildning och Forskning/Läkarutbildningsavtalet research grant in Gothenburg, Lundberg Foundation, Torsten and Ragnar Söderberg Foundation, Novo Nordisk Foundation, Magnus Bergvall Foundation and Åke Wiberg Foundation. The BD FACS Canto II was bought thanks to generous support from the Inga-Britt and Arne Lundberg Foundation. The funders had no role in study design, data collection and analysis, decision to publish, or preparation of the manuscript.

Competing Interests: A pending patent application SE 1351571-3, where the main applicant is Probi AB, who provided the study product, but have not otherwise been involved in the conduction of the experiments and the analyses of data. Name of patent application: SE 1351571-3. Name of study products: Lactobacillus (L.) paracasei DSM13434 (8700:2), L. plantarum DSM 15312 (HEAL 9) and L. plantarum DSM 15313 (HEAL 19). The authors Klara Sjögren and Claes Ohlsson are included in the patent application as inventors but are not further involved in the activities of Probi AB.

* E-mail: Klara.Sjogren@medic.gu.se

Introduction

Fractures caused by osteoporosis constitute a major health concern and result in a huge economic burden on health care systems. In Sweden, the lifetime risk of any osteoporotic fracture is 47% and 24% in women and men, respectively [1]. In USA, the risk has been reported to be 40% and 13% in white women and men, respectively and fractures are associated with significant mortality and morbidity [2]. Cortical bone is the major contributor to non-vertebral fracture risk and comprises more than 80% of the skeleton.

The skeleton is remodeled by bone forming osteoblasts (OBs) and bone resorbing osteoclasts (OCLs). Macrophage colony stimulating factor (M-CSF) increases proliferation and survival of OCLs precursor cells as well as up-regulates expression of receptor activator of nuclear factor-κB (RANK) in OCL. This allows RANK ligand (RANKL) to bind and start the signaling cascade

that leads to OCL formation. The effect of RANKL can be inhibited by Osteoprotegerin (OPG), which is a decoy receptor for RANKL [3].

The association between inflammation and bone loss is well established. In autoimmune diseases, osteoclastic bone resorption is driven by inflammatory cytokines produced by immune cells e.g. activated T cells [4]. In addition, low-grade systemic inflammation, indicated by moderately elevated serum levels of high sensitivity C-reactive protein (hsCRP), associates with low bone mineral density (BMD), elevated bone resorption and increased fracture risk [5-8]. The estrogen deficiency that occurs after menopause results in increased formation and prolonged survival of OCLs. This is suggested to be due to a number of factors including loss of the immunosuppressive effects of estrogen, resulting in increased production of cytokines promoting osteoclastogenesis, and direct effects of estrogen on OCLs [9,10]. In line with these data, blockade of the inflammatory cytokines TNFα and IL-1 leads to a decrease in bone resorption markers in early postmenopausal women [11].

In recent years, the importance of the gut microbiota (GM) for both health and disease has been intensively studied. The GM constitutes of trillions of bacteria which collectively contain 150-fold more genes than our human genome. It is acquired at birth and, although a distinct entity, it has clearly coevolved with the human genome and can be considered a multicellular organ that communicates with and affects its host in numerous ways [12]. The composition of the GM is modulated by a number of environmental factors such as diet and antibiotic treatments. Molecules produced by the gut bacteria can be both beneficial and harmful and are known to affect the host's immune system [13]. Perturbed microbial composition has been postulated to be involved in a range of inflammatory conditions, within and outside the gut including Crohn's disease, ulcerative colitis, rheumatoid arthritis, multiple sclerosis, diabetes, food allergies, eczema and asthma as well as obesity and the metabolic syndrome [13,14]. We recently showed that absence of GM in germ-free (GF) mice leads to increased bone mass associated with reduced bone resorption and altered immune status in bone. Colonisation of GF mice with a normal gut microbiota led to a normalisation of bone mass and immune status in bone marrow [15]. A role of the GM for bone mass is supported by a recent study demonstrating that subtherapeutic antibiotic treatment in early life increases bone mass in young mice [16]. Although the low dose of antibiotics in this study did not cause a significant alteration in bacterial count it caused shifts in the composition of the GM. Furthermore, tetracycline treatment has been shown to prevent bone loss and improve mechanical properties of bone after ovariectomy (ovx) [17,18]. These studies, demonstrate that antibiotic treatment has the capacity to influence both the GM composition and bone mass, supporting the notion that the GM is a regulator of bone homeostasis.

Probiotic bacteria are defined as live microorganisms which when administered in adequate amounts confer a health benefit on the host. Probiotics act by altering the composition or the metabolic activity of the GM [19]. The suggested underlying mechanisms for how probiotics contribute to health are manifold including increased solubility and absorption of minerals, enhanced barrier function and modulation of the immune system [20,21]. Ovx in mice results in bone loss associated with altered immune status, resembling post-menopausal osteoporosis. The purpose of the present study was to determine if probiotic treatment protects mice from ovx-induced bone loss.

Materials and Methods

Ovx Mouse-model and Probiotic Treatment

Six-week-old C57BL/6N female mice were purchased from Charles River (Germany). The mice were housed in a standard animal facility under controlled temperature (22°C) and photoperiod (12-h light, 12-h dark) and had free access to fresh water and soy-free food pellets R70 (Lactamin AB, Stockholm, Sweden). The ovx model for osteoporosis is included in the FDA guidelines for preclinical and clinical evaluation for agents used for the treatment of postmenopausal osteoporosis [22]. Probiotic treatment started two weeks before ovx to study the preventive effect of probiotic treatment on ovx induced bone-loss (Fig. 1A). Mice were treated with either a single *Lactobacillus* (L) strain, *L. paracasei* DSM13434 (L. para) or a mixture of three strains, *L. paracasei* DSM13434, *L. plantarum* DSM 15312 and DSM 15313 referred to as L. mix during 6 weeks. The probiotic strains were selected based on their anti-inflammatory properties in an earlier study [23]. Mice were

randomized into six treatment groups with 10 mice in each as follows: 1. Veh-Ovx, 2. Veh-Sham, 3. L. Para-Ovx, 4. L. Para-Sham, 5. L. Mix-Ovx, 6. L. Mix-Sham (See also Figure 1). The L. strains were given in the drinking water at a concentration of 10^9 colony-forming units (cfu)/ml while control mice received tap water with vehicle (glycerol). Water bottles were changed every afternoon. The survival of the L. strains in the water bottles was checked regularly and after 24 h the concentration dropped one log unit to approximately 10^8 cfu/ml. Each mouse drank on average 4.5 ml water/day. After two weeks of probiotic treatment, the mice were either sham-operated or ovx under inhalation anesthesia with isoflurane (Forene; Abbot Scandinavia, Solna, Sweden). Four weeks after surgery, blood was collected from the axillary vein under anesthesia with Ketalar/Domitor vet, and the mice were subsequently killed by cervical dislocation. Tissues for RNA preparation were immediately removed and snap-frozen in liquid nitrogen for later analysis. Bones were excised and fixed in 4% paraformaldehyde.

A. Study Design

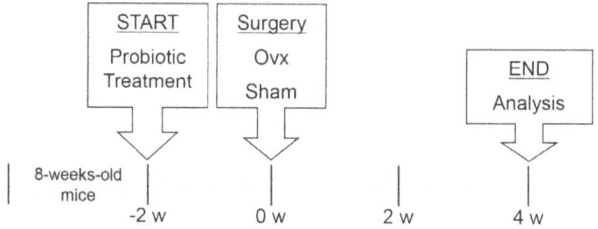

B. Body Weight

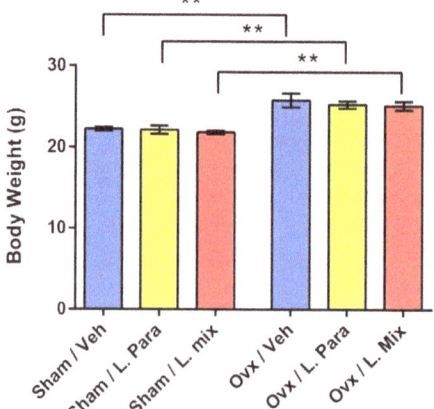

Figure 1. Study Design and Body Weight. Outline of study design (A). Eight-week-old mice were treated with either vehicle (veh), a single *Lactobacillus* (L) strain (L. para) or a mixture of three strains (L. mix) during 6 weeks, starting two weeks before ovx or sham surgery. The L. strains were given in the drinking water at a concentration of 10^9 colony-forming units (cfu)/ml while control mice received tap water with vehicle. Mice were 14-week-old at the end of the study, when tissues were collected for later analysis. Ovx resulted in an expected increased body weight compared to sham mice that was not different after probiotic treatment (B). Results are given as mean±SEM (n = 9–10), ** p≤0.01. Students *t* test ovx vs. sham.

Ethics Statement

All animal experiments had been approved by the local Ethical Committees for Animal Research at the University of Gothenburg.

Peripheral Quantitative Computed Tomography (pQCT)

Computed tomographic scans were performed with the pQCT XCT RESEARCH M (version 4.5B, Norland, Fort Atkinson, WI, USA) operating at a resolution of 70 μm, as described previously [24]. Cortical bone parameters were analyzed *ex vivo* in the mid-diaphyseal region of the femur.

High-resolution μCT

High-resolution μCT analyses were performed on the distal femur by using a 1172 model μCT (Bruker micro-CT, Aartselaar, Belgium). The femurs were imaged with an X-ray tube voltage of 50 kV and current of 201 μA, with a 0.5-mm aluminium filter. The scanning angular rotation was 180° and the angular increment 0.70°. The voxel size was 4.48 μm isotropically. The NRecon (version 1.6.9) was employed to perform the reconstruction following the scans. In the femur, the trabecular bone proximal to the distal growth plate was selected for analyses within a conforming volume of interest (cortical bone excluded) commencing at a distance of 538.5 μm from the growth plate, and extending a further longitudinal distance of 134.5 μm in the proximal direction. To illustrate the effect of probiotics on cortical bone in ovx mice, cortical μCT images of the diaphyseal region of one representative femur from each group were produced and are shown in Figure 2A. These CT images were derived from scans in the diaphyseal region of femur starting at a distance of 3.59 mm from the distal growth plate and extending a further longitudinal distance of 134.5 μm in the proximal direction. For BMD analysis, the equipment was calibrated with ceramic standard samples.

RNA Isolation and Real Time PCR

Total RNA was prepared from cortical bone (femur with the ends removed and bone marrow flushed out with PBS before freezing) and bone marrow using TriZol Reagent (Invitrogen, Lidingö, Sweden). The RNA was reverse transcribed into cDNA using High-Capacity cDNA Reverse Transcription Kit (#4368814, Applied Biosystems, Stockholm, Sweden). RT-PCR analyses were performed using the StepOnePlus Real-Time PCR system (Applied Biosystems). We used predesigned RT-PCR assays from Applied Biosystems (Sweden) for the analysis of IL-6 (Mm00446190_m1), IL-1β (Mm00434228_m1), TNFα (Mm00443258_m1), RANKL (Mm00441908_m1), OPG (Mm00435452_m1), Osterix (Mm04209856_m1), Col1α1 (Mm00801666_g1), osteocalcin (Mm01741771_g1) and TGFβ1 (Mm03024053_m1) mRNA levels. The mRNA abundance of each gene was calculated using the "standard curve method" (User Bulletin 2; PE Applied Biosystems) and adjusted for the expression of 18S (4308329) ribosomal RNA.

Serum and Urine Analysis

Analyses were performed according to the manufacturer's instructions for serum and urine calcium (Ca) (QuantiChrom™-Calcium Assay Kit (DICA-500), Bioassays systems, Hayward, CA, USA), serum and urine creatinine (Mouse Creatinine Kit, Crystal Chem, Downers Grove, IL, USA), serum 25-Hydroxy Vitamin D (EIA, Immunodiagnostic Systems, Herlev, Denmark). As a marker of bone resorption, serum levels of C-terminal telopeptides were assessed using an ELISA kit (Nordic Bioscience Diagnostics, Herlev, Denmark). Serum levels of osteocalcin, a marker of bone formation, were determined with a mouse osteocalcin immunoradiometric assay kit (Immutopics, San Clemente, CA).

Flow Cytometry

Bone marrow cells were harvested by flushing 5 ml PBS through the bone cavity of one femur using a syringe. After centrifugation at 473 g for 5 min, cells were resuspended in Tris-buffered 0.83% NH_4Cl solution (pH 7.29) for 5 min to lyse erythrocytes and then washed in PBS. For flow cytometry analyses, cells were extracellularly stained with BD Horizon v450-conjugated anti-CD4 (Becton Dickinson (BD), Franklin Lakes, NJ, USA) and allophycocyanin (APC) anti-CD25 (BD). By using anti-Mouse Foxp3 Staining Set (eBioscience, Vienna, Austria), Foxp3 was intracellularly stained with Phycoerythrin (PE)-conjugated anti-Foxp3, according to the manufacturer's instructions. Regulatory T cells were defined as $CD4^+CD25^+Foxp3^+$ and results are expressed as frequency of lymphocyte parent gate. Samples were run on a BD FACS Canto II and data was further processed using Flow Jo 8.8.6 software (Three Star Inc, Ashland, USA).

Statistical Analyses

We used GraphPad Prism for all statistical analysis. Results are presented as the means ± SEM. Between-group differences were calculated using unpaired t tests, ovx vs. veh. Comparisons between multiple groups were calculated using a one-way analysis of variance (ANOVA) followed by Dunnett's test to correct for multiple comparisons, within the sham and ovx groups respectively. A two-tailed $p \leq 0.05$ was considered significant.

Results

Probiotic Treatment Protects Mice from Ovx-induced Cortical Bone Loss and Increased Bone Resorption

To determine the preventive effect of probiotic treatment on ovx-induced bone-loss, eight-week-old mice were treated with vehicle (veh), a single *Lactobacillus* (L) strain (L. para) or a mixture of three strains (L. mix) during 6 weeks, starting two weeks before ovx or sham surgery (Figure 1A). Uterus weight can be used as an indicator of estrogen status and ovx resulted in an expected decrease in uterus weight that was similar for all treatments (Table 1). In addition, ovx increased body weight, fat mass and thymus weight in all treatment groups (Figure 1B, Table 1).

In the vehicle treated mice, ovx decreased the cortical bone mineral content (BMC) and cortical cross sectional bone area in the mid-diaphyseal region of femur ($p \leq 0.01$, Figure 2A, C, D). Importantly, ovx did not reduce cortical BMC or cortical cross sectional bone area in the L. para or the L. mix treated mice (Figure 2A, C, D). Cortical BMC was higher in both L. para and L. mix treated ovx mice compared to veh treated ovx mice ($p \leq 0.05$, Figure 2C). We analyzed C-terminal telopeptides to determine if the preventive effect of probiotics on cortical bone was mediated by changes in bone resorption. Ovx increased serum levels of C-terminal telopeptides in veh treated mice (+45±11%, $p \leq 0.05$ over sham) but not in L. para treated (20±9%, non-significant) or L. mix treated (23±9%, non-significant) mice (Table 2). Bone formation, as indicated by serum osteocalcin, was not significantly affected by probiotic treatment (Table 2). Trabecular bone parameters (BV/TV and trabecular BMD) in the distal metaphyseal region of femur were significantly reduced by ovx in all treatment groups ($p \leq 0.05$, Table 2, Figure 2B). These findings demonstrate that probiotic treatment protects mice from ovx-induced cortical bone loss and increased bone resorption.

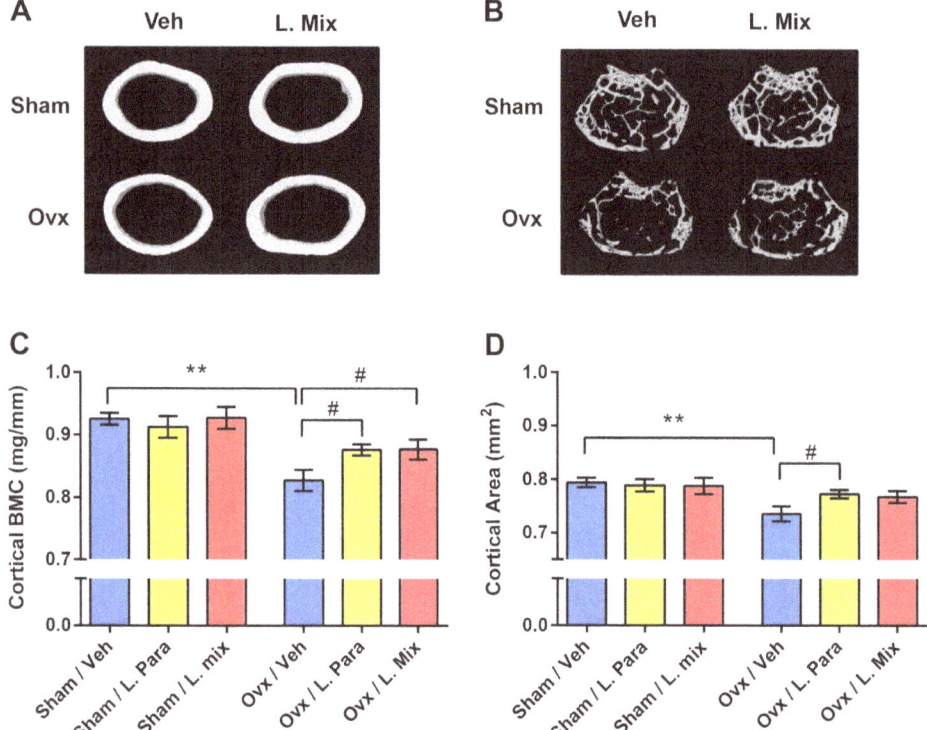

Figure 2. Probiotics Protect Mice from Ovx Induced Cortical Bone-loss. Eight-week-old mice were treated with either vehicle (veh), a single *Lactobacillus* (L) strain (L. para) or a mixture of three strains (L. mix) during 6 weeks, starting two weeks before ovx or sham surgery to study the preventive effect of probiotic treatment on ovx induced bone-loss. At the end of the experiment, dissected femurs were analysed with high-resolution μCT and peripheral quantitative computed tomography (pQCT). Representative μCT images of one cortical section from the veh and L. mix treated sham and ovx groups (A). Representative images of the trabecular bone volume (cortical bone excluded) from the distal metaphyseal region of femur (B). Cortical bone mineral content (BMC) (C) and cortical area (D) were measured by pQCT in the mid-diaphyseal region of femur. Values are given as mean±SEM, (n=9–10). ** p≤0.01, * p≤0.05. Students *t* test ovx vs. sham. # p≤0.05, ANOVA followed by Dunnett's *post hoc* test within the groups, ovx L. Para and L. mix vs. ovx veh.

Probiotic treatment reduces expression of inflammatory cytokines and the RANKL/OPG ratio in cortical bone

To investigate the mechanism for the effect of probiotic treatment on ovx-induced cortical bone loss, we measured bone related mRNA transcripts in cortical bone (Figure 3). The mRNA levels of TNFα, an inflammatory cytokine produced by immune cells that promotes osteoclastogenesis, and IL-1β, a downstream regulator of the effects of TNFα on bone, were significantly decreased by probiotic treatment compared to vehicle treatment in ovx mice (TNFα −46%, p≤0.05; IL-1β −61%, p≤0.05, Figure 3A, B). The expression of IL-6 did not differ between treatments although there was a tendency to decreased expression in the probiotic treatment group (−20%, p = 0.12, Figure 3C).

The RANKL/osteoprotegerin (OPG) ratio is a major determinant of osteoclastogenesis and bone resorption. Importantly, probiotic treatment decreased the RANKL/OPG ratio (−45%, p≤0.05 compared with veh) and this was mainly caused by an increased OPG expression (Figure 3 D-F). In contrast, the mRNA levels of three osteoblast-associated genes, Osterix, Col1α1 and osteocalcin, were not significantly affected by probiotic treatment (Figure 3G-H).

Table 1. Organ Weights.

	Sham			Ovx		
	Veh	**L. Para**	**L. Mix**	**Veh**	**L. Para**	**L. Mix**
Uterus weight (mg)	45.9±4.8	65.9±12.5	65.2±10.4	8.95±0.7**	11.0±3.2**	6.9±0.3**
Gonadal Fat (mg)	371±41	296±40	326±40	597±68*	630±28**	577±58**
Thymus weight (mg)	55.5±3.2	53.6±5.1	47.7±2.6	93.9±4.2**	90.0±5.1**	73.7±5.2** #

Eight-week-old mice were treated with either vehicle (veh), a single *Lactobacillus* (L) strain (L. para) or a mixture of three strains (L. mix) during 6 weeks, starting two weeks before ovx or sham surgery to study the preventive effect of probiotic treatment on ovx induced bone-loss. Mice were 14-weeks-old at the end of the study, when tissues were dissected and weighed. Results are given as mean±SEM, (n=6–10). ** p≤0.01, * p≤0.05, Students *t* test ovx vs. sham. # p≤0.05, ANOVA followed by Dunnett's *post hoc* test within the groups, ovx L. Para and L. mix vs. ovx veh.

Table 2. Trabecular and Cortical Bone Parameters, Serum Bone Markers and Regulatory T cells in Bone Marrow.

	Sham			Ovx		
	Veh	L. Para	L. Mix	Veh	L. Para	L. Mix
Trabecular Bone						
BV/TV (%)	16.2±0.7	16.8±0.8	17.4±0.8	13.2±0.7**	14.4±0.6*	13.8±0.5**
BMD (mg/cm^3)	322±9	331±12	344±8	285±9*	302±7*	298±7**
Tb Th (μm)	45.3±0.7	46.0±0.7	47.9±1.0	43.2±0.8	42.7±0.8**	44.6±1.0*
Tb N (mm^{-1})	3.6±0.1	3.6±0.2	3.6±0.1	3.1±0.1*	3.4±0.1	3.1±0.1*
Tb Sp (μm)	124±1	124±1	124±1	126±1	124±1	127±1
Cortical Bone						
Crt Thk (μm)	181±2	180±3	186±3	168±2**	173±2	176±2#*
Tt Ar (mm^2)	1.96±0.02	1.96±0.04	1.86±0.04	1.92±0.07	2.00±0.03	1.92±0.03
Serum Bone Markers						
C-terminal telopeptides (ng/ml)	12.6±2.2	17.6±1.6	18.4±1.7	18.2±1.4*	21.1±1.6	22.6±1.7
Osteocalcin (ng/ml)	90.9±10.4	97.1±6.7	105.6±6.1	159.9±11.8**	142.1±7.9**	136.9±6.5**
Treg (%CD4+Foxp3+CD25+)	0.117±0.023	0.109±0.017	0.090±0.017	0.054±0.004*	0.069±0.008	0.070±0.014

Eight-week-old mice were treated with either vehicle (veh), a single *Lactobacillus* (L) strain (L. para) or a mixture of three strains (L. mix) during 6 weeks, starting two weeks before ovx or sham surgery to study the preventive effect of probiotic treatment on ovx induced bone-loss. Mice were 14-weeks-old at the end of the study, when tissues were dissected and weighed. Trabecular bone parameters were analysed by high resolution μCT in the distal metaphyseal region of femur; Trabecular bone volume as a percentage of tissue volume (BV/TV); Trabecular bone mineral density (BMD); Trabecular thickness (Tb Th); Trabecular number (Tb N); Trabecular separation (Tb Sp). Cortical bone was measured by pQCT in the mid-diaphyseal region of femur; Cortical thickness (Crt Thk); Total cross-sectional area inside the periosteal envelope (Tt Ar). The resorption marker, C-terminal telopeptides and the formation marker, osteocalcin were measured in serum. Femur bone marrow cells were stained with antibodies recognizing CD4, Foxp3 and CD25. Values represent the percentage of Treg (CD4+ Foxp3+ CD25+) of gated lymphocytes. Results are given as mean±SEM, (n = 6–10). ** p≤0.01, * p≤0.05, Students *t* test ovx vs. sham. # p≤0.05, ANOVA followed by Dunnett's *post hoc* test within the groups, ovx L. Para and L. mix vs. ovx veh.

Regulatory T cells in Bone Marrow

Some of the anti-inflammatory effects exerted by probiotic bacteria are thought to be mediated via the induction of regulatory T (Treg) cells [25]. FACS analysis of bone marrow showed that the frequency of Treg (CD4$^+$CD25$^+$Foxp3$^+$) cells was decreased by ovx in veh treated but not in probiotic treated mice (Table 1). Treg cells are dependent on TGFβ for their induction and maintenance and the expression of TGFβ1 in bone marrow was increased in bone marrow in probiotic compared to veh treated ovx mice (+77±19%, p≤0.01, Figure 3J).

Mineral Metabolism

The urinary fractional excretion of Ca (FECa = (urine Ca × serum creatinine)/(serumserum Ca × urine creatinine)) was increased by ovx in veh treated mice (+86%, p≤0.05, Figure 4). Interestingly, the ovx-induced increase in FECa was completely prevented by probiotic treatment (Figure 4). Serum levels of Ca were increased after ovx in veh but not probiotic treated mice (+13%, p≤0.05, Table 3). The urine Ca/creatinine ratio was not affected by ovx in any of the treatment groups (Table 3). 25-Hydroxy Vitamin D (25(OH)D$_3$) in serum was not affected by ovx or probiotic treatment (Table 3).

Discussion

The GM regulates bone mass and probiotic treatment can affect the GM composition or the metabolic activity of the GM. In the present study we show that probiotic treatment protects mice from ovx-induced cortical bone loss. Both the L. para and the L. mix treatments protected mice from ovx-induced cortical bone loss and increased bone resorption. The urinary fractional excretion of Ca and the bone resorption marker C-terminal telopeptides in serum

were increased by ovx in veh treated but not in probiotic treated mice, suggesting that the probiotic treatments reduced bone resorption in ovx mice. Probiotic treatment reduced the expression of the two inflammatory cytokines, TNFα and IL-1β, and increased the expression of OPG in cortical bone of ovx mice. These findings indicate that probiotic treatment alters the immune status in bone, resulting in attenuated bone resorption in ovx mice.

We have earlier shown that absence of GM leads to increased bone mass in GF mice and colonisation with a normal GM rapidly normalises bone mass [15]. The increased bone mass in GF mice was associated with an altered immune status reflected by decreased expression of inflammatory cytokines in bone. Estrogen deficiency increases inflammatory cytokines and reduces OPG in bone, resulting in bone loss [26]. Ovx mice have altered immune status and bone loss, resembling post-menopausal osteoporosis. Since probiotic treatment has the capacity to modulate the immune system, we hypothesized that probiotic treatment may attenuate ovx-induced increase in inflammatory cytokines and may, thereby, preserve the bone mass in ovx mice. The probiotic strains used, *L. paracasei* DSM13434 (L. para) or a mixture of three strains, *L. paracasei* DSM13434, *L. plantarum* DSM 15312 and DSM 15313 referred to as L. mix, in the present study were selected based on their anti-inflammatory properties in an earlier study [23]. In the study by Lavasani *et al* the selected lactobacilli strains had a suppressive effect on experimental autoimmune encephalomyelitis in an animal model of multiple sclerosis. L. para and L. mix prevented ovx-induced cortical bone loss to a similar extent. We can, therefore, conclude that treatment with L. para prevents ovx-induced cortical bone loss in mice while the possible independent roles of the two used *L. plantarum* strains for cortical bone remain to be determined. We believe that the main property of the L. para that explains the reversal effect of ovx on cortical

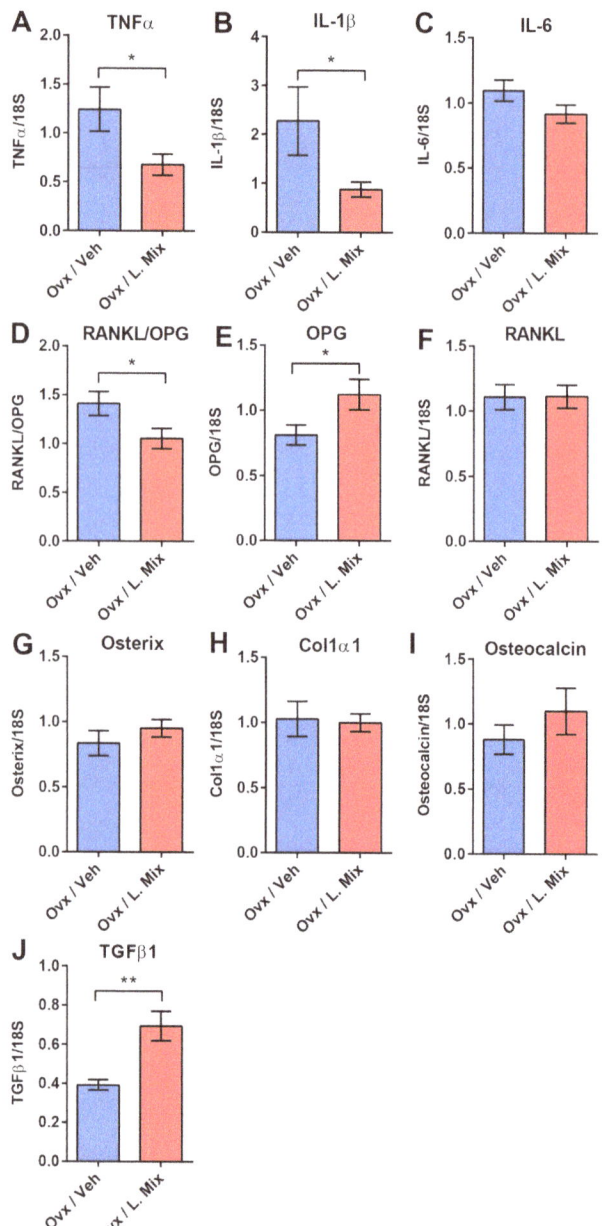

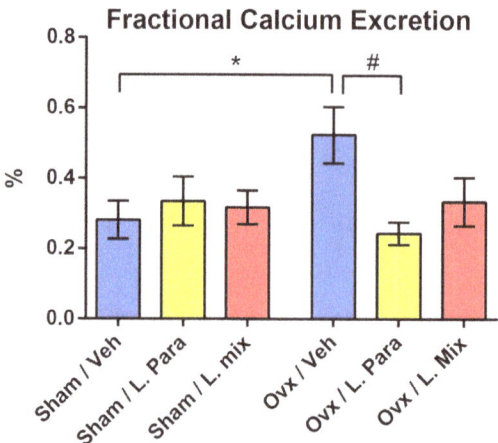

Figure 4. The Fractional Excretion of Ca was Increased by Ovx in the Veh Treated but not in the L. para or the L. mix Treated Mice. Ca and creatinine were measured in serum and urine from 14-week-old mice that had been treated with vehicle (veh), a single *Lactobacillus* (L) strain (L. para) or a mixture of three strains (L. mix) during 6 weeks, starting two weeks before ovx or sham surgery. Urinary fractional Ca excretion was calculated with the formula FECa = (urine Ca × serum creatinine)/(serum Ca × urine creatinine). Values are given as mean ± SEM, n = 5–10 in each group. * $p \leq 0.05$. Students t test ovx vs. sham. # $p \leq 0.05$, ANOVA followed by Dunnett's *post hoc* test within the groups, ovx L. Para and L. mix vs. ovx veh.

Figure 3. Probiotics Reduces Expression of Inflammatory Cytokines and the RANKL/OPG ratio in Cortical Bone and Increases Expression of TGFβ in Bone Marrow. QRT-PCR analysis of the expression of genes known to promote bone resorption; (A) Tumor Necrosis Factor alpha (TNFα), (B) Interleukin-1β (IL-1β), (C) Interleukin-6 (IL-6), (D) Ratio of Receptor activator of nuclear factor kappa-B ligand (RANKL) and Osteoprotegerin (OPG), and individual graphs for (E) OPG, (F) RANKL and genes known to promote bone formation; (G) Osterix, (H) Collagen, type I, α1 (Col1α1) and (I) osteocalcin in cortical bone and (J) transforming growth factor (TGF)β in bone marrow from 14-week-old ovariectomized (ovx) mice treated with either vehicle (veh) or a mixture of three probiotic *Lactobacillus* strains (L. mix) during 6 weeks, starting two weeks before ovx or sham surgery to study the preventive effect of probiotic treatment on ovx-induced bone-loss. Values are given as mean ± SEM, n = 9–10. ** $p \leq 0.01$, * $p \leq 0.05$ versus veh treatment, Student's t-test.

bone is its anti-inflammatory capacity. However, further studies are required to in detail characterize its protective effect on ovx-induced cortical bone loss.

Cortical bone constitutes approximately 80% of the bone in the body and several studies demonstrate that cortical bone is the major determinant of bone strength and, thereby, fracture susceptibility [27–29]. Thus, the substantial cortical bone sparing effect of probiotic treatment in ovx mice in the present study indicates that this treatment might have the capacity to reduce non-vertebral fracture risk in postmenopausal women.

As expected, ovx increased bone resorption, as indicated by elevated serum levels of C-terminal telopeptides, in veh treated mice. In contrast, serum levels of C-terminal telopeptides were not significantly affected by ovx in probiotic treated mice. Bone formation, as indicated by serum osteocalcin was not influenced by probiotic treatment. Collectively, these findings indicate that the bone sparing effect of probiotics in ovx mice might be the result of attenuated bone resorption. A possible inhibitory effect on bone resorption is supported by the finding that the urinary fractional excretion of Ca was increased by ovx in the veh treated but not in the probiotic treated mice. Estrogen therapy to post-menopausal women increases BMD associated with a reduced urinary fractional excretion of Ca [30,31]. We propose that treatment with probiotics supresses bone resorption and as a consequence the urinary fractional excretion of Ca is decreased. However, we cannot exclude other effects of probiotics on urinary fractional excretion of Ca. In addition, it has been proposed that biochemical bone markers might be more influenced by trabecular than cortical bone. Thus, it is possible that effects on the release of biochemical bone markers from cortical bone to serum might be confounded by abundant release from trabecular bone. Therefore, further analyses of other more specific cortical bone parameters, such as osteoclast number in cortical bone, are required to confirm that the protective effect of probiotics on ovx-induced cortical bone loss is mediated via altered cortical bone resorption.

Table 3. Mineral Metabolism.

	Sham			Ovx		
	Veh	L. Para	L. Mix	Veh	L. Para	L. Mix
Serum Ca (mg/dl)	9.1±0.4	9.2±0.4	8.5±0.3	10.3±0.4*	9.3±0.4	8.7±0.3#
Urine Ca/Creatinine Ratio	6.7±0.7	6.3±0.4	5.4±0.6	8.5±1.3	5.9±0.5	5.6±0.7
25(OH)D₃ (ng/ml)	16.5±1.3	17.5±1.7	16.5±1.8	16.7±1.1	17.6±1.0	16.8±1.2

Calcium (Ca) and Creatinine were measured in serum and urine and 25-Hydroxy Vitamin D (25(OH)D$_3$) was measured in serum from 14-week old mice that had been treated with vehicle (veh), a single *Lactobacillus* (L) strain (L. para) or a mixture of three strains (L. mix) during 6 weeks, starting two weeks before ovx or sham surgery. Results are given as mean±SEM, (n = 5–10). ** p≤0.01, * p≤0.05, Students *t* test ovx vs. sham, # p≤0.05, ANOVA followed by Dunnett's *post hoc* test within the groups, ovx L. Para and L. mix vs. ovx veh.

Mechanistic studies of the bone sparing effect of probiotic treatment in ovx mice revealed that the expression of several osteolytic cytokines such as TNFα and IL-1β as well as the RANKL/OPG ratio in cortical bone were suppressed by probiotic treatment. TNFα promotes osteoclastogenesis indirectly by stimulating RANKL expression by marrow stromal cells and osteoblasts and by direct stimulation of OCL precursors exposed to permissive levels of RANKL [32–34]. IL-1 is a downstream regulator of the effects of TNFα on osteoclastogenesis [35,36]. The inhibitory effect of probiotics on the RANKL/OPG ratio in the present study was mainly due to an increased expression of OPG in cortical bone. OPG directly inhibits OCL differentiation at a late stage in a dose dependant manner [37]. Together, these findings indicate that probiotic treatment suppress osteoclastogenesis, resulting in reduced OCL-mediated bone resorption.

Treg cells are critical for maintaining self-tolerance and negatively regulate immune responses. In an earlier study, Lavasani et al demonstrated that the probiotic strains used in the present study induce Treg cells [23]. Several probiotic L. strains have been described to have a therapeutic effect in experimental mouse models of inflammatory bowel disease, atopic dermatitis, and rheumatoid arthritis associated with enrichment of Treg cells in the inflamed regions [25]. This inhibitory effect of probiotic L. strains was recently shown to depend on suppressive motifs in the DNA enriched in these strains that potently prevented dendritic cell activation and maintained Treg cell conversion during inflammation [38]. TGFβ is crucial for the induction and activity of Treg cells [39]. Interestingly, ovx decreased Treg cells in bone marrow in veh but not probiotic treated mice in the present study. Furthermore, the expression of TGFβ1 was increased by probiotic compared to veh treatment

after ovx, suggesting that probiotic treatment prevents down regulation of Treg cells via an induction of TGFβ1. *In vitro* studies have shown that Treg cells directly inhibit OCL differentiation and function and that this effect of Treg cells is stimulated by estrogen and dependent on expression of TGFβ1 [40–42]. In addition, adoptive transfer of Treg cells decreases the number of OCLs and limits bone loss in ovx mice [43]. Collectively these findings may suggest that the suppressive effect of probiotic treatment on inflammatory cytokines and bone resorption involves effects by Treg cells.

In conclusion, treatment with L. para or the L. mix prevents ovx-induced cortical bone loss. Our findings indicate that these probiotic treatments alter the immune status in bone as demonstrated by reduced expression of inflammatory cytokines and increased expression of OPG, resulting in attenuated bone resorption in ovx mice. These data suggest a therapeutic potential for probiotics in the treatment of postmenopausal osteoporosis.

Acknowledgments

We would like to thank Irini Lazou Ahrén and Niklas Larsson at Probi AB, Lund, Sweden for providing the probiotic strains and Lotta Uggla and Biljana Aleksic for excellent technical assistance.

Author Contributions

Conceived and designed the experiments: CO CE FF AA UI KS. Performed the experiments: CO CE FF AA SW HF SMS KS. Analyzed the data: CO CE FF AA SW HF SMS UI KS. Contributed reagents/materials/analysis tools: CO CE FF AA SW HF SMS UI KS. Wrote the paper: CO CE FF AA SW HF SMS UI KS.

References

1. Kanis JA, Johnell O, Oden A, Sembo I, Redlund-Johnell I, et al. (2000) Long-term risk of osteoporotic fracture in Malmo. Osteoporos Int 11: 669–674.
2. Melton LJ, 3rd, Chrischilles EA, Cooper C, Lane AW, Riggs BL (1992) Perspective. How many women have osteoporosis? J Bone Miner Res 7: 1005–1010.
3. Boyle WJ, Simonet WS, Lacey DL (2003) Osteoclast differentiation and activation. Nature 423: 337–342.
4. Kong YY, Feige U, Sarosi I, Bolon B, Tafuri A, et al. (1999) Activated T cells regulate bone loss and joint destruction in adjuvant arthritis through osteoprotegerin ligand. Nature 402: 304–309.
5. Pasco JA, Kotowicz MA, Henry MJ, Nicholson GC, Spilsbury HJ, et al. (2006) High-sensitivity C-reactive protein and fracture risk in elderly women. JAMA 296: 1353–1355.
6. Ding C, Parameswaran V, Udayan R, Burgess J, Jones G (2008) Circulating levels of inflammatory markers predict change in bone mineral density and resorption in older adults: a longitudinal study. J Clin Endocrinol Metab 93: 1952–1958.
7. Schett G, Kiechl S, Weger S, Pederiva A, Mayr A, et al. (2006) High-sensitivity C-reactive protein and risk of nontraumatic fractures in the Bruneck study. Arch Intern Med 166: 2495–2501.
8. Eriksson AL, Moverare-Skrtic S, Ljunggren O, Karlsson M, Mellstrom D, et al. (2013) High sensitive CRP is an independent risk factor for all fractures and vertebral fractures in elderly men: The MrOS Sweden study. J Bone Miner Res.
9. Martin-Millan M, Almeida M, Ambrogini E, Han L, Zhao H, et al. (2010) The estrogen receptor-alpha in osteoclasts mediates the protective effects of estrogen on cancellous but not cortical bone. Mol Endocrinol 24: 323–334.
10. Nakamura T, Imai Y, Matsumoto T, Sato S, Takeuchi K, et al. (2007) Estrogen prevents bone loss via estrogen receptor alpha and induction of Fas ligand in osteoclasts. Cell 130: 811–823.
11. Charatcharoenwitthaya N, Khosla S, Atkinson EJ, McCready LK, Riggs BL (2007) Effect of blockade of TNF-alpha and interleukin-1 action on bone resorption in early postmenopausal women. J Bone Miner Res 22: 724–729.
12. Qin J, Li R, Raes J, Arumugam M, Burgdorf KS, et al. (2010) A human gut microbial gene catalogue established by metagenomic sequencing. Nature 464: 59–65.

13. Maynard CL, Elson CO, Hatton RD, Weaver CT (2012) Reciprocal interactions of the intestinal microbiota and immune system. Nature 489: 231–241.

14. Tremaroli V, Backhed F (2012) Functional interactions between the gut microbiota and host metabolism. Nature 489: 242–249.

15. Sjogren K, Engdahl C, Henning P, Lerner UH, Tremaroli V, et al. (2012) The gut microbiota regulates bone mass in mice. J Bone Miner Res 27: 1357–1367.

16. Cho I, Yamanishi S, Cox L, Methe BA, Zavadil J, et al. (2012) Antibiotics in early life alter the murine colonic microbiome and adiposity. Nature 488: 621–626.

17. Williams S, Wakisaka A, Zeng QQ, Barnes J, Martin G, et al. (1996) Minocycline prevents the decrease in bone mineral density and trabecular bone in ovariectomized aged rats. Bone 19: 637–644.

18. Pytlik M, Folwarczna J, Janiec W (2004) Effects of doxycycline on mechanical properties of bones in rats with ovariectomy-induced osteopenia. Calcif Tissue Int 75: 225–230.

19. Bron PA, van Baarlen P, Kleerebezem M (2012) Emerging molecular insights into the interaction between probiotics and the host intestinal mucosa. Nat Rev Microbiol 10: 66–78.

20. Yan F, Polk DB (2011) Probiotics and immune health. Curr Opin Gastroenterol 27: 496–501.

21. Scholz-Ahrens KE, Ade P, Marten B, Weber P, Timm W, et al. (2007) Prebiotics, probiotics, and synbiotics affect mineral absorption, bone mineral content, and bone structure. J Nutr 137: 838S–846S.

22. Thompson DD, Simmons HA, Pirie CM, Ke HZ (1995) FDA Guidelines and animal models for osteoporosis. Bone 17: 125S–133S.

23. Lavasani S, Dzhambazov B, Nouri M, Fak F, Buske S, et al. (2010) A novel probiotic mixture exerts a therapeutic effect on experimental autoimmune encephalomyelitis mediated by IL-10 producing regulatory T cells. PLoS One 5: e9009.

24. Windahl SH, Vidal O, Andersson G, Gustafsson JA, Ohlsson C (1999) Increased cortical bone mineral content but unchanged trabecular bone mineral density in female ERbeta(−/−) mice. J Clin Invest 104: 895–901.

25. Kwon HK, Lee CG, So JS, Chae CS, Hwang JS, et al. (2010) Generation of regulatory dendritic cells and CD4+Foxp3+ T cells by probiotics administration suppresses immune disorders. Proc Natl Acad Sci U S A 107: 2159–2164.

26. Clowes JA, Riggs BL, Khosla S (2005) The role of the immune system in the pathophysiology of osteoporosis. Immunol Rev 208: 207–227.

27. Zebaze RM, Ghasem-Zadeh A, Bohte A, Iuliano-Burns S, Mirams M, et al. (2010) Intracortical remodelling and porosity in the distal radius and post-mortem femurs of women: a cross-sectional study. Lancet 375: 1729–1736.

28. Holzer G, von Skrbensky G, Holzer LA, Pichl W (2009) Hip fractures and the contribution of cortical versus trabecular bone to femoral neck strength. J Bone Miner Res 24: 468–474.

29. Zheng HF, Tobias JH, Duncan E, Evans DM, Eriksson J, et al. (2012) WNT16 influences bone mineral density, cortical bone thickness, bone strength, and osteoporotic fracture risk. PLoS Genet 8: e1002745.

30. Bansal N, Katz R, de Boer IH, Kestenbaum B, Siscovick DS, et al. (2013) Influence of Estrogen Therapy on Calcium, Phosphorus, and Other Regulatory Hormones in Postmenopausal Women: The MESA Study. J Clin Endocrinol Metab.

31. McKane WR, Khosla S, Burritt MF, Kao PC, Wilson DM, et al. (1995) Mechanism of renal calcium conservation with estrogen replacement therapy in women in early postmenopause – a clinical research center study. J Clin Endocrinol Metab 80: 3458–3464.

32. Hofbauer LC, Lacey DL, Dunstan CR, Spelsberg TC, Riggs BL, et al. (1999) Interleukin-1beta and tumor necrosis factor-alpha, but not interleukin-6, stimulate osteoprotegerin ligand gene expression in human osteoblastic cells. Bone 25: 255–259.

33. Ochi S, Shinohara M, Sato K, Gober HJ, Koga T, et al. (2007) Pathological role of osteoclast costimulation in arthritis-induced bone loss. Proc Natl Acad Sci U S A 104: 11394–11399.

34. Lam J, Takeshita S, Barker JE, Kanagawa O, Ross FP, et al. (2000) TNF-alpha induces osteoclastogenesis by direct stimulation of macrophages exposed to permissive levels of RANK ligand. J Clin Invest 106: 1481–1488.

35. Zwerina J, Redlich K, Polzer K, Joosten L, Kronke G, et al. (2007) TNF-induced structural joint damage is mediated by IL-1. Proc Natl Acad Sci U S A 104: 11742–11747.

36. Wei S, Kitaura H, Zhou P, Ross FP, Teitelbaum SL (2005) IL-1 mediates TNF-induced osteoclastogenesis. J Clin Invest 115: 282–290.

37. Simonet WS, Lacey DL, Dunstan CR, Kelley M, Chang MS, et al. (1997) Osteoprotegerin: a novel secreted protein involved in the regulation of bone density. Cell 89: 309–319.

38. Bouladoux N, Hall JA, Grainger JR, dos Santos LM, Kann MG, et al. (2012) Regulatory role of suppressive motifs from commensal DNA. Mucosal Immunol 5: 623–634.

39. Marie JC, Letterio JJ, Gavin M, Rudensky AY (2005) TGF-beta1 maintains suppressor function and Foxp3 expression in CD4+CD25+ regulatory T cells. J Exp Med 201: 1061–1067.

40. Kim YG, Lee CK, Nah SS, Mun SH, Yoo B, et al. (2007) Human CD4+CD25+ regulatory T cells inhibit the differentiation of osteoclasts from peripheral blood mononuclear cells. Biochem Biophys Res Commun 357: 1046–1052.

41. Zaiss MM, Axmann R, Zwerina J, Polzer K, Guckel E, et al. (2007) Treg cells suppress osteoclast formation: a new link between the immune system and bone. Arthritis Rheum 56: 4104–4112.

42. Luo CY, Wang L, Sun C, Li DJ (2011) Estrogen enhances the functions of CD4(+)CD25(+)Foxp3(+) regulatory T cells that suppress osteoclast differentiation and bone resorption in vitro. Cell Mol Immunol 8: 50–58.

43. Buchwald ZS, Kiesel JR, Yang C, DiPaolo R, Novack DV, et al. (2013) Osteoclast-induced Foxp3+ CD8 T-cells limit bone loss in mice. Bone 56: 163–173.

9

Bifidobacterium breve Attenuates Murine Dextran Sodium Sulfate-Induced Colitis and Increases Regulatory T Cell Responses

Bin Zheng[1]**, Jeroen van Bergenhenegouwen**[1,2]**, Saskia Overbeek**[1]**, Hendrik J. G. van de Kant**[1]**, Johan Garssen**[1,2]**, Gert Folkerts**[1]**, Paul Vos**[2]**, Mary E. Morgan**[1]**, Aletta D. Kraneveld**[1]*

1 Division of Pharmacology, Utrecht Institute for Pharmaceutical Science, Faculty of Science, Utrecht University, Utrecht, The Netherlands, **2** Nutricia Research, Utrecht, The Netherlands

Abstract

While some probiotics have shown beneficial effects on preventing or treating colitis development, others have shown no effects. In this study, we have assessed the immunomodulating effects of two probiotic strains, *Lactobacillus rhamnosus (L. rhamnosus)* and *Bifidobacterium breve (B. breve)* on T cell polarization *in vitro*, using human peripheral blood mononuclear cells (PBMC), and *in vivo*, using murine dextran sodium sulfate (DSS) colitis model. With respect to the latter, the mRNA expression of T cell subset-associated transcription factors and cytokines in the colon was measured and the T helper type (Th) 17 and regulatory T cell (Treg) subsets were determined in the Peyer's patches. Both *L. rhamnosus* and *B. breve* incubations *in vitro* reduced Th17 and increased Th2 cell subsets in human PBMCs. In addition, *B. breve* incubation was also able to reduce Th1 and increase Treg cell subsets in contrast to *L. rhamnosus*. *In vivo* intervention with *B. breve*, but not *L. rhamnosus*, significantly attenuated the severity of DSS-induced colitis. In DSS-treated C57BL/6 mice, intervention with *B. breve* increased the expression of mRNA encoding for Th2- and Treg-associated cytokines in the distal colon. In addition, intervention with *B. breve* led to increases of Treg and decreases of Th17 cell subsets in Peyer's patches of DSS-treated mice. *B. breve* modulates T cell polarization towards Th2 and Treg cell-associated responses *in vitro* and *in vivo*. *In vivo B. breve* intervention ameliorates DSS-induced colitis symptoms and this protective effect may mediated by its effects on the T-cell composition.

Editor: Ivan J. Fuss, National Institutes of Health, United States of America

Funding: This study was performed within the framework of Dutch Top Institute Pharma (project D1-101). The funders had no role in study design, data collection and analysis, decision to publish, or preparation of the manuscript.

Competing Interests: Prof. Johan Garssen, besides being employed as a professor at Utrecht University, is also employed by Danone Research B.V. as a research manager. The research presented in this manuscript, however, was performed independently of his Danone Research B.V. duties. Paul Vos and Jeroen van Bergenhenegouwen are also employed by Danone Research, however they were not directly involved in the analysis of the data. The remaining authors have no competing interests.

* E-mail: A.D.Kraneveld@uu.nl

Introduction

Inflammatory bowel disease (IBD) is a chronic inflammatory disease that affects the gastrointestinal tract and consists of two major forms, Crohn's disease (CD) and ulcerative colitis (UC). Although the exact mechanisms of IBD development still remain to be elucidated, a feature that is common to IBD pathogenesis is a dysregulated effector T cell response to the commensal microflora [1,2]. T cells are important components of the adaptive immune system. Upon activation, T cells expand and differentiate into various effector CD4+ T cells such as Th1, Th2, Th17 cells, and Treg cells. The differentiation of these T cell subsets is induced by the specific transcription factors T-bet [3], GATA3 [4], RORγt [5] and Foxp3 [6,7], respectively.

Until recently, the classical T cell subsets (Th1 and Th2) have been considered the major players during the development of IBD. However, there is an increasing body of evidence showing the importance of the Th17 pathway in IBD [2]. Th17 cells are characterized by RORγt expression and IL17 production [5,8], and increased Th17 cells have been found in IBD patients [9,10].

Although the development of Th17 cells is independent of the Th1 and Th2 program, it shares the same requirement for TGFβ with Treg cells [11]. Treg cells have a unique regulatory function by suppressing the activity of other T cell subsets (Th1, Th2 and Th17 cells) and, thereby, helping control autoimmunity [12]. In contrast to Th17 cells, decreased amounts of Treg cells have been found in the peripheral blood of IBD patients as compared to normal controls [13,14]. In addition, increased apoptosis of Treg cells was found in the inflamed mucosa of IBD patients compared to non-inflamed control colons [15]. Murine models of IBD have further illustrated the protective effects of Treg cells during colitis. Immunodeficient mice that are adoptively transferred with Treg-depleted naïve CD4+ T cells develop spontaneous colitis; in contrast, mice transferred naïve CD4+ T cells combined with Treg cells do not develop colitis [16,17]. Additionally, Mice lacking interleukin (IL)-10, an important anti-inflammatory cytokine needed for both the induction of Treg cells and their effector function, spontaneously develop colitis [18].

In the last decade, products supplemented with live bacteria, called probiotics, have become increasingly popular [19]. The use of probiotics has been proposed to be beneficial for human health and there is increased interest for their use in IBD. This is due to the beneficial effect of probiotic treatment in other intestinal diseases such as traveler's diarrhea and antibiotic-associated diarrhea [20]. However, the working mechanisms of probiotics still need to be elucidated. Gut-derived bacteria from the genera *Lactobacillus* and *Bifidobaterium* are the most studied probiotics. Diverse effects of the probiotics have been demonstrated using human cell culture systems and animal models and one of the most important effects is their ability to modulate immune responses [21]. Studies using human peripheral blood mononuclear cells (PBMC) have demonstrated the abilities of gut-derived bacteria to modulate T cell polarization by inducing different T-cell subsets including Treg cells in a strain dependent manner [22,23]. Moreover, two independent clinical studies using two different *Bifidobacteria* strains have demonstrated their immune modulating capacities by both enhancing the TGFβ signaling and increasing peripheral Treg cells numbers [24,25].

Recently, Plantinga *et al* assessed the cytokine production of PBMC stimulated with two probiotic strains, *L. rhamnosus* and *B. breve*. Exposure to either bacterial strain led to increased IL-10 levels. In addition, exposure to *B. breve* led to a reduction of IFNγ production, a Th1-associated cytokine, as compared to the *L. rhamnosus* [26]. In this study, we further investigated the same probiotic strains by examining their effects on CD4+ T cell differentiation both *in vitro* and *in vivo*. We demonstrated that both strains had the ability to shift CD4+ T cell polarization in stimulated PBMCs away from Th17 cell development towards Th2 differentiation. In addition, *B. breve* induced the development of Treg cells while decreasing the development of Th1 cells. Administering these bacterial strains in the DSS-induced colitis model showed that while *L. rhamnosus* had little effect on disease severity, *B. breve* ameliorated DSS-induced colitis, increased Treg- and Th2-associated responses and locally reduced CD4+RORγt+ Foxp3- T cells while simultaneously increasing CD4+ RORγt- Foxp3+ T cells.

Methods and Materials

Human peripheral blood mononuclear cell stimulations

Human PBMCs were isolated from buffy coats, which were obtained from the Sanquin blood bank (Utrecht, the Netherlands). The cell fraction containing PBMCs was obtained by density centrifugation of 1:3 diluted buffy coats on Ficoll-Paque PLUS (GE Healthcare, Eindhoven, the Netherlands). Subsequently, the obtained cells were washed with phosphate buffered saline (PBS; Lonza Verviers SPRL, Verviers, Belgium) and the erythrocytes were lysed using sterile lysis buffer (0.15M NH_4Cl, 0.01 M $KHCO_3$ and 0.1mM EDTA, pH 7.4). After lysis, the remaining cells (PBMCs) were washed again with PBS supplemented with 2% heat-inactivated Fetal Calf Serum (FCS; Lonza Verviers SPRL, Verviers, Belgium) and resuspended in Roswell Park Memorial Institute (RPMI) 1640 medium (Lonza Verviers SPRL, Verviers, Belgium) supplemented with 2.5% FCS, 1% penicillin/ streptomycin, 1 mM pyruvate and 50 µg/ml gentamicin.

A total of 10^5 PBMCs were incubated either with anti-CD3 (Sanquin, Amsterdam, the Netherlands) alone (at a final concentration of 1:10.000) or in combination with *L. rhamnosus* or *B. breve*. Both *L. rhamnosus* (NutRes 1 formerly known as NumRes 1) and *B. breve* (NutRes 204 formerly known as NumRes 204) were provided by Danone Research BV (Wageningen, the Netherlands) as live bacteria in a 20% glycerol stock. The bacteria:PBMC ratio was 20:1 and incubated in 96-well plates (Greiner bio-one, Stonehouse, UK) at 37°C for 48 hours or 7 days.

Experimental colitis and administration of probiotics

Female C57BL/6 mice were purchased from Charles River Laboratories (Maastricht, the Netherlands). All mice were used at 8–12 weeks of age and were housed under standard conditions in the animal facilities at Utrecht University.

Experimental colitis was induced by adding 1.5% DSS to the drinking water for 5 days. 10^9/dose of *L. rhamnosus* or *B. breve* probiotics were administrated by oral gavage every two days, starting 9 days prior to the DSS treatment and continued to the end of the experiment. Colitis development was monitored by measuring the weight and the fecal condition. The fecal condition was measured on day 0, 3 and 5. On day 6, the mice were sacrificed and the colons and Peyer's patches were isolated for further analysis.

The severity of the colitis was determined by calculating the body weight change, feces condition and the colon length. The body weight change was determined by calculating the percentage of weight change relative to the starting weight before DSS treatment on day 0. The fecal condition score was determined using two parameters: stool consistency (0 = normal, 1 = soft with normal form, 2 = loss of form/diarrhea) and fecal bleeding (0 = no blood, 1 = blood observation using Colo-rectal Test kit (Axon Lab AG, Germany), 2 = blood observation without test).

After sacrificing the mice, the colons were excised between the ileocaecal junction and rectum and were prepared for histological evaluation. The colon was opened longitudinally, placed on a piece of blotting paper, and fixed in 10% formalin. After fixing, the colons were rolled, paraffin-embedded, and sectioned (5 µm). Two researchers assessed general inflammatory features blindly after staining sections with hematoxylin and eosin according the assessment system described below. Assessments included four pathological criteria: the extent of cellular infiltration (0: no infiltration, 1: infiltration between the crypts, 2: infiltration in the submucosa, 3: infiltration in the muscularis externa, 4: infiltration in entire tissue); cover area of cellular infiltration in the region (0: no infiltration, 1: < 25%, 2: 25%–50%, 3: 50%–75%, 4: >75%); loss of crypts (0: no damage, 1: 30% shortening of crypts, 2: 65% shorting of crypts, 3: total loss of crypts, 4: loss of entire epithelial layer); extent of crypts loss in the region (0: no crypt loss, 1: < 25%, 2: 25% – 50%, 3: 50%–75%, 4: > 75%).

Ethics statement

All experiments were performed in accordance with the guidelines issued by the Dutch ethics committee for animal studies. The protocol was specifically approved by the ethics committee for animal studies of Utrecht University (DEC approval number 2009.II.06.046). All efforts were made to minimize suffering.

Immunohistochemical staining

A subset of the mice from each group was examined using immunohistochemistry. After sacrificing the mice, the colons were opened longitudinally and half of each colon was fixed in 10% formalin, rolled, paraffin-embedded, and sectioned (5 µm). The sections were subjected to a heat-induced epitope retrieval step. Slides were washed with PBS and blocked with rabbit or goat serum before an overnight incubation (4°C) with primary antibodies against Ly-6B (AbD Serotec, Dusseldorf, Germany), RORγt (eBioscience San Diego, CA USA) or Foxp3 (eBioscience San Diego, CA USA). For detection, biotinylated goat anti-rat (Dako, Glostrup, DK) secondary antibodies were administered followed by incubation with peroxidase-labeled streptavidin (Vectastain EliteABC kit, Vector, Burlingame, CA USA). The peroxidase activity was visualized using the substrate, DAB

(Sigma, Gillingham, UK). The cell nuclei were visualized by a short incubation with Mayer's hematoxylin (Klinipath, Duiven, the Netherlands). Background staining was determined by substituting the primary antibody with a rat IgG isotype control (Abcam, Cambrige, UK).

The number of Foxp3+ cells was quantified by counting positive cells within the lamina propria area excluding the induced and tertiary lymphoid follicle regions. The density of RORγt+ cell was determined as follows: RORγt+ cells were counted in colonic patches and quantified as a function of $0.01mm^2$ colonic patch area.

MPO measurement

A subset of the mice from each group was used to determine the MPO concentration in the colon. After sacrificing the mice, the colons were opened longitudinally and half of each colon was transferred into RIPA buffer (Thermo Scientific, Rockford, IL USA) and homogenized using a Precellys 24-Dual homogenizer (Precellys, Villeurbanne, France). The homogenates were centrifuged at 14000 rpm for 10 minutes at 4°C and the MPO concentration in the supernatant was measured using an ELISA kit according to the manufacturer's protocol (Hycult biotech, Uden, the Netherlands).

Real-time PCR

A subset of the mice from each group was used to determine the mRNA expression of a selection of genes in the colon. After sacrificing the mice, Total RNA of 1 cm distal colon pieces was isolated using the RNAeasy kit (Qiagen, Germantown, MD USA) and, subsequently, reverse transcribed into cDNA using the iScript cDNA synthesis kit (BioRad, Hercules, CA USA). Real-time PCR quantification was performed using the iQ SYBR Green super mix kit (BioRad, Hercules, CA USA) with the CFX 96 Real-time system (BioRad, Hercules, CA USA) and the RNA expression was determined using built-in detection system of CFX 96 Real-time system (BioRad, Hercules, CA USA). The RNA expression value and normalized gene expression ($\Delta\Delta C_T$) value was calculated using the built-in gene expression analysis module in CFX Manager software version 1.6. The sequence of specific primers for T cell transcription factor genes and the gene for the household protein *ribosomal protein S13 (Rps13)* are listed in Table 1. The primers for the cytokines: *interferon gamma (Ifnγ), Il12p35, Il4, Il5, Il13, Il23p19, Il17, Il6, Tgfβ* and *Il10* were purchased from SABioscience (Frederick, MD USA). The final data for the target samples were normalized against the internal control *Rps13*.

Intracellular staining for cytokines and transcription factors

The isolated human PBMCs were incubated for 48 hour or 7 days as described. The PBMCs, which were incubated for 48

hours, were stained first extracellularly with antibodies for CD4 and CD69, followed by intracellular staining for GATA3, RORγt, FOXP3 and T-bet. The Fluorescent Minus One (FMO) control of each marker was determined by taking out the indicated marker antibody during the staining of control human PBMCs. In addition, the possible background of each marker antibody within CD4 cells was determined by substituting the indicated antibody with an appropriate isotype antibody with a matching fluorescent label.

The PBMCs, which were incubated for 7 days, were provided with fresh culture medium for 24 hours and then subsequently stimulated with PMA (50ng/ml) and ionomycin (750ng/ml) in the presence of Brefeldin A (eBioscience, San Diego, CA USA) for 4 hours. After stimulation, PBMCs were first stained extracellularly with anti-CD4, followed by intracellular staining for IL-4, IL-17, IL-10 and IFNγ. The FMO controls and isotype controls of these marker antibodies were also assessed as described in the previous paragraph.

Peyer's patches isolated from the mice of experimental colitis study were prepared as single-cell suspensions by passing through a 0.75 μm cell strainer. Cells were first stained extracellularly with antibodies for CD4, followed by intracellular staining for Foxp3 and RORγt. The FMO controls and isotype controls of these marker antibodies were examined in mLN cells obtained from non-treated mice. All antibodies and intracellular staining buffers were obtained from eBioscience (San Diego, CA USA). All samples were read on a BD FACSCanto II (BD Biosciencse, Franklin Lakes, NJ USA) and the data were analyzed using BD FACSDiva software (BD Biosciences). The (activated) T cells were determined by gating on CD4+ (CD69+) cells. Subsequently, the different T cells subsets were defined on the found T cell subset associated transcription factors and the specific cytokine producing T cells were found by gating on appropriated cytokines.

Statistical analysis

Means with SEM are represented in each graph. Statistical analysis was performed using GraphPad Prism version 5.0 for windows (GraphPad Software, San Diego, CA USA). P-values were calculated using either the 2-way ANOVA followed by Bonferroni post-tests or a Mann-Whitney test. P-values considered as significant are indicated as ***<0.001, **<0.01, and *<0.05.

Results

L. rhamnosus and *B. breve* reduce Th17 differentiation in PBMCs

To assess the immunomodulatory capacity of the bacterial strains, PBMCs were stimulated with a combination of anti-CD3 together with *L. rhamnosus* or *B. breve*, and the different T cells subsets were analyzed using flow cytometry. Differences were

Table 1. qPCR primer sequences.

Primer Sequence 5'->3'		
	Forward primer	**Reverse primer**
Gata3	GCGGTACCTGTCTTTTTCGT	CACACAGGGGCTAACAGTCA
Foxp3	CACTGGGCTTCTGGGTATGT	AGACAGGCCAGGGGATAGTT
Rorc	TGCAAGACTCATCGACAAGG	AGGGGATTCAACATCAGTGC
Rps13	GTCCGAAAGCACCTTGAGAG	AGCAGAGGCTGTGGATGACT

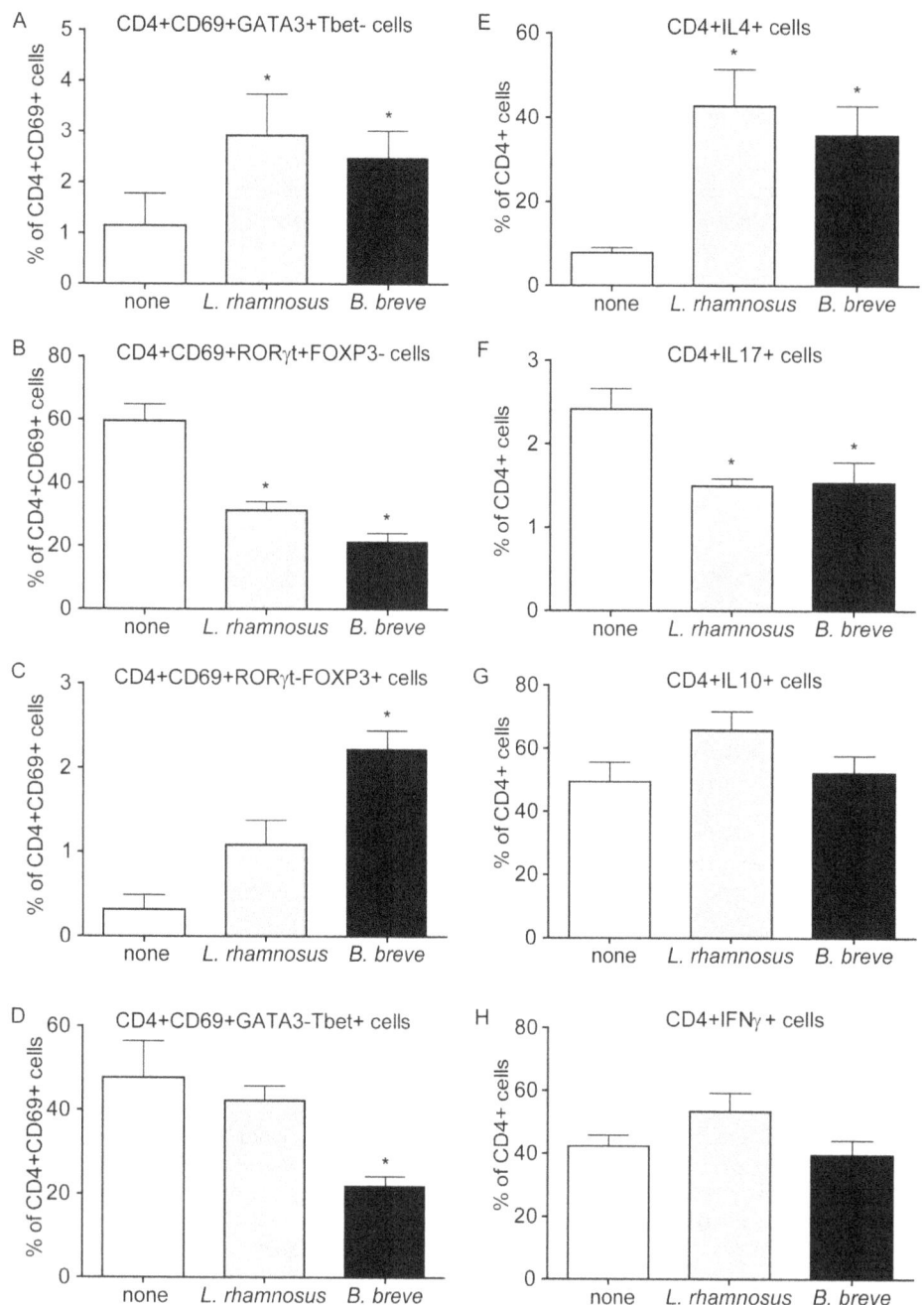

Figure 1. *L. rhamnosus* and *B. breve* alter T cell differentiation in human PBMCs. PBMCs were stimulated with anti-CD3 alone (white bars), with a combination of anti-CD3 and *L. rhamnosus* (grey bar) or a combination of anti-CD3 and *B. breve* (black bar) for 48 hours or 7 days. A–D) The percentages of Th2 (GATA3+Tbet-), Th17 (RORγ+FOXP3-), Treg (RORγ-FOXP3+) or Th1 (GATA-Tbet+) cells within the activated T cells (CD4+CD69+) in the PBMCs were determined after 48 hours of incubation. Percentages within activated CD4+CD69+ T cell population are shown. E–H) The percentages of cytokines (IL10, IL17, IL4 or IFNγ) producing CD4+ T cells in the PBMCs were determined after 7 days of incubation. Percentages within CD4+ T cell population are shown. The Results are expressed as mean + SEM, n = 3, * p<0.05.

found in the T cell subtype composition within the activated CD4+CD69+ T cells (Figure S1 and Figure 1A – D). Both strains significantly increased Th2 (CD4+CD69+GATA3+ Tbet-; Figure 1A) and decreased Th17 (CD4+CD69+RORγ+FOXP3-; Figure 1B) cell subsets. Incubation with *B. breve*, but not *L. rhamnosus*, led to a significantly increased Treg (CD4+CD69+ RORγ-FOXP3+; Figure 1C) and decreased Th1 cell (CD4+ CD69+ GATA3+Tbet-; Figure 1D) subsets. To further confirm

the changes in T cell subsets, the IL-4, IL-17, IL-10 and IFNγ producing CD4+ T cells within total PBMCs were analyzed after 7 days of stimulation (Figure S2 and Figure 1E – H). Cytokine expression of CD4+ T cells stimulated with anti-CD3 and the bacteria mirrored the results seen with the transcription factors. Both *L. rhamnosus* and *B. breve* significantly increased the population of CD4+IL-4+ T cells and decreased the population of CD4+ IL17+ T cells (Figure 1E and F). No changes were observed in the

CD4+IL-10+ and CD4+IFNγ+ T cell populations for both bacteria (Figure 1G and H). The results of FMO controls and isotype controls indicate that the staining antibodies were working sufficiently and that we used proper gate-settings (Figure S1 and S2).

These data indicate that *L. rhamnosus* as well as *B. breve* are able to limit the differentiation of CD4+ T cells *in vitro* towards Th17 cells. Additionally, *B. breve* induced *de novo* Treg induction and reduced Th1 cells.

Intervention with *B. breve*, but not *L. rhamnosus*, ameliorates DSS-induced colitis

To study the effect of the *L. rhamnosus* and *B. breve* strains *in vivo*, the murine DSS-induced colitis model was used. Mice received *L. rhamnosus* or *B. breve* 9 days prior to colitis induction and the bacterial administration was continued until the end of the experiment. Control mice receiving bacteria did not display any clinical changes (data not shown). DSS treatment increased feces condition score, histology score and mildly reduced body weight and the colon length. Intervention with *B. breve*, but not *L. rhamnosus* led to improvement of feces condition and to a significant reduction of DSS-induced colon shortening, colon epithelial damage and cellular infiltration as compared to mice with DSS treatment alone (Figure 2A–D).

In order to visualize changes in infiltrating inflammatory cells in the colon after DSS treatment, immunohistochemistry was employed to determine the number of cells expressing Ly-6B, which is expressed on the surface of neutrophils and inflammatory macrophages [27]. DSS treatment significantly enhanced the infiltration of Ly-6B+ cells. Mice treated with DSS and *B. breve* intervention tended to have reduced amounts of Ly-6B+ cells in the colon (Figure 2E). Consistent with the Ly-6B staining, quantification of MPO concentration (an indicator for neutrophil influx) in the colon showed that DSS treatment significantly increased the MPO concentration in colon of colitis mice. Intervention with *B. breve* reduced the MPO expression by approximately 35%, although no significant different was determined (Figure 2F).

These data indicate that *B. breve* intervention leads to improvements in the outcome of DSS-induced colitis in mice.

B. breve intervention enhances the mRNA expression of Th2- and Treg-associated cytokines in distal colon

As both *L. rhamnosus* and *B. breve* were able to alter T cell differentiation *in vitro*, we investigated if *L. rhamnosus* and *B. breve* induced similar changes *in vivo* during colitis. DSS-induced colitis, on its own, significantly increased the mRNA expression of *Ifnγ*, *Il6*, *Il17* and *Tgfβ* as compared to controls. *L. rhamnosus* intervention did not modulate the transcription of cytokines in healthy control mice nor DSS-treated mice, except for a significant increase of *Il5* in DSS-treated mice (data not shown).

B. breve administration in healthy control mice, on the other hand, significantly increased mRNA transcription of Th2- (*Il4*, *Il5* and *Il13*) and Treg- (*Il10* and *Tgfβ*) associated cytokines as well as *Il23* in the colon. In contrast, Th1- (*Ifnγ* and *Il12*) associated cytokines (Figure 3A) were unaffected. *B. breve* intervention of DSS-treated mice induced a similar mRNA cytokine expression pattern in the colon as healthy control mice with *B. breve intervention*. However, the expression was more pronounced and significantly increased *Il6* and *Il17* mRNA expression levels were observed (Figure 3B).

These results demonstrate that *B. breve* intervention alters mRNA expression patterns in the colon and increased the mRNA expression of *Il6* and *Il17*, and Th2 and Treg-associated cytokines.

B. breve intervention leads to increased numbers of Foxp3+ cells in the colon and altered Treg and Th17 cell populations in the Peyer's patches during colitis

As intervention with *B. breve* led to significant changes in cytokine transcription that were indicative for skewing in the T cell response towards a Th2 and Treg response combined with a Th17 response, we assessed the mRNA expression of Th17-, Th2- and Treg-associated transcription factors; *Rorc*, *Gata3* and *Foxp3*, respectively, in the colon. Significantly increased *Gata3* and *Foxp3* mRNA expression levels were detected in both healthy and DSS-treated mice receiving *B. breve*, while no different was detected for *Rorc* expression in both healthy and DSS-treated mice receiving *B. breve* (Figure 4A).

To determine whether increases in the regulatory T cell response caused by *B. breve* intervention were also reflected by an increased number of Foxp3+ cells in the colon, we visualized and quantified colon Foxp3+ cells using immunohistochemistry. Indeed, increased numbers of Foxp3+ cells were found in the colon of DSS-treated mice with *B. breve* intervention (Figure 4B).

It has been shown that the conditions, which favor Treg development, naturally antagonize Th17 polarization [28]. Since Th17 cells express the transcription factor RORγt [5], we also examined the numbers of RORγt+ cells in the colon using immunohistochemistry. RORγt+ cells were found primarily in the lymphoid follicles and we analyzed the number of these cells per 0.01 mm^2 colonic patch. When analyze the effect of *B. breve* intervention on the number of RORγt+ cells in colonic patches, taking the *B. breve* intervention and exposure to water or DSS together, a trend of decreasing the amount of RORγt+ cells was observed in mice with *B. breve* (Two way ANOVA: $F_{1,8} = 4,29$ $p = 0.07$, Figure 4 C).

Although only a trend in reducing the number of RORγt+ cells by *B. breve* was observed in the colonic patches, analysis of CD4+ T cells within GALT, namely the Peyers patches of the small intestine, using flow cytometry revealed that *B. breve* intervention significantly decreased the Th17 (CD4+RORγt+Foxp3-) cell subset in Peyer's patches and significantly increased the Treg (CD4+RORγt-Foxp3+) cell subset (Figure 4D and Figure S3).

These results indicate that *B. breve* intervention is capable of increasing the Treg cell population and decreasing Th17 cells in the GALT during colitis.

Discussion

While the incidence rate of IBD has increased [29], there is still no curative therapy for IBD, and the treatments that do exist focus mainly on relieving symptoms and often lead to unwanted side-effects [30]. In the last decade, probiotics, defined as "live microorganisms that when administrated in adequate amounts, confer a health benefit on the host", have been proposed as potential candidates for IBD treatment. The increased interest in the immunomodulatory properties of specific probiotic strains stems from the success of using probiotics to treat a varied number of intestinal diseases [20]. Since a dysregulated T cell response is a common feature in IBD [31], we assessed the capability of two probiotic strains, *L. rhamnosus* and *B. breve* to modulate the development of different T cell subsets *in vitro*, using PBMCs isolated from healthy volunteers. In addition, the effect of these specific bacterial strains on the experimental colitis and the development of different T cell subset *in vivo* have been assessed.

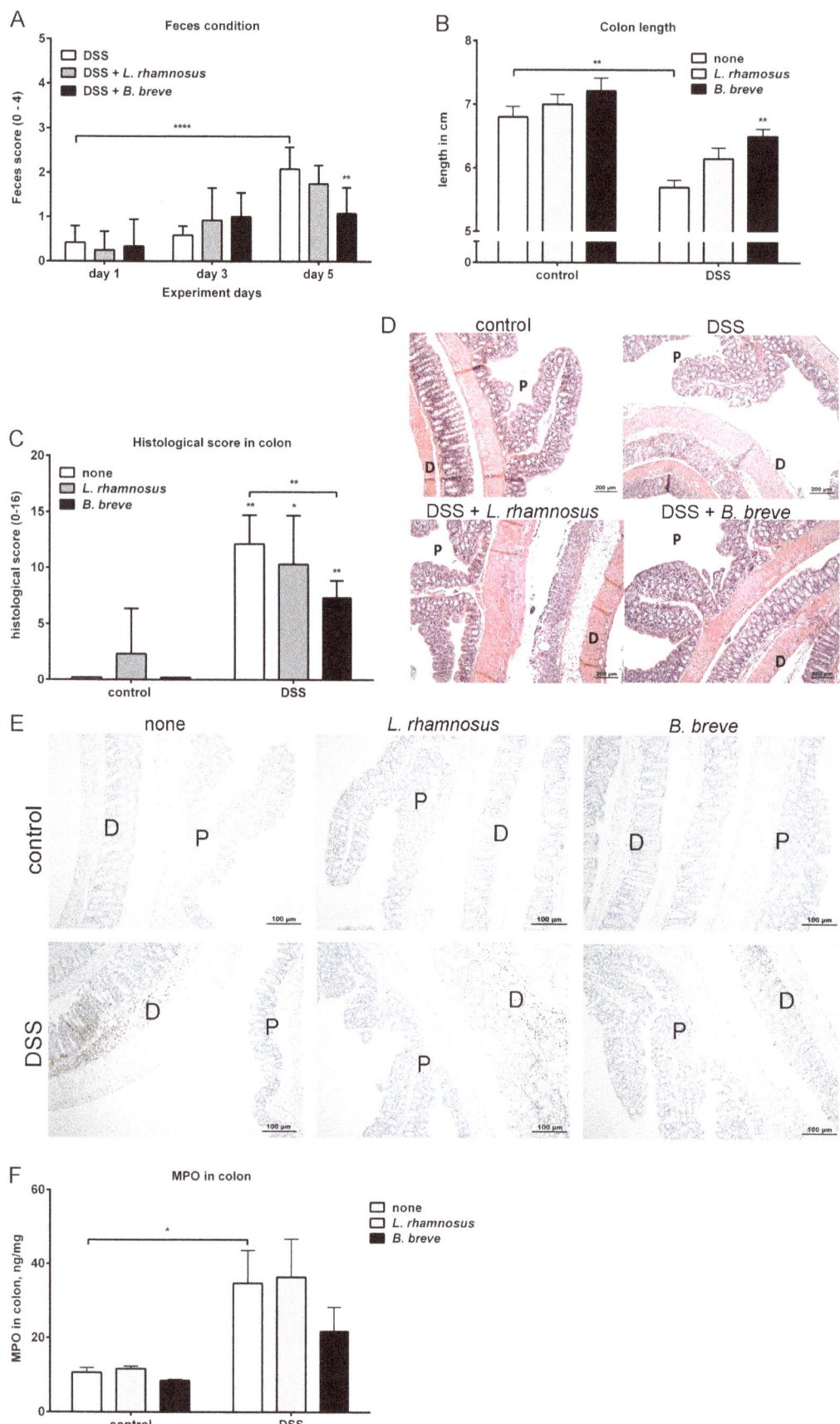

Figure 2. *B. breve*, but not *L. rhamnosus*, ameliorates DSS-induced colitis. C57BL/6 mice with or without probiotics treatment received either normal drinking water or drinking water with DSS for 5 days. A) The fecal condition was calculated on day 0, day 3 and day 5 after DSS treatment. On day 6, the mice were sacrificed and B) the colon length of each mouse was measured. Results are expressed as mean ± SEM, n = 6 mice per group, pooled from two independent experiments. Colons were collected and examined for histological score as described in materials and methods. C) The histological scoring graph and D) representative H&E staining photos are shown. Results are expressed as mean + SEM, n = 3 mice per group, pooled from two independent experiments. E) The presence of Ly-6B+ cells was visualized in the proximal (p) and distal (d) colons using immunohistochemistry. The pictures are representative of 3 separate mice per group obtained from two experiments. F) The concentration of MPO was measured in colon homogenates of each group. Results are expressed as mean + SEM, n = 4 mice per group, pooled from two independent experiments. * p<0.05; ** p<0.01.

We hypothesize that these specific gut-derived bacterial strains could have protective effects on experimental colitis via their capability to modulate the development of different T cell subsets.

Our results were generally consistent with the results from previous study by Plantinga *et al* concerning the same bacteria [26]. In our study, significantly decreased CD4+CD69+RORγt+ FOXP3 and CD4+IL-17+ T cell subsets were observed in PBMCs stimulated with both bacteria, however, only *B. breve* stimulation led to a reduction of the Th1 cell subset. In addition, we found *B. breve* stimulation significantly increased the FOXP3+ Treg cell subset, which is often associated with anti-inflammatory effects [12], suggesting an anti-inflammatory property of this bacterial strain.

The increased Th2 cell subset found in PBMCs stimulated with *B. breve* may contribute to the decreased Th1 cell subset due to the mutual antagonizing effects of Th1 and Th2 cells on each other [32]. Increased Th2 cells and CD4+IL4+ T cell subsets were also observed in PBMCs stimulated with *L. rhamnosus*, but no change in the Th1 cell subset was seen suggesting that *L. rhamnosus* may have a different T cell modulating mechanism.

The capability of *B. breve* to alter T cell differentiation by inducing Treg cell and reducing Th17 cell development *in vitro* indicates that using this specific bacterial strain *in vivo* may have a protective function in IBD. Murine colitis models are a useful tool to examine the clinical efficacy and possible working mechanism of probiotics in the development of IBD. A multitude of *Bifidobacteria* strains have shown protective effect in colitis models. For example, a mixture of probiotics including *Bifidobacterium longum* induces Treg cell expansion and prevents trinitrobenzene sulfonic acid (TNBS)-induced colitis in mice [33]. In addition, prior adminis-

tration of a probiotic mixture, including four *Bifidobacteria* strains, to DSS-treated mice also demonstrated protective effects [34,35].

Here, we tested the effect of *B. breve* administration in a DSS-induced colitis model. For a long time, the DSS-induced acute colitis model was regarded by some as an erosive, self-limiting model of colonic injury and inflammation. A previous study has demonstrated that T cells are not necessary in DSS-induced colitis [36]. However, recent studies show that bacteria penetrated the mucosa layer before inflammation in acute colitis model [37] and microflora is necessary during the development of DSS colitis [38]. Penetration of bacteria in the mucosal layer will lead to the activation of resident innate immune cells that in turn can lead to an adaptive immune response where T cells are involved. Indeed, we have recently demonstrated that antigen-specific T cells develop during the acute stage of DSS-induced colitis [39]. In addition, it has been shown that transient Treg depletion leads to increased severity of DSS colitis [40]. We hypothesized that the induction of Treg cells in the intestinal mucosa by intervention with *B. breve* could induce protective effect during DSS-induced colitis. The DSS colitis model is, thus, a valuable model to investigate the rise of T cell associated responses during intestinal inflammation mimicking early IBD. Altogether, T cells can affect the development of acute colitis, although the specific mechanism still needs to be investigated.

In this study, our data show that intervention with *B. breve* is beneficial in DSS-induced colitis by improving the weight loss, fecal condition, colon histology score which includes epithelial damage and cellular infiltration and colon shortening. DSS-induced colitis is often associated with increased MPO activity, which is indicative for an increased number of infiltrating neutrophils [41]. In line with this finding, increased numbers of

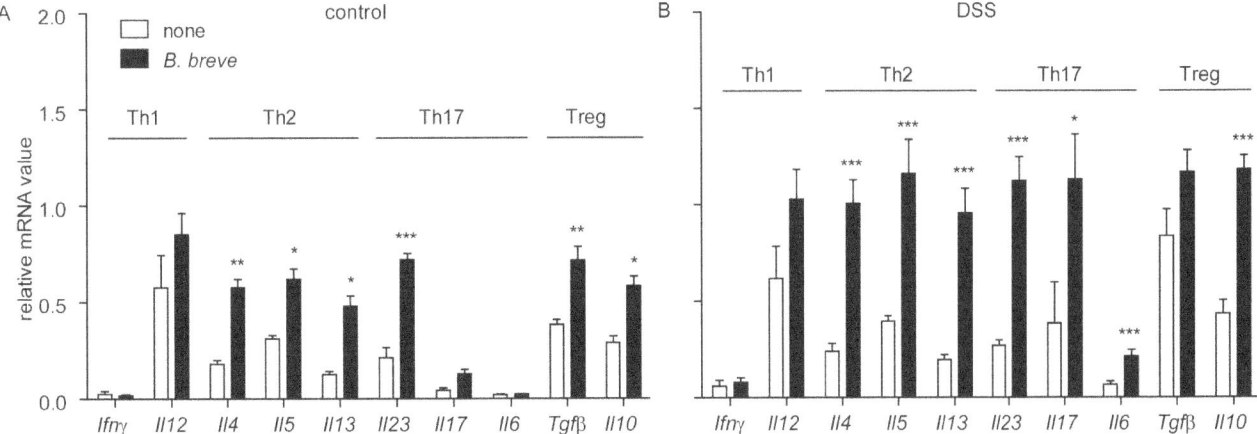

Figure 3. *B. breve* intervention changes mRNA expression of Th2-, Th17- and Treg- associated cytokines in the colon. The mRNA expression of Th1- (*Ifnγ* and *Il12*), Th2- (*Il4, Il5* and *Il13*), Th17- (*Il23* and *Il17*) and Treg- (*Tgfβ* and *Il10*) associated cytokines was quantified in the distal colons of both A) healthy and B) DSS-treated mice with or without *B. breve* intervention. Results are expressed as mean + SEM, n = 5 mice per group, pooled from two independent experiments. * p<0.05; ** p<0.01; *** p<0.001.

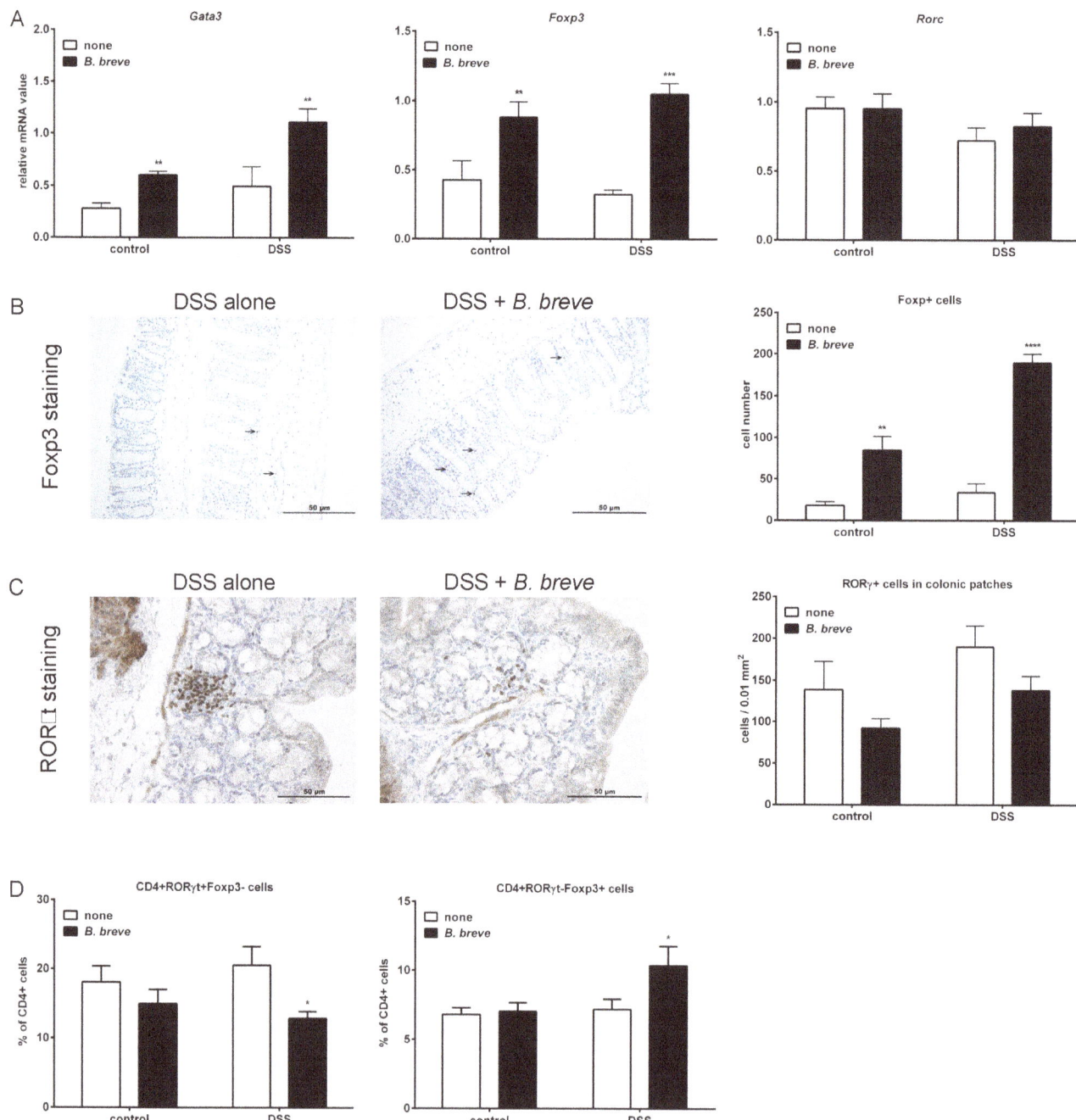

Figure 4. *B. breve* **intervention leads to increased numbers of Foxp3+ cells in the colon and Peyer's patches.** A) The mRNA expression of Th2- (*Gata3*), Th17- (*Rorc*) and Treg- (*Foxp3*) associated transcription factors was quantified in the distal colons of both healthy and DSS-treated mice with or without *B.breve* intervention. Results are expressed as mean + SEM, n = 5 mice per group, pooled from two independent experiments. B) Foxp3+ cells were visualized in the colon of DSS-treated mice with or without *B. breve* intervention using immunohistochemistry. The number of Foxp3+ cells was determined as described in the materials & methods and shown in the graph, The pictures are representative of n = 3 mice per group obtained from two independent experiment. C) RORγt+ cells were visualized in the colon of DSS-treated mice with or without *B. breve* intervention using immunohistochemistry. The number of RORγt+ cells was determined as described in the materials & methods and depicted in the graph. D) The percentage of Th17 cells (CD4+RORγt+Foxp3-) and Treg cells (CD4+RORγt-Foxp3+) was determined in the Peyer's patches obtained from both healthy and DSS-treated mice with or without *B. breve* intervention. Percentages within CD4+ T cell population are shown. Results are expressed as mean + SEM, n = 6 mice per group, pooled from two independent experiments. * p<0.05, ** p<0.01; *** p<0.001.

LyB6+ cells and increased MPO levels were found in the colons of DSS-treated mice. DSS-induced enhancement of MPO expression was decreased by 35% due to *B. breve* intervention, although this did not reach significance. Intervention with *L. rhamnosus* did not affect the DSS-induced colitis, which is similar to results found in experiments using *Lactobacillus rhamnosus GG* performed by Mileti *et al* [42]. A possible explanation could be that *L. rhamnosus* is less able to modify the T cell composition as compared to *B. breve*. it has been postulated that it is essential to target both Th1 and Th17 cells for treatment for CD, the major form of IBD [43]. The fact that *L. rhamnosus* is not as protective for DSS-induced colitis as *B. breve* could be explained by data from the *in vitro* experiments that show exposure to *B. breve* reduced both Th1 and Th17 cell subsets, whereas exposure to *L. rhamnosus* only reduced the Th17 cell subset. In addition, *B. breve* increased the expression of Treg cell–associated cytokines and transcription factors *in vivo*, while *L. rhamnosus* did not induce any of these changes.

Analysis of mRNA expression in the colon showed increased expression of Th2 (*Il4, Il5 and Il13*)- and Treg (*Il10*)-associated cytokines in both healthy and DSS-treated mice with B. breve intervention. An increased Th2 response often results in a decreased Th1 response due to the mutual antagonizing effects of Th1 and Th2 cells on each other [32]. Treg cells are able to repress the activity of other T cell subsets to induce an anti-inflammatory effect [12]. There are two major regulatory T cell populations, namely Foxp3+ Treg and IL10-producing type 1 regulatory T (Tr1) cells that are know to maintain intestinal homeostasis [44]. Therefore, it can be concluded that besides Foxp3+ Treg cells, Tr1 cells could also be involved. Interestingly, the increased *Il10* expression in the colon is in line with recent findings, which demonstrated an increased number of IL10 producing Tr1 cells in the colon after *B. breve* intervention [35].

Next to the increased Th2- and Treg-associated cytokines, we also observed an increased mRNA expression of Th17 associated cytokines including the effector cytokine IL17 in DSS treated mice with *B. breve* intervention as compared to DSS treatment alone. It should be noted that Th17 cells are not the only source of IL17 production as it was demonstrated that also innate lymphoid cells can produce IL17 upon activation by IL23 derived from macrophages and dendritic cells [45] [46]. RORγt is the master transcription factor of Th17 cells [5], but is also expressed in IL17-producing innate lymphoid cells (ILC) [47]. We did not observe an effect of either DSS or treatment with *B.* breve on the expression of *Rorc* in the colon. The increased expression of *Il17* in the colon observed after *B. breve* intervention might be the result of IL23 mediated activation of resident ILC that are mainly found in the lamina propria in close proximity of epithelial cells. However, another possibility is that another, RORγt independent, IL17 producing source is present in the colon, such as B cells [48]. Although IL17 is often thought to promote the development of IBD [49], a recent study has demonstrated a protective function of IL17 in intestinal inflammation [50]. The exact role of IL17 during the IBD development still needs to be elucidated in additional studies.

Treg cells are associated with anti-inflammatory and tolerance inducing mechanisms [12,51]. Although it is not totally clear how Treg cells effect the development of IBD, lack of Treg cells are often found in IBD patients [14,15]. Animal models of IBD have further demonstrated the importance of Treg cells during the development of colitis [17,33]. Foxp3 expression is associated with Treg cell development [6,7] and the anti-inflammatory properties of Foxp3+ Treg cells have been demonstrated by a number of studies in both mice and humans [10,52]. In this study, *B. breve* stimulation induced Treg cell differentiation *in vitro* and *in vivo*. Moreover, *B. breve* intervention ameliorated DSS-induced colitis symptoms and increased Foxp3+ T cells in Peyer's patches. Recent studies have demonstrated that the home of Peyer's patches, the small intestine, is involved in DSS colitis [53,54]. Peyer's patches, like other lymphoid organs, contain dendritic cells that taken up antigens and present them to T cells, leading to T cell activation and differentiation. The increased Foxp3+ T cells in Peyer's patches indicate T cell differentiation that favors anti-inflamma-

tory response. The resulting activated and expanded T cells have the potential to travel to the colon and induce immune regulation [55,56,57]. In line with this hypothesis, an increased amount of Foxp3+ cells was found in the colon of mice with *B. breve* intervention. These data suggest that the protective effects of *B. breve* on DSS colitis might lie with its capability to induce Treg cells

In conclusion, *B. breve* NutRes 204 stimulation *in vitro* leads to T cell skewing toward Treg cells in human PBMCs. Additionally, intervention with *B. breve* NutRes 204 ameliorated DSS-induced colitis symptoms *in vivo* with an increased amount of Treg cells and a reduced amount of Th17 cells in the GALT. This suggests that patients suffering from IBD could potentially benefit from *B. breve* intervention.

Supporting Information

Figure S1　FACS dot plots of T cell composition in human PBMCs with or without bacteria intervention. FACS dot plots of A) fluorescence minus one (FMO) controls and B) isotype controls of FOXP3, RORγ, GATA3 and Tbet staining antibodies within CD4+ T cells are shown. Representative FACS dot plots of Th2 (GATA3+Tbet-) and Th1 (GATA-Tbet+), Th17 (RORγ+FOXP3-) and Treg (RORγ-FOXP3+) cells in the PBMCs after 48 hours incubation with either C) anti-CD3 stimulation alone, D) a combination of anti-CD3 and *L. rhamnosus*, or E) a combination of anti-CD3 and *B. breve* are illustrated. The percentage of activated CD4+CD69+ T cells is calculated within total live cells and the percentage of Th2, Th1, Th17 and Treg cells are shown within activated CD4+CD69+ T cell population.

Figure S2　FACS dot plots of T cell – associated cytokine producing T cells in human PBMCs with or without bacteria intervention. A) FACS dot plots of A) fluorescence minus one (FMO) controls and B) isotype controls of IL4, IL17, IL10 and IFNγ staining antibodies within CD4+ T cells are shown. Representative FACS dot plot of CD4+ T cells in the PBMCs after 7 days stimulation with anti-CD3 alone, or a combination of anti-CD3 with either *L. rhamnosus* or *B. breve* are shown in C). Gated on the CD4+ T cells, the percentages of D) IL4+, E) IL17+, F) IL10+ and G) IFNγ+ CD4+ T cells were determined. The percentage of CD4+ T cells is calculated within total live cells and the percentages of IL4+, IL17+, IL10+ and IFNγ+ CD4+ T cells are presented within CD4+ T cell population.

Figure S3　FACS dot plots of Treg cells and Th17 cells in the mice with or without *B. breve* intervention. A) FACS dot plots of A) fluorescence minus one (FMO) controls and B) isotype controls of Foxp3 and RORγt staining antibodies within CD4+ T cells are shown. Representative FACS dot plots of CD4+ cells in the Peyer's patches obtained from both healthy and DSS-treated mice, with or without *B. breve* intervention, are shown in C). Gated on CD4+ T cells, the percentages of D) Th17 (CD4+RORγt+Foxp3-) and Treg (CD4+RORγt-Foxp3+) cells were determined. The percentage of CD4+ T cells is calculated within total live cells and the percentages of Th17 and Treg cells are determined within CD4+ T cell population.

Acknowledgments

We would like to thank Ms Kostadinova for her help with FACS analysis and Ms Dingjan, Ms Fernstrand en Mr van Aalst for their help preparing and assessing histology slides.

Author Contributions

Conceived and designed the experiments: BZ MEM ADK. Performed the experiments: BZ MEM JvB HJGvdK SO. Analyzed the data: BZ MEM ADK GF JG. Contributed reagents/materials/analysis tools: HJGvdK SO JvB PV. Wrote the paper: BZ MEM ADK GF JG.

References

1. Xavier RJ, Podolsky DK (2007) Unravelling the pathogenesis of inflammatory bowel disease. Nature 448: 427–434.
2. Maynard CL, Weaver CT (2009) Intestinal effector T cells in health and disease. Immunity 31: 389–400.
3. Szabo SJ, Kim ST, Costa GL, Zhang X, Fathman CG, et al. (2000) A novel transcription factor, T-bet, directs Th1 lineage commitment. Cell 100: 655–669.
4. Zheng W, Flavell RA (1997) The transcription factor GATA-3 is necessary and sufficient for Th2 cytokine gene expression in CD4 T cells. Cell 89: 587–596.
5. Ivanov, II, McKenzie BS, Zhou L, Tadokoro CE, Lepelley A, et al. (2006) The orphan nuclear receptor RORgammat directs the differentiation program of proinflammatory IL-17+ T helper cells. Cell 126: 1121–1133.
6. Hori S, Nomura T, Sakaguchi S (2003) Control of regulatory T cell development by the transcription factor Foxp3. Science 299: 1057–1061.
7. Fontenot JD, Gavin MA, Rudensky AY (2003) Foxp3 programs the development and function of CD4+CD25+ regulatory T cells. Nat Immunol 4: 330–336.
8. Harrington LE, Hatton RD, Mangan PR, Turner H, Murphy TL, et al. (2005) Interleukin 17-producing CD4+ effector T cells develop via a lineage distinct from the T helper type 1 and 2 lineages. Nat Immunol 6: 1123–1132.
9. Rovedatti L, Kudo T, Biancheri P, Sarra M, Knowles CH, et al. (2009) Differential regulation of interleukin 17 and interferon gamma production in inflammatory bowel disease. Gut 58: 1629–1636.
10. Eastaff-Leung N, Mabarrack N, Barbour A, Cummins A, Barry S (2010) Foxp3+ Regulatory T Cells, Th17 Effector Cells, and Cytokine Environment in Inflammatory Bowel Disease. Journal of Clinical Immunology 30: 80–89.
11. Bettelli E, Carrier Y, Gao W, Korn T, Strom TB, et al. (2006) Reciprocal developmental pathways for the generation of pathogenic effector TH17 and regulatory T cells. Nature 441: 235–238.
12. Shevach EM (2011) Biological functions of regulatory T cells. Adv Immunol 112: 137–176.
13. Maul J, Loddenkemper C, Mundt P, Berg E, Giese T, et al. (2005) Peripheral and intestinal regulatory CD4+ CD25(high) T cells in inflammatory bowel disease. Gastroenterology 128: 1868–1878.
14. Wang Y, Liu XP, Zhao ZB, Chen JH, Yu CG (2011) Expression of CD4+ forkhead box P3 (FOXP3)+ regulatory T cells in inflammatory bowel disease. J Dig Dis 12: 286–294.
15. Veltkamp C, Anstaett M, Wahl K, Moller S, Gangl S, et al. (2011) Apoptosis of regulatory T lymphocytes is increased in chronic inflammatory bowel disease and reversed by anti-TNFalpha treatment. Gut 60: 1345–1353.
16. Powrie F, Leach MW, Mauze S, Caddle LB, Coffman RL (1993) Phenotypically distinct subsets of CD4+ T cells induce or protect from chronic intestinal inflammation in C. B-17 scid mice. Int Immunol 5: 1461–1471.
17. Read S, Malmstrom V, Powrie F (2000) Cytotoxic T lymphocyte-associated antigen 4 plays an essential role in the function of CD25(+)CD4(+) regulatory cells that control intestinal inflammation. J Exp Med 192: 295–302.
18. Kuhn R, Lohler J, Rennick D, Rajewsky K, Muller W (1993) Interleukin-10-Deficient Mice Develop Chronic Enterocolitis. Cell 75: 263–274.
19. Markets Ma (2014) Probiotics Market by Products (Functional Foods, Dietary Supplements, Specialty Nutrients, Animal Feed), Applications (Regular, Therapeutic, Preventive Health Care) & Ingredients (Lactobacilli, Bifidobacteria, Yeast) - Global Trends & Forecasts to 2019. Available: http://www.marketsandmarkets.com/Market-Reports/probiotic-market-advanced-technologies-and-global-market-69html. Accessed 2014, April 1.
20. Ringel Y, Quigley EMM, Lin HC (2012) Using Probiotics in Gastrointestinal Disorders. Am J Gastroenterol Suppl 1: 34–40.
21. Boirivant M, Strober W (2007) The mechanism of action of probiotics. Curr Opin Gastroenterol 23: 679–692.
22. de Roock S, van Elk M, van Dijk ME, Timmerman HM, Rijkers GT, et al. (2010) Lactic acid bacteria differ in their ability to induce functional regulatory T cells in humans. Clin Exp Allergy 40: 103–110.
23. de Roock S, van Elk M, Hoekstra MO, Prakken BJ, Rijkers GT, et al. (2011) Gut derived lactic acid bacteria induce strain specific CD4(+) T cell responses in human PBMC. Clin Nutr 30: 845–851.
24. Fujii T, Ohtsuka Y, Lee T, Kudo T, Shoji H, et al. (2006) Bifidobacterium breve enhances transforming growth factor beta1 signaling by regulating Smad7 expression in preterm infants. J Pediatr Gastroenterol Nutr 43: 83–88.
25. Konieczna P, Groeger D, Ziegler M, Frei R, Ferstl R, et al. (2012) Bifidobacterium infantis 35624 administration induces Foxp3 T regulatory cells in human peripheral blood: potential role for myeloid and plasmacytoid dendritic cells. Gut 61: 354–366.
26. Plantinga TS, van Maren WW, van Bergenhenegouwen J, Hameetman M, Nierkens S, et al. (2011) Differential Toll-like receptor recognition and induction of cytokine profile by Bifidobacterium breve and Lactobacillus strains of probiotics. Clin Vaccine Immunol 18: 621–628.
27. Rosas M, Thomas B, Stacey M, Gordon S, Taylor PR (2010) The myeloid 7/4-antigen defines recently generated inflammatory macrophages and is synonymous with Ly-6B. J Leukoc Biol 88: 169–180.
28. Zhou L, Lopes JE, Chong MM, Ivanov, II, Min R, et al. (2008) TGF-beta-induced Foxp3 inhibits T(H)17 cell differentiation by antagonizing RORgammat function. Nature 453: 236–240.
29. Lakatos PL (2006) Recent trends in the epidemiology of inflammatory bowel diseases: up or down? World J Gastroenterol 12: 6102–6108.
30. Brown SJ, Mayer L (2007) The immune response in inflammatory bowel disease. Am J Gastroenterol 102: 2058–2069.
31. Monteleone G, Caprioli F (2010) T-cell-directed therapies in inflammatory bowel diseases. Clin Sci (Lond) 118: 707–715.
32. Kidd P (2003) Th1/Th2 balance: the hypothesis, its limitations, and implications for health and disease. Altern Med Rev 8: 223–246.
33. Roselli M, Finamore A, Nuccitelli S, Carnevali P, Brigidi P, et al. (2009) Prevention of TNBS-induced colitis by different Lactobacillus and Bifidobacterium strains is associated with an expansion of gammadeltaT and regulatory T cells of intestinal intraepithelial lymphocytes. Inflamm Bowel Dis 15: 1526–1536.
34. Nanda Kumar NS, Balamurugan R, Jayakanthan K, Pulimood A, Pugazhendhi S, et al. (2008) Probiotic administration alters the gut flora and attenuates colitis in mice administered dextran sodium sulfate. J Gastroenterol Hepatol 23: 1834–1839.
35. Jeon SG, Kayama H, Ueda Y, Takahashi T, Asahara T, et al. (2012) Probiotic Bifidobacterium breve Induces IL-10-Producing Tr1 Cells in the Colon. PLoS Pathog 8: e1002714.
36. Axelsson LG, Landstrom E, Goldschmidt TJ, Gronberg A, Bylund-Fellenius AC (1996) Dextran sulfate sodium (DSS) induced experimental colitis in immuno-deficient mice: effects in CD4(+) -cell depleted, athymic and NK-cell depleted SCID mice. Inflamm Res 45: 181–191.
37. Johansson ME, Gustafsson JK, Sjoberg KE, Petersson J, Holm L, et al. (2010) Bacteria penetrate the inner mucus layer before inflammation in the dextran sulfate colitis model. PLoS One 5: e12238.
38. Hudcovic T, Stepankova R, Cebra J, Tlaskalova-Hogenova H (2001) The role of microflora in the development of intestinal inflammation: acute and chronic colitis induced by dextran sulfate in germ-free and conventionally reared immunocompetent and immunodeficient mice. Folia Microbiol (Praha) 46: 565–572.
39. Morgan ME, Zheng B, Koelink PJ, van de Kant HJG, Haazen LCJM, et al. (2013) New Perspective on Dextran Sodium Sulfate Colitis: Antigen-Specific T Cell Development during Intestinal Inflammation. PLoS ONE 8: e69936.
40. Boehm F, Martin M, Kesselring R, Schiechl G, Geissler EK, et al. (2012) Deletion of Foxp3+ regulatory T cells in genetically targeted mice supports development of intestinal inflammation. BMC Gastroenterol 12: 97.
41. Yan Y, Kolachala V, Dalmasso G, Nguyen H, Laroui H, et al. (2009) Temporal and spatial analysis of clinical and molecular parameters in dextran sodium sulfate induced colitis. PLoS One 4: e6073.
42. Mileti E, Matteoli G, Iliev ID, Rescigno M (2009) Comparison of the immunomodulatory properties of three probiotic strains of Lactobacilli using complex culture systems: prediction for in vivo efficacy. PLoS One 4: e7056.
43. Brand S (2009) Crohn's disease: Th1, Th17 or both? The change of a paradigm: new immunological and genetic insights implicate Th17 cells in the pathogenesis of Crohn's disease. Gut 58: 1152–1167.
44. Barnes MJ, Powrie F (2009) Regulatory T cells reinforce intestinal homeostasis. Immunity 31: 401–411.
45. Takatori H, Kanno Y, Watford WT, Tato CM, Weiss G, et al. (2009) Lymphoid tissue inducer-like cells are an innate source of IL-17 and IL-22. J Exp Med 206: 35–41.
46. Geremia A, Arancibia-Carcamo CV, Fleming MP, Rust N, Singh B, et al. (2011) IL-23-responsive innate lymphoid cells are increased in inflammatory bowel disease. J Exp Med 208: 1127–1133.
47. Walker JA, Barlow JL, McKenzie AN (2013) Innate lymphoid cells—how did we miss them? Nat Rev Immunol 13: 75–87.
48. Leon B, Lund FE (2013) IL-17-producing B cells combat parasites. Nat Immunol 14: 419–421.
49. Fujino S, Andoh A, Bamba S, Ogawa A, Hata K, et al. (2003) Increased expression of interleukin 17 in inflammatory bowel disease. Gut 52: 65–70.
50. O'Connor W, Jr., Kamanaka M, Booth CJ, Town T, Nakae S, et al. (2009) A protective function for interleukin 17A in T cell-mediated intestinal inflammation. Nat Immunol 10: 603–609.
51. Sakaguchi S, Sakaguchi N, Shimizu J, Yamazaki S, Sakihama T, et al. (2001) Immunologic tolerance maintained by CD25+ CD4+ regulatory T cells: their common role in controlling autoimmunity, tumor immunity, and transplantation tolerance. Immunol Rev 182: 18–32.
52. Sakaguchi S, Miyara M, Costantino CM, Hafler DA (2010) FOXP3+ regulatory T cells in the human immune system. Nat Rev Immunol 10: 490–500.

53. Yazbeck R, Howarth GS, Butler RN, Geier MS, Abbott CA (2011) Biochemical and histological changes in the small intestine of mice with dextran sulfate sodium colitis. J Cell Physiol 226: 3219–3224.

54. Geier MS, Smith CL, Butler RN, Howarth GS (2009) Small-intestinal manifestations of dextran sulfate sodium consumption in rats and assessment of the effects of Lactobacillus fermentum BR11. Dig Dis Sci 54: 1222–1228.

55. du Pre MF, Samsom JN (2011) Adaptive T-cell responses regulating oral tolerance to protein antigen. Allergy 66: 478–490.

56. Hauet-Broere F, Unger WW, Garssen J, Hoijer MA, Kraal G, et al. (2003) Functional CD25- and CD25+ mucosal regulatory T cells are induced in gut-draining lymphoid tissue within 48 h after oral antigen application. Eur J Immunol 33: 2801–2810.

57. Koboziev I, Karlsson F, Grisham MB (2010) Gut-associated lymphoid tissue, T cell trafficking, and chronic intestinal inflammation. Ann N Y Acad Sci 1207 Suppl 1: E86–93.

10

Comparison of Microbiological, Histological, and Immunomodulatory Parameters in Response to Treatment with Either Combination Therapy with Prednisone and Metronidazole or Probiotic VSL#3 Strains in Dogs with Idiopathic Inflammatory Bowel Disease

Giacomo Rossi[1], Graziano Pengo[2], Marco Caldin[3], Angela Palumbo Piccionello[1], Jörg M. Steiner[4], Noah D. Cohen[5], Albert E. Jergens[6], Jan S. Suchodolski[4]*

1 School of Veterinary Medical Sciences, University of Camerino, Camerino, Italy, 2 Clinic "St. Antonio", Cremona, Italy, 3 San Marco Laboratories, Padova, Italy, 4 Gastrointestinal Laboratory, Department of Small Animal Clinical Sciences, College of Veterinary Medicine and Biomedical Sciences, Texas A&M University, College Station, Texas, United States of America, 5 Department of Large Animal Clinical Sciences, College of Veterinary Medicine and Biomedical Sciences, Texas A&M University, College Station, Texas, United States of America, 6 Department of Veterinary Clinical Sciences, College of Veterinary Medicine, Iowa State University, Ames, Iowa, United States of America

Abstract

Background: Idiopathic inflammatory bowel disease (IBD) is a common chronic enteropathy in dogs. There are no published studies regarding the use of probiotics in the treatment of canine IBD. The objectives were to compare responses to treatment with either combination therapy (prednisone and metronidazole) or probiotic strains (VSL#3) in dogs with IBD.

Methodology and Principal Findings: Twenty pet dogs with a diagnosis of IBD, ten healthy pet dogs, and archived control intestinal tissues from three euthanized dogs were used in this open label study. Dogs with IBD were randomized to receive either probiotic (D-VSL#3, n = 10) or combination drug therapy (D-CT, n = 10). Dogs were monitored for 60 days (during treatment) and re-evaluated 30 days after completing treatment. The CIBDAI (P<0.001), duodenal histology scores (P< 0.001), and CD3+ cells decreased post-treatment in both treatment groups. FoxP3+ cells (p<0.002) increased in the D-VSL#3 group after treatment but not in the D-CT group. TGF-β+ cells increased in both groups after treatment (P = 0.0043) with the magnitude of this increase being significantly greater for dogs in the D-VSL#3 group compared to the D-CT group. Changes in apical junction complex molecules occludin and claudin-2 differed depending on treatment. *Faecalibacterium* and *Turicibacter* were significantly decreased in dogs with IBD at T0, with a significant increase in *Faecalibacterium* abundance observed in the animals treated with VSL#3 strains.

Conclusions: A protective effect of VSL#3 strains was observed in dogs with IBD, with a significant decrease in clinical and histological scores and a decrease in CD3+ T-cell infiltration. Protection was associated with an enhancement of regulatory T-cell markers (FoxP3+ and TGF-β+), specifically observed in the probiotic-treated group and not in animals receiving combination therapy. A normalization of dysbiosis after long-term therapy was observed in the probiotic group. Larger scale studies are warranted to evaluate the clinical efficacy of VSL#3 in canine IBD.

Editor: Mathias Chamaillard, INSERM, France

Funding: There was no external funding support for this research. VSL Pharmaceuticals supplied the probiotic strains used in this study, but otherwise did not provide any financial support and had no role in study design, data collection, analysis, interpretation or preparation of the manuscript.

Competing Interests: VSL Pharmaceuticals supplied the probiotic strains used in this study. Graziano Pengo is employed by Clinic "St. Antonio" and Marco Caldin by San Marco Laboratories. There are no further patents, products in development or marketed products to declare.

* E-mail: jsuchodolski@cvm.tamu.edu

Introduction

Similar to human inflammatory bowel disease (IBD), three main factors are considered to be fundamental in the pathogenesis of canine idiopathic IBD: the interactions between the mucosal immune system, host genetic susceptibility, and environmental factors (e.g., microbiota, nutrition) [1–3]. Experimental evidence supports a role for commensal bacteria in the pathogenesis of IBD;

for example, spontaneous colitis develops in mice deficient in interleukin (IL)-2 [4] and IL-10 [5] when colonized with a complex microbiota, but not in mice raised under germ-free conditions. Recent studies suggest involvement of the intestinal microbiota in the pathogenesis of canine and feline IBD [2,6–9]. Also, antibiotics such as metronidazole are useful in the treatment of IBD in humans [10] and dogs [11], and there is evidence that children with IBD respond to probiotic administration [12]. Collectively, these findings suggest that the intestinal microbiota plays a crucial role in the pathogenesis of IBD and modulation of intestinal microbiota may be beneficial in the treatment of mucosal inflammation. While probiotics are used frequently in small animal practice, there are only few published studies regarding their efficacy in dogs with chronic enteropathies. In one investigation, a probiotic cocktail was shown to reduce clinical severity in a prospective, placebo-controlled trial in dogs with food-responsive diarrhea treated with an elimination diet [13], but studies evaluating idiopathic IBD have not been reported.

VSL#3 is a high-dose, multi-strain probiotic product containing viable lyophilized bacteria consisting of 4 strains of *Lactobacillus* (*L. casei*, *L. plantarum*. *L. acidophilus*, and *L. delbrueckii* subsp. *bulgaricus*), 3 strains of *Bifidobacterium* (*B. longum*, *B. breve*, and *B. infantis*), and 1 strain of *Streptococcus sulivarius* subsp *thermophilus*. The VSL#3 strains have shown efficacy in humans for the prevention, treatment, and maintenance of remission of both pouchitis and ulcerative colitis in adults and children [12,14,15].

The purpose of the present study was to perform a randomized open-label trial to compare the microbiological, histological, and immunomodulatory effects between the commercial multi-strain probiotic SIVOY, a probiotic product formulated with VSL#3 strains for pets (VSL Pharmaceuticals, Inc., Gaithersburg, MD, USA) and combination therapy with prednisone and metronidazole in canine IBD.

Our results suggest a protective effect of the probiotic mixture in dogs with IBD, with a significant decrease in clinical and histological scores, and a decrease in CD3+ T-cell infiltration. Protection was associated with an enhancement of regulatory T-cell markers (FoxP3+ and TGF-β+), specifically observed in the probiotic-treated group and not in animals receiving combination therapy. The protective effect of the probiotic VSL#3 strains was also associated with normalization of dysbiosis, specifically increases in *Faecalibacterium* spp.

Materials and Methods

Animals

The study was approved by the Camerino University Institutional Animal Care and Use Committee protocol and all owners of the IBD dogs gave informed written consent before enrollment. Twenty pet dogs (Table 1) with a long-time diagnosis of IBD according to published criteria [16] were evaluated at the Veterinary Teaching Hospital, Camerino University, for chronic gastroenteritis. Inclusion criteria included recurrence of clinical signs and absence of any immunomodulating drug therapy (e.g., corticosteroids, metronidazole, and sulfasalazine) within a month before referral. Diagnostic criteria for IBD included: persistent (>3 weeks) gastrointestinal signs, failed responses to dietary (hydrolysate or commercial intact protein elimination diet) or symptomatic therapies (anthelminthics, antibiotics, anticholinergics, gastrointestinal protectants) alone, a thorough diagnostic evaluation with failure to document other causes for gastroenteritis, and histopathologic evidence of intestinal inflammation. The minimum diagnostic evaluation in all dogs included a complete blood count, serum biochemistry, urinalysis, direct (wet mount) and indirect (flotation) examination of feces for endoparasites, and survey abdominal radiographs. In some instances, additional tests including contrast radiography, abdominal ultrasound (performed in 16 of the 20 dogs) and measurement of serum concentrations of trypsin-like immunoreactivity and/or folate and cobalamin were performed. Additional inclusion criteria were the absence of extra-alimentary tract inflammation based on results obtained from initial diagnostic testing. Dogs with hypoproteinemia or a suspicion of intestinal lymphangiectasia were excluded from the study.

Ten pet dogs (Table 1), living in home environments and free of gastrointestinal signs for at least four months, were enrolled as control group (D H) for comparison of fecal microbiota between healthy dogs and dogs with IBD. Control dogs were judged to be healthy based on normal results on physical examination, complete blood count, serum biochemistry, urinalysis, repeated fecal examinations, and dirofilarial antigen assay.

Study design

The trial was a 90 day open-label evaluation to compare the effects of VSL#3 strains versus combination drug therapy on histological, microbiological, and immunological markers. Dogs were randomized into two groups using a computer-generated randomization list. The VSL#3 group (D-VSL#3; n = 10) received between 112 and 225 billion (112 to 225×10^9) lyophilized bacteria per 10 kg daily for 60 consecutive days; the D-CT group (n = 10) received a combination protocol of metronidazole at 20 mg/Kg q12 h and prednisone at 1 mg/kg body weight/day. The clinical disease activity (CIBDAI score) was assessed at baseline (T0) and after 90 days (T1) of enrollment, which was 30 days following completion of either treatment. The CIBDAI is based on 6 criteria, each scored on a scale from 0 3: attitude/activity, appetite, vomiting, stool consistency, stool frequency, and weight loss. After summation, the total composite score is determined to be clinically insignificant (score 0 3), mild (score 4 5), moderate (score 6 8) or severe (score 9 or greater) [17].

Fecal samples were also collected at each visit then immediately stored at −80°C, until microbiota analysis. The evaluation time point 30 days post-treatment was chosen to determine whether individual dogs would relapse within 30 days following completion of either treatment regimen.

Tissue sampling

After enrollment (time point T0) and after 90 days (T1), multiple (10 15 specimens) mucosal biopsy specimens were procured endoscopically from the small and/or large intestine of all dogs with IBD (n = 20, 10 dogs per treatment group). Fifteen dogs having predominantly upper gastrointestinal signs (i.e., vomiting, small bowel diarrhea, anorexia, and/or weight loss) underwent esophagogastroduodenoscopy, whereas upper and lower endoscopic examinations were performed in 5 dogs having mixed signs of enterocolitis (i.e., GI signs associated with tenesmus, hematochezia, mucoid feces, and/or frequent defecation). Biopsy specimens were obtained directly from mucosal lesions of increased granularity, friability, or erosions as well as areas of normal-appearing mucosa. Tissues for histopathology were placed in 10% neutral buffered formalin, then paraffin embedded and serial 3 µm thick sections were prepared. For ethical considerations, no endoscopic examinations were performed in healthy dogs. Histopathology was performed by a single pathologist, who was blinded regarding history, clinical signs, or endoscopic observations. A severity score was assigned for each dog, by using a standardized and previously described histologic grading system, based on the extent of architectural disruption and mucosal

Table 1. Summary characteristics of enrolled dogs.

	Treatment groups		
	VSL#3 (n = 10)	CT (n = 10)	Healthy Control (n = 10)
Breed	Golden Retriever, Husky, Boxer, Rottweiler, Jack Russell Terrier, WHW Terrier, German shepherd (2), Shih Tzu, Yorkshire Terrier	Golden retriever (2), Cocker Spaniel, Boxer, Bull Terrier, Carlino, WHW Terrier, German shepherd, Shar Pei, Yorkshire Terrier	Golden Retriever, Epagneul Breton, Chow Chow, Rottweiler, Border collie, German shepherd, Bolognese, Miniature Schnauzer, Yorkshire Terrier (2)
Sex	m = 5, mn = 1, f = 1, fs = 3	m = 5, f = 2, fs = 3	m = 5, f = 5
Median age (range) in years	5.8 (2.5–11)	5.5 (1.5–9)	6.5 (1–12)
Body weight (range) in kg	18.9 (2–36)	18.7 (1.5–30)	20.6 (2.8–45)
Median (range) time to remission (days)	10.6 (5–15)	4.8 (2.5–7)	n/a

m = male, mn = neutered male, f = female, fs = spayed female; CT = combination therapy; n/a-not applicable.

epithelial changes [17,18], as recently been proposed by the WSAVA for diagnosis of gastrointestinal inflammation [19].

Tissues were also evaluated for expression patterns of apical junction complex (AJC) molecules in both dog groups after end of the therapy. To obtain control tissue from healthy dogs for this analysis, archived formalin-fixed and paraffin-embedded colonic tissues from three male dogs with no clinical signs of intestinal disease were retrieved from the University of Camerino Veterinary Pathology Unit archives. These samples had been obtained immediately post-mortem from dogs that were presented for euthanasia (euthanized dogs, ED) for old age (n = 1), nasal carcinoma (n = 1), or splenic haemangiosarcoma (n = 1). Ages ranged from 7 years to 14 years and histopathological examination of full-thickness intestinal biopsies was normal in all these ED cases.

Immunohistochemical evaluation

Paraffin sections were rehydrated and neutralized for endogenous peroxidases with 3% hydrogen peroxide for 5 minutes followed by rinsing for 5 minutes in distilled water. For antigen retrieval, slides were incubated in three antigen retrieval solutions: citrate buffer (pH 6.0) for TGF-β, EDTA (pH 8.0) for CD3 and FoxP3, and 0.01 M Tris-EDTA buffer (pH 9.0) for claudin 2, occludin and E-cadherin in a steamer (Black & Decker, Towson, MD, USA) for 20 minutes. Non-specific binding was blocked by incubation of slides for 10 minutes with a protein-blocking agent (Protein-blocking agent, Dako, Carpinteria, CA, USA) before application of the primary antibody. Slides were incubated overnight in a moist-chamber with the following primary antibodies: monoclonal (mAb) rat anti-human CD3 (Monoclonal rat anti-human CD3 clone MCA1477, Serotec abD, Biorad Laboratories, Hercules, CA, USA) diluted 1:50, mAb anti-mouse/rat FoxP3 antibodies (Monoclonal anti-mouse/rat FoxP3 antibodies clone FJK-16s, eBioscience, San Diego, CA, USA) diluted 1:400, and mAb mouse anti-TGF-β (Monoclonal mouse anti-TGF-β, clone 1D11, Serotec abD, Biorad Laboratories, Hercules, CA, USA) diluted 1:25 [19,20]. Polyclonal rabbit anti-claudin-2 (Polyclonal rabbit anti-claudin-2 (PAD: MH44), Invitrogen Ltd.,

Paisley, UK) and anti-occludin (anti-occludin PAD: Z-T22, Invitrogen Ltd., Paisley, UK) antibodies and monoclonal mouse anti-E-cadherin IgG2α (Monoclonal mouse anti-E-cadherin IgG2α (clone: 36), BD Biosciences, Oxford, UK) were used as described previously [21].

The immunoreaction with streptavidin–immunoperoxidase (Streptavidin–immunoperoxidase, Black & Decker, Towson, MD, USA) was visualized with 3,3'-diaminobenzidine substrate (3,3'-diaminobenzidine substrate, Vector, Burlingame, UK). Tissues were counterstained with Mayer's hematoxylin. For negative immunohistochemical controls the primary antibodies were omitted. Sections of canine spleen and tonsil served as positive control tissues for CD3 and FoxP3 cell staining and sections of canine placenta for that of TGF-β expression. Positive control tissues for claudin/occludin and E-cadherin staining consisted of canine lung and kidney sections, respectively.

For scoring of intestinal CD3+ T-lymphocytes, FoxP3+ cells, and TGF-β+ cells, these cells were quantified in select compartments of the GI tract (small intestine: villi, basal crypt area, villus-crypt junction; large intestine: apical crypt area, basal crypt area). All cellular types were evaluated using a light microscope (Carl Zeiss, Jena, Germany), a ×40 objective, a ×10 eyepiece, and a square eyepiece graticule (10×10 squares, having a total area of 62,500 μm²). Ten appropriate fields were chosen for each compartment and arithmetic means were calculated for each intestinal region. Results were expressed as IHC positive cells per 62,500 μm². For all parameters, cells on the margins of the tissue sections were not considered for evaluation to avoid inflation of positive cell numbers.

For the evaluation of different lymphocytes subsets in the same histological sections, consecutive 3-μm-thick bioptic cross sections were cut. Sections were placed consecutively on each of eight separate slides, after which the ninth section was placed on the first slide, next to the first section, continuing for 48 sections. A single slide, upon which were six bioptic cross sections from each dog, was analyzed for any given immunostain. Numbers of CD3+ T-lymphocytes, FoxP3+ cells, and TGF-β+ cells, were quantified by using an image-analysis system consisting of a light microscope

Table 2. Oligonucleotides primers/probes used in this study.

qPCR primers/probe	Sequence (5'-3')	Target	Annealing (°C)	Reference
Forward	CCGGAWTYATTGGGTTTAAAGGG	Bacteroidetes	60	[45]
Reverse	GGTAAGGTTCCTCGCGTA			
Forward	GAAGGCGGCCTACTGGGCAC	*Faecalibacterium*	60	[47]
Reverse	GTGCAGGCGAGTTGCAGCCT			
Forward	ACTGAGAGGTTGAACGGCCA	Family Ruminococcaceae	59	[47]
Reverse	CCTTTACACCCAGTAAWTCCGGA			
Forward	CGCATAACGTTGAAAGATGG			
Reverse	CCTTGGTAGGCCGTTACCC	*C. perfringens* 16S	58	[48]
Probe	TCATCATTCAACCAAAGGAGCAATCC			
Forward	KGGGCTCAACMCMGTATTGCGT	Fusobacteria	51	[9]
Reverse	TCGCGTTAGCTTGGGCGCTG			
Forward	TCTGATGTGAAAGGCTGGGGCTTA	*Blautia*	56	[9]
Reverse	GGCTTAGCCACCCGACACCTA			
Forward	CCTACGGGAGGCAGCAGT	Universal Bacteria	59	[49]
Reverse	ATTACCGCGGCTGCTGG			
Forward	CAGACGGGGACAACGATTGGA	*Turicibacter*	63	[9]
Reverse	TACGCATCGTCGCCTTGGTA			
Forward	TCGCGTCCGGTGTGAAAG	*Bifidobacterium*	60	[50]
Reverse	CCACATCCAGCATCCAC			
Forward	CCCTTATTGTTAGTTGCCATCATT	*Enterococcus*	61	[46]
Reverse	ACTCGTTGTACTTCCCATTGT			
Forward	AGCAGTAGGGAATCTTCCA[a]	*Lactobacillus*	58	[46]
Reverse	CACCGCTACACATGGAG[b]			
Forward	TTATTTGAAAGGGGCAATTGCT	*Streptococcus*	54	[51]
Reverse	GTGAACTTTCCACTCTCACAC			

[a] Originally described by [52].
[b] Originally described by [53].

(Carl Zeiss, Jena, Germany) attached to a Javelin JE3462 high-resolution camera and a personal computer equipped with a Coreco-Oculus OC-TCX frame grabber and high-resolution monitor. Computerized color-image analysis was performed by using Image-Pro Plus software (Media Cybernetics). The area of each biopsy in all six cross sections in every dog was recorded, as was the total number of T-lymphocytes determined by immuno-staining as previously described. For each dog, the total bioptic area was calculated as the sum of the areas of all fields in all six bioptic cross sections on one slide. CD3+ T-lymphocytes, FoxP3+ cells, and TGF-β+ cells were counted per section, and stained cell densities were expressed as the number of lymphocytes/cells per square millimeter of analyzed bioptic area [22].

To assess AJC expression (claudin-2, occludin, and E-cadherin) in biopsies sampled after treatment in both groups (at T1 for D-VSL#3 and D-CT), and to compare data to the AJC expression in non-IBD control dogs (ED group), stained tissue sections were evaluated at ×200 and ×630 (oil immersion) magnification to identify areas of consistent staining and acceptable orientation. Immunostaining was evaluated along the length of multiple enteric/colonic crypts and in areas of intact luminal epithelium. Stain intensity was subjectively graded as absent (−), weak (+), moderate (++), or strong (+++), and the localization and distribution of chromogen were noted. For evaluation, intestinal epithelium was divided into luminal, proximal, and distal gland/crypt regions, and the intercellular junction was divided into apical and basolateral compartments. Finally, the scoring of intestinal AJC molecules expression was calculated as previously described for CD3+ T-cells, FoxP3+ cells, and TGF-β+ cells. The AJC molecules were assessd only at T1 (following treatment intervention), because at T0 all dogs had endoscopically visible lesions of intestinal inflammation including erosions, friability, and increased mucosal granularity. Also, dogs had histopathologic lesions of intestinal inflammation of varying severity. Intestinal inflammation was associated with different degrees of epithelial infiltration by lymphocytes (i.e., intraepithelial lymphocytes) in all dogs of both groups. In these instances, it was not considered useful to evaluate AJCs as they were assumed to be altered, but instead AJC molecules were evaluated at T1 when the previously observed endoscopic lesions of inflammation had resolved.

Plasma citrulline

Plasma concentrations of citrulline were measured in the D-VSL#3 treated group only. Plasma samples were taken at baseline (T0) and after 90 days (T1) and stored at −80°C until evaluation. Samples were precipitated with organic solvents and quantified by MS/MS mass spectrometer equipped with electrospray ionization (ESI) interface in positive ion mode (Waters TQ Detector, Water Corp., Milford, MA, USA). All assays were performed in duplicate fashion. All data collected in centroid mode were processed using commercial software (MassLynx 4.1 software, Water Corp., Milford, Ma, USA).

Microbiota analysis

Fresh naturally voided samples were collected from all 20 diseased dogs (at T0 and T1) and 10 healthy dogs (one time point), flash-frozen in liquid nitrogen and stored at $-80°C$. DNA was extracted using a bead-beating method (PowerSoil DNA Isolation Kit, MoBio Laboratories, Carlsbad, CA) according to the manufacturer's protocol. Selected bacterial groups within the fecal microbiota were analyzed by quantitative PCR (qPCR) assays as described previously for canine fecal samples (Table 2) [9,23]. Amplified DNA from each bacterial group was normalized for total amplified bacterial DNA (log_{10} amplified DNA for each bacterial group divided by log_{10} of amplified bacterial DNA) as described previously [23].

Data analysis (statistics)

To evaluate differences at baseline as well as post-treatment between both treatment groups, a combined statistical analysis model was used. This model takes into account differences between the treatment groups at T0 as well as post-treatment at T1. The effects of time (i.e., T0 or T1), treatment, and their 2-way interaction on the various outcome parameters were measured (viz., histology, CIBDAI, TGF-β+, CD3+, and FoxP3+ cells). Dog was modeled as a random effect to account for repeated measures (before and after treatment) for individual dogs; time, treatment, and their 2-way interaction were modeled as fixed, categorical variables. Datasets for TGF-β and FoxP3+ were log_{10}-transformed to meet distributional assumptions underlying the statistical modeling. Confidence intervals (CIs) were estimated using maximum likelihood methods. The correlation structure for mixed-effects modeling was that of compound symmetry. Model fit was assessed visually by examining plots of standardized residuals versus fitted values, and by examining the AIC and BIC values for models. A significance level of $P<0.05$ was used for all analyses (S-PLUS, Version 8.2, TIBCO, Inc., Palo Alto, CA, USA). Changes in plasma citrulline concentrations were compared in the D-VSL#3 group between T0 and T1 using a Wilcoxon matched pairs test. The expression of AJC molecules expression (claudin-2, occludin, and E-cadherin) were compared at T1 between the dogs in the D-VSL#3, the D-CT, and the ED group using a Kruskal-Wallis-Test. The microbiota data obtained by qPCR were compared between healthy dogs and both treatment groups at T0 using an ANOVA or Kruskal-Wallis test where appropriate after evaluating for normal distribution using the Kolmogorov–Smirnov test. Changes in bacterial groups between T0 and T1 were compared using Wilcoxon matched pairs tests. Resulting P-values were corrected for multiple comparisons using the false discovery rate as described by Benjamini & Hochberg, and a $P<0.05$ was considered significant [9].

Results

Table 1 summarizes the signalment of the dogs enrolled into the study. No significant differences for age, sex, or body weights were identified ($P>0.05$ for each) between the dog groups. Table 3 and Figure 1 summarize the changes in histology scores, CIBDAI, and TGF-β+, FoxP3+, and CD3+ T-cell expression in both treatment groups.

Histology scores

Although there was a residual inflammatory infiltrate present (Figure 2), histology scores were significantly ($P<0.0001$) reduced at T1 relative to T0 in both treatment groups (Table 3 and Figure 1). There were no significant differences in the magnitude of this reduction between treatments ($P=0.1452$).

CIBDAI

Despite computer randomization, the severity of clinical signs, as judged by the CIBDAI score, was significantly higher at baseline in the D-CT group (median 9, range 7–13) compared to the D-VSL#3 group (median 7, range 5–10; $P<0.0001$). This was, in part, related to the fact that the all 3 dogs with severe disease were randomly allocated to the D-CT group. Another explanation was the presence of active (versus quiescent) clinical disease at presentation in some dogs. Both groups, however, had overall moderate-to-severe median disease activity at presentation. Clinical scores decreased significantly in both treatment groups over time ($P<0.0001$). As reported by the owners, recovery was more rapid in the D-CT group compared to the D-VSL#3 group ($P=0.0011$). The median time of clinical remission of the main clinical sign (i.e., diarrhea or vomiting) of the dogs in the D-CT group was 4.8 days (range, 2.5 to 7.0 days); while in D-VSL#3 group an improvement was observed in a median of 10.6 days (range, 5.0 to 15.0 days).

TGF-β+

While the TGF-β expression increased significantly in both treatment groups between T0 and T1 ($P=0.0043$), the magnitude of this increase was significantly greater for dogs in the D-VSL#3 group than those for the dogs in the D-CT group at Time T1 ($P=0.0008$) without any obvious preferential localization throughout the small or large intestine (Figure 1).

CD3+ T-cells

The number of CD3+ lymphocytes was increased in dogs with IBD in both treatment groups at T0 (before treatment), with small or large intestinal involvement depending of intestinal tract involved in the inflammatory process. CD3+ T-cells were significantly ($P<0.0001$) reduced at T1 relative to T0 in both treatment groups (Figure 1 and 2), and there were no significant differences in the magnitude of this reduction between both treatments ($P=0.7527$).

FoxP3+ cells

At T0, there were no significant differences in the number of cells between the two treatment groups (Figure 1). No significant increase in FoxP3+ cells was observed in the D-CT group ($P=0.3296$). However, a significant increase in FoxP3+ cells between T0 and T1 ($P=0.0001$) was observed in the D-VSL#3 group.

Expression of AJC proteins

Mucosal biopsies were evaluated in both treatment groups at T1 (Figure 3 and 4). Additionally, samples from 3 ED dogs were utilized as controls. Occludin was significantly lower in the D-CT group ($P<0.0001$) compared to the D-VSL#3 and ED groups. In contrast, Claudin-2 in the large intestine was significantly higher in the D-CT group ($P<0.0001$; Table 3, Figure 4) compared to the D-VSL#3 and ED groups. No significant differences were observed for the other AJC proteins.

Stain intensity was subjectively graded as absent (−), weak (+), moderate (++), or strong (+++), and the localization and distribution of chromogen was noted. Occludin-specific labeling was most intense at the epithelial cell AJC (Fig. 3), with fainter labeling observed along the basolateral membranes. This staining

Table 3. Summary statistics for evaluated markers.

	VSL#3		CT		ED	
	T0	T1	T0	T1	-	
Histology score	11.5 (7–14)	4 (3–7)	9 (3–14)	3 (0–9)	—	P<0.0001*
CIBDAI score	7 (5–10)	0 (0–2)	9 (7–13)	0 (0–3)	—	P<0.0001*
CD3+ cells[†]	3318 (±447.1)	1204 (±240.4)	3427 (±1813)	845 (±849)	—	P<0.0001*
FoxP3+ cells[†]	26.9 (±26.9)	353.6 (±175.1)	11.1 (±9.5)	51.5 (±32.2)	—	P = 0.0001**
TGF-β+ cells[†]	35.4 (±30.3)	791.8 (±771.9)	32.6 (±21.8)	136.7 (±122)	—	P = 0.0043*
Citrulline (µg/ml)	3.46 (±1.82)	4.66 (±2.34)	—	—	—	P = 0.0113
E-caderin[†]	—	4767 (±2288)	—	4735 (±1319)	4877 (±971)	P = 0.9467
Occludin[†]	—	4523 (±1366)	—	814 (±387)	6511 (±1239)	P<0.0001
Claudin-2 (SI) [†]	—	4274 (±1201)	—	4421 (±1293)	4994 (±1183)	P = 0.7944
Claudin-2 (LI) [†]	—	525 (±264)	—	5771 (±1588)	680 (±305)	P<0.0001

Numerical data are expressed as median (range) for histology and CIBDAI and as mean (± SD) for remaining data.
*significant differences between T0 and T1 in both treatment groups.
**significant differences between T0 and T1 for the VSL#3 group only.
[†]cells per 62,500 µm².

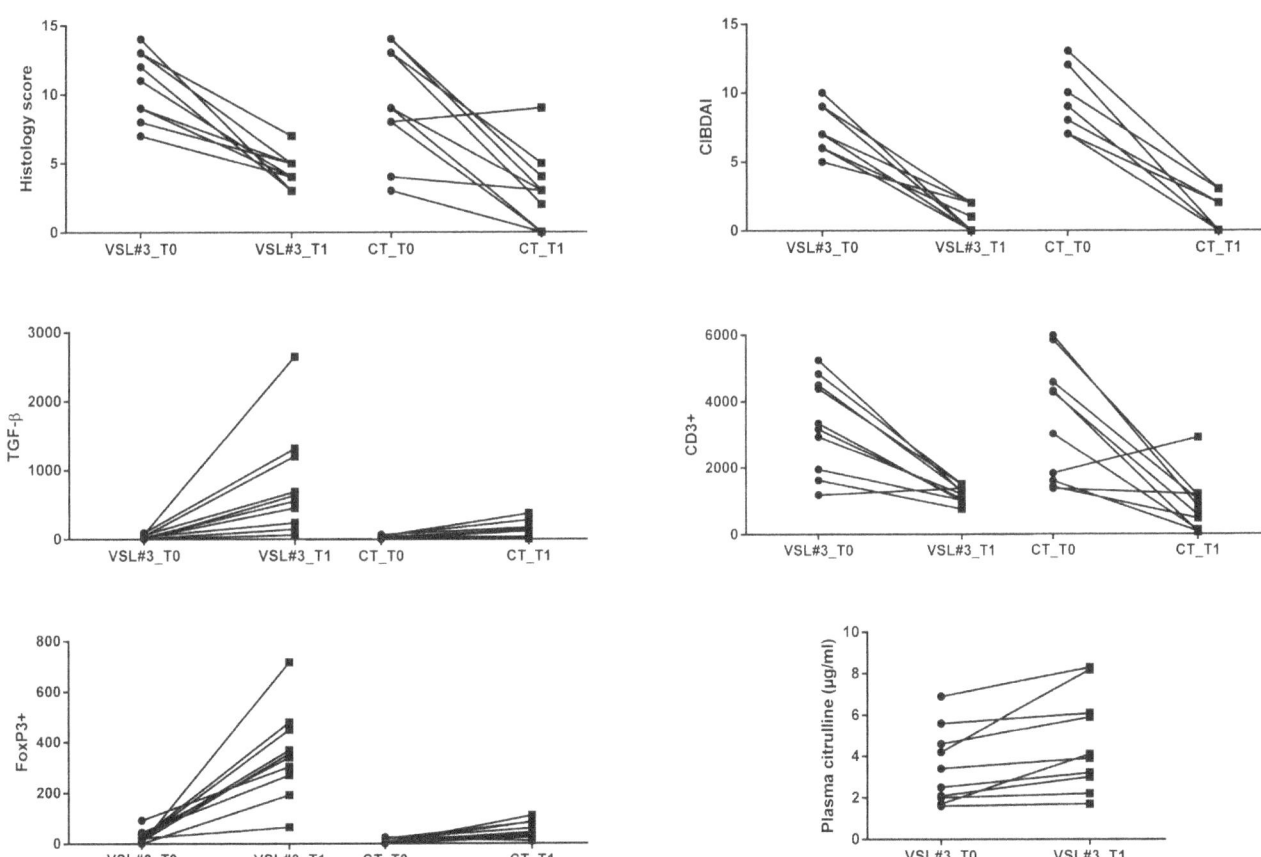

Figure 1. Results for histology scores, CIBDAI, CD3+ cells, FoxP3+ cells, TGF-β cells, and plasma citrulline concentrations. Significant differences between baseline (T0) and 30 days after the end of therapy (T1) were observed for all parameters in both treatment groups except the expression of FoxP3+ T-cells in the CT group (P = 0.3296). While TGF-β increased significantly in both treatment groups, the magnitude of the increase was significantly higher in dogs treated with VSL#3 (P = 0.0008). Data for CD3+ cells, FoxP3+ cells, TGF-β cells expressed as cells per 62,500 µm².

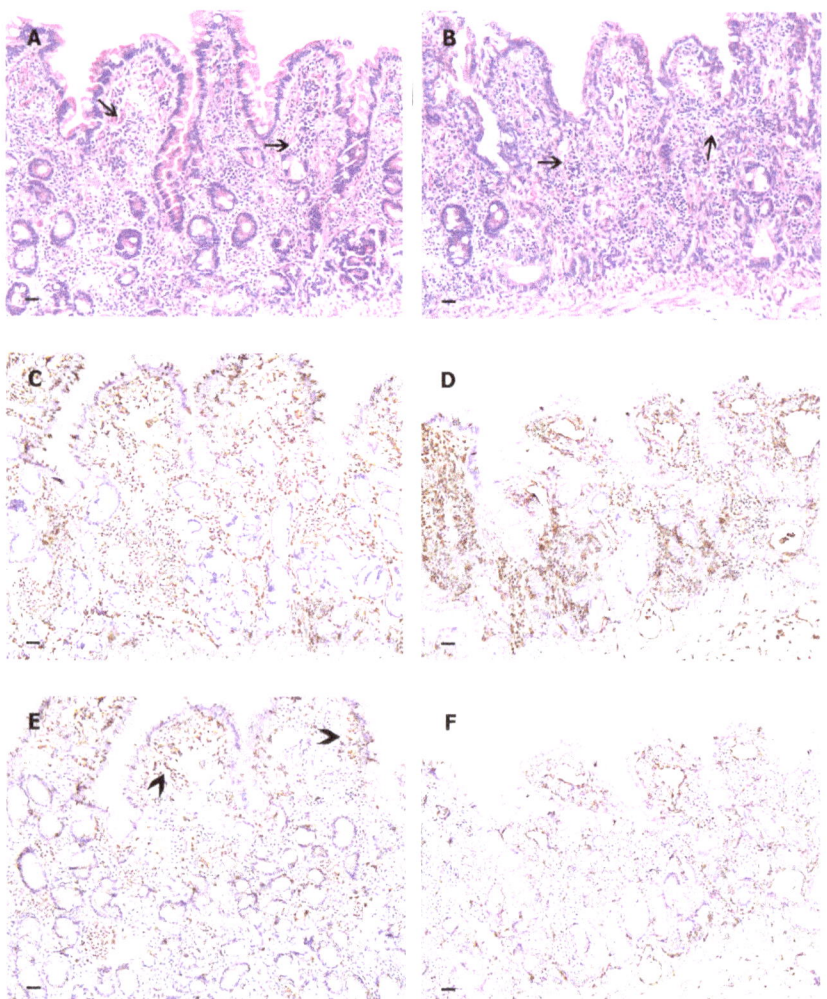

Figure 2. Histology of intestinal mucosa of dogs with IBD after treatment with VSL#3 (A, C, E) and CT (B, D, F). A residual inflammatory infiltrate with lymphocytic-plasmacytic cells (arrows) is evident after the therapy in both samples (H&E, 40X). In both treatment groups ssimilar patterns of mucosal infiltrations with CD3+ T-lymphocytes are evident (C and D). Infiltration with Fox-P3+ cells are proportionally increased in a sample belonging to a VSL#3 treated dog (E) compared to the sample from CT treated dog (F). Note the particular Fox-P3+ T-cells concentrations at the apical portion of villi in the VSL#3 treated dog (E) (arrow-heads) (IHC, ABC method, Harris haematoxylin nuclear counterstain, 40X).

appeared to be uniformly expressed throughout the epithelium of both ED and D-VSL#3 groups. On the contrary, weak to absent expression was observed in the luminal epithelium and in the small intestinal glands of some dogs in the D-CT group. No discernible difference in the distribution or staining intensity of E-cadherin was observed between normal and affected dogs; as, the overall intensity of E-cadherin expression decreased from the luminal epithelium to the distal crypts. At the luminal epithelium, labeling was uniform along the length of the intercellular junction, while the expression was becoming polarized toward the AJC in the distal glands/crypts. E-cadherin-specific labeling was restricted to the AJC and basolateral membranes of intestinal epithelial cells. Moreover, there was little evidence of specific labeling outside the epithelium. In ED and D-VSL#3 groups, claudin-2 was readily detectable in the duodenal epithelium and glands and in colonic crypt epithelium. Immunostaining decreased in intensity from the distal to the proximal crypt and was minimally detectable at the luminal surface of the colon. Claudin-2-specific labeling was largely restricted to the epithelial cell AJC, with some punctate basolateral labeling noted. However, claudin-2 expression was increased in the proximal crypt and luminal epithelium in all CT dogs.

Citrulline

Plasma citrulline concentrations increased significantly in dogs in the D-VSL#3 group between T0 and T1 (P = 0.0113; Figure 1 and Table 3).

Microbiota analysis

The qPCR results (Figure 5) showed that at T0 dogs with IBD (in both treatment groups) had significantly decreased abundance of *Faecalibacterium* spp. (p = 0.008) and *Turicibacter* spp. (p = 0.0078) when compared to healthy dogs. No other bacterial groups evaluated were significantly different compared to the healthy dogs. The qPCR analysis, revealed that the abundance of *Faecalibacterium* spp. increased significantly in the D-VSL#3 group (T1 vs T0; p = 0.03) but not in the D-CT group (T1. vs. T0; p = 0.46). No significant changes were observed for any other bacterial groups in response to treatment.

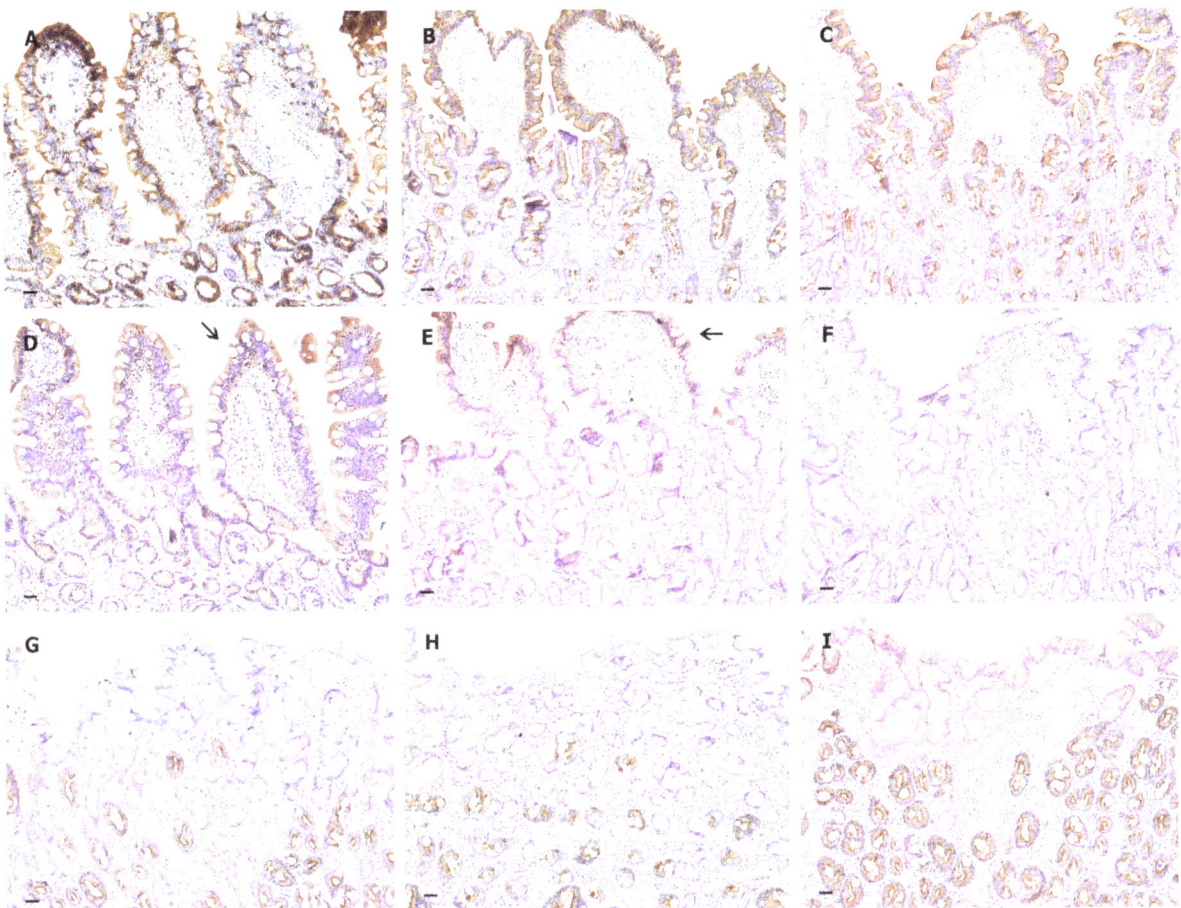

Figure 3. Expression of AJC proteins in the intestinal mucosa of control dogs (ED group) (A, D, G) and dogs treated with VSL#3 (B, E, H) or CT (C, F, I). No discernible differences in the distribution or staining intensity of E-cadherin are observed between normal mucosa (A) and IBD samples (B and C); the overall intensity of E-cadherin staining decreased from the luminal epithelium to the distal crypts. Occludin-specific labelling is most intense at the epithelial cell AJC (arrows) of the luminal epithelium covering the apical portion of villi in ED (D) and VSL#3 (E); a weak to absent expression is observed in the luminal epithelium and in some intestinal glands of the small intestine of the CT sample (F). In colonic samples belonging to ED (G) and VSL#3 (H) groups, claudin-2 is readily detectable only in the colonic crypt epithelium, decreasing in intensity from the distal to the proximal crypt and becoming barely detectable at the luminal surface of the colon. In contrast, claudin-2 expression is increased in the proximal crypt and luminal epithelium of all samples from CT dogs (I).

Discussion

In this study, 20 dogs with long standing IBD were randomized to receive either a probiotic containing VSL#3 strains (SIVOY) or a combination therapy of prednisone and metronidazole. Using a statistical analysis that takes into account differences between the treatment groups at enrollment (T0) as well as post-treatment (T1), we observed differences in some of the evaluated variables depending on the treatment regimen. Histology scores, CIBDAI, and infiltration with mucosal CD3+ T-cells decreased significantly in both treatment groups, and there was no significant effect between the two treatments. FoxP3+ T-cells increased in dogs treated with VSL#3 but not in the D-CT group. While TGF-β+ cells increased significantly in both treatment groups, the magnitude of the increase was significantly greater in dogs treated with VSL#3. The expression of occludin and claudin-2 was also significantly different between dogs treated with probiotic VSL#3 compared to combination therapy.

Although the etiology of canine IBD is poorly understood, there is evidence from clinical observations, studies in humans, and animal models to incriminate the intestinal microbiota as one factor influencing aberrant host responses. Evidence for the role of enteric microbiota in the pathogenesis of IBD in humans is supported by clinical responses to fecal stream diversion in patients with Crohn's disease (CD) and antimicrobial therapy in both CD and ulcerative colitis (UC) patients [3,10]. Furthermore, genetic mutations in NOD2/CARD15 and TLR-4 (Toll-like-receptor-4) in IBD patients make them less able to respond to bacterial components, resulting in defective innate immune responses to enteric microbiota [24]. Dietary factors also appear to play a role in mediating mucosal inflammation in dogs based on the beneficial clinical response to elimination or "hypoallergenic" diets in many of these animals [16]. All 20 patients enrolled in this study were diagnosed as having long-standing idiopathic IBD, and in the past had undergone unsuccessful dietary trials (e.g., elimination diets to exclude adverse food events). During the study period, all dogs remained on their pre-trial diets, and no dietary changes were performed as part of the here presented study. These diets were similar in nutritional composition across both treatment groups. In the D-VSL#3 group, 4 dogs were on an Adult Dry Maintenance Diet, 4 dogs were on a novel protein diet, and 2 dogs were on an elimination diet. The dogs in the D-CT group had a similar diet

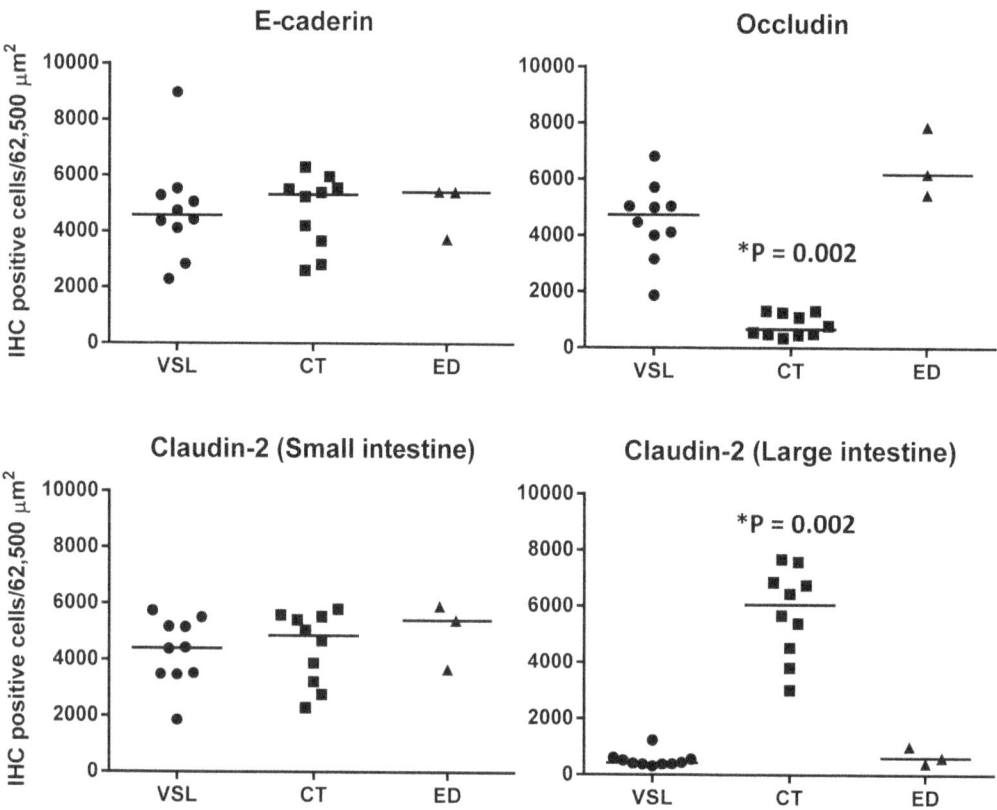

Figure 4. Expression of AJC proteins. Mucosal biopsies were evaluated after the end of treatment (T1) either with the probiotic (VSL) or combination drug therapy (CT), and compared to archived mucosal samples from dogs euthanized for non-gastrointestinal disorders (ED). (*significantly different to the other 2 groups; line denotes median).

distribution, with 4 dogs receiving an Adult Dry Maintenance Diet, 5 dogs receiving a novel protein diet, and 1 dog receiving an elimination diet. Therefore, it is unlikely that the diets were a significant confounding factor in this study since they were broadly similar in both treatment groups.

Probiotic therapy is becoming increasingly popular in veterinary medicine, and has been recommended for the treatment or prevention of a variety of gastrointestinal disorders. However, few objective studies attesting clinical efficacy of probiotics for gastroenteritis are available. The administration of probiotics to dogs with IBD represents warrants further investigation. It has been demonstrated that colitis in both humans and mice is associated with increased levels of cytokines such as TNF-α, IL-6, IL-12p70 and IL-23 [25,26]. Thus, a proper selection of probiotic strains for the treatment of IBD is crucial and should be based on their potential ability to induce an anti-inflammatory pattern of cytokines (IL- 10^{high}, TGF-β^{high}, IL-12p70^{low}, IL-23^{low}, TNF-α^{low}) and attenuate intestinal inflammation. Apart from their immuno-modulatory effects, it has been suggested that probiotics have an effect on the gut microbiome by their antimicrobial activities directed toward intestinal pathogens [3]. In humans, VSL#3 showed efficacy for maintenance of remission of ulcerative colitis [14]. To our knowledge, this is the first study to investigate the microbiological, histological, and immunomodulatory effects of VSL#3 in dogs with IBD and to compare these effects to a commonly used combination therapy with prednisone and metronidazole.

Based on qPCR analysis, only the bacterial genera *Faecalibacterium* and *Turicibacter* were found to be significantly decreased in dogs

with IBD at baseline relative to healthy dogs. These results are consistent with recent findings [9], where *Faecalibacterium* was also the predominant bacterial group decreased in fecal samples of dogs with IBD. *Faecalibacterium prausnitzii* is also consistently decreased in human IBD patients and considered an important bacterial group for maintaining microbial homeostasis [27]. A suggested direct immunomodulatory mechanism of action of *F. prausnitzii* is the secretion of metabolites with anti-inflammatory effects, due to blocking NF-κB activation and IL-8 production [27]. In contrast to previous findings, Fusobacteria were not significantly different in dogs with IBD relative to healthy dogs in the current study [9]. Neither treatment with VSL#3 nor with conventional therapy led to major changes in the overall microbial abundance of bacterial phyla (Bacteroidetes, Firmicutes, Fusobacteria) as assessed 30 days following discontinuation of treatment (at T1). Figure 5 illustrates that there were no significant luminal increases in the administered probiotic genera (i.e., *Bifidobacterium*, *Lactobacillus*, and *Streptococcus*) in dogs receiving VSL#3. This is in line with some studies demonstrating that the administration of probiotics do appear to have only minor and transient detectable effects on fecal microbial communities as assessed by qPCR assays or sequencing of 16S rRNA genes [23,28,29]. In the VSL#3 group, however, *Faecalibacterium* spp. increased significantly after treatment, although a trend for an increase in this bacterial group was also observed in the CT group. These results are in line with a previous study, in which *Faecalibacterium* increased after 4 months of conventional treatment in dogs with IBD and this increase correlated with the improvement in clinical disease activity [9]. This would suggest that the significant increase in fecal

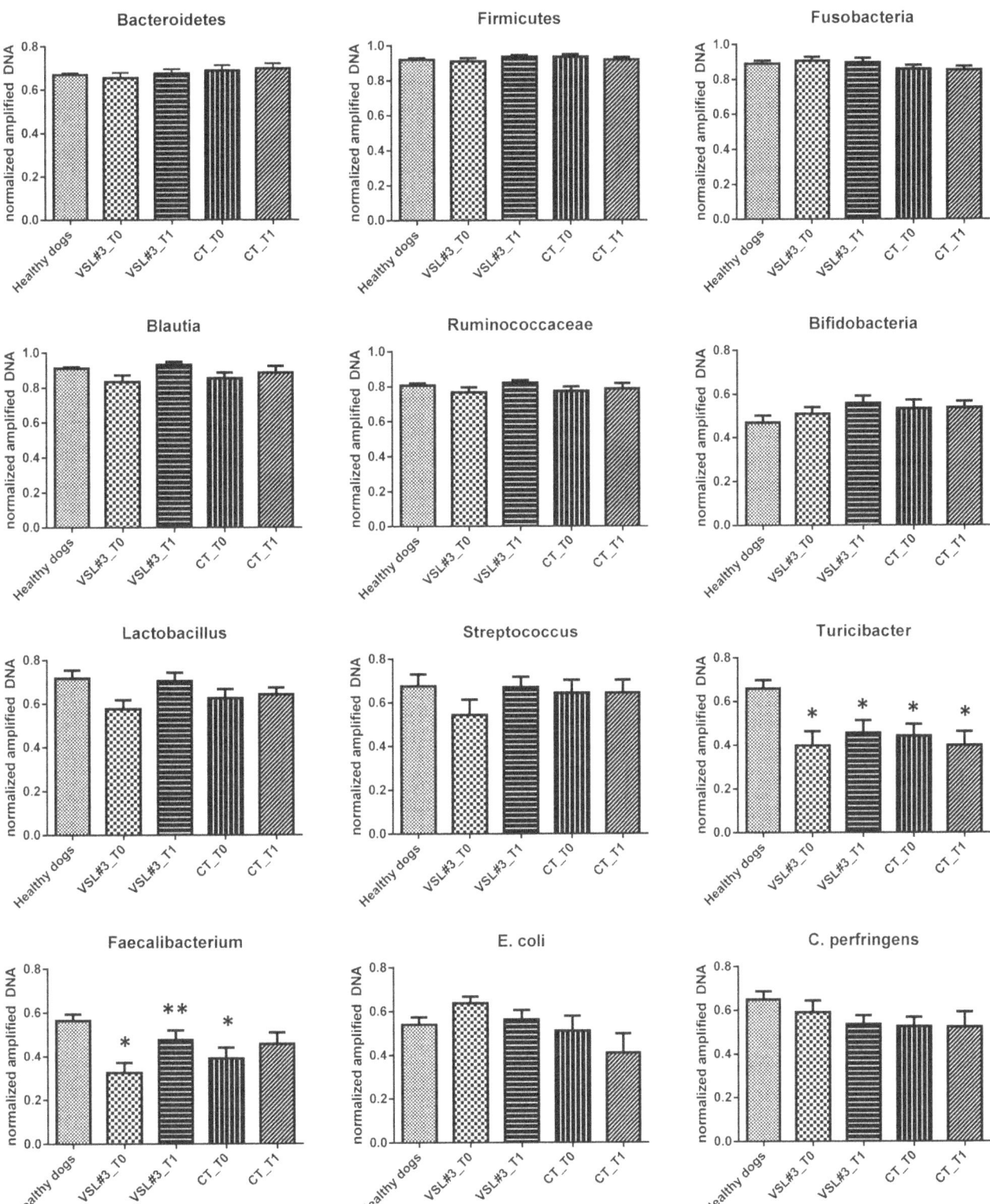

Figure 5. Results of quantitative PCR assays for selected bacterial groups. Dogs with IBD (in both treatment groups) had significantly decreased abundance of *Faecalibacterium* spp. (p = 0.008) and *Turicibacter* spp. (p = 0.0078) compared to the healthy dogs. *Faecalibacterium* spp. increased significantly in the VSL#3 treated dogs at T1 but not in the CT group. (*significantly different compared to healthy dogs; **significantly different after treatment compared to pre-treatment).

Faecalibacterium is not necessarily specific for the probiotic treatment, but may be a general indicator for normalization of fecal dysbiosis after long-term therapy. The *Faecalibacterium–Subdoligranulum* group is a major bacterial group in the canine gastrointestinal tract, comprising 16% of total bacterial counts in feces of healthy dogs and is believed to be of importance in canine gastrointestinal health [30]. Therefore, more-in depth studies evaluating the functional properties of canine *Faecalibacterium* strains are warranted. Some limitation of the microbiota analysis performed in this study need to be noted. Analyzing the fecal microbiota using sequencing of 16S rRNA genes may have revealed potential changes either in microbial diversity indices or in bacterial groups that were not covered by our qPCR assays. For technical reasons, a sequencing approach was not possible in this study. However, we have utilized qPCR assays targeting the microbiota on various phylogenetic levels and also targeting bacterial groups that are major bacterial groups in the canine intestine and that have been shown to be important in canine IBD [9]. Furthermore, in the current study, only fecal samples were analyzed, and the potential impact of treatment on the composition of the small intestinal mucosa-associated microbiota may have been missed. Previous studies have revealed that dogs with IBD have significant differences in small intestinal microbiota compared to controls, and future studies should evaluate the effect of probiotics on the small intestinal microbiota of these dogs [8]. Also, in this study we assessed the fecal microbiota 30 days after the discontinuation of therapy, and it is possible that a transient change in the fecal microbiota during the administration period may have remained undetected and/or changed during the 30 days post-treatment.

It has been speculated that IBD is associated with a loss of intestinal barrier function, as multiple genes encoding for proteins responsible for maintenance of intestinal barrier function (i.e., those encoding for claudin-8, metallothionein, and matrix metalloproteinases) were down-regulated in dogs with IBD in a previous study [31]. The observation that the expression and distribution of occludin and claudin-2 in the large intestine were not significantly different between dogs treated with VSL#3 and the non-IBD control dogs (ED group), but were significantly different compared to the D-CT group, suggests potential effects of VSL#3 on intestinal barrier function, warranting further studies [32]. Similar changes in the distribution of claudin-2 expression have been observed in humans with active UC, where claudin-2 was detected at the surface epithelium [33]. Similarly, down regulation of occludin has been observed in the intestinal mucosa of patients with both UC and CD [34]. Here we compared the expression patterns of AJC proteins between healthy dogs (euthanized dogs; group ED) and dogs with IBD after the two different types of treatment (VSL # 3 or CT treated dogs). The expression pattern of AJC proteins in the ED group was similar to that described by Ohta *et al.* in healthy dogs [35]. In contrast, based on our results it seems that dogs in the CT-group had a greater deviation from the physiological conditions in expression of Claudin-2 in the colon. This particular expression pattern resembles that observed in samples from the colon of dogs with colitis [21]. While we cannot conclusively state that there was an improvement in the expression pattern after probiotic treatment, as samples were not evaluated at T0, we speculate that the expression pattern of AJC proteins in dogs treated with VSL#3 appears to resembles more the physiological state as observed in healthy dogs [35]. Future studies are warranted to confirm this observation. At this point it remains also unclear why claudin-2 is increased in the large intestine of dogs treated with drug therapy,

and further work is needed to elucidate the mechanism behind this increased expression of claudin-2.

Dogs treated with VSL#3 showed significantly increased plasma citrulline concentrations 30 days after end of administration, suggesting restitution of the mucosal barrier. Plasma citrulline concentrations are a marker of global enterocyte mass in humans, rodents, and pigs [36], and have recently been shown to reflect intestinal mucosal recovery in response to severe injury in dogs [37]. Unfortunately, we were able to statistically evaluate the blood levels of citrulline only in the D-VSL#3 group, as plasma citrulline concentrations were not available for all dogs in the D-CT group. Because of the small samples size in the D-CT group, we decided not to perform any statistical analysis to compare plasma citrulline concentrations between treatments. Therefore, it is currently unknown whether the observed increase in plasma citrulline concentrations was specific for the treatment with VSL#3 strains, or would also be present in dogs treated with conventional therapy.

The immunohistochemical results showed cross-reactivity for canine tissues of all antibodies used in this study. This is in line with results from previous studies which have shown that these antibodies are useful for immunohistochemical assessment of canine tissues. In particular, cross-reactivity of the rat anti-human CD3 antigen, clone MCA1477, for canine CD3 positive T-lymphocytes has been shown previously on gastric tissue of dogs [20]. Cross-reactivity of the clone FJK-16s used to stain canine FoxP3-lymphocytes has been reported in another study [38]. Similarly, other authors have successfully used the monoclonal antibody against TGF-β positive dog lymphocytes (clone 1D11) [39]. Finally, the specificities of the antibodies used for canine AJC proteins (i.e., pAb anti-claudin-2 (PAD: MH44), anti-occludin (PAD: Z-T22), and mAb anti-E-cadherin (IgG2α, clone: 36) were, similarly to our study, also reported on sections of intestinal tissue in dogs with IBD [21].

The evaluation of immunomorphological variables suggests a potential anti-inflammatory effect of VSL#3 strains, as decreased mucosal CD3+ T-lymphocytes, and increased FoxP3+ and TGF-β+ positive cells were observed 30 days after the end of administration. Immunohistochemistry results showed a difference in the predominant immunophenotype of infiltrating cells in intestinal lamina propria of biopsies from VSL#3 treated dogs. More specifically, the VSL#3 treated dogs showed increases in CD3+/FoxP3+ cells (Figure 2) in the intestinal mucosa, while dogs treated with prednisone and metronidazole displayed an overall decrease in all inflammatory cell populations that was accompanied by a decrease of FoxP3+ lymphocytes and TGF-β expressing cells (Figure 2). These findings are consistent with a previous study in a mouse model, where VSL#3 also led to increased FoxP3+ expressing T-cells in intestinal lymphoid follicles [40]. In clinical studies with human IBD patients as well as studies on rodent models of IBD, VSL#3 has shown various other anti-inflammatory mechanisms. For example, VSL#3 was shown to induce heat-shock-proteins in intestinal epithelial cells (IEC) [41] or enhance proliferation of IL-10-dependent TGF-β-bearing regulatory T-cells in Th1-dependent murine colitis [42]. These variables have not been examined in the current study, and it would be useful to evaluate these markers in future clinical studies. Furthermore, qPCR quantification of both pro-inflammatory (i.e., TNF-α, IL1-β, IL-8) as well as regulatory genes (FoxP3, IL-10) would have been useful to perform since canine probes have already been published [43] and these studies showed increases in IL-8 in colorectal inflammation [44].

As limitations to this study it should be noted that only a small number of dogs was evaluated, and the power to detect differences in some of the evaluated variables may have been insufficient to detect differences between treatment groups. Furthermore, this was an open-label study and no placebo group was included. Ideally, the clinical effect of the treatment with probiotic strains should be evaluated in a double-blinded placebo controlled trial and compared to a non-treated group. However, in the case of chronic IBD, it is difficult to enroll a non-treated group as these dogs show chronic signs of disease, and therefore we chose in this study to compare the effects of VSL#3 strains to the commonly used combination therapy with prednisone and metronidazole. Our study results suggest that probiotic treatment induces differential anti-inflammatory immune responses when compared

to routine combination therapy as evidenced by significant increases in FoxP3+ cells and a significantly larger increase in TGF-β. The findings lay the foundation for future larger scale placebo controlled clinical studies to evaluate clinical benefits of probiotic VSL#3 strains in the treatment of dogs with IBD.

Author Contributions

Conceived and designed the experiments: GR GP AEJ JSS. Performed the experiments: GR GP MC APP JSS. Analyzed the data: GR NDC AEJ JMS JSS. Contributed reagents/materials/analysis tools: GR GP MC APP. Wrote the paper: GR JMS NDC AEJ JSS.

References

1. Suchodolski JS (2011) Companion animals symposium: Microbes and gastrointestinal health of dogs and cats. J Anim Sci 89: 1520–1530.
2. Allenspach K, House A, Smith K, McNeill FM, Hendricks A, et al. (2010) Evaluation of mucosal bacteria and histopathology, clinical disease activity and expression of Toll-like receptors in German shepherd dogs with chronic enteropathies. Vet Microbiol 146: 326–335.
3. Rioux KP, Madsen KL, Fedorak RN (2005) The role of enteric microflora in inflammatory bowel disease: Human and animal studies with probiotics and prebiotics. Gastroenterol Clin North Am 34: 465–482.
4. Contractor NV, Bassiri H, Reya T, Park AY, Baumgart DC, et al. (1998) Lymphoid hyperplasia, autoimmunity, and compromised intestinal intraepithelial lymphocyte development in colitis-free gnotobiotic IL-2-deficient mice. J Immunol 160: 385–394.
5. Sellon RK, Tonkonogy S, Schultz M, Dieleman LA, Grenther W, et al. (1998) Resident enteric bacteria are necessary for development of spontaneous colitis and immune system activation in interleukin-10-deficient mice. Infect Immun 66: 5224–5231.
6. Xenoulis PG, Palculict B, Allenspach K, Steiner JM, Van House AM, et al. (2008) Molecular-phylogenetic characterization of microbial communities imbalances in the small intestine of dogs with inflammatory bowel disease. FEMS Microbiol Ecol 66: 579–589.
7. Suchodolski JS, Xenoulis PG, Paddock CG, Steiner JM, Jergens AE (2010) Molecular analysis of the bacterial microbiota in duodenal biopsies from dogs with idiopathic inflammatory bowel disease. Vet Microbiol 142: 394–400.
8. Suchodolski JS, Dowd SE, Wilke V, Steiner JM, Jergens AE (2012) 16S rRNA Gene Pyrosequencing Reveals Bacterial Dysbiosis in the Duodenum of Dogs with Idiopathic Inflammatory Bowel Disease. Plos ONE 7: e39333.
9. Suchodolski JS, Markel ME, Garcia-Mazcorro JF, Unterer S, Heilmann RM, et al. (2012) The fecal microbiome in dogs with acute diarrhea and idiopathic inflammatory bowel disease. Plos ONE 7: e51907.
10. Sutherland L, Singleton J, Sessions J, Hanauer S, Krawitt E, et al. (1991) Double-Blind, Placebo Controlled Trial of Metronidazole in Crohns-Disease. Gut 32: 1071–1075.
11. Jergens AE, Crandell J, Morrison JA, Deitz K, Pressel M, et al. (2010) Comparison of oral prednisone and prednisone combined with metronidazole for induction therapy of canine inflammatory bowel disease: a randomized-controlled trial. J Vet Intern Med 24: 269–277.
12. Turner D, Levine A, Escher JC, Griffiths AM, Russell RK, et al. (2012) Management of pediatric ulcerative colitis: joint ECCO and ESPGHAN evidence-based consensus guidelines. J Pediatr Gastroenterol Nutr 55: 340–361.
13. Sauter SN, Benyacoub J, Allenspach K, Gaschen F, Ontsouka E, et al. (2006) Effects of probiotic bacteria in dogs with food responsive diarrhoea treated with an elimination diet. J Anim Physiol Anim Nutr (Berl) 90: 269–277.
14. Bibiloni R, Fedorak RN, Tannock GW, Madsen KL, Gionchetti P, et al. (2005) VSL#3 probiotic-mixture induces remission in patients with active ulcerative colitis. Am J Gastroenterol 100: 1539–1546.
15. Tursi A, Brandimarte G, Papa A, Giglio A, Elisei W, et al. (2010) Treatment of Relapsing Mild-to-Moderate Ulcerative Colitis With the Probiotic VSL#3 as Adjunctive to a Standard Pharmaceutical Treatment: A Double-Blind, Randomized, Placebo-Controlled Study. Am J Gastroenterol: 2218–2227.
16. Simpson KW, Jergens AE (2011) Pitfalls and progress in the diagnosis and management of canine inflammatory bowel disease. Vet Clin North Am Small Anim Pract 41: 381–398.
17. Jergens AE, Schreiner CA, Frank DE, Niyo Y, Ahrens FE, et al. (2003) A scoring index for disease activity in canine inflammatory bowel disease. J Vet Intern Med 17: 291–297.
18. German AJ, Helps CR, Hall EJ, Day MJ (2000) Cytokine mRNA expression in mucosal biopsies from German Shepherd dogs with small intestinal enteropathies. Dig Dis Sci 45: 7–17.
19. Day MJ, Bilzer T, Mansell J, Wilcock B, Hall EJ, et al. (2008) Histopathological Standards for the Diagnosis of Gastrointestinal Inflammation in Endoscopic Biopsy Samples from the Dog and Cat: A Report from the World Small Animal Veterinary Association Gastrointestinal Standardization Group. J Comp Pathol 138 S1–S43.
20. Rossi G, Fortuna D, Pancotto L, Renzoni G, Taccini E, et al. (2000) Immunohistochemical study of lymphocyte populations infiltrating the gastric mucosa of beagle dogs experimentally infected with Helicobacter pylori. Infect Immun 68: 4769–4772.
21. Ridyard AE, Brown JK, Rhind SM, Else RW, Simpson JW, et al. (2007) Apical junction complex protein expression in the canine colon: differential expression of claudin-2 in the colonic mucosa in dogs with idiopathic colitis. J Histochem Cytochem 55: 1049–1058.
22. Engel AG, Arahata K (1986) Mononuclear cells in myopathies: quantitation of functionally distinct subsets, recognition of antigen-specific cell-mediated cytotoxicity in some diseases, and implications for the pathogenesis of the different inflammatory myopathies. Hum Pathol 17: 704–721.
23. Garcia-Mazcorro JF, Lanerie DJ, Dowd SE, Paddock CG, Grutzner N, et al. (2011) Effect of a multi-species synbiotic formulation on fecal bacterial microbiota of healthy cats and dogs as evaluated by pyrosequencing. FEMS Microbiol Ecol 78: 542–554.
24. Franchimont D, Vermeire S, El Housni H, Pierik M, Van Steen K, et al. (2004) Deficient host-bacteria interactions in inflammatory bowel disease? The toll-like receptor (TLR)-4 Asp299gly polymorphism is associated with Crohn's disease and ulcerative colitis. Gut 53: 987–992.
25. Becker C, Dornhoff H, Neufert C, Fantini MC, Wirtz S, et al. (2006) Cutting edge: IL-23 cross-regulates IL-12 production in T cell-dependent experimental colitis. J Immunol 177: 2760–2764.
26. Fuss IJ, Becker C, Yang Z, Groden C, Hornung RL, et al. (2006) Both IL-12p70 and IL-23 are synthesized during active Crohn's disease and are down-regulated by treatment with anti-IL-12 p40 monoclonal antibody. Inflamm Bowel Dis 12: 9–15.
27. Sokol H, Pigneur B, Watterlot L, Lakhdari O, Bermudez-Humaran LG, et al. (2008) Faecalibacterium prausnitzii is an anti-inflammatory commensal bacterium identified by gut microbiota analysis of Crohn disease patients. PNAS 105: 16731–16736.
28. Vitali B, Ndagijimana M, Cruciani F, Carnevali P, Candela M, et al. (2010) Impact of a synbiotic food on the gut microbial ecology and metabolic profiles. BMC Microbiol 10: 4.
29. Larsen N, Vogensen FK, Gobel R, Michaelsen KF, Abu Al-Soud W, et al. (2011) Predominant genera of fecal microbiota in children with atopic dermatitis are not altered by intake of probiotic bacteria Lactobacillus acidophilus NCFM and Bifidobacterium animalis subsp. lactis Bi-07. FEMS Microbiol Ecol 75: 482–496.
30. Garcia-Mazcorro JF, Dowd SE, Poulsen J, Steiner JM, Suchodolski JS (2012) Abundance and short-term temporal variability of fecal microbiota in healthy dogs. MicrobiologyOpen 1: 340–347.
31. Wilke VL, Nettleton D, Wymore MJ, Gallup JM, Demirkale CY, et al. (2012) Gene expression in intestinal mucosal biopsy specimens obtained from dogs with chronic enteropathy. Am J Vet Res 73: 1219–1229.
32. Madsen K, Cornish A, Soper P, McKaigney C, Jijon H, et al. (2001) Probiotic bacteria enhance murine and human intestinal epithelial barrier function. Gastroenterol 121: 580–591.
33. Prasad S, Mingrino R, Kaukinen K, Hayes KL, Powell RM, et al. (2005) Inflammatory processes have differential effects on claudins 2, 3 and 4 in colonic epithelial cells. Lab Invest 85: 1139–1162.
34. Gassler N, Rohr C, Schneider A, Kartenbeck J, Bach A, et al. (2001) Inflammatory bowel disease is associated with changes of enterocytic junctions. Am J Physiol Gastrointest Liver Physiol 281: G216–228.
35. Ohta H, Yamaguchi T, Rajapakshage BK, Murakami M, Sasaki N, et al. (2011) Expression and subcellular localization of apical junction proteins in canine duodenal and colonic mucosa. Am J Vet Res 72: 1046–1051.
36. Curis E, Nicolis I, Osowska S, Zerrouk N, Benazeth S, et al. (2005) Almost all about citrullin in mammals. Amino Acids 29: 177–205.
37. Dossin O, Rupassara SI, Weng HY, Williams DA, Garlick PJ, et al. (2011) Effect of Parvoviral Enteritis on Plasma Citrulline Concentration in Dogs. J Vet Intern Med 25: 215–221.
38. Pinheiro D, Singh Y, Grant CR, Appleton RC, Sacchini F, et al. (2011) Phenotypic and functional characterization of a CD4(+) CD25(high) FOX-P3(high) regulatory T-cell population in the dog. Immunol 132: 111–122.

39. Colitz CM, Malarkey D, Dykstra MJ, McGahan MC, Davidson MG (2000) Histologic and immunohistochemical characterization of lens capsular plaques in dogs with cataracts. Am J Vet Res 61: 139–143.

40. Bassaganya-Riera J, Viladomiu M, Pedragosa M, De Simone C, Carbo A, et al. (2012) Probiotic bacteria produce conjugated linoleic acid locally in the gut that targets macrophage PPAR gamma to suppress colitis. Plos ONE 7: e31238.

41. Petrof EO, Kojima K, Ropeleski MJ, Musch MW, Tao Y, et al. (2004) Probiotics inhibit nuclear factor-kappaB and induce heat shock proteins in colonic epithelial cells through proteasome inhibition. Gastroenterol 127: 1474–1487.

42. Di Giacinto C, Marinaro M, Sanchez M, Strober W, Boirivant M (2005) Probiotics ameliorate recurrent Th1-mediated murine colitis by inducing IL-10 and IL-10-dependent TGF-beta-bearing regulatory cells. J Immunol 174: 3237–3246.

43. Ohta H, Takada K, Torisu S, Yuki M, Tamura Y, et al. (2013) Expression of CD4+ T cell cytokine genes in the colorectal mucosa of inflammatory colorectal polyps in miniature dachshunds. Vet Immunol Immunopathol 155: 259–263.

44. Tamura Y, Ohta H, Torisu S, Yuki M, Yokoyama N, et al. (2013) Markedly increased expression of interleukin-8 in the colorectal mucosa of inflammatory colorectal polyps in miniature dachshunds. Vet Immunol Immunopathol 156: 32–42.

45. Muhling M, Woolven-Allen J, Murrell JC, Joint I (2008) Improved group-specific PCR primers for denaturing gradient gel electrophoresis analysis of the genetic diversity of complex microbial communities. ISME J 2: 379–392.

46. Malinen E, Rinttila T, Kajander K, Matto J, Kassinen A, et al. (2005) Analysis of the fecal microbiota of irritable bowel syndrome patients and healthy controls with real-time PCR. Am J Gastroenterol 100: 373–382.

47. Garcia-Mazcorro JF, Suchodolski JS, Jones KR, Clark-Price SC, Dowd SE, et al. (2012) Effect of the proton pump inhibitor omeprazole on the gastrointestinal bacterial microbiota of healthy dogs. FEMS Microbiol Ecol 80: 624–636.

48. Wise MG, Siragusa GR (2005) Quantitative detection of Clostridium perfringens in the broiler fowl gastrointestinal tract by real-time PCR. Appl Environ Microbiol 71: 3911–3916.

49. Lubbs DC, Vester BM, Fastinger ND, Swanson KS (2009) Dietary protein concentration affects intestinal microbiota of adult cats: a study using DGGE and qPCR to evaluate differences in microbial populations in the feline gastrointestinal tract. J Anim Physiol Anim Nutr (Berl) 93: 113–121.

50. Rinttila T, Kassinen A, Malinen E, Krogius L, Palva A (2004) Development of an extensive set of 16S rDNA-targeted primers for quantification of pathogenic and indigenous bacteria in faecal samples by real-time PCR. J Appl Microbiol 97: 1166–1177.

51. Furet JP, Quenee P, Tailliez P (2004) Molecular quantification of lactic acid bacteria in fermented milk products using real-time quantitative PCR. Int J Food Microbiol 97: 197–207.

52. Walter J, Hertel C, Tannock GW, Lis CM, Munro K, et al. (2001) Detection of Lactobacillus, Pediococcus, Leuconostoc, and Weissella species in human feces by using group-specific PCR primers and denaturing gradient gel electrophoresis. Appl Environ Microbiol 67: 2578–2585.

53. Heilig HG, Zoetendal EG, Vaughan EE, Marteau P, Akkermans AD, et al. (2002) Molecular diversity of Lactobacillus spp. and other lactic acid bacteria in the human intestine as determined by specific amplification of 16S ribosomal DNA. Appl Environ Microbiol 68: 114–123.

11

Comparative Genome Analysis of *Lactobacillus rhamnosus* Clinical Isolates from Initial Stages of Dental Pulp Infection: Identification of a New Exopolysaccharide Cluster

Mangala A. Nadkarni[1,2]*, **Zhiliang Chen**[3], **Marc R. Wilkins**[3], **Neil Hunter**[1,2]

1 Institute of Dental Research, Westmead Centre for Oral Health and Westmead Millennium Institute, Westmead, New South Wales, Australia, **2** Faculty of Dentistry, The University of Sydney, Sydney, New South Wales, Australia, **3** Systems Biology Initiative, School of Biotechnology and Biomolecular Sciences, The University of New South Wales, Sydney, New South Wales, Australia

Abstract

The human oral microbiome has a major role in oral diseases including dental caries. Our studies on progression of caries infection through dentin and more recently, the invasion of vital dental pulp, detected *Lactobacillus rhamnosus* in the initial stages of infection of vital pulp tissue. In this study employing current high-throughput next generation sequencing technology we sought to obtain insight into genomic traits of tissue invasive *L. rhamnosus*, to recognise biomarkers that could provide an understanding of pathogenic potential of lactobacilli, generally regarded as safe. Roche GS FLX+ technology was used to generate whole genome sequences of two clinical isolates of *L. rhamnosus* infecting vital pulp. Detailed genome-wide comparison of the genetic profiles of tissue invasive *L. rhamnosus* with probiotic *L. rhamnosus* was performed to test the hypothesis that specific strains of *L. rhamnosus* possessing a unique gene complement are selected for the capacity to invade vital pulp tissue. Analysis identified 264 and 258 genes respectively, from dental pulp-invasive *L. rhamnosus* strains LRHMDP2 and LRHMDP3 isolated from two different subjects that were not present in the reference probiotic *L. rhamnosus* strain ATCC 53103 (GG). Distinct genome signatures identified included the presence of a modified exopolysaccharide cluster, a characteristic confirmed in a further six clinical isolates. Additional features of LRHMDP2 and LRHMDP3 were altered transcriptional regulators from RpoN, NtrC, MutR, ArsR and zinc-binding Cro/CI families, as well as changes in the two-component sensor kinase response regulator and ABC transporters for ferric iron. Both clinical isolates of *L. rhamnosus* contained a single *Spa*FED cluster, as in *L. rhamnosus* Lc705, instead of the two *Spa* clusters (*Spa*CBA and *Spa*FED) identified in *L. rhamnosus* ATCC 53103 (GG). Genomic distance analysis and SNP divergence confirmed a close relationship of the clinical isolates but segregation from the reference probiotic *L. rhamnosus* strain ATCC 53103 (GG).

Editor: Christiane Forestier, Université d'Auvergne Clermont 1, France

Funding: This study was supported partially by NIDCR grant R01 DE015272-07, by the Westmead Centre for Oral Health, by the Australian Government Super Science Scheme, the New South Wales State Government Science Leveraging Fund (SLF) and by the University of New South Wales. The funders had no role in study design, data collection and analysis, decision to publish, or preparation of the manuscript.

Competing Interests: The authors have declared that no competing interests exist.

* E-mail: mangala.nadkarni@sydney.edu.au

Introduction

The human oral microbiome has a major causal association with dental caries. Organic acids produced by microbial hydrolysis of dietary sugars cause demineralization of hydroxyapatite crystals leading to dental caries. The Human Microbe Identification Microarray (HOMIM) for identification of named and unnamed taxa associated with oral infections [1], in combination with pyrosequencing, has enabled the complex uncultured microbial ecosystem of the oral cavity to be more precisely defined [2–5]. Other techniques such as scanning electron microscopy (SEM), fluorescence *in-situ* hybridization (FISH) and confocal microscopy using fluorochrome-labelled 16S rRNA probes, have given insights into the community structure and spatial distribution of polymicrobial consortia in infected tissues [6,7]. More recently, these analyses have provided evidence indicating that a restricted group

of lactobacilli, particularly *Lactobacillus rhamnosus*, are implicated in the initial stages of infection of vital pulp tissue [6].

While many current techniques have allowed partial profiling of the human microbiome, important questions regarding functional attributes, metabolic specialization and virulence traits can only be answered by comprehensive post-genomic analysis of the ecosystems of the human body [8]. Current high-throughput Next Generation Sequencing has the power to provide detailed knowledge of the diverse array of genes that influence biochemical and physiological functions responsible for host-microbe associations. This enables the identification of bacterial traits that contribute to adhesion, invasion and evasion of host defences. In the present study, whole genome sequencing of two clinical isolates of *L. rhamnosus* using Roche GS FLX+ technology combined with detailed genome-wide analysis of the genetic profiles of tissue invasive *L. rhamnosus* compared with probiotic *L. rhamnosus*, was undertaken. The hypothesis to be tested was that invasive *L.*

rhamnosus possess a unique complement of genes that could facilitate invasion of dental pulp tissue.

We have generated draft genome sequences of two clinical isolates of *L. rhamnosus* strains, LRHMDP2 and LRHMDP3 [9] isolated from infected dental pulps from carious teeth. These represent the initial stages of infection of pulp tissue. Here we report comprehensive comparative genomic analysis of our two clinical isolates of *L. rhamnosus*, LRHMDP2 and LRHMDP3 (isolated from two different subjects) with probiotic *L. rhamnosus* GG (ATCC 53103) and explore the relevance of the genomic features of the clinical isolates to invasion of dental pulp. Both clinical isolates showed absence of typical exopolysaccharide (*eps*) and *Spa*CBA pilus clusters. Newly identified genes from the clinical isolates included a modified *eps* cluster, a two-component sensor kinase, a response regulator and ABC transporters for ferric iron. Additional genes detected in the clinical isolates included putative open reading frames encoding RNA polymerase sigma 54 factor RpoN, NtrC, MutR, ArsR and zinc-binding Cro/CI family transcriptional regulators. PTS system mannose, galactitol, mannitol, cellobiose and beta-glucoside-specific II components were also identified. This analysis provided evidence that clinical isolates of invasive *L. rhamnosus* differ from probiotic reference strains of this organism.

Results and Discussion

Genome sequencing

L. rhamnosus clinical isolates (LRHMDP2 and LRHMDP3) were sourced from infected dental pulps from carious teeth categorised as representing the initial stages of infection of vital pulp tissue from two different subjects [6] (see Methods section for detail description for selection criteria; **Tables S1, S2a, S2b**). These isolates were speciated by PCR analysis of 16S rRNA gene sequences using *Lactobacillus* genus-specific primers as described in an earlier report [10]. LRHMDP2 and LRHMDP3 16S rRNA gene sequences were 99% homologous to *L. rhamnosus*. Genomic DNA from both the clinical isolates was sequenced by a whole genome shotgun strategy using Roche GS FLX+ pyrosequencing at the Ramaciotti Centre for Gene Function Analysis, University of New South Wales (see Methods section for detailed sequencing protocol).

Genome properties

A total of 77,329 reads from *L. rhamnosus* LRHMDP2 and 75,985 reads from *L. rhamnosus* LRHMDP3 were generated to reach a depth of ~17-fold genome coverage, and assembled by Newbler Assembler 2.6 into 51 and 47 contigs longer than 200 base pairs, respectively (**Table 1**, [9]). Gene definition and annotation was performed by merging the result from the RAST (Rapid Annotation using Subsystem Technology) server and tRNAscan-SE. This showed that sequences of the two strains, *L. rhamnosus* LRHMDP2 and *L. rhamnosus* LRHMDP3, included 2,911,290 base pairs and 2,911,934 base pairs, respectively. The genomic sequence of *L. rhamnosus* LRHMDP2 comprised 2,910 coding sequences (CDSs), 7 rRNA loci and 56 tRNAs. In comparison, the genome sequence of *L. rhamnosus* LRHMDP3 contained 2,925 CDSs, 6 rRNA loci and 57 tRNAs. The G+C content of both clinical strains of *L. rhamnosus*, LRHMDP2 and LRHMDP3 was 46.6%, similar to *L. rhamnosus* GG and *L. rhamnosus* Lc705 (**Table 1, Figure 1** [9]). By pair-wise genome comparisons Lukjancenko *et al.* have reported 93.3% similarity between two *L. rhamnosus* strains [11].

Table 1. General Genome sequencing[a] features and annotation[b] features of *L. rhamnosus* clinical isolates and *L. rhamnosus* GG and *L. rhamnosus* Lc705 genomic features.

Organism	Total Reads	genome coverage	Number of contigs[c] (>200 bp)	Genome size (Mbp)	Coding sequences (CDS)	rRNA operons	tRNA genes	GC content (%)	Hypothetical proteins (total)
LRHMDP2[d]	77,329	17 X	51	2.91	2,910	7	56	46.6	815
LRHMDP3[d]	75,985	17 X	47	2.91	2,925	6	57	46.6	819
GG[e]	-	-	-	3.01	2944	5	57	47	734
Lc705[e]	-	-	-	3.03	2992	5	61	47	760

[a]Using Roche GS(FLX+) pyrosequencing.
[b]Based on results merged from the RAST (Rapid Annotation using Subsystem Technology) server and tRNAscan-SE.
[c]assembled by Newbler Assembler 2.6.
[d]Both *L.rhamnosus* clinical isolates (LRHMDP2 and LRHMDP3) were sourced from infected dental pulps from carious teeth categorised as representing the initial stages of infection of pulp tissue.
[e][13].

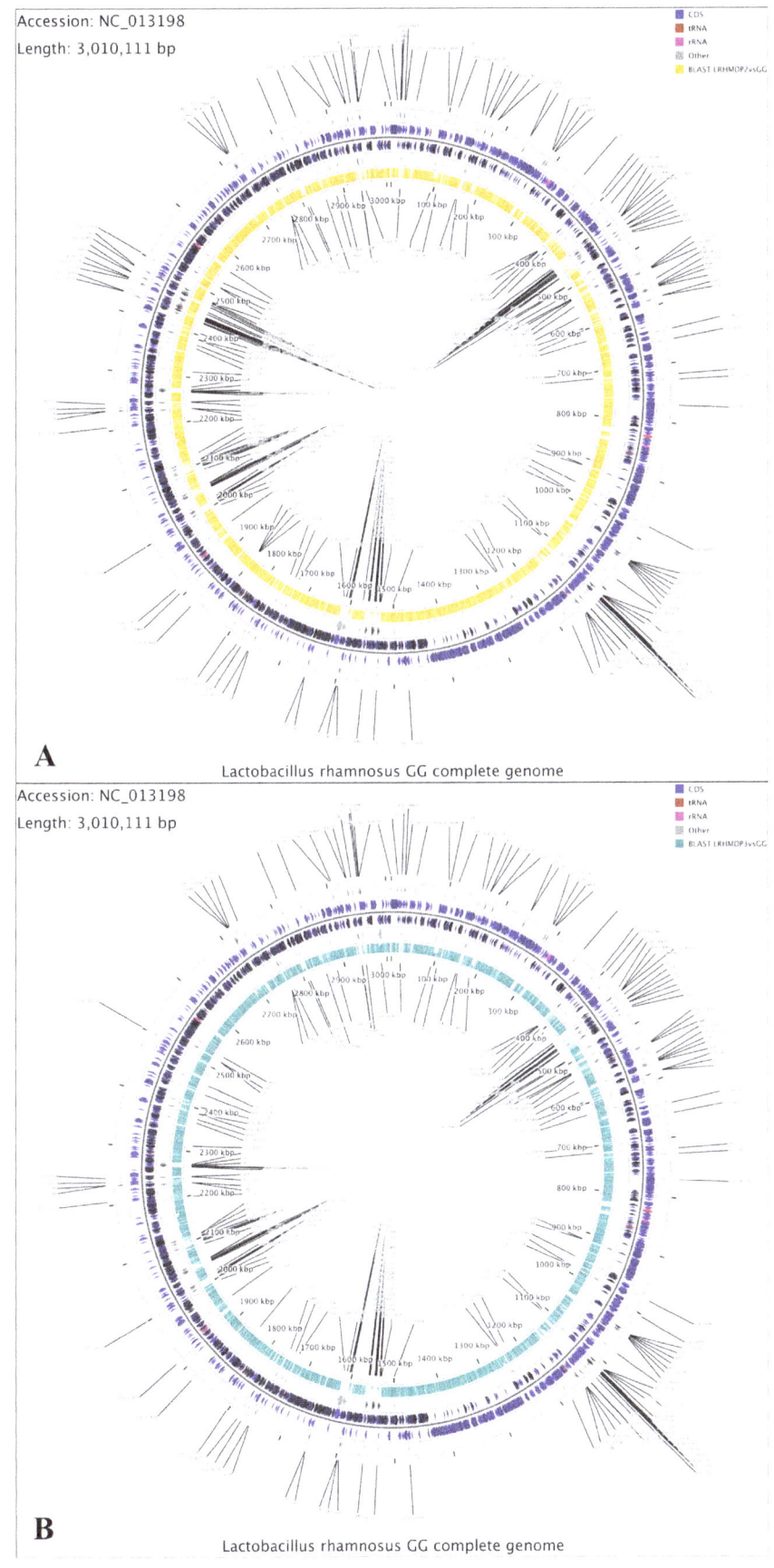

A   Lactobacillus rhamnosus GG complete genome

B   Lactobacillus rhamnosus GG complete genome

Figure 1. (A) Difference in gene content between *L. rhamnosus* **GG and** *L. rhamnosus* **LRHMDP2; (B) Difference in gene content between** *L. rhamnosus* **GG and** *L. rhamnosus* **LRHMDP3.** In both figures, genes from positive and negative strands of *L. rhamnosus* GG are shown in rings II and III (from outside). Protein coding genes, tRNA and rRNA genes are shown by different colours. Directions of the arrows indicate the translational directions. Genes found to be absent in *L.rhamnosus* LRHMDP2/LRHMDP3 are highlighted in rings I and IV, and are labelled using gene symbols. The innermost rings (V) represent the homology of *L.rhamnosus* LRHMDP2/LRHMPD3 against *L. rhamnosus* GG, regions. Gaps indicate lower BLAST scores between the genomes of *L. rhamnosus* LRHMDP2/LRHMDP3 and *L. rhamnosus* GG [48].

Comparative genome analysis

Genome annotation using the RAST server assigns functions to genes, enables assembly of metabolic pathways and makes connections into sub-systems represented in the genome to achieve comparative analysis with annotated genomes maintained in the SEED (http://www.theseed.org/) [12]. Biological roles could be assigned to 72.0% of CDSs for the genomic sequences of *L. rhamnosus* LRHMDP2 and *L. rhamnosus* LRHMDP3. A comparative annotation summary indicated 264 of the 2,910 coding sequences from *L. rhamnosus* strain LRHMDP2 and 258 of the 2,925 coding sequences from *L. rhamnosus* strain LRHMDP3 (~9% of coding sequences) in the clinical isolates were newly identified genes (**Table 1, Table 2**) where homologues could not be detected in the probiotic reference strain of *L. rhamnosus* ATCC 53103 (GG). New genes are defined as no BLAST hits found or the normalised BLAST score less than 2 where normalised BLAST score is calculated as BLAST score divided by alignment length. Some 9% of coding sequences attributed to genes from the probiotic *L. rhamnosus* strain GG as well as the other probiotic *L. rhamnosus* strain Lc705 were not detected in either clinical isolate (**Table 3**).

Comparisons based on sub-system category distribution highlighted distinctive features that discriminated the two clinical isolates of *L. rhamnosus*, (LRHMDP2 and LRHMDP3) and *L. rhamnosus* GG. Comparative genomic analysis between *L. rhamnosus* GG and the two strains, *L. rhamnosus* LRHMDP2 and *L. rhamnosus* LRHMDP3, based on 27 sub-system category distributions, showed differences in 6 distribution categories, namely; co-factors, vitamins, prosthetic groups and pigments; cell wall and capsule; virulence, disease and defence traits; phage, prophage, transposable elements, plasmids, iron acquisition and metabolism and carbohydrates (data not shown).

Compared to the *L. rhamnosus* GG genome, components not detected in either of the clinical isolate assemblies were mainly assigned to genes encoding transposases (IS5, IS30, IS3/IS911, IS150/IS3, IS4 family proteins) and phage-related products. Also, not detected in the clinical isolates were characteristic exopolysaccharide (*eps*) cluster genes, pilus-specific genes (the *spa*CBA cluster), mannose/fructose/sorbose- specific PTS system genes and genes encoding MerR and XRE family transcriptional regulators. Further omissions included type III restriction modification systems and

CRISPRs and CRISPR-associated (*cas*) genes, in addition to genes coding for hypothetical proteins (**Table S3**). Similarly, genes present in the other *L. rhamnosus* probiotic strain, *L. rhamnosus* Lc705 that were absent in both clinical isolate assemblies were mainly assigned to transposases (IS111A, IS1328, IS1533, IS116, IS110, IS902, IS4, IS1480, IS30, IS5, tnp4), phage-related products, exopolysaccharide (*eps*) clusters; also fructose-specific, sorbose-specific, galactitol-specific, lactose/cellobiose-specific and mannose-specific PTS systems. Also noted as absent from the clinical isolates were other carbohydrate metabolism genes and the type I restriction modification system in addition to genes encoding hypothetical proteins (**Table S4**).

The clinical isolates, *L. rhamnosus* LRHMDP2 and *L. rhamnosus* LRHMDP3 were also distinct from the other *L. rhamnosus* probiotic strain Lc705 as no plasmid genes could be identified in either as compared to *L. rhamnosus* Lc705 that harbours a 64.5 kb plasmid [13]. However, the two clinical isolates from infected dental pulp resembled *L. rhamnosus* Lc705 and other lactobacilli such as *L. gasseri*, *L. johnsonii* and *L. iners* in the absence of CRISPRs and *cas* detected in *L. rhamnosus* GG [13,14]. CRISPRs and *cas* comprise a distinct defence system present in ~47% bacterial genomes (http://crispr.u-psud.fr/crispr/). This system provides resistance against foreign genetic elements and bacteriophage predation by inactivating incoming DNA. Similarly, both clinical isolates resembled *L. rhamnosus* Lc705 in the absence of the *spa*CBA cluster but presence of a homologous *spa*FED cluster also present in *L. rhamnosus* GG. In Gram-positive bacteria the three subunits of the *spa*CBA gene cluster are flanked by transposable elements and in *L. rhamnosus* strain GG this cluster is located on a genomic island GGISL2 [13]. Absence of the *spa*CBA gene cluster in the clinical isolates could be related to the paucity in these strains of transposable elements that regulate horizontal gene transfer.

Both clinical isolates were characterised by a modified *eps* cluster, a two-component sensor kinase, a response regulator and ABC transporters for ferric iron, RNA polymerase sigma 54 factor RpoN, MutR, NtrC, ArsR and zinc-binding Cro/CI family transcriptional regulators, ThiJ/PfpI family protein and PTS system mannose, galactitol, mannitol, cellobiose and beta-glucoside-specific II components in addition to genes encoding hypothetical proteins (**Table S5a, S5b**). The majority of the coding sequences for newly identified genes had closest homology with genes in bacteria from the phylum *Firmicutes*.

Inter-strain divergence between *L. rhamnosus* strains LRHMDP2 and LRHMDP3

Although the probiotic *L. rhamnosus* strains GG and Lc705 were shown to exhibit 331 and 383 strain-specific genes respectively [13], the inter-strain difference between clinical isolates *L. rhamnosus* LRHMDP2 and *L. rhamnosus* LRHMDP3 was limited. Analysis revealed 26 strain-specific genes for *L. rhamnosus* LRHMDP2 and 28 strain-specific genes for *L. rhamnosus* LRHMDP3 (**Table S6a, S6b**). Many of the strain-specific genes encoded hypothetical proteins. Biological function could be assigned to 4 of these genes from *L. rhamnosus* LRHMDP2 and 8 from *L. rhamnosus* LRHMDP3. Type II secretory pathway/competence component/competence protein ComGC, immunity

Table 2. Number of new genes identified in the assemblies of clinical isolates, *L. rhamnosus* LRHMDP2 and *L. rhamnosus* LRHMDP3[a].

Assembly Name	Number of genes
L. rhamnosus LRHMDP2	264[b]
L. rhamnosus LRHMDP3	258[c]

[a]based on comparative studies with reference probiotic *L. rhamnosus* strains ATCC53103 (GG) and Lc705.
[b]inclusive of 176 genes coding for hypothetical proteins.
[c]inclusive of 166 genes coding for hypothetical proteins.

Table 3. Number of genes from reference strain *L. rhamnosus* ATCC 53103 (GG) and *L. rhamnosus* Lc705 not detected in both assemblies[a].

Reference probiotic *L. rhamnosus* strains	Number of genes
L. rhamnosus ATCC 53103 (GG)	275[b]
L. rhamnosus Lc705	331[c]

[a]L. rhamnosus LRHMDP2 and L. rhamnosus LRHMDP3.
[b]inclusive of 77 genes coding for hypothetical proteins.
[c]inclusive of 106 genes coding for hypothetical proteins and 63 plasmid genes.

protein PlnI, membrane-bound protease CAAX family and adhesion exoprotein were unique to *L. rhamnosus* LRHMDP2. DinG family ATP-dependent helicase YoaA, a putative regulator of the mannose operon ManO, cytochrome d ubiquinol oxidase subunit I, phage lysin, flagellar hook-length control protein FliK, phosphotransferase system, phosphocarrier protein HPr, phosphoenolpyruvate-protein phosphotransferase of PTS system and transcriptional regulator CtsR, were identified only in *L. rhamnosus* LRHMDP3. Genome sequencing has identified CtsR (class III stress gene repressor) in other *L. rhamnosus* strains. However, CtsR is well characterised in *Lactobacillus plantarum* - found in vegetable, meat and dairy products and also as a natural inhabitant of the mammalian gut. CtsR plays an important role in the response to environmental stress conditions [15]. CtsR is also considered to play an important role in stress resistance and virulence in *Listeria monocytogenes*, a firmicute that causes food-borne illness and is associated with invasive gastrointestinal infection [16].

Genomic distance and SNP divergence

The genomic difference between the genomic sequences of two *L. rhamnosus* clinical isolates and eleven other *L. rhamnosus* strains from NCBI was analysed. **Figure S1** shows a dendrogram that clusters the 13 different *L. rhamnosus* strains based on full genome nucleotide-level comparisons. The branches of the dendrogram are scaled to show the distance between genomes. The two clinical isolates show small genomic difference, and are found in a different part of the dendogram compared to the other probiotic *L. rhamnosus* dairy isolates, intestinal isolates from healthy humans and isolates from feces of healthy humans.

A more specific SNP divergence analysis between the probiotic *L. rhamnosus* strain GG and the two *L. rhamnosus* clinical isolates identified 50,459 SNPs with 95% confidence (**Table S7**). Of these, 6,899 SNPs were found to be located in non-coding regions. Both clinical isolates exhibited 50,370 SNPs in common when compared with GG. Only 30 SNPs were common between *L. rhamnosus* GG and *L. rhamnosus* LRHMDP2 compared to *L. rhamnosus* LRHMDP3. *L. rhamnosus* GG and *L. rhamnosus* LRHMDP3 shared only 45 SNPs compared to *L. rhamnosus* LRHMDP2. These findings indicate that SNP divergence between the two clinical isolates is 0.2%. The divergence equivalent to 50,459 SNPs between the probiotic *L. rhamnosus* strain GG and the two *L. rhamnosus* clinical isolates, confirmed the segregation of the clinical isolates.

Exopolysaccharide cluster

Most of the lactic acid bacteria (LAB) produce exopolysaccharide (EPS) and the composition of exopolysaccharide has been shown to influence adhesion properties of probiotic strains [17,18]. Metabolic shunting away from glycolysis is considered to be an important driver of EPS formation [19]. The organization of an exopolysaccharide (*eps*) gene cluster from LAB includes genes encoding glycosyltransferases (key enzymes for biosynthesis of the EPS repeating unit located in the central region of the *eps* cluster), polysaccharide biosynthesis export protein and transcriptional regulator, in addition to chain length determinant protein (*wzd* ORF) at one end of the *eps* gene cluster and protein-tyrosine phosphatase (*wzb* ORF) on the other end of the *eps* gene cluster [17,19,20]. To date, the *eps* cluster genes of only two probiotic *L. rhamnosus* strains GG and ATCC 9595 have been characterised in detail [18,21].

The *eps* cluster from the two clinical isolates, *L. rhamnosus* LRHMDP2 and *L. rhamnosus* LRHMDP3, is composed of 15 ORFs (12.44 kb) and 14 ORFs (11.55 kb) respectively, with notable absence of genes (*rmlACBD*) involved in dTDP-rhamnose biosynthesis (**Figure 2, Table 4, Table 5**). Genes involved in dTDP-rhamnose biosynthesis were identified in the 16.37 kb *eps* gene cluster of *L. rhamnosus* GG consisting of 17 putative ORFs [18] and the 18.7 kb *eps* gene cluster of *L. rhamnosus* ATCC 9595 consisting of 18 putative ORFs [21]. However, insertion of the putative transposase DDE domain from *L. casei* ATCC 334 (62% identity) was found to inactivate *rmlA* from the *eps* cluster of *L. rhamnosus* GG [18]. In both clinical isolates, a degenerate transposase belonging to the TnpB_IS66 superfamily and with 97% identity with the homolog from *L. rhamnosus* LMS2-1 was identified at a similar location within the *eps* cluster (**Table 4, Table 5**). The absence of *rmlACBD* genes involved in the dTDP-rhamnose biosynthetic enzyme system in the *eps* cluster from both clinical isolates implies an alternate route for EPS synthesis employed by the clinical isolates as compared to *L. rhamnosus* GG and *L. rhamnosus* ATCC 9595 [19]. The *eps* cluster of both clinical isolates also comprises 3 glycosyltranferases as opposed to 5 glycosyltranferases in *L. rhamnosus* GG [18] and in *L. rhamnosus* ATCC 9595 [21]. The presence of family 2 glycosyltransferases showing 50% amino acid identity to a homolog from *Lactobacillus pentosus* strain IG1 suggests a different pathway for polymerisation of monosaccharide repeating units. Family 2 glycosyltransferases are inverting glycosyltransferases, implying probable β-glycosidic linkage in the repeating unit of EPS (www.cazy.org/glycosyltransferases.html). A near-identical EPS biosynthetic gene cluster in *L. rhamnosus* strains ATCC 9595 and *L. rhamnosus* Lc705 produces rhamnose-rich EPS as opposed to the galactose-rich EPS from *L. rhamnosus* GG. On a broader scale, however, the organisation of the *eps* gene cluster from all the *L. rhamnosus* strains studied including the two clinical isolates, remains in close agreement. For instance, the *wzb* ORF as part of the *eps* cluster from *L. rhamnosus* strain GG and both clinical isolates are predicted to share 100% amino acid sequence identity. Remaining genes of the *eps* cluster from both clinical isolates showed closest identity to two *eps* cluster genes from *L. rhamnosus* strain LMS2-1, *Lactobacillus zeae* strain KCTC 3804 and *Lactobacillus pentosus* strain IG1 and a single *eps* cluster gene each from *L. rhamnosus* strains GG, Lc705

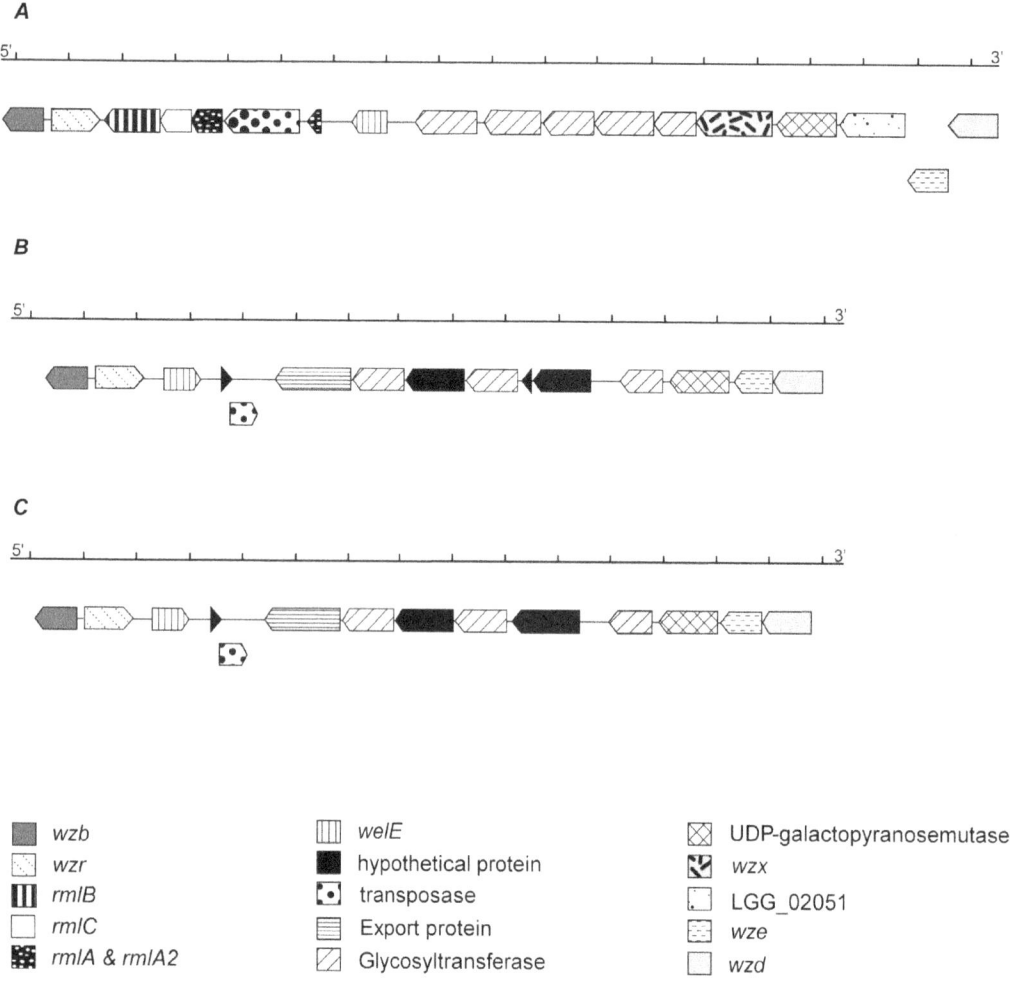

Figure 2. Genomic organisation of EPS gene cluster of *L. rhamnosus* GG (A), *L. rhamnosus* LRHMDP2 (B) and *L. rhamnosus* LRHMDP3 (C). Nomenclature analogy of exopolysaccharide/capsular gene cluster is maintained in accordance to *L rhamnosus* ATCC53103/GG and ATCC 9595 *eps* cluster [18,21].

and HN001. Biological function is unknown for four ORFs from the *L. rhamnosus* LRHMDP2 *eps* gene cluster and 3 ORFs from the *L. rhamnosus* LRHMDP3 *eps* gene cluster whereas in *L. rhamnosus* strains GG, Lc705 and ATCC 9595, biological function could be attributed to most of the predicted translated products.

The genomic organisation of the *eps* cluster genes from *L. rhamnosus* GG showed that, except for one ORF (*wzr*), all other ORFs of the *eps* cluster are in reverse orientation. Whereas 11 ORFs in *L. rhamnosus* LRHMDP2 and 10 ORFs in *L. rhamnosus* LRHMDP3 are in reverse orientation to the remaining ORFs of the *eps* cluster (**Figure 2**). In both clinical isolates, in addition to *wzr* (cell envelope-associated transcriptional attenuator LytR-CpsA-Psr, subfamily F2), priming glycosyltransferase, *welE* (undecaprenyl-phosphate galactosephosphotransferase) also maintained a forward direction. These findings imply that in the clinical isolates, in addition to organisation, transcriptional direction of *wzr* and *welE* could regulate expression of *eps* cluster genes.

Minor variation in structure of the polysaccharide is found to influence adherence and biofilm formation and the nature of immune responses. A deletion mutant for the priming glycosyl-transferase (*welE*) of *L. rhamnosus* GG showed altered heteropoly-saccharide composition with increased adherence and biofilm

capacity compared with the galactose-rich EPS from wild type *L. rhamnosus* GG [18]. A homolog for the priming glycosyltransferase *welE* from *L. rhamnosus* HN001 and a homolog for *wchA* from *L. pentosus* IG1 were detected in the *eps* cluster from both clinical isolates. In *S. pneumoniae*, *wchA* encodes the initial transferase transferring glucose-1-phosphate to undecaprenol phosphate, a similar function to *WbaP*, a galactosyl-1-phosphate transferase of the *Salmonella enterica* O-antigen gene cluster that catalyzes the first step of O-antigen synthesis [22,23].

Phosphotransferase system (PTS) transporters, ATP-binding cassette (ABC) transporters and sugar transporters

Genome-wide studies, using microarray-based expression analysis and also signature-tagged mutagenesis (STM) have shown that, in many bacteria, catabolism of complex carbohydrates could play a crucial role in the pathogenesis of disease [24]. Lactobacilli, including probiotic lactobacilli with a broad spectrum pattern of carbohydrate utilisation, were found to encode PTS transporters as a dominant mechanism for carbohydrate transport [13,25,26]. In *L. rhamnosus* LRHMDP2 and *L. rhamnosus* LRHMDP3, 85 and 86 PTS transporters respectively, were identified (equivalent to

Table 4. Exopolysaccharide cluster genes from *L. rhamnosus* LRHMDP2.

ORF	Size (bp)	Predicted encoded function	Predicted domain(s) present in encoded ORF	Best BLASTx hit (accession no.)		% Amino acid identity
				Protein	Organism	
wzb	768	Manganese-dependent protein-tyrosine phosphatase	No hit in Pfam database	Wzb (ABV54212.1)	*L. rhamnosus* GG	100
wzr	894	Cell envelope-associated transcriptional attenuator LytR-CpsA-Psr, subfamily F2	1 transmembrane domain and LytR_cpsA_psr domain (PF03816)	Wzr (YP_003174725.1)	*L. rhamnosus* Lc705	95
welE	669	Undecaprenyl-phosphate galactosephosphotransferase	Bacterial sugar transferase (PF02397)	WelE (ABS71033.1)	*L. rhamnosus* HN001	85
	198	hypothetical protein	low complexity	conserved hypothetical protein (ZP_04441887.1)	*L. rhamnosus* LMS2-1	95
	516	Degenerate transposase	TnpB_IS66 super family (PF05717)	Transposase (ZP_04441884.1)	*Lactobacillus rhamnosus* LMS2-1	97
	1425	polysaccharide biosynthesis export protein	Polysacc_synt (PF01943) and 6 transmembrane domains	Oligosaccharide translocase (YP_003352680.1)	*Lactococcus lactis* subsp. *lactis* KF147	43
	1011	Glycosyltransferase family 2	Glycosyltransferase family 2 (PF00535)	WchA (CCC15460.1)	*L. pentosus* IG1	51
	1086	hypothetical protein	Polysaccharide pyruvyltransferase (PF04230)	Polysaccharide pyruvyltransferase (YP_005047298.1)	*Clostridium clariflavum* DSM 19732	44
	978	Glycosyltransferase group 2 family protein	Glycosyltransferase family 2 (PF00535)	WelL (CCC15461.1)	*Lactobacillus pentosus* IG1	50
	210	hypothetical protein	2 transmembrane domains	Endoplasmic reticulum metallopeptidase 1 (CCH00546.1)	*Fibrella aestuarina* BUZ 2	34
	1101	hypothetical protein	9 transmembrane domains	hypothetical protein (ZP_06987467.1)	*Bacteroides* sp. 3_1_19	25
	792	Glycosyltransferase	No hit in Pfam database	WciB (ABQ58967.1)	*Streptococcus oralis*	53
	1122	UDP-galactopyranosemutase	UDP-galactopyranosemutase(PF03275)	UDP-galactopyranose mutase (ZP_09453942.1)	*L. zeae* KCTC 3804	88
wze	756	Tyrosine-protein kinase EpsD	ATPase MipZ (PF09140)andCbiA (PF01656)	Wze (ZP_09453943.1)	*L. zeae* KCTC 3804	89
wzd	915	Tyrosine-protein kinase transmembrane modulator EpsC	Chain length determinant protein (PF02706) and 1 transmembrane domain	Wzd (ACN94846.1)	*L. rhamnosus* GG	79

~3% of coding sequences) (**Table 6**). Not all PTS transporters were complete (encoding for all subunits). Both clinical isolates encoded multiple clusters for cellobiose uptake and hydrolysis as observed in *L. casei* [27]. Unlike *L. rhamnosus* GG, maltose/maltodextrin metabolism in the clinical isolates could be associated with the ABC transporter, a mechanism also adopted by *L. casei* and *Lactococcus lactis* [26,28]. The PTS for mannose-specific IIA, IIB, IIC, IID, galactitol-specific IIA, IIC, mannitol-specific IIA, lactose-specific IIB, cellobiose-specific IIB, IIC and PTS system beta-glucoside-specific IIA, IIB, IIC identified in both clinical isolates, were not homologous to the genes in the genome of *L. rhamnosus* GG (**Table S5a, S5b**).

Sixty nine ATP-binding cassette (ABC) transporters identified in clinical isolates included multidrug transporter, antimicrobial peptide transporter, bacteriocin/lantibiotic transporter, phosphate, nitrate/sulfonate/bicarbonate, alkane sulfonates and ferric iron, zinc and manganese transporters, oligopeptide and amino acid, glutamine, proline, glycine, betain transporters, osmotically activated L-carnitine/cholin (OpuCA, CB, CC, CD) transporters, excinuclease, spermidine, putrescine transport, glycerol-3-phosphate (UgpCAE), ribose (RbsDACB), N-acetyl-D-glucosamine,

maltose/maltodextrin transporters and multisugar transporters. Most of the ABC transporters were also detected in *L. rhamnosus* GG except for ferric iron, lantibiotic, alkane sulfonates, L-carnitine/cholin, ribose, N-acetyl-D-glucosamine and maltose/maltodextrin transporters. In both clinical isolates, ferric iron ABC transporters (iron-binding protein; LRHMDP2_1748, LRHMDP3_480 and permease protein; LRHMDP2_1750, LRHMDP3_478) were associated with two-component sensor kinase (LRHMDP2_1746, LRHMDP3_482) and response regulator (LRHMDP2_1747, LRHMDP3_481). Nearest homologs for ferric iron ABC transporters (iron-binding protein and permease protein), two-component sensor kinase and response regulator could be identified in *L. casei* and *L. paracasei* strains.

Genome-wide analysis has shown that virulence attributed to ABC transporters is associated with uptake of nutrients, metal ions such as iron, zinc, and manganese and cell attachment in a given physiological niche [29,30]. However, ABC transport systems, particularly important in bacterial virulence in disease models, were detected in clinical isolates as well as probiotic *L. rhamnosus* GG [29,30]. ABC transporters for antimicrobial peptides were detected in both clinical isolates as well as in probiotic *L. rhamnosus*

Table 5. Exopolysaccharide cluster genes from *L. rhamnosus* LRHMDP3.

ORF	Size (bp)	Predicted encoded function	Predicted domain(s) present in encoded ORF	Best BLASTx hit (accession no.) Protein	Organism	% Amino acid identity
wzb	768	Manganese-dependent protein-tyrosine phosphatase	No hit in Pfam database	Wzb (ABV54212.1)	*L. rhamnosus* GG	100
wzr	894	Cell envelope-associated transcriptional attenuator LytR-CpsA-Psr, subfamily F2	1 transmembrane domain and LytR_cpsA_psr domain (PF03816)	Wzr (YP_003174725.1)	*L. rhamnosus* Lc705	95
welE	669	Undecaprenyl-phosphate galactosephosphotransferase	Bacterial sugar transferase (PF02397)	WelE (ABS71033.1)	*L. rhamnosus* HN001	85
	198	hypothetical protein	low complexity	conserved hypothetical protein (ZP_04441887.1)	*L. rhamnosus* LMS2-1	95
	516	Degenerate transposase	TnpB_IS66 super family (PF05717)	Transposase (ZP_04441884.1)	*Lactobacillus rhamnosus* LMS2-1	97
	1425	polysaccharide biosynthesis export protein	Polysacc_synt (PF01943) and 6 transmembrane domains	Oligosaccharide translocase (YP_003352680.1)	*Lactococcus lactis* subsp. *lactis* KF147	43
	1011	Glycosyltransferase family 2	Glycosyltransferase family 2 (PF00535)	WchA (CCC15460.1)	*L. pentosus* IG1	51
	1086	hypothetical protein	Polysaccharide pyruvyltransferase (PF04230)	Polysaccharide pyruvyltransferase (YP_005047298.1)	*Clostridium clariflavum* DSM 19732	44
	981	Glycosyltransferase group 2 family protein	Glycosyltransferase family 2 (PF00535)	WelL (CCC15461.1)	*Lactobacillus pentosus* IG1	50
	1323	hypothetical protein	9 transmembrane domains	hypothetical protein (ZP_06987467.1)	*Bacteroides* sp. 3_1_19	24
	792	Glycosyltransferase	No hit in Pfam database	WciB (ABQ58967.1)	*Streptococcus oralis*	53
	1122	UDP-galactopyranosemutase	UDP-galactopyranosemutase(PF03275)	UDP-galactopyranose mutase (ZP_09453942.1)	*L. zeae* KCTC 3804	88
wze	756	Tyrosine-protein kinase EpsD	ATPase MipZ (PF09140)andCbiA (PF01656)	Wze (ZP_09453943.1)	*L. zeae* KCTC 3804	89
wzd	915	Tyrosine-protein kinase transmembrane modulator EpsC	Chain length determinant protein (PF02706) and 1 transmembrane domain	Wzd (ACN94846.1)	*L. rhamnosus* GG	79

Table 6. PTS, ABC and sugar transporters in *L. rhamnosus* GG, LRHMDP2, LRHMDP3.

Organism	PTS transporters	ABC transporter	sugar transporters
LRHMDP2[a]	85	69	70
LRHMDP3[a]	86	70	71
GG[b]	75	85	56

[a]Count from annotations.
[b]count from KEGG.

GG. Resistance towards host defence molecules such as antimicrobial peptides is identified as an important virulence phenotype and considered as a target for antimicrobial therapy for infectious diseases of humans [31].

Surface-exposed proteins

Surface-exposed proteins were identified using genome-wide analysis for predicted signal sequences; sortase motif, lipid attachment motif, choline binding motif, YSIRK signal peptide for Gram-positive cell-wall attached proteins, membrane proteins and exported proteins (**Table 7**). In *L. rhamnosus* LRHMDP2 and *L. rhamnosus* LRHMDP3, as well as in *L. rhamnosus* GG, surface proteins with YSIRK signal peptides found in other Gram-positive cell-wall attached proteins and putative choline binding motifs, were absent. However, 12 LPXTG-type cell wall surface anchor proteins could be identified in *L. rhamnosus* LRHMDP2 and 10 in *L. rhamnosus* LRHMDP3 (**Table 8**). As expected, surface-exposed pilus-specific proteins (SpaCBA) of *L. rhamnosus* GG were not detected in either clinical isolate (see comparative genomic analysis). LPXTG-type Gram-positive anchor regions with designated biological functions were identified in both clinical isolates. These included a putative cell wall surface anchor family protein with a repeating unit of a collagen binding domain, a subtilisin-like serine protease, a putative cell wall anchored protein SasC (LPXTG motif), similar to MabA, a modulator of *L. rhamnosus* GG adhesion and biofilm formation and a cell wall surface anchor family protein (internalinJ), renamed as mucus binding factor (MBF) in *L. rhamnosus* GG. Other LPXTG-type Gram-positive anchor regions are currently identified as hypothetical proteins with no assigned biological function.

The putative cell wall surface anchor family protein with a repeat unit of collagen binding protein domain B is unique to clinical isolates and no homolog could be identified in *L. rhamnosus*

GG. Collagen binding domains (CBD) from the bacterial adhesion domain superfamily are commonly detected in Firmicutes [32]. Collagen is one of the most common components of the extracellular matrix (ECM) and binding to ECM frequently precedes the invasion of host tissues. Organic material in dentinal matrix mainly consists of collagen type I fibrils and it has been suggested that adhesion to collagen components within dentinal tubules assists bacterial invasion [33]. In earlier studies we elucidated the microbiome invading dentinal tubules *in situ* [6]. Based on the phases of infection, we could identify lactobacilli consistently in the limited and established stages of infection of dental pulp [6]. Persistence of *Enterococcus faecalis* in root canals of failed endodontic treatment was associated with collagen binding protein and serine protease activity [34]. Similarly, in *S. mutans*, a collagen-binding protein, Cnm, is implicated as a virulence factor that could contribute to infection [35]. Growing evidence for association between Microbial Surface Component Recognizing Adhesive Matrix Molecules (MSCRAMM) and disease processes implies that a cell wall surface anchor family protein with repeat units of a collagen binding domain and subtilisin-like serine protease activity could play a crucial role in the apparent invasive behaviour of *L. rhamnosus*, which could be confirmed by *in-situ* mRNA expression studies in the context of tissue pathophysiology. A subtilisin-like serine protease identified in both clinical isolates was not detected in *L. rhamnosus* GG, but is present in *L. rhamnosus* strain Lc705 [13].

In both clinical isolates the putative cell wall anchored protein SasC (LPXTG motif), a 200 amino acid Gram-positive anchor (LRHMDP2_2569 and LRHMDP3_2533), showed 97% identity to the LPXTG-motif cell wall anchor domain of the extracellular matrix protein, MabA (LGG_01865) from *L. rhamnosus* GG. However, in both clinical isolates, except for the LPXTG motif, no DUF1542 (domain of unknown function 1542) could be identified. MabA containing LPXTG-motif cell wall anchor domain with multiple DUF1542 repeat domains, was characterised as a modulator of adhesion and biofilm formation in probiotic *L. rhamnosus* GG [36]. In *S. aureus*, SasC comprises an N-terminal signal peptide with a YSIRK motif, a C-terminal LPXTG motif and a FIVAR (found in various architectures) motif that includes 17 DUF1542 repeat domains. SasC is identified as a novel factor involved in cell aggregation and biofilm formation, and is implicated in *S. aureus* colonization during infection [37]. The putative cell wall anchored protein in *L. rhamnosus* GG is devoid of the saccharide-binding FIVAR domain that mediates hyaluronate and fibronectin binding [36,38]. In both clinical isolates, a putative homolog of the cell-wall-anchored protein SasC (LPXTG Motif)

Table 7. Surface exposed proteins in *L. rhamnosus* GG, LRHMDP2, LRHMDP3.

Surface exposed proteins	*L. rhamnosus* **GG**	*L. rhamnosus* **LRHMDP2**	*L. rhamnosus* **LRHMDP3**
Putative Signal Peptides[a]	253	235	179
Candidate lipoprotein signal peptides[b]	145	141	143
Putative choline-binding motifs	0	0	0
Signal peptide YSIRK for Gram-positive cell wall-attached proteins[c]	0	0	0
Membrane proteins	756	715	706
Exported proteins	98	135	136

[a]Identified by SignalP4.1 using a cutoff of Y score lower limit of 0.3.
[b]Matches to pattern {DERK}(6)-[LIVMFWSTAG](2)-[LIVMFSTAGCQ]-[AGS]-C by prosite scan:
[c]Identified from proteins predicted to have signal sequences by [YS]-[SA]-[IL]-[RK](2)-x(3)-G-x(2)-S.

Table 8. LPXTG-type Gram-positive anchor regions.

LRHMDP2:

gi|411186888|gb|EKS54010.1|	hypothetical protein LRHMDP2_251
gi|411186173|gb|EKS53298.1|	hypothetical protein LRHMDP2_549
gi|411185004|gb|EKS52134.1|	Subtilisin-like serine protease
gi|411183640|gb|EKS50777.1|	hypothetical protein LRHMDP2_1916
gi|411183327|gb|EKS50466.1|	internalin J
gi|411182566|gb|EKS49713.1|	Adhesion exoprotein
gi|411182429|gb|EKS49578.1|	hypothetical protein LRHMDP2_2346
gi|411181795|gb|EKS48956.1|	putative cell-wall-anchored protein SasC (LPXTG motif)
gi|411181505|gb|EKS48676.1|	Phage tail fiber protein
gi|411181513|gb|EKS48684.1|	hypothetical protein LRHMDP2_2752
gi|411181406|gb|EKS48582.1|	hypothetical protein LRHMDP2_2786
gi|411181363|gb|EKS48543.1|	Cell wall surface anchor family protein putative

LRHMDP3:

gi|411185129|gb|EKS52258.1|	hypothetical protein LRHMDP3_629
gi|411184803|gb|EKS51934.1|	hypothetical protein LRHMDP3_1001
gi|411184385|gb|EKS51518.1|	Subtilisin-like serine protease
gi|411182972|gb|EKS50114.1|	cell surface protein
gi|411182504|gb|EKS49652.1|	hypothetical protein LRHMDP3_2273
gi|411182229|gb|EKS49382.1|	hypothetical protein LRHMDP3_2400
gi|411181925|gb|EKS49083.1|	putative cell-wall-anchored protein SasC (LPXTG motif)
gi|411181547|gb|EKS48716.1|	cell wall surface anchor family protein
gi|411181369|gb|EKS48548.1|	hypothetical protein LRHMDP3_2824
gi|411181377|gb|EKS48556.1|	Phage tail fiber protein
gi|411181289|gb|EKS48474.1|	Cell wall surface anchor family protein putative

locus, LRHMDP2_2895 (comprising 1716 amino acids) and LRHMDP3_2912 (comprising 1744 amino acids) with no putative conserved domains, was also detected, showing 96% identity to LGG_01865.

Altered domain structure of the cell wall anchored proteins in both clinical isolates and in probiotic GG due to the absence of YSIRK signal peptide surface proteins, as well as absence of DUF1542 repeat domains in clinical isolates, requires further investigation.

The cell wall surface anchor family protein (internalinJ) identified in both clinical isolates showed 98% sequence identity to internalin J (LGG_02337) from *L. rhamnosus* GG. This protein was characterised and renamed as mucus binding factor (MBF) [39]. MBF, together with MabA protein, is postulated to provide ancillary support in the adhesion process once *Spa*C has established primary contact with host cells. In *E. faecalis* on the other hand, internalin was shown to confer invasiveness via a "zipper" mechanism similar to that observed during the internalin-E-cadherin-mediated entry of *L. monocytogenes* [40]. A *Spa*CBA cluster could not be detected in either clinical isolate; however, internalin J could have pathogenic attributes.

Progression of caries is associated with changes in nutrient availability from carbohydrate rich to a nutrient limiting tissue environment. Of note, response regulators, RpoN, NtrC, MutR and zinc-binding Cro/CI were among the newly identified genes in both clinical isolates. Of interest was RpoN (σ 54 RNA polymerase), whose nearest homolog (showing ~86%–89% identity) was in *L. casei*. In *L. casei*, *L. plantarum* and *L. monocytogenes*, mannose PTS is controlled by σ 54 RNA polymerase [41]. In addition, ferric iron ABC transporters (iron-binding protein and permease protein) associated with two-component sensor kinase and response regulator, are unique to both clinical isolates. Positive and negative transcriptional regulators as well as two-component regulatory systems, have a significant role in the virulence and pathogenicity of the Gram-positive pathogen, *L. monocytogenes*. Although the aetiology of acute infection differs from chronic infection, virulence/pathogenicity islands recognised in pathogenic bacteria associated with acute infections could not be identified in either clinical isolate within the confidence limits of the draft genomes. Pathogenic Gram-positive bacteria involved in infectious diseases invariably express exopolysaccharide (or capsule), pili, PTS transporters, ABC transporters and surface-exposed proteins with adhesive properties that contribute to virulence. Paradoxically, in addition to the presence of five genomic islands, the health benefit conferred by probiotic *L. rhamnosus* GG is attributed to exopolysaccharide and pilus protein encoded by the *Spa*CBA pilus cluster that is considered to establish primary contact in adhesive processes. This mechanism is supported by ancillary involvement of MBF and MabA proteins for prolonged persistence of *L. rhamnosus* GG in the gastrointestinal tract.

Comparative genomics based on COG analysis of bifidobacteria, lactobacilli and other probiotic bacteria has also indicated that virulence genes are not a separate category, but instead pathogenic strains are characterised by variation in genes assigned to COG categories M (cell wall/membrane biosynthesis) and O (post-translational modifications and chaperones) [11]. Cai *et al.* proposed a model for niche-associated evolution in *L. casei* showing evolution of *L. casei* niche generalists from the *Lactobacillus* ancestor with gain and/or loss of plasmids and genomic islands. They have also proposed that enhanced fitness of dairy specialist *L. casei* occurred by further gene loss from niche generalists capable of adaptation to wider range of environmental conditions [42]. Genetic divergence between *L. rhamnosus* clinical isolates implicated in infection of dental pulp and probiotic *L. rhamnosus* GG provides an indication of possible niche adaptation. This is a phenomenon prevalent in the evolution of lactobacilli associated with mucosal surfaces and also from food-related habitats [43,44].

Conclusions

Dental caries is a progressive chronic infection associated with polymicrobial aetiology. Progression of lactobacilli through dentinal tubules and the presence of apparently encapsulated *L. rhamnosus* in the initial stages of pulp infection are indicative of an important role in advanced stages of the infection. In both clinical isolates, the absence of the *L. rhamnosus* GG *Spa*CBA pilus cluster (**Table S9**), existence of a modified MabA-like protein, a unique repeat unit of a collagen binding protein domain and the presence of an altered exopolysaccharide gene cluster (**Table 4, Table 5, Table S9**), suggests the surface components of the clinical isolates of *L. rhamnosus* may employ a different mechanism for invasion of dental pulp. Surface components of *L. rhamnosus* clinical isolates are postulated to facilitate the observed invasion of dental pulp by adherence both to cellular elements and extracellular matrices. Invasion could also be facilitated by the evasion of host defence mechanisms, through the masking effect of an extracellular polysaccharide layer. The presence of capsular polysaccharide (CPS) or extracellular polysaccharide (EPS) is wide-spread in Gram-positive and Gram-negative pathogens. These serve dual roles in survival and pathogenic action by modulating adaptive immune responses and by regulating phagocytosis [45]. Therefore, transcriptional orientation of the *eps* cluster genes, the presence of two genes homologous to priming glycosytransferases, the absence of *rml*ACBD genes involved in the dTDP-rhamnose biosynthetic pathway and the presence of a family 2 glycosyltransferase in the *eps* cluster of both clinical isolates of *L. rhamnosus*, is predicted to alter EPS composition and could influence pathogenicity. Interestingly, the presence of putative CDSs encoding CtsR (class III stress gene repressor), RpoN (σ54 RNA polymerase), mannose PTS and internalin J as identified in the clinical isolates, correlates with the presence of homologues in *L. monocytogenes*, a firmicute involved in invasive gastrointestinal infection.

The impact of acquisition of new genes by the two clinical isolates as evident in the segregation of the clinical isolates from other probiotic *L. rhamnosus* strains and specifically, the SNP divergence from *L. rhamnosus* GG, will be better understood with future transcriptomic and functional studies.

Materials and Methods

Collection and processing of carious teeth

Carious teeth were collected from adult patients attending the Westmead Centre for Oral Health. The protocol was approved by the Human Ethics Committee, Sydney West Area Health Service. The decision by the clinician, independent from the study, to extract the carious tooth, was based on clinical examination and radiographs; alternatively, extraction of the tooth was the patient's informed decision. The patient signed the consent form to participate in the research. The carious tooth was not included in the study if the patient had antibiotic administered in the two weeks preceding the extraction or if the patient had reported a history of Hepatitis C or HIV. Early stage of pulp infection was inferred based on radiographic image(s) showing integrity of periodontal ligament with no shadow effect for peri-apical infection and with no clinical signs or symptoms of irreversible pulpitis, based on patient descriptors such as thermal sensitivity, pain and postural variation. Stages of infection were subsequently re-assessed during downstream processing of the tissues. Carious teeth meeting the above criteria were transported in sterile and reduced PBS glycerol to the Institute of Dental Research within 2 h of extraction and were processed. Carious dentine was excavated to near totality and the scored tooth was rinsed in sterile 18-MΩ water twice before splitting to avoid contamination from the carious dentine. A longitudinal groove was carefully carved around the carious tooth to bisect the carious lesion using a sterile bur and pressure syringe and sterile 18-MΩ water; the tooth was split sagittaly with sterile hand instruments and the extent and location of the carious lesion and patho-physiology of pulp tissue was noted. Based on these observations, classification of the carious lesion for stages of infection was re-assessed. An intact pulp with mild erythema was provisionally classified as at an early (limited) stage of infection. Exposed pulp localized on one side of the bisected tooth was carefully lifted with sterile tweezers, rinsed in sterile, reduced PBS with 15% glycerol and placed in a sterile screw capped cryo-tube containing 0.4 ml sterile PBS with 15% glycerol. The resuspended pulp was homogenized using a 2 ml sterile glass homogenizer and stored at −80°C after subsequent processing.

PCR analysis of bacterial phyla in infected pulp tissue

Earlier findings based on FISH studies [6] indicating that fewer taxa are detected in early stages of infection was confirmed by PCR analysis of bacterial phyla in infected pulp tissue (**Table S1**).

Isolation and identification of lactobacilli from infected vital pulp

An aliquot (20 µl) of homogenized pulp sample was streaked on freshly prepared Man, Rogosa and Sharpe (MRS) agar and incubated anaerobically at 37°C for 48 h. Each bacterial colonial form with distinct morphology was subcultured in MRS medium (2 ml). Cultures were incubated anaerobicaly at 37°C for 24 h. An aliquot (0.85 ml) of bacterial culture was collected in screw capped cryo-tubes containing 0.15 ml sterile glycerol, homogenized and stored at −80°C for future use. Remaining culture was processed for routine Gram stain to study bacterial morphology and for purification of genomic DNA as described [6]. Briefly, an aliquot (40 µl) of homogenised pulp was digested at 56°C for 40 min in cell lysis buffer consisting of 2 mg ml^{-1} proteinase K (Qiagen), 2 mg ml^{-1}lysozyme (Roche) and 1000 U ml^{-1}mutanolysin (Sigma) in 10 mM sodium phosphate buffer pH 6.7. After the addition of 0.1%SDS (Sigma), DNA was extracted using QIAamp DNA Mini kit (Qiagen) to give a final elution volume of 200 µl as described [6]. Genomic DNA isolated from bacterial cultures showing morphology of Gram-positive rods was subjected to PCR using *Lactobacillus* genus-specific and species-specific primers. Qualitative PCR and sequence identity for the 16S rRNA gene was performed as described previously [10].

Design and evaluation of PCR primers

Primers were designed using Primer Express (Applied Biosystems) for the detection of bacteria from the families, *Streptococcaceae*, *Acidaminococcaceae*, *Lachnospiraceae* and *Coriobacteriaceae*as well as from

P. alactolyticus and *Propionibacterium* strain FMA5 (**Table S2a**). Primers used for the detection of *Lactobacillaceae*, *Prevotellaceae* and *F. nucleatum* have previously been reported (**Table S2a**). Primers for the *L. rhamnosus* exopolysaccharide and pilus cluster gene set (**Table S8**) were designed using OligoExplorer 1.2 (Gene Link) and the specificity of the primers determined by DNA sequencing of the amplicons produced (Australian Genome Research Facility).

DNA amplification for identification of bacteria

Amplicons of 16S rRNA genes for given taxa were obtained from carious dentine or infected pulp using HotStarTaq Master Mix kit (Qiagen), using 100 nM of each of the appropriate primer set (Invitrogen or Sigma; **Table S2**) and 2 µl extracted DNA solution from the infected pulp. The PCR reaction was carried out using a GeneAmp PCR system 9700 (Applied Biosystems) in a 25 µl reaction volume at 95°C for 10 min followed by 40 cycles at 95°C for 15 s and 60°C or 62°C for 1 min or 2 min, respectively. PCR products were subjected to electrophoresis on 1.6% agarose gels containing ethidium bromide and DNA bands visualized by UV illumination.

DNA amplification for identification of *L. rhamnosus* exopolysaccharide and pilus cluster genes

In order to amplify specific genes, PCR was performed with HotStarTaq Master Mix (Qiagen), 200 nM of the appropriate primers (Integrated DNA Technologies; **Table S8**) and 2 µl of extracted DNA from eight *L. rhamnosus* clinical isolates. Amplification was performed at 95°C for 15 min followed by 40 cycles of 95°C for 15 s and 58°C for 1 min or 55°C for 1 min or 56°C for 2 min (**Table S8**) for the detection of exopolysaccharide and pilus cluster genes. DNA isolated from *L. rhamnosus* ATCC 53103 and *L. rhamnosus* ATCC 9595 was included as a positive control. The PCR products were subjected to electrophoresis as described above.

DNA sequencing

PCR amplicons were purified using a MO BIO kit (MO BIO Laboratories Inc.) and sequenced at the Australian Genome Research Facility using the appropriate forward primer. Sequence identity of 16S rDNA PCR amplicons was established using BLAST to search nucleotide collection (nr/nt) databases accessed through NCBI (http://www.ncbi.nlm.nih.gov/).

Whole genome sequencing using Roche GS -FLX+

Two genomic DNA samples (1–2 µg DNA, A260/A280 ratio in the range 1.8–2.0) were delivered to the Ramaciotti Centre for Gene Function Analysis, University of New South Wales, Australia for whole genome sequencing using Roche GS FLX+. The sequencing was undertaken according to standard Roche/FLX procedures.

Genome Assembly and BLAST analysis

The sequence reads were assembled *de novo* with Newbler assembler 2.6 using the default parameters. The contigs assembled by Newbler were then joined by CAP3 [46] using default parameters. The final sets of contigs were cross-checked against the *Lactobacillus rhamnosus* GG genome sequence using blastn as well as blastx with an E-value cut-off of $10e^{-6}$. The absence of a gene was defined as no BLAST hits against the reference or a normalized blastx score of less than 2, where normalized blastx score was calculated as BLAST score divided by alignment length.

Genomic Distance Analysis and SNP Detection

Genomic distance between the eleven probiotic *L. rhamnosus* strains; GG, Lc705, HN001, R0011, LMS2-1, CASL, LOCK900, LOCK908, ATCC 8530, ATCC 21052, MTCC 5462 and the 2 clinical isolates *L. rhamnosus* LRHMDP2 and *L. rhamnosus* LRHMDP3 was analyzed by whole-genome BLAST comparison using the NCBI web-based BLAST, with an E-value threshold of 10^{-6}. The whole genome sequence of *L. rhamnosus* GG was used as a query sequence in the BLAST search against the other genomes. The Tree View option was then selected to generate a dendrogram using the neighbour-joining method that clusters sequences according to their distances from the query sequence.

SNP divergence between the probiotic *L. rhamnosus* strain GG and the two clinical isolates *L. rhamnosus* LRHMDP2 and *L. rhamnosus* LRHMDP3 was investigated using the web-based program SNPs Finder (http://snpsfinder.lanl.gov) [47]. The 3 genomes were uploaded at the same time for SNP detection. SNPs were evaluated in homologous regions of 600 bp that shared sequence similarity of at least 95%. Protein coding information for *L. rhamnosus* GG was used as the gene coordinate file in SNP detection for anchoring gene positions.

Supporting Information

Figure S1 Dendrogram of genomic difference between the 2 clinical isolates, *L. rhamnosus* LRHMDP2 and *L. rhamnosus* LRHMDP3 and 11 other L.rhamnosus strains based on genomic BLAST. Genomic distance between eleven *L. rhamnosus* strains; GG, Lc705, HN001, R0011, LMS2-1, CASL, LOCK900, LOCK908, ATCC 8530, ATCC 21052, MTCC 5462 and the 2 clinical isolates *L. rhamnosus* LRHMDP2 and *L. rhamnosus* LRHMDP3 was analyzed by whole-genome BLAST comparison using the NCBI web-based BLAST, with an E-value threshold of 10^{-6}. The whole genome sequence of *L. rhamnosus* GG was used as a query sequence in the BLAST search against the other genomes. The dendrogram was generated using the neighbour-joining method that clusters sequences according to their distances from the query sequence. Percent relatedness is indicated on the scale.

Table S1 PCR analysis of bacterial taxa in infected dental pulp tissue.

Table S2 Primers used to amplify taxa (a) and Specificity of primer pairs for detecting a given taxon (b).

Table S3 Genes from the reference *L. rhamnosus* strain GG not present in both assemblies.

Table S4 Genes from the reference *L. rhamnosus* strain Lc705 not present in both assemblies.

Table S5 New genes present in *L. rhamnosus* LRHMDP2 (a) and *L. rhamnosus* LRHMDP3 (b).

Table S6 Strain specific genes for *L. rhamnosus* LRHMDP2 (a) and *L. rhamnosus* LRHMDP3 (b).

Table S7 SNP divergence between *L. rhamnosus* GG and *L rhamnosus* LRHMDP2 and LRHMDP3.

Table S8   Primers used to amplify selected exopolysaccharide genes and pilus cluster genes.

Table S9   Targeted genetic profile of exopolysaccharideand pilus cluster genes for clinical isolates from pulp sample.

Acknowledgments

We thank Dr. Nattida Charadram for collection of teeth with carious lesions and extraction of pulp.

References

1. Paster BJ, Dewhirst FE (2009) Molecular microbial diagnosis. Periodontology 2000 51:38–44.
2. Belda-Ferre P, Alcaraz LD, Cabrera-Rubio R, Romero H, Simon-Soro A, et al. (2012) The oral metagenome in health and disease. The ISME journal 6:46–56.
3. Griffen AL, Beall CJ, Campbell JH, Firestone ND, Kumar PS, et al. (2012) Distinct and complex bacterial profiles in human periodontitis and health revealed by 16S pyrosequencing. The ISME journal 6:1176–1185.
4. Ling Z, Kong J, Jia P, Wei C, Wang Y, et al. (2010) Analysis of oral microbiota in children with dental caries by PCR-DGGE and barcoded pyrosequencing. Microbial ecology 60:677–690.
5. Zaura E, Keijser BJ, Huse SM, Crielaard W (2009) Defining the healthy "core microbiome" of oral microbial communities. BMC microbiology 9:259.
6. Nadkarni MA, Simonian MR, Harty DW, Zoellner H, Jacques NA, et al. (2010) Lactobacilli are prominent in the initial stages of polymicrobial infection of dental pulp. J Clin Microbiol 48:1732–1740.
7. Schaudinn C, Carr G, Gorur A, Jaramillo D, Costerton JW, et al. (2009) Imaging of endodontic biofilms by combined microscopy (FISH/cLSM - SEM). Journal of microscopy 235:124–127.
8. Huttenhower C, HMP Consortium (2012) Structure, function and diversity of the healthy human microbiome. Nature 486:207–214.
9. Chen Z, Wilkins MR, Hunter N, Nadkarni MA (2013) Draft Genome Sequences of Two Clinical Isolates of Lactobacillus rhamnosus from Initial Stages of Dental Pulp Infection. Genome Announc 1.
10. Byun R, Nadkarni MA, Chhour KL, Martin FE, Jacques NA, et al. (2004) Quantitative analysis of diverse Lactobacillus species present in advanced dental caries. J Clin Microbiol 42:3128–3136.
11. Lukjancenko O, Ussery DW, Wassenaar TM (2011) Comparative genomics of Bifidobacterium, Lactobacillus and related probiotic genera. Microbial ecology 63:651–673.
12. Aziz RK, Bartels D, Best AA, DeJongh M, Disz T, et al. (2008) The RAST Server: rapid annotations using subsystems technology. BMC genomics 9:75.
13. Kankainen M, Paulin L, Tynkkynen S, von Ossowski I, Reunanen J, et al. (2009) Comparative genomic analysis of Lactobacillus rhamnosus GG reveals pili containing a human- mucus binding protein. Proceedings of the National Academy of Sciences of the United States of America 106:17193–17198.
14. Macklaim JM, Gloor GB, Anukam KC, Cribby S, Reid G (2011) At the crossroads of vaginal health and disease, the genome sequence of Lactobacillus iners AB-1. Proceedings of the National Academy of Sciences of the United States of America 108 Suppl 1:4688–4695.
15. Fiocco D, Capozzi V, Collins M, Gallone A, Hols P, et al. (2009) Characterization of the CtsR stress response regulon in Lactobacillus plantarum. Journal of bacteriology 192:896–900.
16. Hu Y, Raengpradub S, Schwab U, Loss C, Orsi RH, et al. (2007) Phenotypic and transcriptomic analyses demonstrate interactions between the transcriptional regulators CtsR and Sigma B in Listeria monocytogenes. Applied and environmental microbiology 73:7967–7980.
17. Jolly L, Stingele F (2001) Molecular organization and functionality of exopolysaccharide gene clusters in lactic acid bacteria. International Dairy Journal 11:733–745.
18. Lebeer S, Verhoeven TL, Francius G, Schoofs G, Lambrichts I, et al. (2009) Identification of a Gene Cluster for the Biosynthesis of a Long, Galactose-Rich Exopolysaccharide in Lactobacillus rhamnosus GG and Functional Analysis of the Priming Glycosyltransferase. Applied and environmental microbiology 75:3554–3563.
19. Welman AD, Maddox IS (2003) Exopolysaccharides from lactic acid bacteria: perspectives and challenges. Trends in biotechnology 21:269–274.
20. Laws A, Gu Y, Marshall V (2001) Biosynthesis, characterisation, and design of bacterial exopolysaccharides from lactic acid bacteria. Biotechnology advances 19:597–625.
21. Peant B, LaPointe G, Gilbert C, Atlan D, Ward P, et al. (2005) Comparative analysis of the exopolysaccharide biosynthesis gene clusters from four strains of Lactobacillus rhamnosus. Microbiology (Reading, England) 151:1839–1851.
22. Bentley SD, Aanensen DM, Mavroidi A, Saunders D, Rabbinowitsch E, et al. (2006) Genetic analysis of the capsular biosynthetic locus from all 90 pneumococcal serotypes. PLoS genetics 2:e31.
23. Jiang SM, Wang L, Reeves PR (2001) Molecular characterization of Streptococcus pneumoniae type 4, 6B, 8, and 18C capsular polysaccharide gene clusters. Infection and immunity 69:1244–1255.
24. Shelburne SA, Davenport MT, Keith DB, Musser JM (2008) The role of complex carbohydrate catabolism in the pathogenesis of invasive streptococci. Trends in microbiology 16:318–325.
25. Francl AL, Thongaram T, Miller MJ (2010) The PTS transporters of Lactobacillus gasseri ATCC 33323. BMC microbiology 10:77.
26. Monedero V, Maze A, Boel G, Zuniga M, Beaufils S, et al. (2007) The phosphotransferase system of Lactobacillus casei: regulation of carbon metabolism and connection to cold shock response. Journal of molecular microbiology and biotechnology 12:20–32.
27. Broadbent JR, Neeno-Eckwall EC, Stahl B, Tandee K, Cai H, et al. (2012) Analysis of the Lactobacillus casei supragenome and its influence in species evolution and lifestyle adaptation. BMC genomics 13:533.
28. Monedero V, Yebra MJ, Poncet S, Deutscher J (2008) Maltose transport in Lactobacillus casei and its regulation by inducer exclusion. Research in microbiology 159:94–102.
29. Davidson AL, Dassa E, Orelle C, Chen J (2008) Structure, function, and evolution of bacterial ATP-binding cassette systems. Microbiol Mol Biol Rev 72:317–364.
30. Garmory HS, Titball RW (2004) ATP-binding cassette transporters are targets for the development of antibacterial vaccines and therapies. Infection and immunity 72:6757–6763.
31. Nizet V (2006) Antimicrobial peptide resistance mechanisms of human bacterial pathogens. Current issues in molecular biology 8:11–26.
32. Chagnot C, Listrat A, Astruc T, Desvaux M (2012) Bacterial adhesion to animal tissues: protein determinants for recognition of extracellular matrix components. Cellular microbiology 14:1687–1696.
33. Love RM, Jenkinson HF (2002) Invasion of dentinal tubules by oral bacteria. Crit Rev Oral Biol Med 13:171–183.
34. Hubble TS, Hatton JF, Nallapareddy SR, Murray BE, Gillespie MJ (2003) Influence of Enterococcus faecalis proteases and the collagen-binding protein, Ace, on adhesion to dentin. Oral microbiology and immunology 18:121–126.
35. Abranches J, Miller JH, Martinez AR, Simpson-Haidaris PJ, Burne RA, et al. (2011) The collagen-binding protein Cnm is required for Streptococcus mutans adherence to and intracellular invasion of human coronary artery endothelial cells. Infection and immunity 79:2277–2284.
36. Velez MP, Petrova MI, Lebeer S, Verhoeven TL, Claes I, et al. (2010) Characterization of MabA, a modulator of Lactobacillus rhamnosus GG adhesion and biofilm formation. FEMS immunology and medical microbiology 59:386–398.
37. Schroeder K, Jularic M, Horsburgh SM, Hirschhausen N, Neumann C, et al. (2009) Molecular characterization of a novel Staphylococcus aureus surface protein (SasC) involved in cell aggregation and biofilm accumulation. PLoS one 4:e7567.
38. Williams RJ, Henderson B, Sharp LJ, Nair SP (2002) Identification of a fibronectin-binding protein from Staphylococcus epidermidis. Infection and immunity 70:6805–6810.
39. von Ossowski I, Reunanen J, Satokari R, Vesterlund S, Kankainen M, et al. (2010) Mucosal adhesion properties of the probiotic Lactobacillus rhamnosus GG SpaCBA and SpaFED pilin subunits. Applied and environmental microbiology 76:2049–2057.
40. Lecuit M, Ohayon H, Braun L, Mengaud J, Cossart P (1997) Internalin of Listeria monocytogenes with an intact leucine-rich repeat region is sufficient to promote internalization. Infection and immunity 65:5309–5319.
41. Stevens MJ, Molenaar D, de Jong A, De Vos WM, Kleerebezem M (2010) sigma54-Mediated control of the mannose phosphotransferase sytem in Lactobacillus plantarum impacts on carbohydrate metabolism. Microbiology 156:695–707.
42. Cai H, Thompson R, Budinich MF, Broadbent JR, Steele JL (2009) Genome sequence and comparative genome analysis of Lactobacillus casei: insights into their niche-associated evolution. Genome biology and evolution 1:239–257.
43. Makarova K, Slesarev A, Wolf Y, Sorokin A, Mirkin B, et al. (2006) Comparative genomics of the lactic acid bacteria. Proceedings of the National Academy of Sciences of the United States of America 103:15611–15616.

Data Availability: The Whole Genome Shotgun project described in this paper has been deposited at DDBJ/EMBL/GenBank under the accession numbers AMQW00000000 for L. rhamnosus LRHMDP2 and AMQX00000000 for L. rhamnosus LRHMDP3.

Author Contributions

Conceived and designed the experiments: MAN. Performed the experiments: MAN. Analyzed the data: MAN ZC MRW. Contributed reagents/materials/analysis tools: MAN NH ZC MRW. Wrote the paper: MAN NH ZC MRW.

44. Makarova KS, Koonin EV (2007) Evolutionary genomics of lactic acid bacteria. Journal of bacteriology 189:1199–1208.
45. Avci FY, Kasper DL (2010) How bacterial carbohydrates influence the adaptive immune system. Annual review of immunology 28:107–130.
46. Huang X, Madan A (1999) CAP3: A DNA sequence assembly program. Genome research 9:868–877.
47. Song J, Xu Y, White S, Miller KW, Wolinsky M (2005) SNPsFinder—a web-based application for genome-wide discovery of single nucleotide polymorphisms in microbial genomes. Bioinformatics (Oxford, England) 21:2083–2084.
48. Stothard P, Wishart DS (2005) Circular genome visualization and exploration using CGView. Bioinformatics (Oxford, England) 21:537–539.

12

A Systems Biology Approach Investigating the Effect of Probiotics on the Vaginal Microbiome and Host Responses in a Double Blind, Placebo-Controlled Clinical Trial of Post-Menopausal Women

Jordan E. Bisanz[1,2]**, Shannon Seney**[1]**, Amy McMillan**[1,2]**, Rebecca Vongsa**[3]**, David Koenig**[3]**, LungFai Wong**[3]**, Barbara Dvoracek**[3]**, Gregory B. Gloor**[1,4]**, Mark Sumarah**[5]**, Brenda Ford**[6]**, Dorli Herman**[6]**, Jeremy P. Burton**[1,2,7,8]**, Gregor Reid**[1,2,7]*

1 Canadian Centre for Human Microbiome and Probiotic Research, Lawson Health Research Institute, London, Canada, **2** Microbiology and Immunology, The University of Western Ontario, London, Canada, **3** Kimberly Clark Corporation, Corporate Research and Engineering-Microbial Control, Neenah, Wisconsin, United States of America, **4** Biochemistry, The University of Western Ontario, London, Canada, **5** Agriculture and Agri-Food Canada, London, Canada, **6** Springbank Medical Clinic, London, Canada, **7** Surgery, The University of Western Ontario, London, Canada, **8** Division of Urology, The University of Western Ontario, London, Canada

Abstract

A lactobacilli dominated microbiota in most pre and post-menopausal women is an indicator of vaginal health. The objective of this double blinded, placebo-controlled crossover study was to evaluate in 14 post-menopausal women with an intermediate Nugent score, the effect of 3 days of vaginal administration of probiotic *L. rhamnosus* GR-1 and *L. reuteri* RC-14 (2.5×10^9 CFU each) on the microbiota and host response. The probiotic treatment did not result in an improved Nugent score when compared to when placebo. Analysis using 16S rRNA sequencing and metabolomics profiling revealed that the relative abundance of *Lactobacillus* was increased following probiotic administration as compared to placebo, which was weakly associated with an increase in lactate levels. A decrease in *Atopobium* was also observed. Analysis of host responses by microarray showed the probiotics had an immune-modulatory response including effects on pattern recognition receptors such as TLR2 while also affecting epithelial barrier function. This is the first study to use an interactomic approach for the study of vaginal probiotic administration in post-menopausal women. It shows that in some cases multifaceted approaches are required to detect the subtle molecular changes induced by the host to instillation of probiotic strains.

Trial Registration: ClinicalTrials.gov NCT02139839

Editor: Jan S. Suchodolski, GI Lab, United States of America

Funding: This project was funded by Kimberly Clark Corperation who were involved in the study design, analysis and preparation of the manuscript. This work was funded by Kimberly Clark and RV, DK, LW and BD are employees of Kimberly Clark. In order to avoid conflict of interest, all authors, including employees of Kimberly Clark, were blinded to which treatment was received by which subject, until all the sample analysis had been completed.

Competing Interests: In order to avoid conflict of interest, all authors, including employees of Kimberly Clark, were blinded to which treatment was received by which subject, until all the sample analysis had been completed.

* Email: gregor@uwo.ca

Introduction

The vaginal microbiota is a dynamic ecosystem that is usually mono dominated by the *Lactobacillus* genus in times of health but it can transform quickly to a dysbiotic state where a range of microorganisms rise in prominence and cause the polymicrobial bacterial vaginosis (BV) [1]. The pre-menopausal vaginal microbiota and its constituents have been extensively studied using genome sequencing, 16S rRNA community profiling and RNA-seq based meta-transcriptomics [2–5]. However, until recently, few studies have used similar in-depth methods to decipher the vaginal microbiome of post-menopausal women, despite its impact on quality of life [6,7]. In previous studies, we have shown using high-throughput sequencing methods that contrary to conventional knowledge [8], the vaginal microbiome of post-menopausal women is not significantly different from that of pre-menopausal women [9]. Our study showed that the post menopausal vagina appeared to have greater stability than the rapidly fluctuating pre-menopausal microbiome [9,10] presumably due to a lack of hormone cycling. We found that similar to premenopausal women, the healthy post-menopausal vaginal microbiome is dominated by a combination of *Lactobacillus iners* and *L. crispatus* while BV-associated organisms such as *Gardnerella vaginalis* and *Atopobium vaginae* are increased in states of dysbiosis [9,11,12]. Despite some deficiencies, Nugent scoring is still the standard technique in most research and clinical settings to rapidly diagnose bacterial vaginosis [13]. The score is calculated by microscopic examination of a Gram-stained vaginal smear and numeration of cell morphotypes to assign a score from 0 to 10 where 0–3 is considered normal, 4–6 is intermediate and 7–10 is BV [13]. The intermediate scores are particularly interesting as they indicate a risk of transition to BV, or they could be a state that reverts to a healthy lactobacilli dominated microbiota.

Given vaginal contact with sanitary products by menstruating women and the large proportion of postmenopausal women whom suffer urinary incontinence requiring products such as incontinence pads, also in proximity to the vagina, these products may offer a vehicle to deliver a prophylactic probiotic to women that routinely suffer BV or urinary tract infections. We were interested to determine whether the microbiota and Nugent scores were altered by the administration of probiotics with the potential view of perhaps instilling these organisms in such products. *Lactobacillus rhamnosus* GR-1 and *Lactobacillus reuteri* RC-14 used in this prospective study are well-characterized probiotic strains used in combination to prevent and treat BV [14–17]. In addition to clinical and Nugent scoring, 16S rRNA sequencing was used to examine changes in the microbiota, GC-MS profiled metabolome changes, the host transcriptional responses were tested using an Affymetrix microarray and inflammatory mediators by multiplex cytokine analysis. This approach was designed to provide a holistic understanding of the probiotic-microbiota-host interactome [18].

Results

Participant recruitment and demographics

A total of 22 subjects were screened and 14 participants were enrolled into a 129-day prospective double-blind, cross-over placebo controlled study (design in Figure S1). A CONSORT flow diagram is displayed in Figure 1.

Clinical Outcomes

Nugent score and vaginal pH measurements (data not shown) were taken before and after treatment with the probiotic and placebo. Table 1 shows there were no statistically significant improvement in Nugent scores between the treatment groups. Furthermore no statistical differences were observed in vaginal pH measurement.

Microbiome profiling

Sequences were successfully obtained for 104 of 106 samples with the mean number of reads being 33,396 (range 1,475 to 223,629) across 3 separate 316 Ion Torrent Sequencing Chips. A total of 221 Operational Taxonomic Units (OTUs) were defined with clustering at 97% identity (Table S1) at a minimum abundance of 1% in any one sample. A no template control was also sequenced and high-abundance contaminating sequences were filtered out *in silico* before OTU assignment.

Figure 2 shows a heat map of the 50 most abundant OTUs across all samples and time points. We were able to identify 2 OTUs which were assigned presumptive strain designations to *L. rhamnosus* GR-1 and *L. reuteri* RC-14 (OTU_12 and OTU_23 respectively) based upon 100% ID matches to the 16S rRNA gene of private genome assemblies of these strains. A separate heat map was created showing only the abundance of these two OTUs compared to all other OTUs detected (Figure S2). Furthermore with one notable exception (participant 19), these strains only appeared in combination following the probiotic treatment period. In the case of participant 19 however, the proportion of reads is still quite low (2.1% and 0.1%) for both OTUs. The *L. rhamnosus* GR-1 is far more abundant than the *L. reuteri* which may suggest

it was a *L. casei* group member other than the GR-1 used in the study as this group has high sequence homology, especially in the V6 16S rRNA region [19]. In one individual (participant 09) both probiotic strains were detected even though only placebo should have been received. The probiotic strains could be unambiguously detected after probiotic treatment in 7 of 12 cases (04, 10, 15, 16, 17, 18, 21), which we termed responders. While in the non-responders (12, 14, 19, 05 and 07) the probiotic strains were not detected after the probiotic treatment. The abundance of *L. rhamnosus* GR-1 did appear to increase over the study period in participant 12, which may suggest temporary persistence or colonization. In all other cases, the probiotics were only detected in high abundance immediately after the treatment suggesting that the washout period chosen for the study was appropriate.

Samples were clustered based upon unweighted pair group method with arithmetic mean (UPGMA) using weighted UniFrac distance metrics [20] showing that participants were generally most similar to themselves over the study period and as we had previously shown, the post menopausal vaginal microbiome is more stable than that of pre menopausal women (Figure 3).

All samples were plotted on a PCoA using weighted Unifrac distances and overlaid with the contributions of the 20 most abundant genera (Figure S3). This shows 3 weak groupings, one dominated by *Lactobacillus*, another defined by *Atopobium* and *Gardnerella*, and a third defined by a large number of diverse genera.

Data were summarized to genus level and the relative abundance of *Lactobacillus* was considered. Increase in total *Lactobacillus* is significant (Figure 4; FDR<0.05) following probiotic administration though there is no significant difference in Shannon's diversity index (p = 0.25, Wilcoxon signed-rank test). After probiotic administration, the proportion of total lactobacilli increased in all participants except participant 19 who appears to have had an indigenous *L. casei/rhamnosus* present. Even those without significant detection of the probiotic lactobacilli have an increase in indigenous lactobacilli (fold increases for the ambiguous participants are 1.27, 1.21, 2.35, 14.43 and 52.43). Additionally, in some cases, such as participant 17 (Figure 5), it is not only the probiotics that increased in proportional abundance, but also indigenous lactobacillus such as *L. gasseri/johnsonii*, (p = 0.011 across all samples) which is not significant in the placebo group if the case of participant 09 is excluded (where probiotic appears to have been received rather than placebo). In the 7 individuals where the probiotics could be easily detected, the average fold increase was 120 fold (range 3.2 to 389). Significant decreases in the proportion of *Atopobium* were also observed after probiotic treatment with a trend of reduction of *Gardnerella* and *Prevotella*. Interestingly the placebo appears to have increased the proportion of *Streptococcus* (FDR<0.01) while both applications appear to have increased levels of *Staphylococcus*.

Metabolome

Using GC-MS, 68 metabolites were detected and a heat map is displayed in Figure S4. The Pearson's correlation coefficient between lactate and lactate producing bacteria was 0.47 (p< 0.001), indicating a weak positive correlation (Figure S5). Among those who responded to probiotic intervention and had sufficient material for metabolomics analysis (n = 4), there was an increase in lactate levels (p = 0.13), whereas there tended to be a decrease with placebo (p = 0.23). There was no difference in the overall metabolome of all subjects after probiotic or placebo intervention (Figures S6A and S6B respectfully), nor was there a significant difference in any one metabolite. It is worth noting that low yields

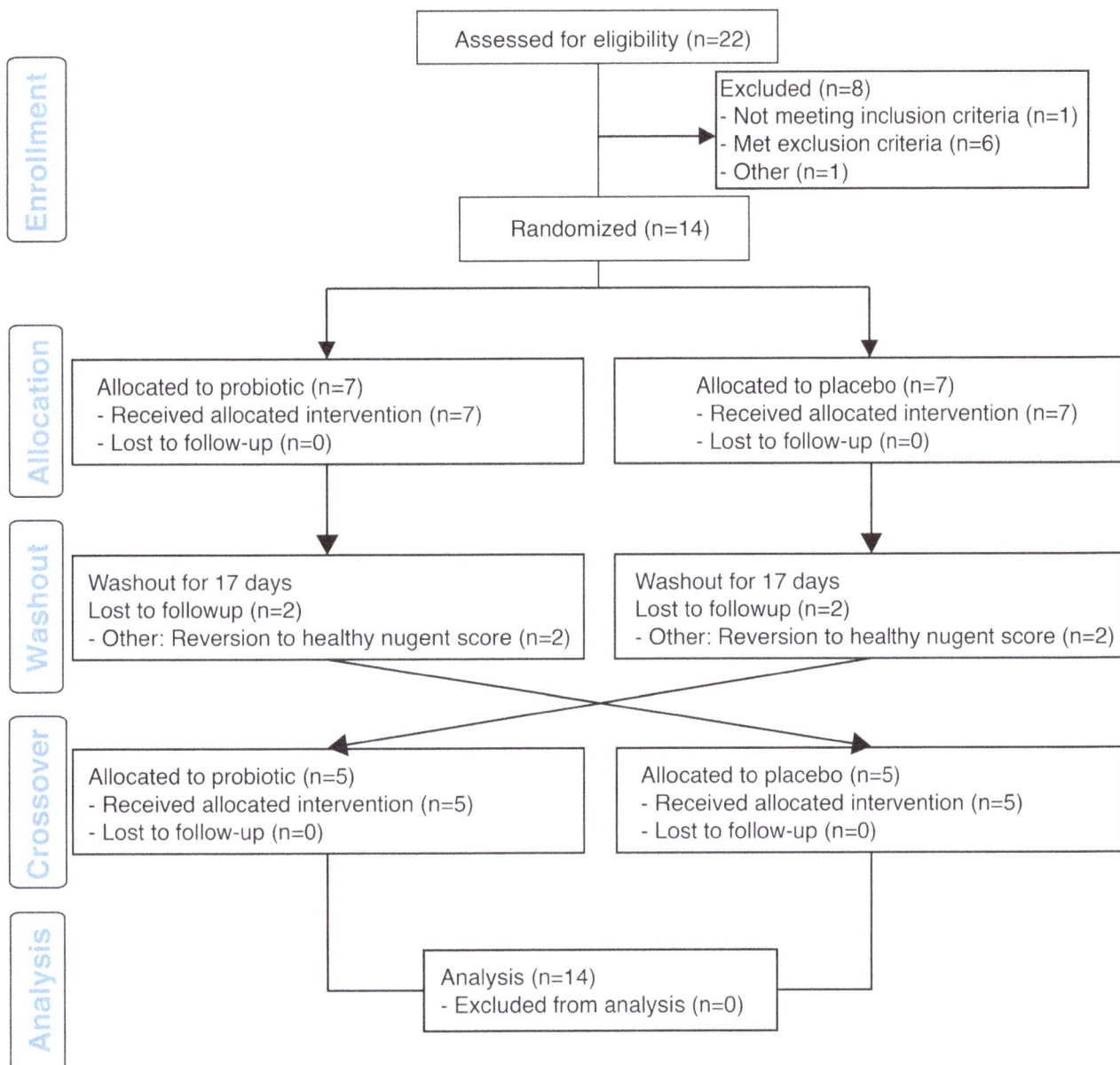

Figure 1. CONSORT flow diagram.

Table 1. Summary of Nugent Score Change - Proportion of Improvement from Baseline.

			Placebo		
			Improved[1]	Not Improved	*p*-value[2]
One-Day after last Test Article Use	Probiotics:	Improved	0	2(20%)	0.1573
		Not Improved	0	8(80%)	
Eight-Days after last Test Article Use	Probiotics:	Improved	0	2(20%)	0.5637
		Not Improved	1(10%)	7(70%)	

[1]Improvement in Nugent Score: indicated by shift of intermediate score (4–6) at baseline to normal score (0–3) at follow-up visits. Baseline Nugent score for study Phase-I was collected at visit 3, and baseline Nugent score for study Phase-II was collected at visit 6.
[2]*p*-value is based on McNemars test to determine results of cross-over treatment regimen.

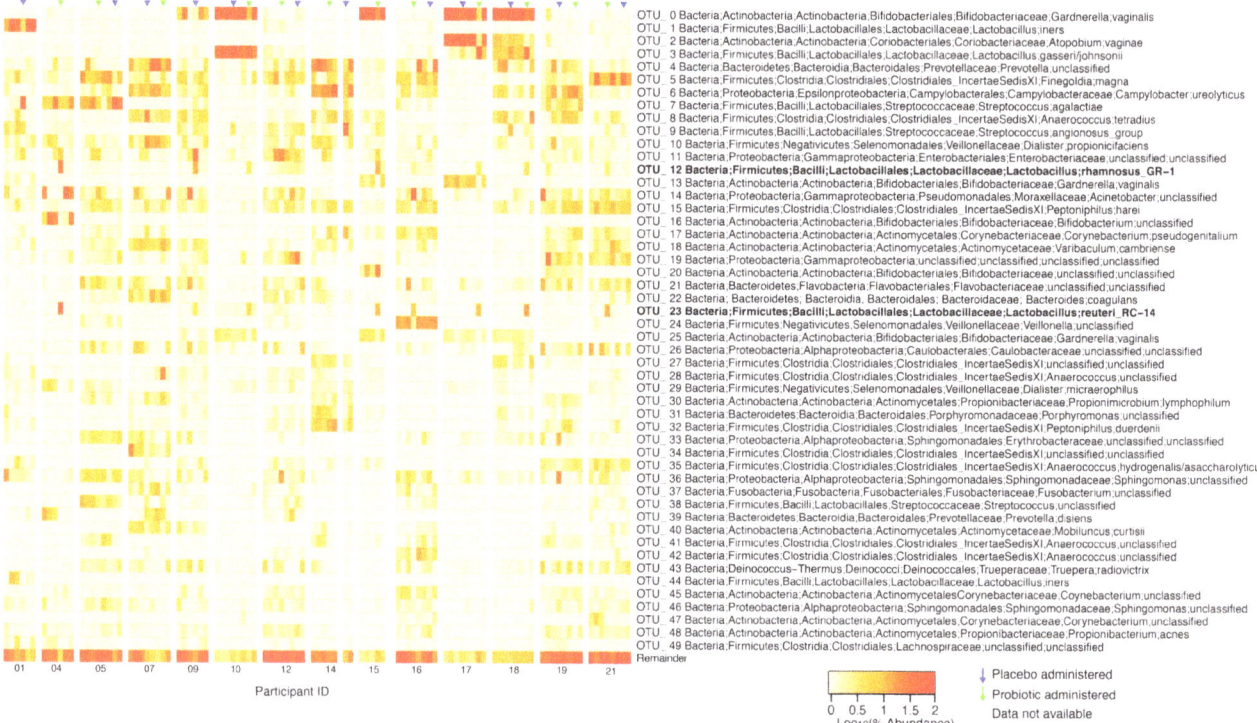

Figure 2. Microbiota heat map. Each column represents the microbiota of a single sample with the 50 most abundant OTUs displayed with their taxonomies and the remainder pooled. Samples are clustered by the participant of origin and organized from first to last visit from left to right. The time points immediately following administration of the placebo or probiotic are indicated with a blue or green arrow (respectively). OTUs representing the putative OTUs for *L. rhamnosus* GR-1 and *L. reuteri* RC-14 have been bolded.

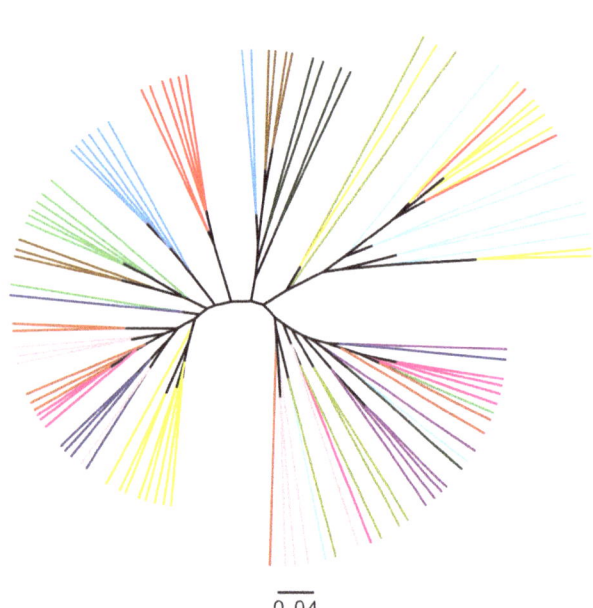

Figure 3. UPGMA clustering of all participants microbiota based upon weighted UniFrac distances. In general, participants cluster most closely with themselves. The sample tips are colored by the participant of origin.

of metabolites were obtained from some of the collected swabs so lower abundance analytes may not be detectable.

Cytokine/Chemokine analysis

Interleukin 5 (IL-5) was found to be significantly up-regulated following probiotic treatment but not following placebo treatment. In other cytokines such as IL-1B, IL-6, TNFa and GM-CSF there were weak positive trends in increase of cytokine levels, however this was also observed in the placebo in some cases (Figure S7).

Microarray

Due to limited yield from RNA samples, only 2 complete sample sets (before and after intervention from the same individual, 4 arrays) were obtained from the probiotic intervention and 3 sets from the placebo intervention (6 arrays). Each sample was analyzed on its own Affymetrix Human Gene 2.0 ST array. Using a paired t-test design to account for paired-samples before and after probiotic administration, 90 probe sets were detected as being differentially expressed (≥ 2 fold change, p<0.05) of which 57 were coding genes or long non-coding RNA (a new feature of the Affymetrix 2.0 ST array) (Table S2A). Two of these were differentially regulated in both conditions (gene list differentially regulated by placebo is displayed in Table S2B), ANKRD20A5P psuedogene (ID cluster 16851230) and a long non-coding RNA (ID cluster 16851230) with no known biological significance.

Many genes of interest were differentially expressed by probiotic treatment including Interleukin 18 (IL18; −2.5 fold after probiotic), CR1-complement receptor (CR1; 10.4 fold), Caspase 14 (CASP14; 5.3 fold, previously observed to function in epithelial development and barrier function [21]) and Toll like receptor 2

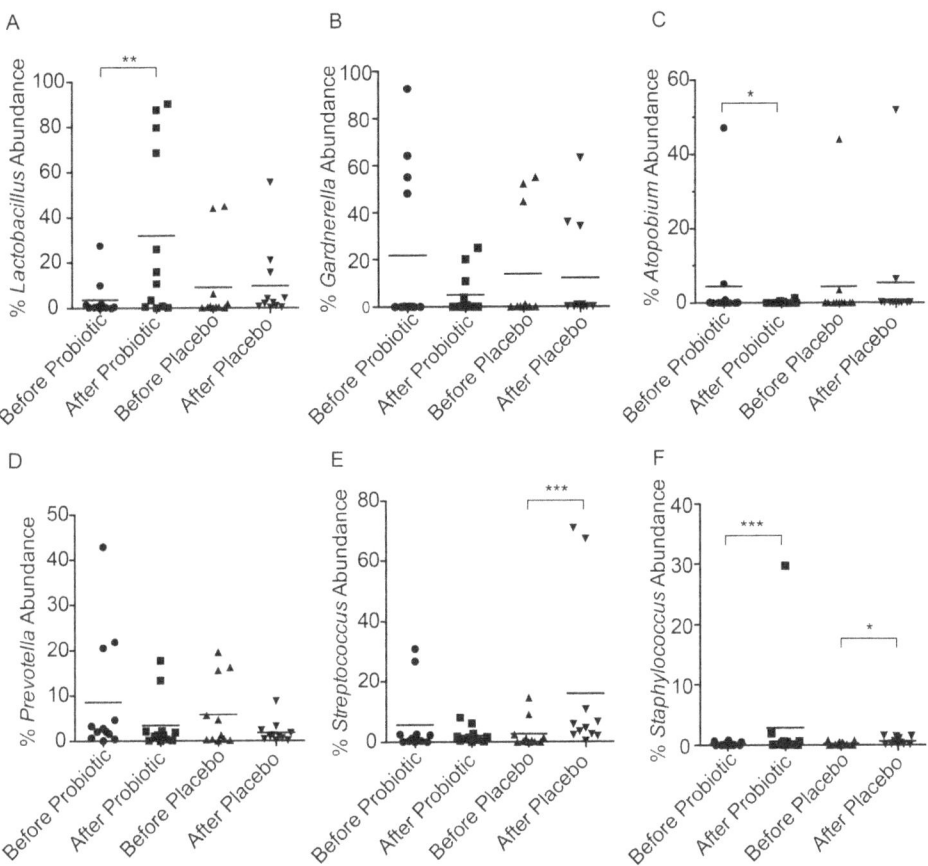

Figure 4. Selected genus relative abundances following probiotic and placebo interventions. (A) *Lactobacillus*, (B) *Gardnerella*, (C) *Atopobium*, (D) *Prevotella*, (E) *Streptococcus*, (F) *Staphylococcus*. Following probiotic administration for 3 days, the proportion of *Lactobacillus* is significantly increased while that of *Atopobium* is decreased and *Staphylococcus* is increased. Placebo interventions increased *Streptococcus* and *Staphylococcus* abundance. *FDR<0.1, **FDR<0.05, ***FDR<0.01.

(TLR2; 3.84 fold) (Table S2A). To look at functional consequences, GO enrichment was applied showing the most highly affected functions to be pattern recognition receptor activity, complement receptor activity, inflammatory response and gram-positive bacterial cell surface binding (Table 2). This was mirrored in the Ingenuity Pathway Analysis (Figure S8). In comparison, the placebo appears to have acted on functions related to epithelial cell differentiation, similar to our previous study [9].

Discussion

In this study, we aimed to evaluate the effect of 3 days of vaginal administration of *L. rhamnosus* GR-1 and *L. reuteri* RC-14 on the nugent score of post-menopausal women as well as effects on the microbiota, metabolome and host. As the subjects under study had no symptoms of disease, not unexpectedly, the probiotic prophylaxis did not cause any clinical changes. There were no adverse effects of the probiotic therapy. The probiotic administration resulted in a temporary increase in the relative proportion of *Lactobacillus* in the vagina, which for women whose intermediate score tends to be a predecessor for BV, may help prevent or delay this transition. The near absence of *L. crispatus* (OTU_198), detected in only 1 of 15 women above a 1% threshold and throughout the study at maximum 1.4%, may indicate that the subjects were at a higher risk of BV, since *L. crispatus* is thought to play a protective role in maintaining the vaginal microbiota [22].

The increase in lactobacilli and potential decreases in BV-associated organisms such as *Atopobium* following probiotics could have been due to displacement of the host's microbes with the total bacterial load remaining the same; or the total bacterial load increased as a result of adding in exogenous lactobacilli. The observation that Shannon's diversity is not affected by probiotic instillation favors the second option. It is interesting to note the increased relative abundance of *Staphylococcus* as a result of placebo treatment, which may be the result of the antimicrobial actions of titanium dioxide used in the preparation. It is important to note that the probiotics may create an environment conducive to indigenous lactobacilli such as *L. gasseri* and *L. johnsonii*, however this effect may be short lived. A longer treatment time and follow-up might have determined if there was an effect on diversity, but we were interested in whether a short term probiotic application that women might find easy to administer monthly, could increase the lactobacilli count.

In some individuals it was extremely clear when the probiotic strains had been applied. The reason for this not being universal is unclear, but could be due to a particularly resilient indigenous microbiota or women who were non-responders for unexplained reasons [23].

We identified a positive correlation between lactate abundance and percentage of lactate producers present $r = 0.43$ ($p = 9.6 \times 10^{-6}$). Although this correlation seems intuitive, to our knowledge it has never been directly shown. Only the relative

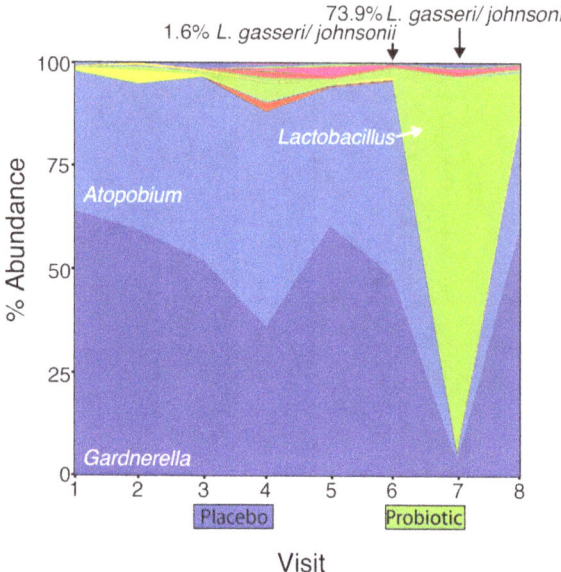

Figure 5. Time series of participant 17. During the administration of probiotic between visits 6 and 7, the abundance of lactobacilli significantly increases, decreasing the proportional abundance of both *Atopobium* and *Gardnerella*. This increase is not due solely to the probiotic strains as a significant increase in indigenous *L. gasseri/ johnsonii* takes place.

abundance of bacterial taxa are measured, and therefore absolute changes affected lactate levels cannot be directly observed. Certain species of *Lactobacillus* have been associated with lower pH and may produce more lactate compared to others [3]. There was a trend towards an increase in lactate after probiotic and decrease after placebo intervention, but as there were only four women whom responded to treatment and from whose samples enough material for metabolite analysis could be extracted, the sample size is likely too small to reach significance. Still, the observation that probiotics could potentially increase lactate levels is promising as

lactate has been shown to have many beneficial properties in the vaginal tract such as HIV inactivation [24].

Overall, the administration of the two lactobacilli strains did not induce any more changes than placebo in the metabolome, but lack of material makes this result inconclusive. In our experience and that of others, the metabolomic patterns differ between health and BV, with the latter showing odorous compounds such as cadaverine and putrescine, but these trends were not clearly observed in this study [25].

In addition to looking at the bacterial communities, we sought to examine host responses to probiotic treatment. It has been well established that probiotic strains can affect host transcription in the gut in a strain specific manor [26] but this is the first time similar studies have been carried out in the vagina. Our findings show that *L. rhamnosus* GR-1 and *L. reuteri* RC-14 had an immunomodulatory effect working on important central inflammatory mediators complement receptor 1, toll like receptor 2 and IL-18. Though limited in the number of subjects available for analysis, the strength of this analysis is the use a paired study design. Interleukin 18 is a proinflammatory cytokine inducing cell-mediated immunity via interferon gamma though it also has effects on B cells and IgE production [27]. In the female reproductive tract it may be of relevance in preterm birth [28] indicating microbial invasion of the amnion, but it also has protective roles against genital herpes simplex 2 [29]. The effect of probiotic treatment on IL-18 expression may not be a direct interaction, but rather indicative of anti-viral actions of the probiotics resulting which has previously been hypothesized and shown *in vitro* [24,30]. Alternatively it may reflect a general up-regulation of the innate immune system by the probiotic strains. Unfortunately, our cytokine/chemokine panel did not cover IL-18 to help support this finding. Alternatively the up-regulation, of complement receptors 1 and 3a, as well as TLR2 both are indicative of increased anti-bacterial innate immune system activity. TLR2 specifically is involved in detection of Gram-positive surface markers (including lactobacilli) [31] while the complement receptors are important for phagocytosis and inflammatory cascades in response to bacterial infections [32]. *L. plantarum* has previously been shown to modulate TLR2 expression which may play a role in modifying infection response [33]. In addition, our cytokine/chemokine multiplex results may

Table 2. Gene Ontology Enrichment in genes differentially expressed by probiotic intervention.

Function	Enrichment Score	Enrichment p-value	# genes in list	% genes in group	GO ID
Pattern Recognition Receptor Activity	12.82	2.71E-06	3	18.75	8329
Complement Receptor Activity	10.9319	1.79E-05	2	50	4875
Inflammatory Response	10.4433	2.91E-05	6	1.70455	6954
Gram-positive Bacterial Cell Surface Binding	10.4222	2.98E-05	2	40	51637
LPS Receptor Activity	10.4222	2.98E-05	2	40	1875
Defense Response	9.64889	6.45E-05	9	0.840336	6952
Regulation of GM-CSF Production	8.92362	1.33E-04	2	20	32645
Positive Regulation of Response to External Stimulus	8.83166	1.46E-04	4	2.5974	32103
Response to Wounding	8.61585	1.81E-04	6	1.22449	9611
Positive Regulation of Inflammatory Response	8.25985	2.59E-04	3	4.22535	50729

Genes differentially expressed (≥2-fold change, p<0.05) were subjected to GO enrichment analysis. Functions heavily relate to innate immune system responses including responses to both Gram-positive and Gram-negative bacteria.

indicate the up-regulation of interleukin 5 which is most commonly associated with eosinophil activation [34]. Though commonly associated with parasites, eosinophils have also been associated with *Actinomyces*-like organisms in the urogenital tract and may display some activity against bacterial infection [35]. These seemingly pro-inflammatory responses may be beneficial in aiding the host immune system to help clear BV-associated pathogens to restore a more normal microbiota of symbionts. Microarray analysis failed to collaborate the finding that interleukin 5 was up-regulated, however the two individuals surveyed by microarray did not show up-regulation by cytokine analysis either. If all participants could have been surveyed this result may have been different.

Interestingly, caspase 14 plays a role in development and barrier function of the epidermis [21] and we had previously observed its differential regulation associated with vaginal dryness and dysbiosis in a subset of post-menopausal women. Its up-regulation following probiotic treatment may be evidence of improvement in strength and integrity of the vaginal epithelium, though further research is needed to verify this claim. Indeed probiotics are thought to positively act on barrier function [36].

In conclusion, this study demonstrated the use of a systems-wide interactomics approach to examine how probiotic application might modulate the vagina in post-menopausal women. The short duration therapy increased the lactobacilli, as well as modulating inflammatory markers. As the transition from healthy to intermediate to BV is such a common occurrence, women may feel that a three day intravaginal probiotic may be beneficial. A larger study is needed to support this, but the present study showed no harm and the potential for such a benefit.

Methods

Participant Recruiting and study design

The protocol for this trial and supporting CONSORT checklist are available as supporting information; see Checklist S1 and Protocol S1. A total of fourteen post-menopausal women were enrolled in the study from a clinical site in Ontario, Canada. Inclusion criteria were individuals with an intermediate Nugent Score (4–6) who were post-menopausal (40 to 80 years old, not having had a menstrual period for the last 12 months, currently in a mutually monogamous sexual relationship or not sexually active, agreeing to be abstinent 72 hours prior to each study visit and agreeing to refrain from intercourse for 48 hours after treatment, agreeing to abstain from use of other intravaginal products throughout the study and in good general health. Exclusion criteria were use of intravaginal products within three months prior to visit 1, a history of immunosuppressive drug therapy, chemotherapy or radiation therapy, a medical condition which might compromise immune system function (ex. cancer, leucopenia, HIV, organ transplant), antibiotic and/or antifungal medication within the last 4 weeks, oral probiotic use within 3 months prior to visit 1, significant changes in diet during the course of the study (based on self report), induced menopause due to surgical/medical intervention (ex. hysterectomy), currently taking estrogen therapy, a history of drug or alcohol abuse, known allergy to either of the probiotic strains or product excipients and participation in a clinical trial involving an investigational product/device within the past three months or who are scheduled to participate in another clinical study concurrently. No formal sample size calculation was carried out and the number of subjects to be randomized was based on feasibility and practical considerations for an exploratory study. The study was approved by Health Canada (Clinical Trial

Application File No. 180061) and is registered with clinicaltrials.gov under accession number (NCT02139839).

The study design is diagramed in Figure S1. Briefly, after obtaining written informed consent, participants were enrolled on day one and subsequently returned for baseline appointments on days 5 and 15 (visits 2 and 3 respectively). On visit 3, the participants were randomized to either receive 3 days of *L. rhamnosus* GR-1/*L. reuteri* RC-14 probiotic (Chr. Hansen, minimum 2.5×10^9 CFU of each strain) or visually identical placebo (49 mg Gelatin and 1 mg titanium dioxide) self-administered by having the participant insert the capsule as far as comfortable into their vagina and then lying in a horizontal position for 15 minutes. Randomization was carried out using a random number generator. Both study staff and study participants were blinding as to the study treatment being given. The capsules were to be administered twice a day for 3 days with the participant returning on the fourth day for sampling. A feminine hygiene pad was applied for several hours after capsule implantation. There was then a 17 day washout period with sampling again at day 26. At day 36 participants were again sampled, individuals who received probiotic in the first treatment period then received placebo and vice versa. On day 40, participants returned for sampling after the treatment period and were followed up with on day 47 and again on day 129, only if they had positive culture results for the presence of the probiotic. At all visits 5 swabs were collected: (i) Dacron swab for bacterial DNA extraction, (ii) Dacron swab for multiplex cytokine/chemokine analysis, (iii) Dacron swab for metabolomics analysis, (iv) a cytobrush (Cytobrush plus GT, Cooper Surgical Inc. USA) brushed against the vaginal wall and stored in 700 µL RNAlater for total RNA extraction (Life Technologies), and (v) Dacron swab for Nugent Score. McNemar's test was applied to examine changes in Nugent score and other symptoms.

Microbiome profiling

Vaginal swabs for microbiome analysis were extracted using the QIAamp DNA stool mini kit (Qiagen). Swabs were vortexed in 1 mL buffer ASL before removal of the swab and addition of 0.1 mm zirconia/silica beads (Biospec Products) with 2, 30 second rounds of bead beating at full speed with cooling on ice in between (Mini-BeadBeater; Biospec Products). Sample amplification for sequencing was carried out using the forward primer (CCATCT-CATCCCTGCGTGTCTCCGACTCAGxxxxxCWACGCGAR-GAACCTTACC) and the reverse primer (CCACTACGCCT-CCGCTTTCCTCTCTATGGGCAGTCGGTGATACRACAC-GAGCTGACGAC) where xxxxx was a sample specific nucleotide barcode, the 5′ end is the adapter sequence for the Ion Torrent sequencer and the sequences following the barcode are complementary to the V6 rRNA region. Amplification was carried out in 42 µL with each primer present at a 10 µL (3.2 pMol/µL stock), 20 µL GoTaq hot start colorless master mix (Promega) and 2 µL extracted DNA. The PCR protocol was as follows: initial activation step at 95°C for 2 minutes and 25 cycles of 1 minute 95°C, 1 minute 55°C and 1 minute 72°C. PCR products were quantified with a Qubit 2.0 Flourometer and the high sensitivity dsDNA specific fluorescent probes (Life Technologies). Samples were mixed at equimolar concentrations and purified with the QIAquick PCR Purification kit (QIAGEN).

All subsequent work was carried out at the London Regional Genomics Centre (LRGC, lrgc.ca, London, Ontario, Canada). Briefly, samples were prepared with an Ion OneTouch System (Life Technologies) and sequenced on an Ion Torrent Personal Genome Machine sequencer on a 316 chip (Life Technologies).

Resulting Reads were extracted and de-multiplexed using modifications of in-house perl and unix-shell scripts [9] with OTUs clustered at 97% identity. Table S3 displays the nucleotide barcodes and it's corresponding sample. Reads were deposited to the Short Read Archive (BioProject ID: PRJNA244441). To control for background contaminating sequences, a no-template control was also sequenced, and any individual sequence unit belonging to an OTU present at $\geq 1\%$ abundance in the NTC was removed before re-clustering with uclust 3.0.612.

Automated taxonomic assignments carried out by examining best hits from comparison the Ribosomal Database Project (rdp.cme.msu.edu) and manually curated as before [9].

Alpha and beta-diversity analysis were made using the OSX distribution of QIIME (MacQIIME 1.70, wernerlab.org/software/macqiime). OTU tables were rarified to 1060 reads for all analysis to control for uneven sampling depth. To avoid inappropriate statistical inferences made from compositional data, log-ratios, a method previously described by Aitchison and adapted to microbiome data was used [37–39] with paired t-tests for comparisons of genus and species level data. The False Discovery Rate (FDR) method was used to control for multiple testing with a significance threshold of FDR = 0.1. All statistical analysis, unless otherwise indicated, was carried out using R (r-project.org).

Microarray analysis

Vaginal cytobrushes were extracted as previously [9]. Briefly a vaginal cytobrush was centrifuged at $5000 \times g$ for 10 min at $4°C$ and the supernatant discarded. The pellet was then extracted with TRIzol (Life Technologies) following the manufacturer's protocol before being DNase treated with Turbo DNA-free kit (Life Technologies). Sample quantity and quality were determined using a Nanodrop 1000 spectrophotometer (Thermo Fisher Scientific, Waltham, USA) and Agilent 2100 Bioanalyzer (Agilent Technologies, Santa Clara, USA). Samples were prepared for analysis on the GeneChip Human 2.0 ST array (Affymetrix, Santa Clara, USA) at the LRGC following standard protocols.

Probe level data was imported into Partek Genomics Suite (Partek, St. Louis, USA) using the RMA algorithm. Gene expression data was deposited in the Gene Expression Omnibus Database with the accession number GSE54363.

To study the effects of the probiotic and placebo treatments on host gene expression, and considering the study design, a paired t-test was used to evaluate gene expression changes in each individual immediately before and after probiotic/placebo administration. This yielded two individuals (14 and 16) with arrays immediately preceding and following the probiotic period treatment and three individuals (01, 14, 05) with arrays surrounding the placebo treatment period. For other participants, insufficient yield and/or quality resulted in a lack of paired samples, and they were thus excluded from downstream analysis.

Sample Preparation GC-MS

Samples were collected from the mid-vaginal wall using the Cytobrush vaginal brush. Vaginal brushes were pre-cut into 1.5 mL tubes and weighed prior to and after sample collection to determine the mass of vaginal fluid collected. Samples were stored at $-80°C$ until analysis. After thawing, brushes were cut and eluted in methanol-water (1:1) in 1.5 mL microcentrifuge tubes. The weight of each sample was divided by the weight of the lightest sample and this fraction was multiplied by 200 μL to determine the volume of methanol-water to add to each sample. This corresponded to a volume of 200–382 μL, depending on the mass of vaginal fluid collected. A blank swab eluted in 200 μL methanol-water was included as a negative control. All samples

were vortexed for 10 s to extract metabolites, centrifuged for 5 min at 10 000 $\times g$, vortexed again for 10 s after which time the brushes were removed from tubes. Samples were centrifuged a final time to pellet cells and 150 μL of the supernatant was transferred to a GC-MS vial. Next, 2.5 μL of 2 mg/mL ribitol was added to each vial as an internal standard. Samples were then dried to completeness using a SpeedVac. After drying, 100 μL of 2% methoxyamine•HCl in pyridine (MOX) was added to each for derivatization and incubated at $50°C$ for 90 min. 100 μL N-Methyl-N-(trimethylsilyl) trifluoroacetamide (MSTFA) was then added to each vial and incubated at $50°C$ for 30 min, then transferred to micro inserts before running on GC-MS (Agilent 7890A GC, 5975 inert MSD with triple axis detector, 30 m DB5-MS column with 10 m duraguard column). Samples were analyzed once for untargeted whole metabolome analysis, and a second time on selective ion monitoring (SIM) mode specific for lactate using the reference ions 117, 147, 191 and 219. Solvent delay was 10 min and sample injection volume was 1 uL. All samples were run in random order and one was run multiple times throughout to ensure machine consistency.

Lactate relative abundance quantification

Chromatograms were deconvoluted in Chemstation (Agilent) and lactate relative abundance determined by peak area using the auto integration function. Lactate relative abundance was plotted against the percentage of lactate producers (*Lactobacillus*, *Bifidobacteria*, *Atopobium*, *Streptococcus*, *Staphylococcus*, *Weisella*) present in each sample to determine the correlation between lactate levels and lactate producing bacteria. The correlation value was determined in R using a Pearson's correlation. Paired t-tests were used to determine if probiotic or placebo intervention had a significant effect on lactate levels using a significance threshold of p = 0.05.

Whole metabolome analysis

Chromatogram files were converted to ELU format using the AMDIS Mass Spectrometry software [40]. Chromatograms were then aligned and abundance of metabolites calculated using the Spectconnect software [41], with the support threshold set to low. In order to determine changes in the metabolome due to probiotic or placebo, Principle component analysis (PCA) was conducted in SIMCA (Umetrics) using the relative abundance matrix (RA) output from Spectconnect. Data were mean centered and pareto scaled prior to PCA. A Mann- Whitney U test was then used to determine metabolites that were significantly altered by intervention (p<0.05).

Cytokine/Chemokine Measurement

Swabs were resuspended in 200 μL extraction buffer (20 mM Tris-HCl ph7.5, 150 mM NaCl, 1 mM PMSF, 0.05% Tween 20 and 1 uL/mL protease inhibitor cocktail (Roche), vortexed, incubated overnight at $4°C$, swab removed and then 50 μL more extraction buffer was added before being stored at $-80°C$. The resulting samples were thawed on ice and loaded onto a Milliplex Human High Sensitivity Cytokine/Chemokine Panel (EMD Millipore, Billerica, Mass, USA) and analyzed for IL-1B, IL-2, IL-4, IL-5, IL-6, IL-7, IL-8, IL-10, IL-12p70, IL-13, IFNg, GM-CSF, and TNFa. Results for IL-2, IL-7, IL-8, IL-10, IL-13, IFNg were not displayed as they were consistently outside the assay range and there was limited material for reanalysis. The plate was analyzed using a Bio-Plex 200 System (Bio-Rad Laboratories, CA, USA) with cytokine/chemokine levels being generated automatically from standard curves using the Bio-Plex Manager software (v.4.1.1 Bio-Rad). In cases where the analyte was detected, but below the limit of detection, ½ the LOD was used for analysis. Results were normalized to total protein as determined with the Qubit Protein Assay Kit and Qubit 2.0 Flourometer (Life Technologies, CA, USA). Paired t-tests were used to make comparisons.

Supporting Information

Figure S1   Study design. The study was designed to have two treatment periods separated by a washout period with a wash-in period and follow up period. Information collected at each visit is outlined in the bottom table. Visit 9 was only necessary if the probiotics were detected by culture methods.

Figure S2   Heat map of probiotic OTUs as compared to the remainder of the microbiota. Each row represents a participants sample at a given time point. The samples are labeled in the following format V(Visit#)_(Participant Identifier).

Figure S3   PCoA of weighted UniFrac distances with overlaid genera. All samples were plotted and colored by the individual of origin, again showing significant clustering by individual. Three broad groups can be determined (i) associated with *Lactobacillus*, (ii) associated with *Gardnerella* and *Atopobium* and (iii) a diverse microbiota including *Prevotella* and *Veillonella*.

Figure S4   Heat map of all detected metabolites across all samples.

Figure S5   Correlation between lactate abundance and percent lactate producers. Each dot represents a different sample and each color a different individual. The coefficient of correlation was 0.43 ($p = 9.6 \times 10^{-6}$).

Figure S6   Principle component analysis (PCA) of metabolites in vaginal fluid before and after probiotic (A) or placebo (B) intervention. Each point represents a different sample. Distribution of samples is based on metabolites alone, where the distance between samples represents how similar the metabolomes of those samples are.

Figure S7   Cytokine levels across groups following before and after probiotic and placebo adjusted by total protein in the sample. (A)IL-1β, (B) IL-4, (C) IL-5, (D) IL-6, (E) GM-CSF, (F) TNFα.

Figure S8   Ingenuity Pathway Analysis of inflammatory network altered by probiotic administration. Genes differentially expressed (red = up-regulated, green = down-regulated) by probiotic treatment were overlaid over a protein interaction network showing TLR2 and IL18 as central nodes in an inflammatory network.

Table S1   Taxonomic assignments of OTUs. The OTU number, the assigned taxonomy and the OTU seed sequence are presented.

Table S2   Filtered gene lists of probes differentially expressed before and after probiotic (A) and placebo (B) administration. Probe sets presented were differentially expressed ≥2-fold with a raw p-value<0.05. Two complete sample sets (before and after intervention from the same individual, 4 arrays) were obtained from the probiotic intervention and 3 sets from the placebo intervention (6 arrays).

Table S3   Nucleotide barcodes for participant sample 16S rRNA sequencing.

Protocol S1   Study protocol.

Checklist S1   CONSORT checklist.

Author Contributions

Conceived and designed the experiments: JEB SS AM RV DK LW BD GBG JPB GR. Performed the experiments: JEB SS AM BF DH. Analyzed the data: JEB SS AM RV DK LW BD GBG MS JPB GR. Contributed reagents/materials/analysis tools: GBG MS. Contributed to the writing of the manuscript: JEB SS AM RV DK GR.

References

1. Ma B, Forney LJ, Ravel J (2012) Vaginal Microbiome: Rethinking Health and Disease. Annu Rev Microbiol 66: 371–389.
2. Macklaim J, Gloor GB, Anukam KC, Cribby S, Reid G (2011) At the crossroads of vaginal health and disease, the genome sequence of Lactobacillus iners AB-1. Proc Natl Acad Sci USA 108: 4688–4695.
3. Ravel J, Gajer P, Abdo Z, Schneider GM, Koenig SSK, et al. (2011) Vaginal microbiome of reproductive-age women. Proc Natl Acad Sci USA 108: 4680–4687.
4. Hummelen R, Fernandes AD, Macklaim JM, Dickson RJ, Changalucha J, et al. (2010) Deep sequencing of the vaginal microbiota of women with HIV. PLoS ONE 5: e12078. doi:10.1371/journal.pone.0012078.
5. Macklaim JM, Fernandes AD, Di Bella JM, Hammond JA, Reid G, et al. (2013) Comparative meta-RNA-seq of the vaginal microbiota and differential expression by Lactobacillus iners in health and dysbiosis. Microbiome 1: 12. doi:10.1186/2049-2618-1-12.
6. Santoro N, Komi J (2009) Prevalence and impact of vaginal symptoms among postmenopausal women. J Sex Med 6: 2133–2142. doi:10.1111/j.1743-6109.2009.01335.x.
7. Nappi RE, Lachowsky M (2009) Menopause and sexuality: Prevalence of symptoms and impact on quality of life. Maturitas 63: 138–141.
8. Cauci S, Driussi S, De Santo D, Penacchioni P, Iannicelli T, et al. (2002) Prevalence of Bacterial Vaginosis and Vaginal Flora Changes in Peri- and Postmenopausal Women. Journal of Clinical Microbiology 40: 2147–2152. doi:10.1128/JCM.40.6.2147-2152.2002.

9. Hummelen R, Macklaim JM, Bisanz JE, Hammond J-A, McMillan A, et al. (2011) Vaginal microbiome and epithelial gene array in post-menopausal women with moderate to severe dryness. PLoS ONE 6: e26602. doi:10.1371/journal.pone.0026602.
10. Gajer P, Brotman RM, Bai G, Sakamoto J, Schütte UME, et al. (2012) Temporal dynamics of the human vaginal microbiota. Sci Transl Med 4: 132ra52. doi:10.1126/scitranslmed.3003605.
11. Burton JP, Reid G (2002) Evaluation of the bacterial vaginal flora of 20 postmenopausal women by direct (Nugent score) and molecular (polymerase chain reaction and denaturing gradient gel electrophoresis) techniques. J Infect Dis 186: 1770–1780.
12. Burton JP, Cadieux PA, Reid G (2003) Improved Understanding of the Bacterial Vaginal Microbiota of Women before and after Probiotic Instillation. Appl Environ Microbiol 69: 97–101.
13. Nugent RP, Krohn MA, Hillier SL (1991) Reliability of diagnosing bacterial vaginosis is improved by a standardized method of gram stain interpretation. Journal of Clinical Microbiology 29: 297–301.
14. Anukam K, Osazuwa E, Ahonkhai I, Ngwu M, Osemene G, et al. (2006) Augmentation of antimicrobial metronidazole therapy of bacterial vaginosis with oral probiotic Lactobacillus rhamnosus GR-1 and Lactobacillus reuteri RC-14: randomized, double-blind, placebo controlled trial. Microbes and Infection 8: 1450–1454.
15. Reid G, Charbonneau D, Erb J, Kochanowski B, Beuerman D, et al. (2003) Oral use of Lactobacillus rhamnosus GR-1 and L. fermentum RC-14

significantly alters vaginal flora: randomized, placebo-controlled trial in 64 healthy women. FEMS Immunol Med Microbiol 35: 131–134.

16. Martinez RCR, Franceschini SA, Patta MC, Quintana SM, Gomes BC, et al. (2009) Improved cure of bacterial vaginosis with single dose of tinidazole (2 g), *Lactobacillus rhamnosus* GR-1, and *Lactobacillus reuteri* RC-14: a randomized, double-blind, placebo-controlled trial. Can J Microbiol 55: 133–138.

17. Hummelen R, Changalucha J, Butamanya NL, Cook A, Habbema JDF, et al. (2010) *Lactobacillus rhamnosus* GR-1 and *L. reuteri* RC-14 to prevent or cure bacterial vaginosis among women with HIV. Int J Gynaecol Obstet 111: 245–248.

18. Bisanz JE, Reid G (2011) Unraveling how probiotic yogurt works. Sci Transl Med 3: 106ps41. doi:10.1126/scitranslmed.3003291.

19. Vásquez A, Molin G, Pettersson B, Antonsson M, Ahrné S (2005) DNA-based classification and sequence heterogeneities in the 16S rRNA genes of *Lactobacillus casei/paracasei* and related species. Syst Appl Microbiol 28: 430–441.

20. Lozupone C, Knight R (2005) UniFrac: a New Phylogenetic Method for Comparing Microbial Communities. Appl Environ Microbiol 71: 8228–8235.

21. Denecker G, Ovaere P, Vandenabeele P, Declercq W (2008) Caspase-14 reveals its secrets. J Cell Biol 180: 451–458.

22. Verstraelen H, Verhelst R, Claeys G, De Backer E, Temmerman M, et al. (2009) Longitudinal analysis of the vaginal microflora in pregnancy suggests that *L. crispatus* promotes the stability of the normal vaginal microflora and that *L. gasseri* and/or *L. iners* are more conducive to the occurrence of abnormal vaginal microflora. BMC Microbiol 9: 116. doi:10.1186/1471-2180-9-116.

23. Reid G, Gaudier E, Guarner F, Huffnagle GB, Macklaim JM, et al. (2010) Responders and non-responders to probiotic interventions: how can we improve the odds? Gut Microbes 1: 200–204.

24. Aldunate M, Tyssen D, Johnson A, Zakir T, Sonza S, et al. (2013) Vaginal concentrations of lactic acid potently inactivate HIV. J Antimicrob Chemother 68: 2015–2025.

25. Yeoman CJ, Thomas SM, Miller MEB, Ulanov AV, Torralba M, et al. (2013) A Multi-Omic Systems-Based Approach Reveals Metabolic Markers of Bacterial Vaginosis and Insight into the Disease. PLoS ONE 8: e56111. doi:10.1371/journal.pone.0056111.

26. van Baarlen P, Troost F, van der Meer C, Hooiveld G, Boekschoten M, et al. (2011) Human mucosal in vivo transcriptome responses to three lactobacilli indicate how probiotics may modulate human cellular pathways. Proc Natl Acad Sci USA 108: 4562–4569.

27. Biet F, Locht C, Kremer L (2002) Immunoregulatory functions of interleukin 18 and its role in defense against bacterial pathogens. J Mol Med 80: 147–162.

28. Jacobsson B, Holst R-M, Mattsby-Baltzer I, Nikolaitchouk N, Wennerholm U-B, et al. (2003) Interleukin-18 in cervical mucus and amniotic fluid: relationship to microbial invasion of the amniotic fluid, intra-amniotic inflammation and preterm delivery. BJOG 110: 598–603.

29. Harandi AM, Svennerholm B, Holmgren J, Eriksson K (2001) Interleukin-12 (IL-12) and IL-18 Are Important in Innate Defense against Genital Herpes Simplex Virus Type 2 Infection in Mice but Are Not Required for the Development of Acquired Gamma Interferon-Mediated Protective Immunity. J Virol 75: 6705–6709.

30. Bolton M, van der Straten A, Cohen CR (2008) Probiotics: potential to prevent HIV and sexually transmitted infections in women. Sex Transm Dis 35: 214–225.

31. Schwandner R, Dziarski R, Wesche H, Rothe M, Kirschning CJ (1999) Peptidoglycan- and lipoteichoic acid-induced cell activation is mediated by toll-like receptor 2. J Biol Chem 274: 17406–17409.

32. Zipfel PF, Skerka C (2009) Complement regulators and inhibitory proteins. Nat Rev Immunol 9: 729–740.

33. Rizzo A, Losacco A, Carratelli CR, Domenico MD, Bevilacqua N (2013) *Lactobacillus plantarum* reduces *Streptococcus pyogenes* virulence by modulating the IL-17, IL-23 and Toll-like receptor 2/4 expressions in human epithelial cells. Int Immunopharmacol 17: 453–461.

34. Molfino NA, Gossage D, Kolbeck R, Parker JM, Geba GP (2012) Molecular and clinical rationale for therapeutic targeting of interleukin-5 and its receptor. Clin Exp Allergy 42: 712–737.

35. Kaya D, Demirezen S, Beksaç MS (2012) The presence of eosinophil leucocytes in cervicovaginal smears with Actinomyces-like organisms: Light microscopic examination. J Cytol 29: 226–229.

36. Sultana R, McBain AJ, O'Neill CA (2013) Strain-dependent augmentation of tight-junction barrier function in human primary epidermal keratinocytes by *Lactobacillus* and *Bifidobacterium* lysates. Appl Environ Microbiol 79: 4887–4894.

37. Aitchison J (1981) A new approach to null correlations of proportions. Math Geol 13: 175–189. doi:10.1007/BF01031393.

38. Aitchison J, Egozcue J (2005) Compositional Data Analysis: Where Are We and Where Should We Be Heading? Math Geol 37: 829–850.

39. Faust K, Sathirapongsasuti JF, Izard J, Segata N, Gevers D, et al. (2012) Microbial co-occurrence relationships in the human microbiome. PLoS Comput Biol 8: e1002606. doi:10.1371/journal.pcbi.1002606.

40. Stein SE (1999) An integrated method for spectrum extraction and compound identification from gas chromatography/mass spectrometry data. J Am Soc Mass Spectrom 10: 770–781.

41. Styczynski MP, Moxley JF, Tong LV, Walther JL, Jensen KL, et al. (2007) Systematic identification of conserved metabolites in GC/MS data for metabolomics and biomarker discovery. Anal Chem 79: 966–973.

13

Interacting Symbionts and Immunity in the Amphibian Skin Mucosome Predict Disease Risk and Probiotic Effectiveness

Douglas C. Woodhams[1,2*¤a], **Hannelore Brandt**[1], **Simone Baumgartner**[1¤b], **Jos Kielgast**[3¤c], **Eliane Küpfer**[1,4], **Ursina Tobler**[1,5], **Leyla R. Davis**[1], **Benedikt R. Schmidt**[1,5], **Christian Bel**[1], **Sandro Hodel**[1], **Rob Knight**[6], **Valerie McKenzie**[2]

1 Institute of Evolutionary Biology and Environmental Studies, University of Zurich, Zurich, Switzerland, 2 Department of Ecology and Evolutionary Biology, University of Colorado, Boulder, Colorado, United States of America, 3 Section for Freshwater Biology, Department of Biology, University of Copenhagen, Copenhagen, Denmark, 4 Department of Evolutionary Biology, Technical University of Braunschweig, Braunschweig, Germany, 5 KARCH, Neuchâtel, Switzerland, 6 Howard Hughes Medical Institute and Department of Chemistry and Biochemistry, BioFrontiers Institute, University of Colorado, Boulder, Colorado, United States of America

Abstract

Pathogenesis is strongly dependent on microbial context, but development of probiotic therapies has neglected the impact of ecological interactions. Dynamics among microbial communities, host immune responses, and environmental conditions may alter the effect of probiotics in human and veterinary medicine, agriculture and aquaculture, and the proposed treatment of emerging wildlife and zoonotic diseases such as those occurring on amphibians or vectored by mosquitoes. Here we use a holistic measure of amphibian mucosal defenses to test the effects of probiotic treatments and to assess disease risk under different ecological contexts. We developed a non-invasive assay for antifungal function of the skin mucosal ecosystem (mucosome function) integrating host immune factors and the microbial community as an alternative to pathogen exposure experiments. From approximately 8500 amphibians sampled across Europe, we compared field infection prevalence with mucosome function against the emerging fungal pathogen *Batrachochytrium dendrobatidis*. Four species were tested with laboratory exposure experiments, and a highly susceptible species, *Alytes obstetricans*, was treated with a variety of temperature and microbial conditions to test the effects of probiotic therapies and environmental conditions on mucosome function. We found that antifungal function of the amphibian skin mucosome predicts the prevalence of infection with the fungal pathogen in natural populations, and is linked to survival in laboratory exposure experiments. When altered by probiotic therapy, the mucosome increased antifungal capacity, while previous exposure to the pathogen was suppressive. In culture, antifungal properties of probiotics depended strongly on immunological and environmental context including temperature, competition, and pathogen presence. Functional changes in microbiota with shifts in temperature provide an alternative mechanistic explanation for patterns of disease susceptibility related to climate beyond direct impact on host or pathogen. This nonlethal management tool can be used to optimize and quickly assess the relative benefits of probiotic therapies under different climatic, microbial, or host conditions.

Editor: Carlos A. Navas, University of Sao Paulo, Brazil

Funding: Financial support came from the Zoological Institute and the Forschungskredit of the University of Zurich, Vontobel Stiftung, Janggen-Pöhn Stiftung, Basler Stiftung für biologische Forschung, Stiftung Dr. Joachim De Giacomi, Zoo Zürich, Grün Stadt Zürich, European Union of Aquarium Curators, Schweizer Tierschutz, Zürcher Tierschutz, Claraz Foundation, the environment departments of the cantons St. Gallen and Zurich, Swiss National Science Foundation (31-125099 to DCW), and U.S. National Science Foundation Population and Community Ecology Section (DEB 1146284 to VJM and RK). The funders had no role in study design, data collection and analysis, decision to publish, or preparation of the manuscript.

* E-mail: dwoodhams@gmail.com

¤a Current address: Department of Biology, University of Massachusetts Boston, Boston, Massachusetts, United States of America
¤b Current address: Department of Aquatic Ecology, Duebendorf, Switzerland, and Institute of Integrative Biology, ETH-Zürich, Zurich, Switzerland
¤c Current address: Section for Freshwater Biology, Department of Biology, University of Copenhagen, Copenhagen, Denmark and Center for Macroecology, Evolution and Climate Natural History Museum of Denmark, Copenhagen, Denmark

Introduction

Probiotic therapies often aim to extend or shape the immune function of hosts by altering the symbiotic microbial community. Probiotics are used in human and veterinary medicine, agriculture and aquaculture, and have been proposed for treatment of emerging wildlife diseases such as those occurring on corals and amphibians [1,2]. Microbiota can mediate pathogenesis through a range of mechanisms [3,4], and disease ecology studies demonstrate that parasitic and non-parasitic microbes interact with each other and with the host immune system such that pathogenicity is often influenced by environmental conditions [5–8]. Thus, the environment affects the risk of disease to individuals, populations, and species, and assessing disease risk under changing conditions is vital to conservation and infectious disease mitigation and can direct the allocation of resources for most effect [9–12].

The microbiota inhabiting skin and mucosal surfaces has a profound impact on host health and immunity [7,13,14], and may be predictive of risk for some diseases [15–17]. Amphibian skin is a model system for diseases affecting vertebrate mucosa. The mucosome, or micro-ecosystem of the mucus, as defined here contains interdependent host factors (mucosal antibodies, antimicrobial peptides, lysozyme, alkaloids) and microbial-community factors (microbiota, antibiotic metabolites). The mucosome has various functions potentially including communication, and predator and pathogen defense. Here, we develop a non-lethal assay and holistic measure referred to as "mucosome function" to describe the effect of amphibian skin mucus on pathogen viability. We examine how environmental and immunological contexts may impact the outcome of host-microbe symbioses, and how mucosome function captures the *in vivo* complexity of the micro-ecosystem and can thus accurately predict susceptibility to infection. We focus on probiotic bacteria and fungi applied to the skin mucosome as biocontrol agents against the emerging amphibian disease chytridiomycosis.

Chytridiomycosis is a major cause of global amphibian population declines and species extinctions [18,19]. The disease is caused by the chytrid fungus *Batrachochytrium dendrobatidis*, or *Bd*, and is strongly influenced by climatic conditions [20]. Climate-linked changes to the entire microbiota, not just *Bd*, may influence disease susceptibility [5]. Current efforts to mitigate chytridiomycosis in wildlife populations have turned to bioaugmentation, or the use of probiotic therapies [1,21]. The successful prophylactic use of *Janthinobacterium lividum* was demonstrated against chytridiomycosis in mountain yellow-legged frogs, *Rana muscosa* [22]. However, when tested on the endangered Panamanian golden frog, *Atelopus zeteki*, the probiotic survived briefly on the skin, but did not protect the amphibians from disease [23]. Similarly, the probiotic *Pedobacter cryoconitis* temporarily reduced infection loads of heavily infected *R. muscosa* [24]. Each target host may thus require probiotic therapy tailored to that species, population, or life-history stage. Screening the various bacteria associated with hosts or their environment to identify effective probiotics is challenging [25,26]. Thus, probiotic therapies for amphibians must be optimized, and an understanding of which candidate bacteria can establish and persist on the host in its natural environmental context is urgently needed.

To date, all attempts to apply probiotic therapy against chytridiomycosis have used simple selection criteria for choosing candidate probiotics. Selection of the most efficient probiotic is challenging because there are hundreds of culturable phylotypes to choose from, either from environmental sources, or more typically, from tolerant host populations that can persist with nonlethal *Bd* infections [1]. However, simple co-culture assays to determine antifungal capacity have been insufficient to ensure probiotic effectiveness [23,24]. Co-factors including interactions of the probiotic with the microbial community already present on the amphibian skin, as well as interactions with host immune defenses, and effects of environmental conditions, may complicate the outcome of biotherapy. Here, we experimentally test the impact of immunological and environmental context on potential probiotic bacteria both *in vitro* and *in vivo*. The tested conditions are illustrative rather than comprehensive for potential environmental conditions, community and immunological interactions. Because it is impractical to test all potential interactions before testing probiotics on amphibians for a disease resistance effect, we suggest a protocol for selecting probiotics with the highest potential benefit, and to test whether the probiotics will likely be effective in the range of foreseeable conditions on the host. Our non-lethal susceptibility assay of mucosome function can help assess disease risk and treatment effects in rare amphibians including relict populations or captive populations of endangered species intended for reintroduction.

Typical approaches to compare species susceptibility and to assess disease risk include pathogen exposure experiments [27], or field surveys to compare infection prevalence and monitor disease and population trajectories [28], or modeling environmental and biogeographic risk factors [10,29]. Deficits of conventional pathogen exposure experiments include lack of environmental context when amphibians are exposed under clean laboratory conditions. Biodiversity including microbiota and macrobiota can influence disease outcome [30], and bacterial community diversity is reduced through time in captivity without natural sources such as soil for re-inoculating the skin [31]. The exposure history, population genetics, and life-history stage of the amphibians used in the experiment, as well as the strain and dose of the pathogen can all affect experimental outcomes, and many threatened species are not suitable for such experiments. In addition, growth of *Bd* is often inhibited by skin microbiota of amphibians [32,33]. However, little is known about how protective microbiota differs among host populations or regions, or how mucosome function is altered by enrichment with potential probiotics.

Our aims in this study were (1) to develop a holistic, simple, non-invasive, and non-lethal method to measure mucosome function against *Bd*. Using this tool, we aimed (2) to test whether mucosome function can predict *Bd* infection prevalence of amphibians in the field and survival in *Bd* exposure experiments. While we show that probiotics are influenced by a variety of factors including competition, temperature, and innate immunity when tested *in vitro*, we aimed (3) to use mucosome function as an ecologically-integrated predictor of probiotic therapy effect so that future research can test probiotic strategies for conservation and not lose hope in the potential of probiotic therapy in the face of immunological and ecological complexity. We provide a detailed protocol for measuring mucosome function in File S1.

Materials and Methods

Ethics statement

Permits to conduct fieldwork were obtained from the Swiss cantonal conservation authorities, and from Germany - German federal licence (Rheinland-Pfalz) no. 425-104.143.0904 Struktur- und Genehmigungsdirektion Nord, Koblenz. All animal procedures were approved by the Veterinary Authority of Zurich (110/2007 and 227/2007) and the Federal Office for the Environment. Fieldwork conformed to standard decontamination practices to avoid transport of pathogens between sites. All animals in experiments were monitored daily for animal welfare and to ameliorate suffering. During experiments, any individual demonstrating clinical signs of disease including lethargy, abnormal skin shedding, and loss of righting reflex were humanely euthanized. At the end of the experiment, all animals were humanely euthanized by overdose of tricaine methanesulfonate.

Survey of *Bd* infection prevalence

To compare *Bd* infection prevalence among species and life-history stages, we combine previously unpublished results from field studies in Switzerland with *Bd* surveys from amphibians across Europe collated by Bd-Maps (www.bd-maps.net, accessed September 1, 2013). In addition to data from 5939 sampled amphibians available from Bd-maps, skin swabs were collected from 2591 amphibians from 12 species and from 66 *Bd*-positive populations from the northern parts of Switzerland and tested for *Bd* between 2007 and 2009 (Table 1). Amphibians were caught by dip-netting and swabbed with a sterile cotton swab (Copan Italia S.p.A., Brescia, Italy). Field material was cleaned and disinfected before moving between different sites to avoid contamination and spread of *Bd* and other pathogens. Extraction and analysis for *Bd*-DNA were done following the qPCR protocol by Boyle *et al.* [34] using *Bd*-specific primers and standards to quantify the amount of DNA. We ran each sample twice and the PCR was repeated if the two wells returned dissimilar results. Reactions below 1 genomic equivalent were scored *Bd*-negative to avoid false positives. Mean infection prevalence with 95% binomial confidence interval was calculated for each species and life stage sampled, and calculated for both Europe and Switzerland.

Bd infection prevalence predicted by skin defenses

Skin defense peptides and mucosome samples were tested against *Bd* for comparison of anti-*Bd* activity with infection prevalence in natural populations by logistic regression in R. Amphibians sampled for skin peptides and mucosome function (Table 1) were sampled in Switzerland and compared to field infection prevalence from Switzerland and across Europe in separate analyses. Skin peptides were collected upon induction by subcutaneous injection of metamorphosed amphibians with 40 nmole/g body mass norepinephrine (bitartrate salt, Sigma) or immersion of larval amphibians in 100 μM norepinephrine, and tested for *Bd* growth inhibition as previously described [35,36]. Skin peptide samples from post-metamorphic amphibians only were used in the logistic regression analyses because different methods of peptide induction were used on larval stages. Mucosome samples from multiple life-history stages of the same species were included and matched to life-history stages sampled for *Bd* diagnostics (Table 1). Detailed methods for measuring mucosome function against *Bd* using a fluorescence assay of *Bd* viability adapted from Stockwell *et al.* [37] (Fig. S1 in File S1) and comparisons of mucosome function and skin peptide defenses against *Bd* are presented in Supporting Information (Figs. S4, S5 in File S1).

Survival predicted by mucosome function

To examine the relationship between mucosome function against *Bd* and susceptibility to infection and subsequent survival we performed experimental exposures to *Bd* on four species. All animals were exposed to zoospores from Swiss lineage *Bd* TG 739 isolated from a moribund *A. obstetricans* in Gamlikon, Switzerland in 2007 [38] and cryopreserved until use. Egg clutches were obtained from *P. esculentus* (n = 8), *B. variegata* (n = 8), *R. temporaria* (n = 45), and *A. obstetricans* (n = 13) in northern Switzerland or southern Germany. *Rana temporaria* were raised in outdoor mesocosms through metamorphosis before experimental exposure of metamorphs to *Bd* (N = 92 exposed, 94 control). Other species were exposed to *Bd* as tadpoles (N = 80 exposed, 40 control per species). All animals were kept in the same laboratory at 19°C during the experiments. We measured the proportion of infected metamorphs by qPCR, and determined relative survival (survival of infected/survival of uninfected controls) at the end of the experiments (50–90 d after metamorphosis). Kaplan Meier curves are presented in the Supporting Information for each species. We examined the relationship between mucosome function against *Bd* and relative survival and proportion infected using logistic regression analyses in R.

Host ecological context and skin defenses

The *in vivo* effects of ambient temperature and skin microbiota on mucosome function against *Bd* and skin peptide defenses were tested on a focal amphibian species, *A. obstetricans*. In Europe, the common midwife toad, *A. obstetricans*, is a species of conservation concern [39] and is particularly sensitive to *Bd* early in life-history [40]. Host-associated bacteria and fungi were surveyed by culturing from populations of midwife toads near Basel, Switzerland in May, 2009, including samples from 19 adults, 32 larvae, and 9 egg clutches. Although many diverse antifungal bacteria have been described in association with skin of some amphibian hosts [32,33], we chose eleven bacterial residents isolated from *A. obstetricans* for the environmental context experiments described below based on potency against *Bd* in culture and high prevalence in the populations sampled (L. Davis, unpublished). Two bacterial isolates with high *in vitro* potency against *Bd* and the ability to withstand host skin defense peptides, and one fungal isolate, were chosen for applications on recently metamorphosed *A. obstetricans*.

All metamorphs used in the study were raised in captivity from wild-caught tadpoles that were naturally exposed to the fungus in their pond of origin, near Zunzgen, Switzerland, but negative for *Bd* by qPCR at the time of the experiment. Toadlets were of similar size (mean±SD: 2.1±0.3 g; ANOVA $F_5 = 1.179$, $P = 0.332$) and treated at the same time with one exception. Toadlets in the *Bd*-exposure group were exposed to *Bd* approximately 2 months prior to the microbial exposure treatments, and the toadlets were smaller (1.5±0.3 g), and no longer infected at the time of sampling based on qPCR.

We treated recently metamorphosed common midwife toads, *Alytes obstetricans* (N = 70: 10 per treatment group, 7 treatments), by housing them individually at 5°C, 18°C, or 25°C with no microbes added, or at 18°C with exposure to *Bd* zoospores (8.5×10^6 zoospores of global panzootic lineage isolated from a *Bufo bufo* in the UK [38]), a probiotic fungus *Penicillium expansum*, or a probiotic bacterium *P. fluorescens* or *F. johnsoniae*. Toadlets were bathed individually for one hour in water containing the microbes and after 2 weeks, toadlets from all treatments were sampled on the same day for mucosome function and subsequently skin peptides, sampled as described above.

Temperature, competition of probiotic strains, and co-culture with *Bd*

To determine the effects of competitive interactions and temperature on probiotic potential, 11 common host-associated isolates were chosen. These included two isolates of *Serratia plymuthica* and one isolate of *Janthinobacterium lividum* from egg clutches of midwife toads, three isolates of *Flavobacterium johnsoniae* and five species of *Pseudomonas* isolated from the skin of adults. Based on 16S rRNA gene sequences, all 11 isolates were considered unique operational taxonomic units (OTUs) at 99%, but clustered into 7 OTUs at 97% similarity as determined by the UCLUST algorithm in QIIME. The 16S rRNA gene sequences of all isolates were deposited in the European Nucleotide Archive (Table S1 in File S1).

In one set of experiments, bacterial isolates were freshly grown at 18°C on RIIA agar media supplemented with 1% tryptone then transferred to experimental conditions. Bacteria and *Bd* (Swiss isolate TG 739) readily grew on the same media. Plate experiments were performed in duplicate. Both isolates of *Serratia plymuthica* were grown separately at 18 and 25°C, or at 18°C on media inoculated with *Bd* and allowed to dry before streaking the bacteria. Two isolates of *F. johnsoniae* were grown separately or combined on media inoculated with *Bd*, and grown at 18°C. When combined, each isolate was streaked across the entire plate. Three *Pseudomonas* isolates were grown either separately, combined, or combined on media inoculated with *Bd*, and grown at 18°C. Control plates of sterile media or *Bd*-only were also tested. All plates were incubated for 3 days, and then rinsed with 2 ml sterile

Table 1. Amphibians from Switzerland sampled for skin peptide effectiveness and mucosome function against *Bd*, and *Bd* infection prevalence at different life-history stages.

Species	Life-history stage[#]	Peptide effectiveness* (N)	SE	Mean mucosome function against Swiss *Bd* (N)	SE	Switzerland: Percent infected (N)	95% binomial confidence interval	Europe: Percent infected (N)	95% binomial confidence interval
Alytes obstetricans	Adult/Subadult	15.92 (8)	6.21	0.012 (10)	0.000	4.9 (41)	0.6–16.5	29.7 (209)[$]	23.5–36.4
Alytes obstetricans	Metamorph	37.75 (5)	12.15						
Alytes obstetricans	Larvae	48.76 (5)	24.23	2.963 (10)	0.681	45.4 (2111)	43.3–47.6	38.0 (3008)	36.3–39.8
Bombina variegata	Adult/Subadult			1.075 (4)	0.081	20.0 (150)	13.9–27.3	21.1 (227)	16.0–27.0
Bufo bufo	Adult	16.34 (15)	3.37753	0.117 (9)	0.082	0.0 (22)	0.0–15.4	0.9 (3606)	0.6–1.2
Bufo bufo	Larvae			1.284 (5)	0.404			6.7 (45)	1.4–18.3
Hyla arborea	Adult	11.42 (7)	2.15210			3.8 (26)	0.1–19.6	12.5 (32)	3.5–29.0
Ichthyosaura alpestris	Adult	0.94 (7)	.52546	1.361 (20)	0.062	24.8 (629)	21.5–28.4	21.5 (775)	18.7–24.6
Lissotriton vulgaris	Adult	1.85 (4)	1.02506			27.3 (22)	10.7–50.2	17.0 (47)	7.7–30.8
Pelophylax lessonae/esculentus	Adult	27.27 (10)	3.18			22.4 (170)	16.3–29.4	15.6 (275)	11.6–20.5
Pelophylax lessonae/esculentus	Metamorph	5.34 (5)	1.88685	0.545 (10)	0.042	13.0 (69)	6.1–23.3	13.2 (76)	6.5–22.9
Rana temporaria	Adult/Subadult	1.97 (13)	.62111	0.251 (10)	0.128	0.0 (10)	0.0–30.9	3.1 (129)	0.9–7.8
Rana temporaria	Larvae			0.220 (5)	0.120	0.0 (20)	0.0–16.8	0.0 (23)	0.0–14.8
Salamandra salamandra	Adult	4.92 (9)	1.32654					11.1 (9)	0.3–48.3
Salamandra salamandra	Larvae	42.78 (5)	13.35528					23.2 (69)	13.9–34.9

Skin peptide effectiveness is the percent inhibition of *Bd* zoospore growth caused by 50 µg/ml peptide multiplied by the quantity of peptides (mg) per g amphibian according to Woodhams et al. [11]. The mucosome function against *Bd* (Swiss isolate TG 739) is a measure of zoospore viability quantified by the ratio of green:red fluorescence as described above. Infection prevalence is the mean from all amphibians in each group from multiple sites and seasons.

[#] Larval and post-metamorphic skin peptide samples extracted by different methods.

*Peptide effectiveness = % inhibition of *Bd* growth at 50 µg/ml * mg peptides/g frog mass.

[$] Includes samples from chytridiomycosis outbreak sites in Spain (S. Walker, unpubl.), not included in logistic regression.

Mili-Q water. Rinse water was then filtered through a 0.22 μm syringe filter.

Bacteria were also grown in liquid RIIA media for 4 d at 14, 19, and 22°C, and metabolites filtered as above. Metabolites from liquid cultures were added to *Bd* zoospores (Global panzootic lineage VMV 813 from a bullfrog, *Lithobates catesbeianus* tadpole) to test for inhibitory effects on pathogen growth. To determine the effect of bacterial filtrate on *Bd* growth, *Bd* zoospores were harvested in 1% tryptone and counted under a hemocytometer. Wells of a 96-well plate were inoculated with 50 μl zoospores at 8×10^6 zoospores per ml. Then, 50 μl of filtrate (or filtrate diluted 1:10) from each of the experimental or control plates, or liquid cultures, was added to the wells in replicates of four. In addition, 6 positive control wells contained *Bd* and 50 μl sterile water or RIIA media, and 6 negative control wells contained heat-killed *Bd* and 50 μl sterile water or RIIA media. The change in optical density measured at 490 nm absorbance over 7 days growth at 19°C was recorded using a Victor3 multilabel plate reader (PerkinElmer). Standard statistical testing was carried out in IBM SPSS Statistics 22. Significant *Bd* growth inhibition (or enhancement) caused by bacterial filtrate was determined by t-test, and a repeatable result (Table S2 in File S1). Percent inhibition depended on filtrate dose (see Results) and was not considered comparable among bacterial isolates.

Effects of host skin peptides and *Bd* metabolites on probiotics in culture

To test for the response of bacterial growth upon culture with either *Bd* filtrate or host skin peptides, bacteria were grown in RIIA liquid media on 96 well plates. Supernatant from a 2-week old culture of *Bd* (type isolate JEL 197) growing in 0.5% tyrptone was filtered through a 0.22 μm syringe filter. An equal volume of *Bd* filtrate or sterile media was added to bacterial cultures. To test effects of peptides, we added an equal volume of sterile water or natural mixtures of partially-purified skin peptides from *A. obstetricans* metamorphs at a final concentration of 100 μg/ml in water. Growth after 48 hr was measured as change in optical density measured at 480 nm. Differences between experimental and control bacterial growth were tested by t-tests using a Bonferroni correction for multiple comparisons.

Results

Survey of *Bd* infection prevalence

Surveys of approximately 8500 amphibians (http://www.bd-maps.net/; this study) at different life-history stages for *Bd* infection based on qPCR indicated high prevalence in larval midwife toads, *A. obstetricans* (45% infected in Switzerland) and aquatic adult newts *Ichthyosaura alpestris* (26%), and *Lissotriton vulgaris* (27%). Low infection prevalence (<5%) was detected in populations of adult *A. obstetricans*, *Bufo bufo*, *Rana temporaria*, and *Hyla arborea* (Table 1, Fig. 1).

Bd infection prevalence predicted by skin defenses

We examined two non-lethal measures of susceptibility to infection in pathogen-free Swiss amphibians acclimated to laboratory conditions. These included testing *Bd* growth or viability upon exposure to natural mixtures of partially purified skin defense peptides, and a holistic functional measure of the skin mucosal ecosystem (mucosome function) including ambient skin defenses: peptides, alkaloids, lysozymes, mucosal antibodies, microbiota and microbial metabolites [41]. Both antifungal skin peptides and mucosome function were correlated with infection prevalence in natural populations across Europe (Fig. 1a,c) and

within Switzerland (Fig. 1b,d). Prevalence of infection with *Bd* decreased with peptide efficiency (Fig. 1c,d, logistic regressions: Europe, P = 0.0015; Switzerland, P = 0.0079). While induced peptide defenses stored in granular glands were measured here, ambient peptides (not induced by norepinephrine) are a natural component of the mucosome [42,43]. Mucosome function was tightly correlated to *Bd* prevalence in natural populations of Swiss amphibians (Fig. 1b, P<0.0001) and in amphibians across Europe (Fig. 1a, P = 0.0020). The odds ratios of *Bd* colonization in Swiss amphibians was 1.950 (Europe, 2.969) with each unit change in mucosome function, and 0.839 (Europe, 0.811) with each unit decrease in skin peptide efficiency. Correlations of mucosome function and induced skin peptide efficiency are presented in Figure S4 in File S1 and suggest that both host and microbial factors contribute to mucosome function against *Bd*.

Survival predicted by mucosome function

Pathogen exposure experiments were conducted on four host species with a Swiss isolate of *Bd*, and relative survival post-metamorphosis of infected tadpoles differed among species (% relative survival, mean±SD days survived): *A. obstetricans* (0%, 24±17.5 d), *Bombina variegata* (39.0%, 32±23.9 d), and *Pelophylax esculentus* (30.4%, 12±12.8 d; Fig. S2 in File S1). Relative survival of recently metamorphosed *Rana temporaria* exposed to *Bd* was 100% (Fig. S3 in File S1), and no colonization by *Bd* was detected by qPCR (n = 92). Success of *Bd* colonization of tadpoles also differed among species (Pearson χ^2_3 = 13.102, P = 0.004): *A. obstetricans* (13.9% infected, n = 36), *B. variegata* (10.7%, n = 75), and *P. esculentus* (7.9%, n = 76). Mucosome function predicted survival (logistic regression, P<0.0001; Fig. 2a) and infection with *Bd* in these species (P = 0.0106; Fig. 2b). The odds of infection increased by 1.751 with each unit change in mucosome function, and the odds of survival decreased by 0.0454.

Host ecological context and skin defenses

Midwife toads, *A. obstetricans*, were treated with various temperature and probiotic therapies and tested for mucosome function. Host context significantly affected mucosome permissiveness or lethality towards *Bd* (Fig. 3a; ANOVA, F_6 = 41.606, P< 0.001). *Bd* viability was similar following incubation with mucosome samples from toads at temperatures ranging from 5–25°C. Mucosome samples from toads previously exposed to *Bd* were least effective at killing *Bd* zoospores, while those from toads treated with probiotics *Flavobacterium johnsoniae* and *Penicillium expansum* were most effective at killing zoospores (Fig. 3a). While *Pseudomonas* in general, and the *P. fluorescens* isolate (76.5c) used in this study were often effective at inhibiting *Bd* in co-culture and produced antifungal metabolites across a range of temperatures ideal for *Bd* growth (Fig. 4a, Table S2 in File S1), there was no significant benefit of this probiotic when applied on hosts in terms of increasing mucosome function and reducing *Bd* viability (Fig. 3a).

Because one significant antimicrobial component of *A. obstetricans* skin mucus is antimicrobial peptides (AMPs) [44], we collected peptide skin secretions, quantified them per surface area of the toads and measured their ability to inhibit *Bd* growth at a standardized concentration of 100 μg/ml. On average, toads produced 0.25 mg peptide per cm^2 surface area, and at 100 μg/ml these peptides inhibited *Bd* growth by 48.7%. These values did not differ significantly among treatment groups, nor did a combined measure of skin peptide effectiveness against *Bd* (% * mg/cm^2, Fig. 3b; Kruskal-Wallis tests, P's>0.05). Thus, skin peptides stored in granular glands were not significantly affected by the 2-week temperature and microbe treatments including previous exposure

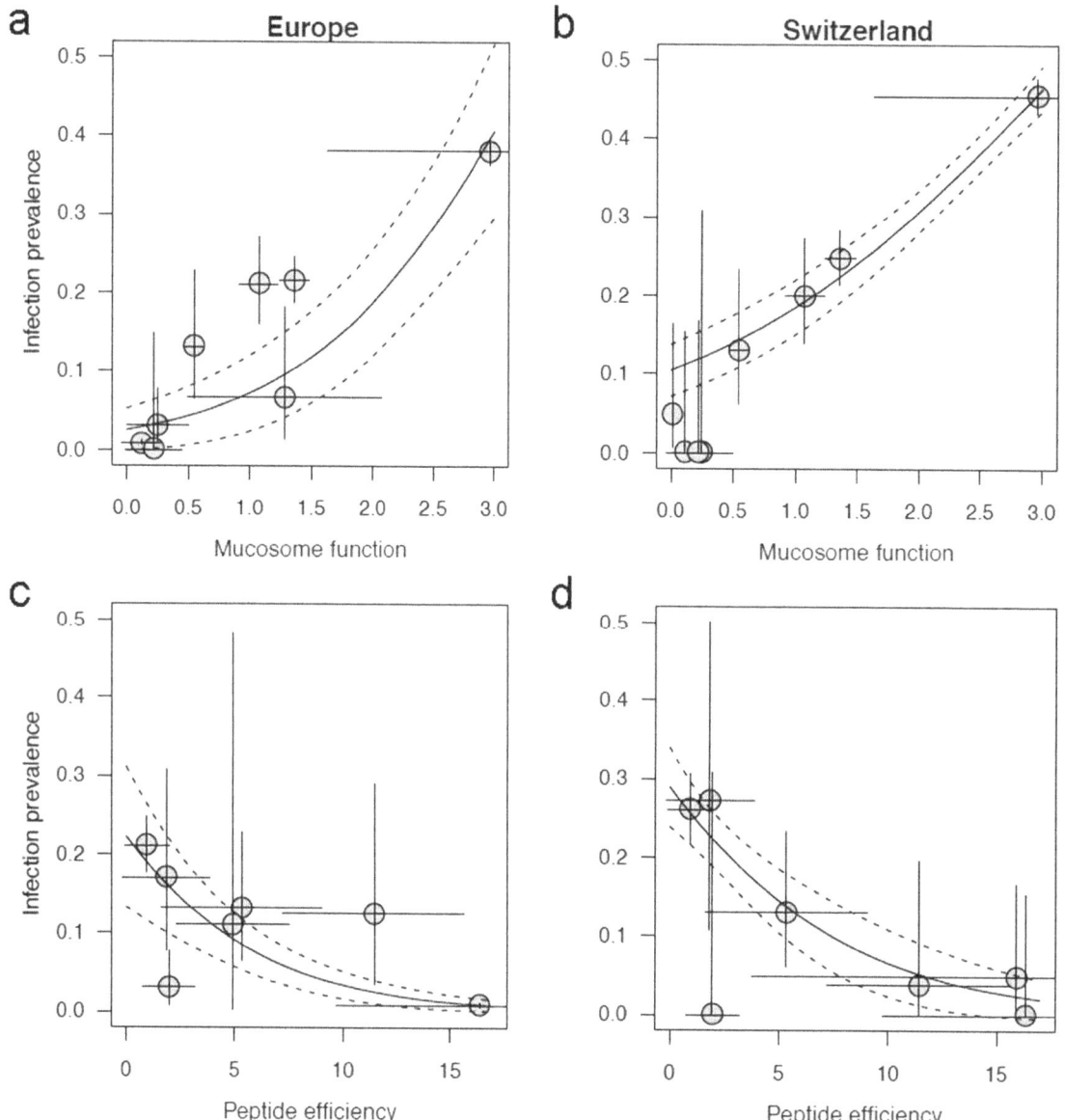

Figure 1. Infection prevalence (mean, 95% binomial CI) of amphibians sampled across Europe and within Switzerland predicted by mucosome function and skin defense peptide activity against *Batrachochytrium dendrobatidis* **(Bd) zoospores.** Mucosome function (mean, SE) indicates *Bd* viability after a 1 hr exposure to amphibian mucus (a,b) and units represent green:red fluorescence. Peptide efficiency (mean, SE) indicates quantity of natural mixtures of skin peptides induced from granular glands multiplied by activity of a standard concentration of peptides against *Bd* zoospore growth. Only post-metamorphic amphibians sampled upon subcutaneous injection with norepinephrine are plotted in (c) and (d). Amphibian skin mucosome function is a better predictor of infection prevalence than induced skin peptide efficiency (logistic regression, see text). Summary data for all species and life-history stages are presented in Table 1.

to *Bd*. There was not a significant correlation between peptide effectiveness and mucosome function against *Bd* (Fig. S5 in File S1; Pearson, $\chi^2 = -0.102$, $P = 0.827$). Zoospore viability after exposure to mucosome samples was significantly higher in the *Bd*-exposure treatment compared to other treatments (Fig. 3a). However, skin peptides induced from hosts in the *Bd*-exposure treatment were effective at inhibiting *Bd* growth, and not significantly different than peptides from toads in other treatments (Fig. 3b).

Temperature, competition of probiotic strains, and co-culture with *Bd*

Environmental conditions affected the capacity of probiotic bacteria to inhibit the fungal pathogen *Bd* (Table S2 in File S1). Two *Serratia plymuthica* isolates (isolates 27 and 28) were capable of inhibiting *Bd* growth when incubated at 18°C. Isolate 27 was inhibitory under all tested conditions: 18°C, 25°C, and 18°C co-cultured with *Bd*. Isolate 28 significantly enhanced *Bd* growth at 25°C, and was neither enhancing nor inhibitory at 18°C when co-cultured with *Bd* (Fig. 4c, Table S2 in File S1). A dose-response of *Bd* growth inhibition was found such that filtrate diluted 1/10 was

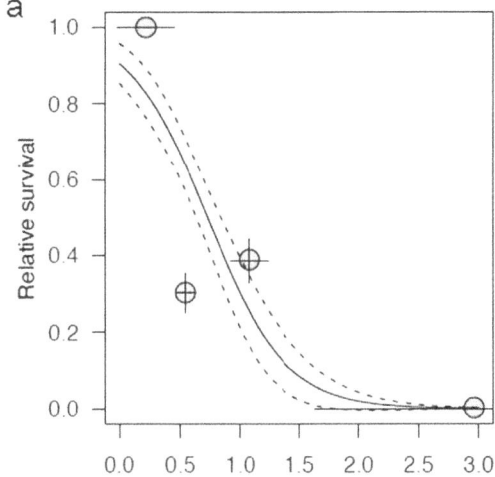

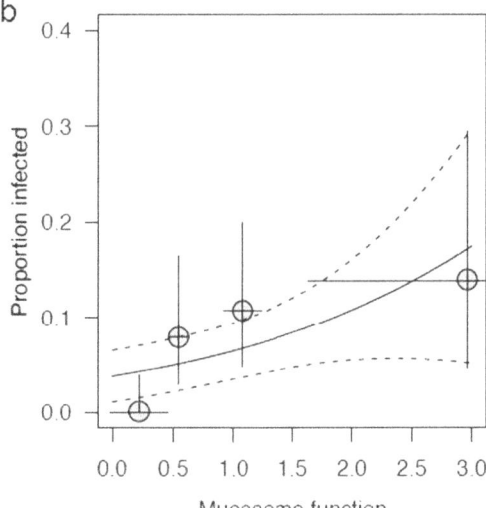

Figure 2. Relative survival (95% binomial CI; a) and Proportion of infected frogs (95% binomial CI; b) predicted by Mucosome function. Post-metamorphosis survival was measured from four Swiss amphibian species after exposure to zoospores of a Swiss *Bd* isolate, TG 739. Survival curves for each species are presented in Supporting Information (Figs. S2, S3 in File S1) and relative survival was calculated as the proportion of infected frogs surviving/proportion of unexposed control frogs surviving. *Alytes obstetricans* showed the highest infection and mortality, and *Rana temporaria* the lowest, with *Bombina variegata* and *Pelophylax esculentus* intermediate. All frogs were raised in captivity from egg clutches and had no history of natural exposure to *Bd*. Mucosome function (mean, SE) indicates *Bd* viability after exposure to amphibian mucus and is a significant predictor of both survival (binomial logistic regressions, P<0.0001) and infection prevalence (P = 0.0106).

significantly less inhibitory than undiluted filtrate (paired t-test, $t_{35} = 9.836$, $P<0.001$), and filtrate from control plates with or without *Bd* significantly enhanced *Bd* growth (Table S2 in File S1). Testing metabolites of the bacteria growing at 14, 19, and 22°C in liquid culture against the global panzootic lineage of *Bd* showed similar results including a dose-response (Fig. 4a,b, paired t-test, $t_{31} = -10.607$, $P<0.001$). In several cases, *Bd* growth was enhanced with addition of diluted bacterial metabolites in comparison to positive control growth with RIIA media only

(>100%, Fig. 4b). Most cultures were more inhibitory of *Bd* at the lower temperatures, except for *J. lividum*, (isolate 77.5b) which was most inhibitory at 22°C (Fig. 4a,b).

While all bacteria were unique based on 16S rRNA gene sequencing when clustered at 99% similarity, probiotic physiology and function against *Bd* did not always correspond to OTU clustering at 97% similarity (Table S1 in File S1). In other words, bacterial isolates considered to be the same "species" based on 16S rRNA could have different antifungal function. Here, only one of two *Flavobacterium johnsoniae* isolates inhibited *Bd* growth. When grown together, the filtrate remained inhibitory. However, when grown together and co-cultured with *Bd*, the filtrate was no longer inhibitory. Three *Pseudomonas* isolates were capable of inhibiting *Bd* growth, and were inhibitory when combined with or without co-culture with *Bd*. The above mentioned growth inhibition of *Bd* caused by bacterial filtrate was significantly different from control bacterial growth with water only added (independent t-tests, *P*s< 0.05 and replicated result; all data shown in Table S2 in File S1). These conditions represent infected or uninfected hosts and are illustrative rather than comprehensive of all possible environmental conditions and competitive interactions.

Effects of host skin peptides and *Bd* metabolites on probiotics in culture

Amphibian skin defense peptides may regulate the skin microbiota. We found that natural mixtures of skin peptides from *A. obstetricans* at a concentration of 100 μg/ml significantly inhibited growth of *Pseudomonas migulae* (73b1) and significantly enhanced growth of *P. filiscindens* (73c1), *Flavobacterium johnsoniae* (70d1), and *Janthinobacterium lividum* (76.5c; t-test, Bonferroni corrected P's<0.05; Fig. S6 in File S1).

We tested for a direct effect of *Bd* metabolites on bacterial growth, and found that filtrate from two-week old cultures of *Bd* in 0.5% tryptone significantly inhibited the growth of *Serratia plymuthica* (5/27b2, 5/28a3), *F. johnsoniae* (81a1, 70d1), and *P. filiscindens* (73c1), while significantly enhancing the growth of *J. lividum* (77.5b1; t-test, Bonferroni corrected P's<0.05; Fig. S6 in File S1).

Discussion

We found that a holistic measure of mucosome function against *Bd* is predictive of infection risk in natural populations of amphibians and survival in laboratory exposure experiments. While induced antimicrobial peptides may explain some variation in infection risk (Fig. 1b,d), mucosome function can be altered through probiotic therapy (Fig. 3a), and thus microbial communities play a major role in determining susceptibility to infection with *Bd*. In particular, tadpoles of the endangered midwife toad, *A. obstetricans* may be most at risk of both infection and subsequent disease-induced mortality upon metamorphosis (Fig. 2), even though adult toads are well protected by the mucosome and perhaps resistant to colonization with *Bd*. Similarly, the common frog *R. temporaria* has strong mucosome activity against *Bd*, shows *Bd* colonization resistance, but has relatively poor skin defense peptides. This suggests that this common species has protective microbial communities. Adaptive defenses are not suspected because frogs were raised from eggs and had no history of exposure to *Bd*.

In this study, we provide several striking examples showing that probiotic capacity depends on immunological and environmental context. These examples lead to recommendations for choosing probiotics based on predictable host conditions. Temperature is known to influence amphibian host immune function [41] and

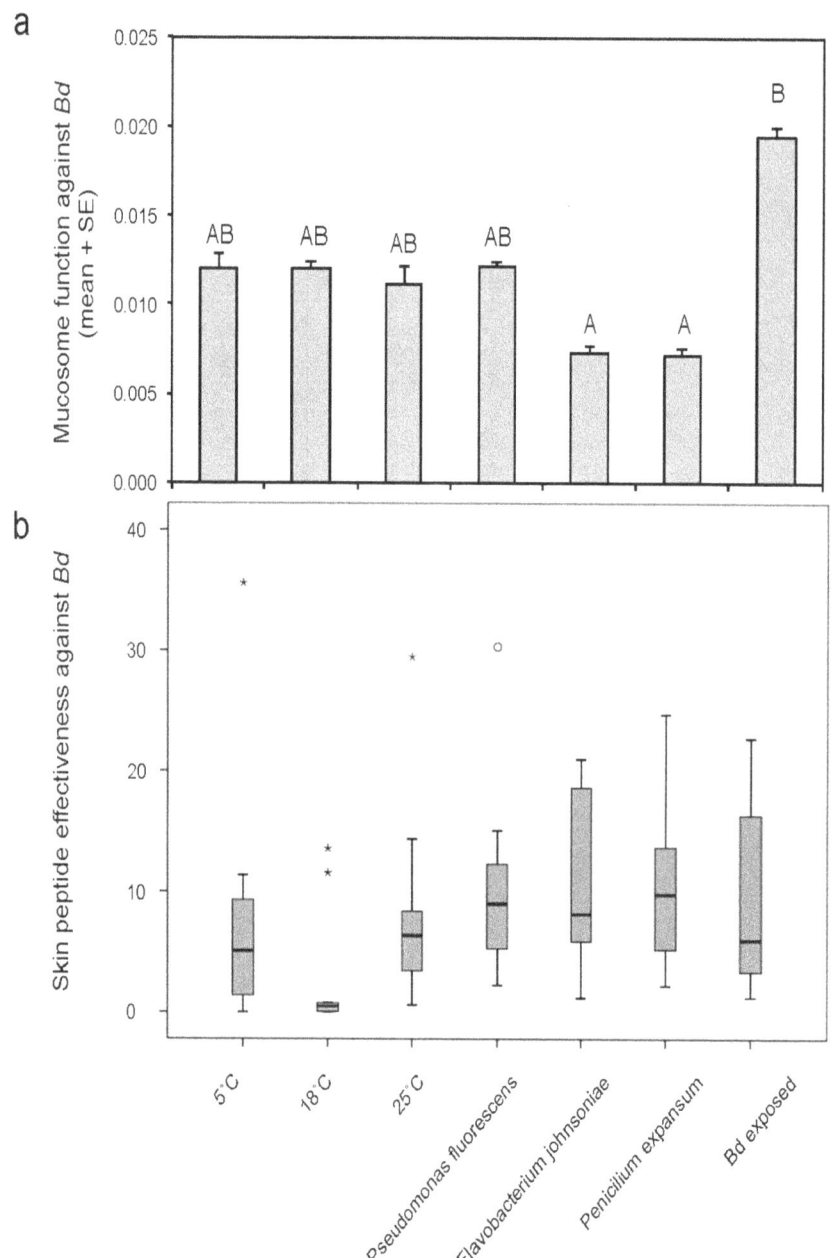

Figure 3. Temperature and probiotic treatments of recently metamorphosed midwife toads, _A. obstetricans_, influence skin mucosome function (a) but not induced skin peptide defenses (b). (a) Mucosome function indicates _B. dendrobatidis_ (_Bd_) viability after exposure to amphibian mucus quantified by green: red fluorescence. Significantly different subsets are indicated by letters above bars (Tukey post-hoc test). _Bd_ zoospore viability was reduced after exposure to mucus from frogs treated with the bacterium _F. johnsoniae_ and the fungus _P. expansum_, and zoospore viability was highest after exposure to mucus from toads previously exposed to _Bd_. (b) Skin peptide effectiveness against _Bd_ did not differ significantly among treatments (ANOVA, $F^6 = 0.952$, $P = 0.466$).

bacterial growth, metabolism, pigment and antibiotic production [45]. However, it was surprising that a shift from 18 to 25°C, a typical natural range for midwife toads, caused a common bacterial symbiont of the eggs and skin, _Serratia plymuthica_, to change from inhibiting _Bd_ to enhancing _Bd_ growth (Fig. 4c). Testing metabolites of the bacteria growing at 14, 19, and 22°C in liquid culture against the global panzootic lineage of _Bd_ showed similar results (Fig. 4b). Functional changes in probiotic activity with shifts in temperature have not previously been reported. Our results provide an alternative mechanistic explanation for patterns

of susceptibility related to climate, which have previously been limited to empirical observation and pathogen-centered effects [46–49].

The microbial interactions we tested also altered antifungal effects relative to what would be predicted from individual isolates. For example, co-culture of _Flavobacterium johnsoniae_ with _Bd_ caused cultures of the bacterium that normally produce antifungal metabolites to switch off antifungal activity: when grown together with _Bd_, _F. johnsoniae_ filtrate was benign, and indeed _Bd_ filtrate inhibited the growth of two out of three _F. johnsoniae_ isolates (Fig.

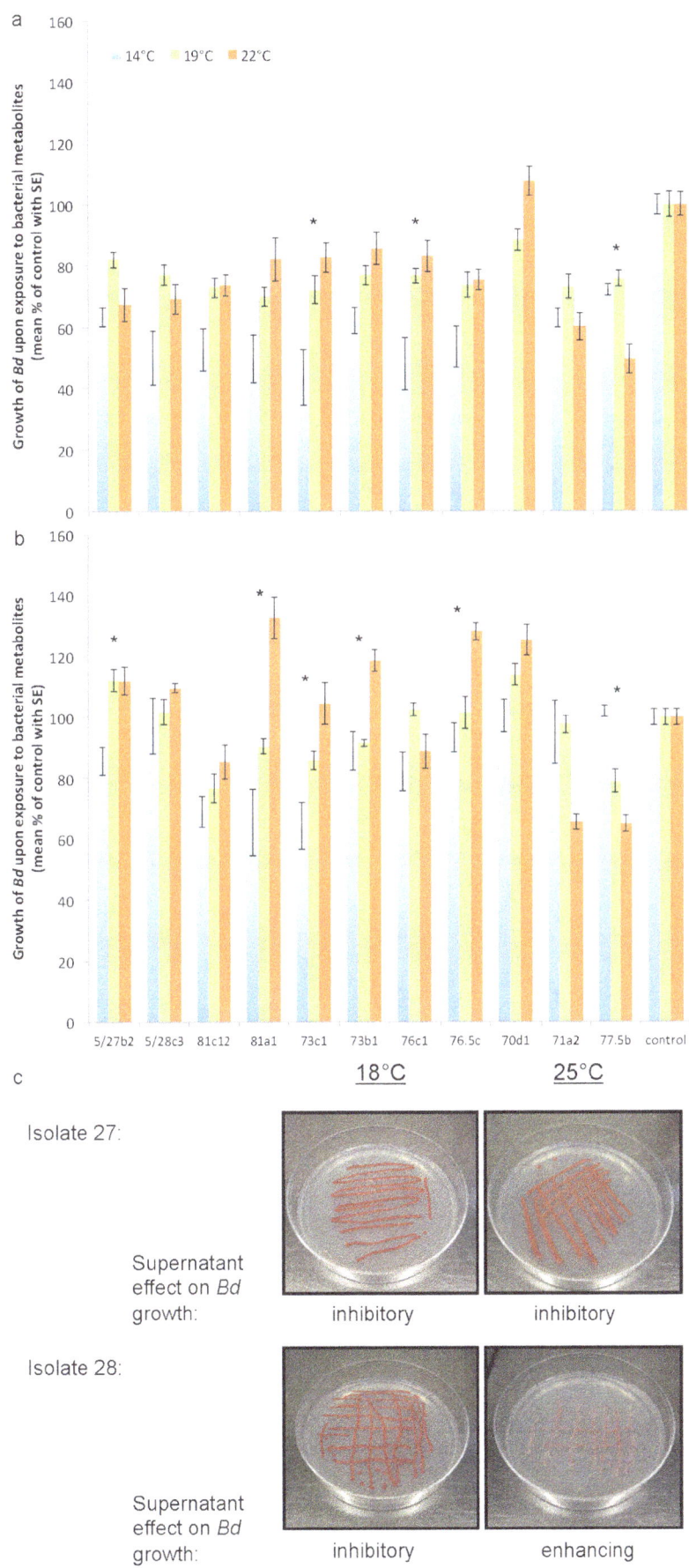

a

b

c

Isolate 27:

Supernatant effect on *Bd* growth: inhibitory   inhibitory

Isolate 28:

Supernatant effect on *Bd* growth: inhibitory   enhancing

Figure 4. Environmental context determines antifungal capacity of probiotics. Tested temperatures (14, 19, 22°C) significantly affected the production of bacterial metabolites in liquid media that could inhibit *B. dendrobatidis* (*Bd*; GPL isolate VMV 813) zoospore growth in a dose-dependent fashion (a = full strength metabolites, b = 1:10 dilution). * indicates that *Bd* growth differed among metabolite temperature treatments (ANOVA, Bonferroni-corrected P's<0.05). (c) Representative replicates are shown of two isolates of *Serratia plymuthica* isolated from egg clutches of common midwife toads, *Alytes obstetricans*, grown on solid media under different temperature conditions. Filtrate from isolate 27 always inhibited growth of *Bd*, but filtrate from isolate 28 inhibited *Bd* growth at 18°C, and enhanced *Bd* growth at 25°C. Filtrate from sterile media (R2A agar supplemented with 1% tryptone) caused enhanced growth of *Bd*. Note that colony color can be an indication of antifungal metabolites such as prodiginines from red *Serratia spp.* [45,67], but are produced only under certain growth conditions.

S6 in File S1). Co-evolution of *Bd* and amphibian hosts is a postulated driver of pathogenicity factors including compounds suppressing host immune defenses [43,50,51]. These factors may extend to inhibiting certain antifungal symbionts or altering their function.

Myriad microbial and immune interactions occur once probiotics are added to living hosts. Thus, testing probiotics *in vivo* is critical for testing the intended antifungal effect of probiotic therapy under realistic environmental conditions. We found that previous exposure to *Bd* may have a negative effect on host immunity or the ability of the mucosome to kill zoospores (Fig. 3A). This result is consistent with a study on Australia green-eyed tree frogs, *Litoria serrata*, showing inhibition of ambient skin peptides with *Bd* infection but no inhibition of inducible stored skin peptides [43]. Because stored skin defense peptides can have potent activity against *Bd*, yet not be active on the skin, induced skin peptides may not accurately predict infection susceptibility. This mystery of how seemingly well-defended species can be affected by chytridiomycosis [52] deserves careful study on the conditions under which host skin defense peptides are activated. Induced skin defense peptides were previously used to predict disease susceptibility in Panama [11] and New Zealand [53]. In Panama, most species had weak peptide defenses and declined after disease emergence while only two species had strong peptide defenses against *Bd* compared to reference species of known disease resistance. Of these two species, the one with the highest levels of skin peptide defenses persisted at the field site (*Espadarana prosoblepon*) [54], and the other species (*Agalychnis lemur*) disappeared, but a relict population has been detected nearby (Julie Ray, pers. comm.). In New Zealand, all native species demonstrated high levels of skin peptide defenses and appear to resist chytridiomycosis [53], although populations are in decline [55].

We found that a bacterium *F. johnsoniae* and a fungal probiotic *P. expansum* can increase the *Bd* killing function of the mucosome. The bacterium *P. fluorescens* did not show this effect. Because host AMPs did not appear to be affected by these treatments (Fig. 3B), the observed effects are most likely caused by antifungal metabolites produced by the microbes growing on the amphibian skin [56]. Upregulation of host mucosal immunity excluding AMPs is an untested alternative mechanism, and potentially a beneficial host response to probiotics. A non-responsive immune system when given probiotics may be preferred from a conservation management standpoint in order for the probiotics to colonize the host, establish within the microbiota and persist. However, this in not necessarily common and immune stimulation in response to probiotics occurs in other systems [57,58].

An ideal probiotic would produce metabolites that inhibit *Bd* growth as shown above, and also be uninhibited by host skin defense peptides. A literature review demonstrates that skin peptides can inhibit the growth of some bacteria, but not others, and suggests that skin defense peptides may be critical in structuring the symbiont community on amphibian skin [52]. Rollins-Smith *et al.* [35] showed that *Aeromonas hydrophila*, a common resident on amphibian skin and also an opportunistic pathogen, could tolerate high levels of host antimicrobial peptides.

This organism shows antifungal characteristics including activity against *Bd* growth [33]. The ability of extracellular products of *A. hydrophila* to inhibit amphibian antimicrobial peptides indicates a co-evolutionary relationship between host and symbionts [59]. In addition, *Pseudomonas mirabilis* and *Serratia liquefaciens* were found to be resistant to antimicrobial peptides from several host frog species [60]. Here we used probiotics that largely resisted low concentrations of natural mixtures of host defense peptides (Fig. S6 in File S1). Thus, to increase the likelihood of probiotic establishment, use of probiotics with a co-evolutionary relationship with the target host may be advantageous.

While easily cultured, the isolates tested here may not be dominant community members based on culture-independent analyses [31,61,62]. Therefore, future studies will benefit by examining the effects of probiotic treatments on the natural microbial communities on host amphibians using culture-independent techniques such as next-generation sequencing. While community interactions are difficult to test in vitro and before probiotics are applied to a host, our results affirm that testing probiotics under certain foreseeable contexts may increase the pace of biotherapy development.

Because potential probiotics that inhibit the growth of *Bd* only do so under certain conditions, we recommend the following screening criteria (Fig. 5): (1) Candidates for probiotic development should be chosen from among the culturable microbiota locally present on tolerant hosts or populations that are able to persist with *Bd* [32,33]. (2) Candidates should have the capacity to inhibit *Bd* growth when grown in isolation, in co-culture with *Bd*, and in an environmental context relevant to the amphibian life-cycle, and (3) the ability to resist immune defenses on host skin, establish within the microbiota, and contribute to antifungal defenses *in vivo*. Resistance to mucosal immune defenses may be critical for establishment within the microbial community associated with the skin, and critical for long-term persistence. Some symbionts appear to be assisted in surviving on the host by thriving on skin mucosal products. Mucosal oligosaccharides, for example, differ among hosts and life-history stages, and may be a selective force in structuring the microbiota [63,64]. Amphibian skin provides a useful model of host-microbiota interactions to better understand mechanisms of microbial community assembly and maintenance within vertebrate mucosa. Indeed, these mechanisms underlie strategies to promote human health by manipulating microbial communities - a long-term goal of the Human Microbiome Project [7,65].

While screening for candidate probiotics, some beneficial organisms may be inadvertently discarded based on tests of bacterial filtrate on *Bd* growth. Microbes producing antifungal metabolites such as bacteriocins [66] or small molecule antibiotics [56,67] will be detected by this method. However, microbes may also compete directly for space or resources, and may exclude pathogenic fungi by other mechanisms [26,68]. Furthermore, microbial secondary metabolites such as prodiginines produced by *Serratia spp.* can be immunosuppressive [67]. Probiotics may strongly influence host immunity through interactions with host Toll-like receptors or NOD-like receptors, or through interactions

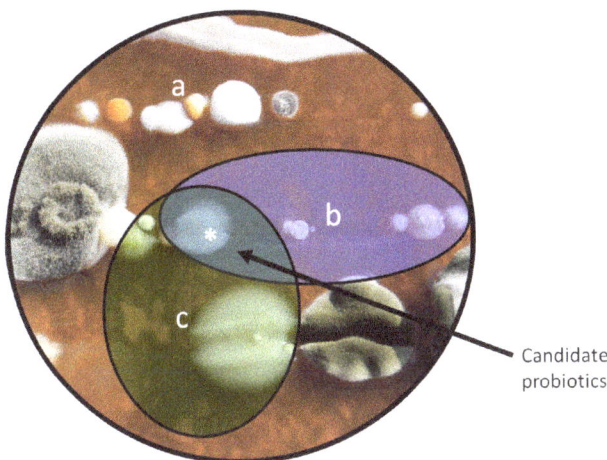

Candidate probiotics

Figure 5. Choosing probiotics with the greatest potential against amphibian chytridiomycosis. Candidate probiotic bacteria (or fungi) are isolated from populations of amphibians that are able to persist in the presence of *B. dendrobatidis* (*Bd*) [1]. To increase the chances of successful prophylactic biotherapy, candidate probiotics should be tested for at least three characteristics: (a) capacity to inhibit *Bd* growth as a pure isolate without specific competitive interactions to induce antifungal metabolites, (b) capacity to inhibit *Bd* at a temperature range consistent with host habitat, and (c) resistance to host skin immune defenses that would complicate probiotic establishment. Remedial biotherapy of already infected individuals should maintain antifungal capacity when grown in competition with *Bd* and withstand the sometimes lethal effects of *Bd* metabolites (Fig. S6 in File S1). Testing probiotic effect in vivo can be accomplished without resorting to pathogen exposure experiments by using the mucosome function assay described here.

with epithelial cells and immune system cells modulating both local and systemic immune responses [69]. The immunomodulatory effect of probiotics cannot be tested with *in vitro Bd* growth assays and host trials are necessary to test for these emergent properties of probiotics.

Antimicrobial peptides and a range of other defenses protect amphibian skin by synergizing or interacting with microbes [41,70]. Thus, a better indication of antifungal effect of probiotics was obtained by testing the mucosome directly on zoospore viability. *In vitro* screening cannot incorporate every factor and eventually *in vivo* trials, both in the lab and under natural conditions are necessary to determine if an overall health benefit is provided. However, beginning with a probiotic that is not likely to become an opportunistic pathogen with changing climatic conditions may be a consideration. Transmissible probiotics would aid disease control at the population level [33], and if able to persist through metamorphosis when applied to tadpoles, disease presentation at this critical developmental stage could be avoided for *A. obstetricans* and other susceptible amphibians [40]. Additionally, *Bd* metabolites are known to be toxic to amphibian lymphocytes [50], and in this study were toxic to certain bacteria such as *Serratia plymuthica* (Fig. S6 in File S1), perhaps prohibiting the use of certain probiotics intended as remedial biotherapy for

infected individuals. The potential for negative biodiversity-function relationships, especially among mixtures of closely related bacteria, cautions against the use of probiotic mixtures that may cause interference competition and reduce host protection [71]. Further refinements to the probiotic screening and discovery process will incorporate next-generation sequencing analyses to target rare or as yet uncultured microbes of interest, and testing microbial consortia that appear linked to disease resistance function. Measuring the effectiveness of applied probiotics is a second step in managing disease risk.

No previous studies have attempted to relate skin microbiota or a holistic measure of skin defense function against *Bd* with disease susceptibility. Given the extreme complexity of the skin micro-ecosystem and interactions described above, the holistic measure of mucosome function presents a significant advance in our capacity to predict relative disease susceptibility, and to measure the success of managed treatments without resorting to infection trials. Here, we examined overall prevalence of infection in Switzerland and Europe and test for correlations at these broad scales with innate defenses from selected life-stages and species (Fig. 1). We found a very strong correlation between mucosome function against *Bd* and infection prevalence in the field and upon experimental exposure. Since Bd-naïve amphibians were sampled for mucosome function, adaptive immunity such as mucosal antibodies is not indicated and antifungal function can be attributed primarily to innate defenses including the microbiota. Indeed, altering the microbiota through probiotic treatments affected mucosome function against *Bd*. In addition to assessing infection risk in natural amphibian assemblages, mucosome functional assays can now be used to assess risk in relict populations or in captive colonies slated for reintroduction. While the efficacies of human probiotics are under scrutiny [2], quantifying the effectiveness of amphibian probiotic treatments under scenarios of changing environmental conditions is a tangible goal.

Supporting Information

File S1 Protocol for determining *Bd* viability, supplementary tables and figures.

Acknowledgments

We thank the V. McKenzie lab, J. Van Buskirk, M. Becker, S. Bell, J. Daskin, and J. Walke for their thoughtful discussion, T. Garner, V. Vasquez, L. Reinert, and L. Rollins-Smith for donation of *Bd* isolates, S. Röthlisberger for performing microsatellite analyses on *Pelophylax* embryos, S. Lötters and M. Veith from University of Trierand for help with field work, and the S.K. Schmidt lab for use of equipment. This work partially emerged from the advanced ecology course led by H.-U. Reyer at the University of Zurich.

Author Contributions

Conceived and designed the experiments: DCW SB JK EK UT. Performed the experiments: DCW HB SB JK EK UT LRD CB SH. Analyzed the data: DCW SB JK EK UT BRS. Wrote the paper: DCW BRS RK VM. Performed field work: DCW JK UT LRD.

References

1. Bletz MC, Loudon AH, Becker MH, Bell SC, Woodhams DC, et al. (2013) Mitigating amphibian chytridiomycosis with bioaugmentation: characteristics of effective probiotics and strategies for their selection and use. Ecol Lett 16: 807–820.
2. Hoffmann DE, Fraser CM, Palumbo FB, Ravel J, Rothenberg K, et al. (2013) Probiotics: Finding the right regulatory balance. Science 342: 314–315.
3. Fukuda S, Toh H, Hase K, Oshima K, Nakanishi Y, et al. (2009) Bifidobacteria can protect from enteropathogenic infection through production of acetate. Nature 469: 543–547.
4. Britton RA, Young VB (2012) Interaction between the intestinal microbiota and host in *Clostridium difficile* colonization resistance. Trends Microbiol 20: 313–319.

5. Belden LK, Harris RN (2007) Infectious diseases in wildlife: the community ecology context. Front Ecol Environ 5: 533–539.

6. Lafferty KD (2010) Interacting parasites. Science 330: 187–188.

7. Clemente JC, Ursell LK, Parfrey LW, Knight R (2012) The impact of the gut microbiota on human health: An integrative view. Cell 148: 1258–1270.

8. Daskin JH, Alford RA (2012) Context-dependent symbioses and their potential roles in wildlife diseases. Proc Roy Soc B 279: 1457–1465.

9. Burgman MA, Ferson S, Akçakaya HR (1993) Risk assessment in conservation biology. London: Chapman & Hall.

10. Murray KA, Skerratt LF (2012) Predicting wild hosts for amphibian chytridiomycosis: integrating host life-history traits with pathogen environmental requirements. Human and Ecological Risk Assessment 18: 200–224.

11. Woodhams DC, Voyles J, Lips KR, Carey C, Rollins-Smith LA (2006) Predicted disease susceptibility in a Panamanian amphibian assemblage based on skin peptide defenses. J Wildl Dis 42: 207–218.

12. Gascon C, Collins JP, Moore RD, Church DR, McKay JE, et al, editors. (2007) Amphibian Conservation Action Plan. Gland: IUCN/SSC Amphibian Specialist Group.

13. Grice EA, Segre JA (2011) The skin microbiome. Nat Rev Microbiol 9: 244–253.

14. Rosenthal M, Goldberg D, Aiello A, Larson E, Foxman B (2011) Skin microbiota: Microbial community structure and its potential association with health and disease. Infect Gen Evol 11: 839–848.

15. Stecher B, Chaffron S, Käppeli R, Hapfelmeier S, Freedrich S, et al. (2010) Like will to like: Abundances of closely related species can predict susceptibility to intestinal colonization by pathogenic and commensal bacteria. PLoS Patho. 6: e1000711.

16. Karlsson FH, Tremaroli V, Nookaew I, Bergström G, Behre CJ, et al. (2012) Gut metagenome in European women with normal, impaired and diabetic glucose control. Nature 498: 99–103.

17. Ross EM, Moate PJ, Marett LC, Cocks BG, Hayes BJ (2013) Metagenomic predictions: From microbiome to complex health and environmental phenotypes in humans and cattle. PLoS ONE 8: e73056.

18. Skerratt LF, Berger L, Speare R, Cashins S, McDonald KR, et al. (2007) Spread of chytridiomycosis has caused the rapid global decline and extinction of frogs. Ecohealth 4: 125–134.

19. Fisher MC, Henk DA, Briggs CJ, Brownstein JS, Madoff LC, et al. (2012) Emerging fungal threats to animal, plant and ecosystem health. Nature 484: 186–194.

20. Murray KA, Skerratt LF, Garland S, Kriticos D, McCallum H (2013) Whether the weather drives patterns of endemic amphibian chytridiomycosis: A pathogen proliferation approach. PLoS ONE 8: e61061.

21. Vredenburg VT, Briggs CJ, Harris RN (2011) Host-pathogen dynamics of amphibian chytridiomycosis: the role of the skin microbiome in health and disease. In: Olsen L, Choffnes E, Relman DA, Pray L, editors.Fungal diseases: an emerging challenge to human, animal, and plant health. Washington, DC: The National Academies Press. pp. 342–355.

22. Harris RN, Brucker RM, Walke JB, Becker MH, Schwantes CR, et al. (2009) Skin microbes on frogs prevent morbidity and mortality caused by a lethal skin fungus. ISME J 3: 818–824.

23. Becker MH, Harris RN, Minbiole KP, Schwantes CR, Rollins-Smith LA, et al. (2011) Towards a better understanding of the use of probiotics for preventing chytridiomycosis in Panamanian golden frogs. Ecohealth 8: 501–506.

24. Woodhams DC, Geiger CC, Reinert LK, Rollins-Smith LA, Lam B, et al. (2012) Treatment of amphibians infected with chytrid fungus: learning from failed trials with itraconazole, antimicrobial peptides, bacteria, and heat therapy. Dis Aquat Org 98: 11–25.

25. Dunne C, O'Mahony L, Murphy L, Thornton G, Morrissey D, et al. (2001) In vitro selection criteria for probiotic bacteria of human origin: correlation with in vivo findings. Am J Clin Nutr 73: 386S–392S.

26. Kesarcodi-Watson A, Kaspar H, Lategan MJ, Gibson L (2008) Probiotics in aquaculture: The need, principles and mechanisms of action and screening processes. Aquaculture 274: 1–14.

27. Searle CL, Gervasi SS, Hua J, Hammond JI, Relyea RA, et al. (2011) Differential host susceptibility to Batrachochytrium dendrobatidis, an emerging amphibian pathogen. Cons Bio 25: 965–974.

28. Balaz V, Vörös J, Civiš P, Vojar J, Hettyey A, et al. (2014) Assessing risk and guidance on monitoring of Batrachochytrium dendrobatidis in Europe through identification of taxonomic selectivity of infection. Cons Bio 28: 213–223.

29. Rödder D, Kielgast J, Bielby J, Schmidtlein S, Bosch J, et al. (2009) Global amphibian extinction risk assessment for the panzootic chytrid fungus. Diversity 1: 52–66.

30. Johnson PTJ, Preston DL, Hoverman JT, Richgels KLD (2013). Biodiversity reduces disease through predictable changes in host community competence. Nature 494: 230–234.

31. Loudon AH, Woodhams DC, Parfrey LW, Archer H, Knight R (2013) Microbial community dynamics and the effect of environmental reservoirs on red-backed salamanders (Plethodon cinereus). ISME J doi: 10.1038/ismej.2013.200.

32. Woodhams DC, Vredenburg VT, Simon MA, Billheimerd D, Shakhtourd B, et al. (2007) Symbiotic bacteria contribute to innate immune defenses of the threatened mountain yellow-legged frog, Rana muscosa. Biol Cons 138: 390–398.

33. Walke JB, Harris RN, Reinert LK, Rollins-Smith LA, Woodhams DC (2011) Social immunity in amphibians: Evidence for vertical transmission of innate defenses. Biotropica 43: 396–400.

34. Boyle DG, Boyle DB, Olsen V, Morgan JAT, Hyatt AD (2004) Rapid quantitative detection of chytridiomycosis (Batrachochytrium dendrobatidis) in amphibian samples using real-time Taqman PCR assay. Dis Aquat Org 60: 141–148.

35. Rollins-Smith LA, Doersam JK, Longcore JE, Taylor SK, Shamblin JC, et al. (2002) Antimicrobial peptide defenses against pathogens associated with global amphibian declines. Dev Comp Immunol 26: 63–72.

36. Daum JM, Davis LR, Bigler L, Woodhams DC (2012) Hybrid advantage in skin peptide immune defenses of water frogs (Pelophylax esculentus) at risk from emerging pathogens. Infect Genet Evol 12: 1854–1864.

37. Stockwell MP, Clulow J, Mahony MJ (2010). Efficacy of SYBR 14/propidium iodide viability stain for the amphibian chytrid fungus Batrachochytrium dendrobatidis. Dis Aquat Org 88: 177–181.

38. Farrer RA, Weinert LA, Bielby J, Garner TWJ, Balloux F, et al. (2011) Multiple emergences of genetically diverse amphibian-infecting chytrids include a globalized hypervirulent recombinant lineage. Proc Natl Acad Sci USA 108: 18732–18736.

39. Barrios V, Olmeda C, Ruiz E, compilers (2012). Action plan for the conservation of the common midwife toad (Alytes obstetricans) in the European Union. Brussels: European Commission.

40. Tobler U, Schmidt BR (2010) Within- and among-population variation in chytridiomycosis-induced mortality in the toad Alytes obstetricans. PLoS ONE 5: e10927 doi:10.1371/journal.pone.0010927

41. Rollins-Smith LA, Woodhams DC (2011) Amphibian immunity: staying in tune with the environment. In: Demas GE, Nelson RJ, editors.Ecoimmunology.New York: Oxford University Press. pp. 92–143.

42. Pask J, Woodhams DC, Rollins-Smith LA (2012) The ebb and flow of antimicrobial skin peptides defends northern leopard frogs, Rana pipiens, against chytridiomycosis. Global Change Bio 18: 1231–1238.

43. Woodhams DC, Bell SC, Kenyon N, Alford RA, Rollins-Smith LA (2012) Immune evasion or avoidance: Fungal skin infection linked to reduced defence peptides in Australian green-eyed treefrogs, Litoria serrata. Fun Bio 116: 1203–1211.

44. Conlon JM, Demandt A, Nielsen PF, Leprince J, Vaudry H, et al. (2009) The alyteserins: Two families of antimicrobial peptides from the skin secretions of the midwife toad Alytes obstetricans (Alytidae). Peptides 30: 1069–1073.

45. Schloss PD, Allen HK, Klimowicz AK, Mlot C, Gross JA, et al. (2010) Psychrotrophic strain of Janthinobacterium lividum from a cold Alaskan soil produces prodigiosin. DNA and Cell Biology 29: 533–541.

46. Kiesecker JM, Blaustein AR, Belden LK (2001) Complex causes of amphibian population declines. Nature 410: 681–684.

47. Pounds JA, Bustamante MR, Coloma LA, Consuegra JA, Fogden MPL, et al. (2006) Widespread amphibian extinctions from epidemic disease driven by global warming. Nature 439: 161–167.

48. Woodhams DC, Alford RA, Briggs CJ, Johnson M, Rollins-Smith LA (2008) Life-history trade-offs influence disease in changing climates: Strategies of an amphibian pathogen. Ecology 89: 1627–1639.

49. Doddington BJ, Bosch J, Oliver JA, Grassly NC, Garcia G, et al. (2013) Context-dependent amphibian host population response to an invading pathogen. Ecology 94: 1795–1804.

50. Fites JS, Ramsey JP, Holden WM, Collier SP, Sutherland DM, et al. (2013) The invasive chytrid fungus of amphibians paralyzes lymphocyte responses. Science 342: 366–369.

51. Voyles J, Rosenblum EB, Berger L (2011) Interactions between Batrachochytrium dendrobatidis and its amphibian hosts: a review of pathogenesis and immunity. Microbes & Infect 13: 25–32.

52. Conlon JM (2011) Structural diversity and species distribution of host-defense peptides in frog skin secretions. Cell Molec Life Sci 68: 2303–2315.

53. Meltser S, Bishop PJ (2009) Skin peptide defences of New Zealand frogs against chytridiomycosis. Anim Cons 13: 44–52.

54. Crawford AJ, Lips KR, Bermingham E (2010) Epidemic disease decimates amphibian abundance, species diversity, and evolutionary history in the highlands of central Panama. Proc Natl Acad Sci USA 107: 13777–13782.

55. Newman DG, Bell BD, Bishop PJ, Burns R, Haigh A, et al. (2010) Conservation status of New Zealand frogs, 2009. N Zeal J Zool 37: 121–130.

56. Brucker RM, Harris RN, Schwantes CR, Gallaher TN, Flaherty DC, et al. (2008) Amphibian chemical defense: Antifungal metabolites of the microsymbiont Janthinobacterium lividum on the salamander Plethodon cinereus. J Chem Ecol 34: 1422–1429.

57. Salinas I, Abelli L, Bertoni F, Picchietti S, Roque A, et al. (2008) Monospecies and multispecies probiotic formulations produce different systemic and local immunostimulatory effects in the gilthead seabream (Sparus aurato L.). Fish & Shellfish Immunol 25: 114–123.

58. Küng D, Bigler L, Davis LR, Gratwicke B, Griffith E, et al. (2013) Stability of microbiota on a Panamanian frog: insights for probiotic therapy against disease. PLoS ONE 9(1): e87101. doi:10.1371/journal.pone.0087101.

59. Schadich E, Cole ALJ (2009) Inhibition of frog antimicrobial peptides by extracellular products of the bacterial pathogen Aeromonas hydrophila. Lett Appl Microbiol 49: 384–387.

60. Schadich E (2009) Skin peptide activities against opportunistic bacterial pathogens of the African clawed frog (Xenopus laevis) and three Litoria frogs. J Herpetol 43: 173–183.

61. McKenzie VJ, Bowers RM, Fierer N, Knight R, Lauber CL (2012) Co-habiting amphibian species harbor unique skin bacterial communities in wild populations. ISME J 6: 588–596.

62. Kueneman JG, Parfrey LW, Woodhams DC, Archer HM, Knight R, McKenzie VJ (2013) The amphibian skin microbiome across species, space and life history stages. Mol Ecol doi: 10.1111/mec.12510.

63. Delplace F, Maes E, Lemoine J, Strecker G (2002) Species specificity of O-linked carbohydrate chains of the oviducal mucins in amphibians: structural analysis of neutral oligosaccharide alditols released by reductive beta-elimination from the egg-jelly coats of *Rana clamitans*. Biochem J 363: 457–471.

64. Varki A (1993) Biological roles of oligosaccharides – all of the theories are correct. Glycobiol 3: 97–130.

65. NIH HMP Working Group, Peterson J, Garges S, Giovanni M, McInnes P, et al (2009) The NIH Human Microbiome Project. Genome Res 19: 2317–2323.

66. Cotter PD, Ross RP, Hill C (2013) Bacteriocins - a viable alternative to antibiotics? Nat Rev Microbiol 11: 95–105.

67. Williamson NR, Fineran PC, Gristwood T, Chawrai SR, Leeper FJ, et al. (2007) Anticancer and immunosuppressive properties of bacterial prodiginines. Future Microbiol 2: 605–618.

68. Wilson M (2005) Microbial inhabitants of humans: their ecology and role in health and disease. Cambridge: Cambridge University Press.

69. Hancock REW, Nijnik A, Philpott DJ (2012) Modulating immunity as a therapy for bacterial infections. Nat Rev Microbiol 10: 243–254.

70. Myers JM, Ramsey JP, Blackman AL, Nichols AE, Minbiole KPC, et al. (2012) Synergistic inhibition of the lethal fungal pathogen *Batrachochytrium dendrobatidis*: The combined effect of symbiotic bacterial metabolites and antimicrobial peptides of the frog *Rana muscosa*. J Chem Ecol 38: 958–965.

71. Becker J, Eisenhauer N, Scheu S, Jousset A (2012) Increasing antagonistic interactions cause bacterial communities to collapse at high diversity. Ecol Lett 15: 468–474.

14

Antioxidants Keep the Potentially Probiotic but Highly Oxygen-Sensitive Human Gut Bacterium *Faecalibacterium prausnitzii* Alive at Ambient Air

M. Tanweer Khan, Jan Maarten van Dijl, Hermie J. M. Harmsen*

Department of Medical Microbiology, University of Groningen, University Medical Center Groningen, Groningen, The Netherlands

Abstract

The beneficial human gut microbe *Faecalibacterium prausnitzii* is a 'probiotic of the future' since it produces high amounts of butyrate and anti-inflammatory compounds. However, this bacterium is highly oxygen-senstive, making it notoriously difficult to cultivate and preserve. This has so far precluded its clinical application in the treatment of patients with inflammatory bowel diseases. The present studies were therefore aimed at developing a strategy to keep *F. prausnitzii* alive at ambient air. Our previous research showed that *F. prausnitzii* can survive in moderately oxygenized environments like the gut mucosa by transfer of electrons to oxygen. For this purpose, the bacterium exploits extracellular antioxidants, such as riboflavin and cysteine, that are abundantly present in the gut. We therefore tested to what extent these antioxidants can sustain the viability of *F. prausnitzii* at ambient air. The present results show that cysteine can facilitate the survival of *F. prausnitzii* upon exposure to air, and that this effect is significantly enhanced the by addition of riboflavin and the cryoprotectant inulin. The highly oxygen-sensitive gut bacterium *F. prausnitzii* can be kept alive at ambient air for 24 h when formulated with the antioxidants cysteine and riboflavin plus the cryoprotectant inulin. Improved formulations were obtained by addition of the bulking agents corn starch and wheat bran. Our present findings pave the way towards the biomedical exploitation of *F. prausnitzii* in redox-based therapeutics for treatment of dysbiosis-related inflammatory disorders of the human gut.

Editor: Gunnar Loh, German Institute of Human Nutrition Potsdam-Rehbrücke, Germany

Funding: M.T.K. was supported by a grant from the Graduate School GUIDE of the University of Groningen. The funders had no role in study design, data collection and analysis, decision to publish, or preparation of the manuscript.

Competing Interests: The authors have declared that no competing interests exist.

* E-mail: h.j.m.harmsen@umcg.nl

Introduction

Diabetes, obesity, inflammatory bowel diseases and various other metabolic disorders are intimately linked to changes in the human gut microbiota [1–4]. It is thus important to develop strategies and tools to revert the condition of 'dysbiosis', where the human gut microbiota is substantially altered, to situations as typically encountered in healthy individuals [1], [5–9]. Therefore, the so-called 'probiotic' supplementation of gut microorganisms is being considered as a highly promising novel therapeutic approach for treating dysbiosis [7], [10], [11]. Such probiotics are historically defined as 'live microbial food supplements that have beneficial effects on the host by improving its intestinal microbial balance' [12]. Importantly, the probiotics required for treating dysbiosis do not only have to compete with pathogenic bacteria and suppress them, but they should also help to positively modulate the gut-microbial fermentation pathways [13]. This calls for the development of novel probiotic formulations that include microorganisms from the human gut microbiome, a formidable resource of beneficial microbes.

Traditionally, the bacteria *Lactobacillus acidophilus*, *Lactobacillus casei*, *Bifidobacterium bifidum* and *Bifidobacterium longum*, and the yeast *Saccharomyces boulardii* have been employed as probiotics in humans. These microorganisms are generally well-accepted and tolerated amongst consumers [13]. Alternatively, the growth and/or activity of particular gut microbes can be stimulated with non-digestible food constituents, such as inulin-type fructans [14]. Such food supplements are often referred to as 'prebiotics' [11], [15]. Generally, prebiotic supplementations are aimed to modulate the composition of the gut microbiota in such a way that the potentially health-promoting bacteria, such as lactobacilli and bifidobacteria are stimulated [15], [16]. 'Synbiotics' are combinations of probiotics and prebiotics that have a beneficial impact on the host by improving the survival and establishment of beneficial microbes in the gastrointestinal tract [11], [17]. Such synbiotics may selectively stimulate microbial growth by activating the metabolism of one, or a limited number of health-promoting microbes. Importantly, the ingestion of probiotics, prebiotics or synbiotics can modulate the short-chain fatty acid (SCFA) profiles in the human colon [11]. Here, the most abundant SCFAs are acetate, propionate and butyrate, which are generally detectable in human feces at ratios of 60:20:20, respectively [18], [19]. The butyrate level is of particular importance, because this SCFA represents the most preferred energy source for colonocytes [19],[20], stimulates cell proliferation [21], and promotes mucus secretion from colonic mucosa [22]. In view of these benefits of butyrate, butyrogenic bacteria are highly promising future probiotics.

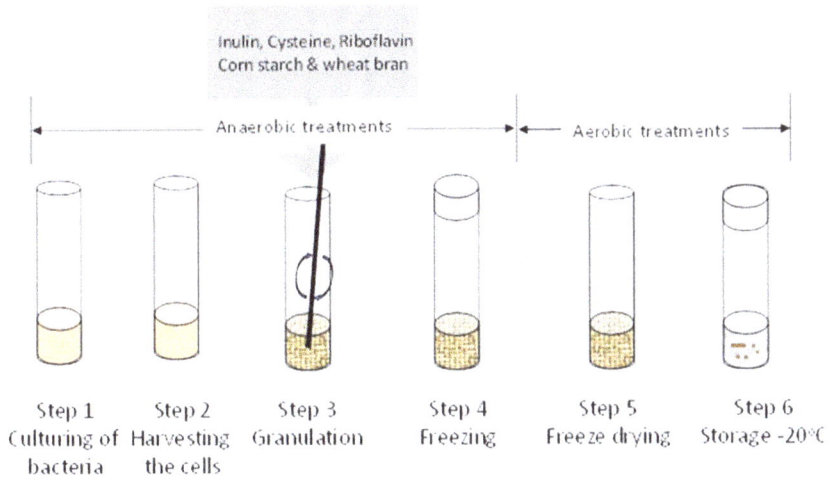

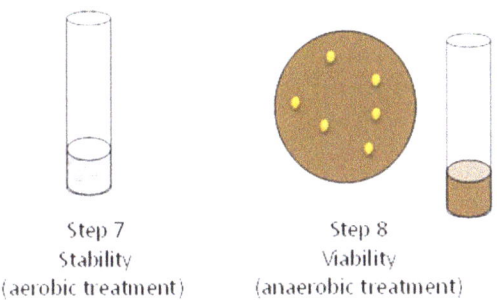

Figure 1. Schematic representation of the methodology employed to investigate the survival of *F. prausnitzii* in different formulations upon exposure to air. Steps 1–4 were conducted in an anaerobic chamber, and steps 5–7 were conducted aerobically in ambient air. The final step 8, involving rehydration and viability assays, was conducted in an anaerobic chamber.

One of the most important butyrate-producing bacteria in the human colon is *Faecalibacterium prausnitzii*. This bacterium and its close relatives, which are amongst the most abundant gut bacteria, belong to the clostridial cluster IV [23]. Notably, faecalibacteria are present at low numbers in patients with inflammatory-bowel disease while, on the other hand, they display significant anti-inflammatory effects in mouse colitis models [24]. These observations imply that faecalibacteria contribute substantially to gut health, which makes them prime candidates for inclusion in probiotic or synbiotic formulations [25]. However, in contrast to the fairly oxygen-tolerant probiotics that are commercialized

today, these probiotics of the future are highly oxygen-sensitive. In fact, faecalibacteria cannot even withstand a few minutes of exposure to ambient air [26]. This extreme oxygen-sensitivity is a huge challenge for the development of probiotic or synbiotic formulations that include *F. prausnitzii* or its close relatives.

Freeze-drying or lyophilization is commonly used to preserve microorganisms since lyophilates can be stored for decades [27]. Accordingly, many microbes are stored as lyophilates in important strain collections like the American Type Culture Collection (ATCC) and the National Collection of Type Cultures (NCTC). To overcome the potential viability loss during formulation, it is

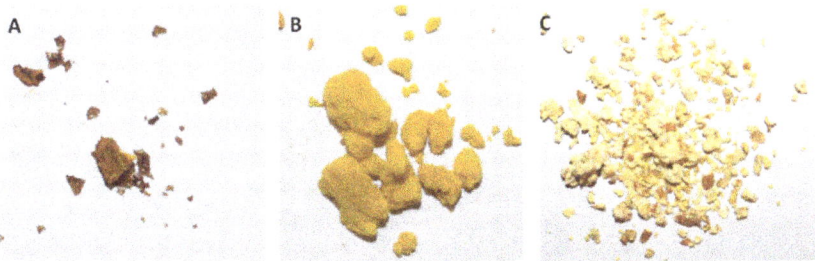

Figure 2. Physical appearance of freeze-dried granules containing *F. praunitzii*. A, Formulation with cysteine and riboflavin results in compact and pellet-like granules. B, Formulation with inulin, riboflavin and cysteine results in a foam-like matrix with a high bulk volume. C, Formulation with wheat bran, inulin, corn starch, cysteine and riboflavin results in hard and compact granules with considerable bulk volume.

Table 1. Survival of *F. prausnitzii* in different formulations.

Combinations					Survival percentage (+/−5%) Exposure to ambient air (h)		
					6	11	24
Inu					nd	nd	0
	Rb				nd	nd	0
	Rb	Cys			nd	nd	10
Inu	Rb				nd	nd	0
Inu	Rb	Cys			nd	nd	59
	Rb	Cys			nd	nd	60
Inu	Rb	Cys			nd	nd	70
			Cs		0	0	0
			Cs	Wb	0	0	0
Inu			Cs	Wb	6	0	0
Inu	Rb	Cys	Cs	Wb	93	100	57
Inu	Rb		Cs	Wb	75	3	0
	Rb	Cys	Cs	Wb	nd	45	0
Inu	Rb	Cys	Cs		nd	40	22
	Rb	Cys	Cs		nd	53	0
Inu				Wb	nd	nd	0

Inu, Inulin; Rb, Riboflavin; Cys, Cysteine; Cs, Corn starch; Wb, Wheat bran; nd, not determined.
Please note that the data shown in Table 1 result from a single experiment. In this experiment we focused predominantly on 24 h survival at ambient air, because this period of time is probably needed for future formulation processes.

Figure 3. Scanning electron micrographs of *F. prausnitzii*-**containing formulations.** All formulations shown contained cysteine and riboflavin supplemented either with corn starch (A), inulin (B), or wheat bran, corn starch and inulin (C and D). Corn starch has a discrete bead-like appearence (A). In contrast, inulin forms a flake-like matrix (B), that can form a coating around corn starch and/or wheat bran, thereby entrapping the bacterial cells (C and D). Images A, B and C were recorded at 500× magnification, and image D was recorded at 1500× magnification. The arrows indicate entrapped *F. prausnitzii* cells.

generally recommended to lyophilize concentrated cultures containing $>10^7$ cells [27–29]. This ensures the presence of sufficient viable cells after lyophilization, long-term storage and reconstitution [27], [29]. For strain preservation purposes, the survival of ~0.1% of the original cell population is in principle sufficient [29]. In contrast, for probiotic fomulation it is important that as many bacteria as possible survive lyophilization and subsequent storage. Therefore, glycerol, mannitol, sorbitol, inulin, dextrin and Crystalean have been frequently applied as cryopreservants to protect probiotic formulations during lyophilization

and to enhance their shelf-life [30], [31]. However, the selection of appropriate and compatible cryopreservants remains a major challenge and no suitable protectants were so far reported for faecalibacteria.

Inulin-type fructans are generally well-known prebiotics, which have been shown to increase the number of bifidobacteria both *in vitro* and *in vivo* [14], [32]. In addition, inulin has been widely employed as a cryopreservative in lyophilization procedures [27], [30], [31], [33]. Similarly, starch and wheat bran can serve as prebiotics [34] that possibly enhance the efficacy of synbiotic formulations [35], [36]. Antioxidants such as glutathione, ascorbate and cysteine can further enhance the viability of probiotics during storage [37]. Recently, we have shown that *F. prausnitzii* can exploit flavins and oxidized thiols as redox mediators to shuttle electrons to oxygen [38]. In this way, *F. prausnitzii* can survive and thrive under moderately oxygenized conditions as encountered in the human gut, where oxygen diffuses in from the epithelial cell layers. In the present study, we therefore investigated whether and how riboflavin and cysteine can be used to develop formulations that preserve viable *F. prausnitzii* cells under oxygenized conditions. Our findings show that this is indeed a feasible strategy, which paves the way for the development of synbiotics containing faecalibacteria.

Materials and Methods

Bacterial strains and culturing conditions

F. prausnitzii strain A2-165 (DSM 17677), previously described [26] was maintained at 37°C on yeast extract, casitone, fatty acid and glucose medium (YCFAG) in an anaerobic chamber. To maintain anaerobic conditions a gas mixture of N2 (81%), H2 (11%) and CO2 (8%) was used. For formulation experiments, the

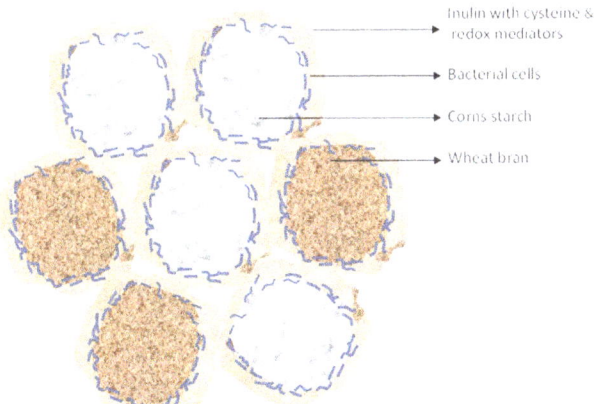

Inulin with cysteine & redox mediators

Bacterial cells

Corn starch

Wheat bran

Figure 4. Tentative schematic representation of a formulation in which *F. prausnitzii* **cells were adhered on cornstarch or wheat bran, and entrapped by an inulin matrix containing riboflavin and cysteine as antioxidants and redox mediators.**

bacterial cells were grown in 50 ml of YCFAG broth to an optical density at 600 nm of ~1. The average yield of bacterial cells on a wet-weight basis was around 0.1 g per 50 ml of broth. Cells were harvested by centrifugation (3300 g) for 10 min. The resulting pellet was then washed in PBS and re-centrifuged to obtain a cell pellet that was used for further experiments.

Formulation procedure

Bacterial cell pellets from 50 ml cultures were re-suspended in 400 μl of inulin solution (10%) with or without 0.2% cysteine or phosphate-buffered saline (PBS). To these mixtures was added either 200 μl of a 16.5 mM riboflavin solution in PBS, or 200 μl PBS without further additions. In some experiments, the bacterial slurry was later mixed with 0.5 g of wheat bran, with or without 1 g of corn starch. The mixtures were then homogenized with a sterile spatula and frozen at −20°C. Upon freezing, the mixtures were lyophilized for 3 h and stored at −20°C until further analysis. Notably, the culturing of bacteria, harvesting, granulation, and freezing were conducted anaerobically, while freeze-drying and storage were performed aerobically.

Stability and viability testing

For stability analyses, freeze-dried granules were exposed to atmospheric air at ambient temperatures for 0 h, 6 h, 11 h or 24 h. Next, the granules were placed in an anaerobic chamber, and rehydrated in PBS to yield 1:10 dilutions. After rehydration, 100 μl or 200 μl of the diluent was plated on YCFAG agar and incubated anaerobically at 37°C for 24 to 36 h. Colony forming units (CFU) per ml were estimated by counting individual colonies formed on the agar plates. The relative stability of lyophilized mixtures was calculated by dividing the CFU/ml determined after exposure to ambient air for a particular period of time by the CFU/ml of the rehydrated mixture immediately after formulation (T_0). All experiments have been repeated at least twice on different dates.

Scanning electron microscopy

Lyophilized granules of different composition were mounted on aluminum stubs, sputtered with gold particles and examined with a JEOL 6301F scanning Electron Microscope (EM), as previously described [39].

Results and Discussion

To investigate whether it is possible to formulate *F. prausnitzii* cells in such a way that they can withstand ambient air for several hours, a formulation protocol was established as schematically depicted in Figure 1. The initial formulation steps including bacterial cultivation, harvesting and mixing with particular compounds were carried out under anaerobic conditions. After lyophilization, the resulting granules were frozen (−20°C) and stored aerobically at −20°C until further use. Depending on their composition, the lyophilized mixtures were recovered as granules with different morphologies. For example, Figure 2A shows the compact flake-like appearance of a lyophilized preparation containing cysteine, riboflavin and *F. prausnitzii* cells; Figure 2B shows the granules that were obtained when inulin was present in addition to cysteine, riboflavin and *F. prausnitzii* cells; and Figure 2C shows the hard and compact granules that were obtained when *F. prausnitzii* was formulated with corn starch, wheat bran, inulin, cysteine and riboflavin. In the latter type of formulation, the bacterial cells were found to adhere to the wheat bran and/or cornstarch. Table 1 shows the relative viability of *F. prausnitzii* in the different tested formulations upon exposure to

ambient air for 6, 11 or 24 h as compared to the bacterial counts obtained immediately after the formulation (T_0). Notably, the established cryopreservant inulin did not detectably enhance faecalibacterial survival by itself. The same was true for riboflavin, despite its known antioxidant activity. On the other hand, cysteine had a mildly protective effect, leading to about 10% bacterial survival upon 24 h exposure to air. When combined, inulin and riboflavin gave no detectable protection, but inulin and cysteine were highly effective leading to almost 60% bacterial survival. Intriguingly, the same level of protection was achieved with a combination of riboflavin and cysteine. The highest level of protection of around 70% was achieved when cysteine was combined with riboflavin and inulin (Table 1). It should be noted that the foamy material that was obtained when the faecalibacteria were formulated with cysteine, riboflavin and inulin is difficult to handle (Fig. 2B). Therefore, we tested the possibility to use corn starch and wheat bran as bulking agents. Notably, these two compounds had no protective effects on the faecalibacteria, neither by themselves nor in combination with inulin (Table 1 and data not shown). However, close to maximum faecalibacterial survival (~60%) was observed when corn starch and wheat bran were combined with inulin, riboflavin and cysteine. This finding is important, because the respective mixture yields hard and compact granules with a considerable bulk volume (Fig. 2C). Importantly, also in the mixture containing inulin, riboflavin, cysteine, corn starch and wheat bran, the inclusion of cysteine is crucial for bacterial protection against air (Table 1). Somewhat unexpectedly, when corn starch was present, also inulin and wheat bran did contribute to the faecalibacterial survival, but a mix of only these two compounds did not support faecalibacterial survival at ambient air (Table 1). Altogether, we conclude that a formulation of *F. prausnitzii* with inulin, riboflavin, cysteine, corn starch and wheat bran is optimal both in terms of faecalibacterial survival and the type of granules that are obtained.

As shown by scanning EM, corn starch forms discrete bead like structures during our formulation procedure (Fig. 3A) while inulin forms a flake-like matrix (Fig. 3B). When combined, corn starch, wheat bran and inulin form a matrix upon lyophilization that seems to facilitate faecalibacterial entrapment on corn starch and wheat bran particles as shown in Figure 3 (C and D) and schematically represented in Figure 4. Keeping highly oxygen-sensitive bacteria, such as *F. prausnitzii*, alive in probiotic or synbiotic formulations has been considered a major challenge [27]. In this study, we describe a methodology for formulating *F. prausnitzii*, which allows cells of this bacterium to survive the exposure to ambient air for at least 24 h. The principal concept underlying our novel formulation is that the faecalibacterial cells are coated with antioxidants, which protects them from the lethal effects of oxygen when exposed to ambient air (Fig. 4). Bulking agents, such as wheat bran and corn starch provide a substratum for microbial cell adherence and inulin serves as a matrix to capture the faecalibacterial cells on the corn starch and wheat bran particles (Fig. 3D). Furthermore, the inulin can form an outer matrix that incorporates the antioxidants cysteine and riboflavin (Fig. 3, C and D). Notably, these compounds will not only protect the faecalibacteria from oxidative damage, but they will also facilitate faecalibacterial growth upon rehydration. The latter relates to the fact that *F. prausnitzii* can use riboflavin and oxidized cysteine to shuttle electrons to oxygen, which allows them to grow to higher cell densities in moderately oxygenized environments [38]. Inulin is well known for its cryopreservative properties [27], [30], [33] and our data suggest that it protects the faecalibacteria during lyophylization. However, inulin is not protective by itself and it can only sustain the viability of faecalibacteria in combination with cysteine or, even better, cysteine plus riboflavin. Furthermore, the addition of inulin also increases the bulk volume of the formulation (Fig. 2B). While the different bulking agents used in the present study result in variable granule sizes as shown in Figure 2, we do not believe that this will result in major differences in the surface areas exposed to oxygen since the different granules are porous and exposed to oxygen on a microscale regardless their macro size (Fig. 3). Instead, our present findings imply that the oxygen protection is provided mainly through the coating with antioxidants and inulin, which creates a reducing environment in which F. prausnitzii can survive. At this stage we do not exactly know why the survival of F. prausnitzii in the presence of bulking agents is somewhat lower than when these bulking agents are absent, but it could be due to the fact that the bulking agents wheat bran and corn starch quickly attract moisture that could lead to decreased survival upon exposure to ambient air. Notably, the bulking agents corn starch, wheat bran and inulin used in this study do not only facilitate the formulation procedure, but they are also known to confer prebiotic effects when ingested by the host [14], [34], [36]. For instance, prebiotic administration of inulin is known to result in increased numbers of F. prausnitzii and other beneficial gut microbes [14]. Wheat bran is a rich source of arabinoxylan, which has been suggested to lower low-grade chronic or systemic inflammation in obese mouse models [34]. In addition, food supplementation with arabinoxylan is known to

counteract the dysbiosis induced by high fat diets [35]. Corn starch or resistant starch can also modulate the gut microbiota, leading to altered SCFA production profiles [40–43]. Lastly, cysteine is known to counteract oxidative stress at the gut mucosal lining [44], while F. prausnitzii is believed to have significant anti-inflammatory effects [24], [45].

Altogether, the beneficial properties of the butyrogenic bacterium F. prausnitzii and the compounds used for its formulation indicate that our present approach has a strong potential to deliver a novel and highly potent synbiotic. This synbiotic formulation could be beneficial for the treatment of the patients with colitis, a disease characterized by severe inflammation of the colon as well as low counts of F. prausnitzii. Here, it would be clearly beneficial to avail of an effective synbiotic formulation to restore high counts of F. prausnitzii. Our present findings represent a major step forward towards the development of such a synbiotic, as we now know how to keep the highly oxygen-sensitive bacterium F. prausnitzii and possibly other anaerobic gut microbes alive at ambient air.

Author Contributions

Conceived and designed the experiments: MTK JMvD HH. Performed the experiments: MTK. Analyzed the data: MTK HH. Wrote the paper: MTK JMvD HH.

References

1. Chassard C, Dapoigny M, Scott KP, Crouzet L, Del'homme C, et al. (2012) Functional dysbiosis within the gut microbiota of patients with constipated-irritable bowel syndrome. Aliment. Pharmacol. Ther. 35: 828–838.
2. Koren O, Goodrich JK, Cullender TC, Spor A, Laitinen K, et al. (2012) Host remodeling of the gut microbiome and metabolic changes during pregnancy. Cell 150: 470–480.
3. Nicholson JK, Holmes E, Kinross J, Burcelin R, Gibson G, et al. (2012) Host-gut microbiota metabolic interactions. Science 336: 1262–1267.
4. Tremaroli V, Backhed F (2012) Functional interactions between the gut microbiota and host metabolism. Nature 489: 242–249.
5. Kayama H, Takeda K (2012) Regulation of intestinal homeostasis by innate and adaptive immunity. Int. Immunol. 24: 673–680.
6. Klaenhammer TR, Kleerebezem M, Kopp MV, Rescigno M (2012) The impact of probiotics and prebiotics on the immune system. Nat. Rev. Immunol. 12: 728–734.
7. Butterworth AD, Thomas AG, Akobeng AK (2008) Probiotics for induction of remission in Crohn's disease. Cochrane Database Syst. Rev. 3: CD006634.
8. Toh ZQ, Anzela A, Tang ML, Licciardi PV (2012) Probiotic therapy as a novel approach for allergic disease. Front. Pharmacol. 3: 171.
9. Videlock EJ, Cremonini F (2012) Probiotics for antibiotic-associated diarrhea. JAMA 308: 665; author reply 665-6.
10. Floch MH, Walker WA, Madsen K, Sanders ME, Macfarlane GT, et al. (2011) Recommendations for probiotic use-2011 update. J. Clin. Gastroenterol. 45 Suppl: S168-71.
11. Vyas U, Ranganathan N (2012) Probiotics, prebiotics, and synbiotics: gut and beyond. Gastroenterol Res Pract 2012: 872716.
12. Fuller R (1989) Probiotics in man and animals. J. Appl. Bacteriol. 66: 365–378.
13. Goldin BR (1998) Health benefits of probiotics. Br. J. Nutr. 80: S203-7.
14. Ramirez-Farias C, Slezak K, Fuller Z, Duncan A, Holtrop G, et al. (2009) Effect of inulin on the human gut microbiota: stimulation of Bifidobacterium adolescentis and Faecalibacterium prausnitzii. Br. J. Nutr. 101: 541–550.
15. Gibson GR, Roberfroid MB (1995) Dietary modulation of the human colonic microbiota: introducing the concept of prebiotics. J. Nutr. 125: 1401–1412.
16. Gibson GR, Beatty ER, Wang X, Cummings JH (1995) Selective stimulation of bifidobacteria in the human colon by oligofructose and inulin. Gastroenterology 108: 975–982.
17. Kolida S, Gibson GR (2011) Synbiotics in health and disease. Annu. Rev. Food Sci. Technol. 2: 373–393, 2011.
18. Cummings JH (1981) Short chain fatty acids in the human colon. Gut 22: 763–779.
19. Mortensen PB, Clausen MR (1996) Short-chain fatty acids in the human colon: relation to gastrointestinal health and disease. Scand. J. Gastroenterol. Suppl. 216: 132–148.
20. Hague A, Singh B, Paraskeva C (1997) Butyrate acts as a survival factor for colonic epithelial cells: further fuel for the in vivo versus in vitro debate. Gastroenterology 112: 1036–1040.
21. Sakata T (1987) Stimulatory effect of short-chain fatty acids on epithelial cell proliferation in the rat intestine: a possible explanation for trophic effects of

fermentable fibre, gut microbes and luminal trophic factors. Br. J. Nutr. 58: 95–103.
22. Shimoyodome A, Meguro S, Hase T, Tokimitsu I, Sakata T (2000). Decreased colonic mucus in rats with loperamide-induced constipation. Comp. Biochem. Physiol. A. Mol. Integr. Physiol. 126: 203–212.
23. Hold GL, Schwiertz A, Aminov RI, Blaut M, Flint HJ (2003) Oligonucleotide probes that detect quantitatively significant groups of butyrate-producing bacteria in human feces. Appl. Environ. Microbio.l 69: 4320–4324.
24. Sokol H, Pigneur B, Watterlot L, Lakhdari O, Bermudez-Humaran LG, et al. (2008) Faecalibacterium prausnitzii is an anti-inflammatory commensal bacterium identified by gut microbiota analysis of Crohn disease patients. Proc. Natl. Acad. Sci. U.S.A. 105: 16731–16736.
25. Miquel S, Martin R, Rossi O, Bermúdez-Humarán LG, Chatel JM, et al. (2013) Faecalibacterium prausnitzii and human intestinal health. Curr. Opin. Microbiol. 16:255–61.
26. Duncan SH, Hold GL, Harmsen HJ, Stewart CS, Flint HJ (2002) Growth requirements and fermentation products of Fusobacterium prausnitzii, and a proposal to reclassify it as Faecalibacterium prausnitzii gen. nov., comb. nov. Int. J. Syst. Evol. Microbiol. 52: 2141–2146.
27. Morgan CA, Herman N, White PA, Vesey G (2006) Preservation of micro-organisms by drying; a review. J. Microbiol. Methods 66: 183–193.
28. Miyamoto-Shinohara Y, Imaizumi T, Sukenobe J, Murakami Y, Kawamura S, et al. (2000) Survival rate of microbes after freeze-drying and long-term storage. Cryobiology 41: 251–255.
29. Bozoglu TF, Ozilgen M, Bakir U (1987) Survival kinetics of lactic acid starter cultures during and after freeze drying. Enzyme Microbiology Technology 9: 531–537.
30. Hubalek Z (2003) Protectants used in the cryopreservation of microorganisms. Cryobiology 46: 205–229.
31. Savini M, Cecchini C, Verdenelli MC, Silvi S, Orpianesi C, et al. (2010) Pilot-scale production and viability analysis of freeze-dried probiotic bacteria using different protective agents. Nutrients 2: 330–339.
32. Jedrzejczak-Krzepkowska M, Bielecki S (2011) Bifidobacteria and inulin-type fructans which stimulate their growth. Postepy. Biochem. 57: 392–400.
33. de Jonge J, Amorij JP, Hinrichs WL, Wilschut J, Huckriede A, et al. (2007) Inulin sugar glasses preserve the structural integrity and biological activity of influenza virosomes during freeze-drying and storage. Eur. J. Pharm. Sci. 32: 33–44.
34. Neyrinck AM, Van Hee VF, Piront N, De Backer F, Toussaint O, et al. (2012) Wheat-derived arabinoxylan oligosaccharides with prebiotic effect increase satietogenic gut peptides and reduce metabolic endotoxemia in diet-induced obese mice. Nutr. Diabetes 2: e28.
35. Neyrinck AM, Possemiers S, Druart C, Van de Wiele T, De Backer F, et al. (2011) Prebiotic effects of wheat arabinoxylan related to the increase in bifidobacteria, Roseburia and Bacteroides/Prevotella in diet-induced obese mice. PLoS One 6: e20944.
36. Ganzle MG, Follador R (2012) Metabolism of oligosaccharides and starch in lactobacilli: a review. Front. Microbiol. 3: 340.

37. Ross RP, Desmond C, Fitzgerald GF, Stanton C (2005) Overcoming the technological hurdles in the development of probiotic foods. J. Appl. Microbiol. 98: 1410–1417.

38. Khan MT, Duncan SH, Stams AJ, van Dijl JM, Flint HJ, et al. (2012) The gut anaerobe Faecalibacterium prausnitzii uses an extracellular electron shuttle to grow at oxic-anoxic interphases. ISME J. 6: 1578–1585.

39. van Drooge DJ, Hinrichs WL, Dickhoff BH, Elli MN, Visser MR, et al. (2005) Spray freeze drying to produce a stable Delta(9)-tetrahydrocannabinol containing inulin-based solid dispersion powder suitable for inhalation. Eur. J. Pharm. Sci. 26: 231–240.

40. Backhed F (2012) Host responses to the human microbiome. Nutr. Rev. 70 Suppl 1: S14–7.

41. Paturi G, Nyanhanda T, Butts CA, Herath TD, Monro JA, et al. (2012) Effects of potato fiber and potato-resistant starch on biomarkers of colonic health in rats fed diets containing red meat. J. Food Sci. 77: H216–23.

42. Tachon S, Zhou J, Keenan M, Martin R, Marco ML (2013) The intestinal microbiota in aged mice is modulated by dietary resistant starch and correlated with improvements in host responses. FEMS Microbiol. Ecol. 83: 299–309.

43. Ze X, Duncan SH, Louis P, Flint HJ (2012) Ruminococcus bromii is a keystone species for the degradation of resistant starch in the human colon. ISME J., 2012.

44. Kim CJ, Kovacs-Nolan J, Yang C, Archbold T, Fan MZ, et al. (2009) L-cysteine supplementation attenuates local inflammation and restores gut homeostasis in a porcine model of colitis. Biochim. Biophys. Acta. 1790: 1161–1169.

45. Sokol H, Seksik P, Furet JP, Firmesse O, Nion-Larmurier I, et al. (2009) Low counts of Faecalibacterium prausnitzii in colitis microbiota. Inflamm. Bowel Dis. 15: 1183–1189.

15

Yeast Modulation of Human Dendritic Cell Cytokine Secretion: An *In Vitro* Study

Ida M. Smith[1,2], Jeffrey E. Christensen[1], Nils Arneborg[2], Lene Jespersen[2]*

1 Health & Nutrition Division Discovery, Chr. Hansen A/S, Hørsholm, Denmark, **2** Department of Food Science, University of Copenhagen, Frederiksberg, Denmark

Abstract

Probiotics are live microorganisms which when administered in adequate amounts confer a health benefit on the host. The concept of individual microorganisms influencing the makeup of T cell subsets via interactions with intestinal dendritic cells (DCs) appears to constitute the foundation for immunoregulatory effects of probiotics, and several studies have reported probiotic strains resulting in reduction of intestinal inflammation through modulation of DC function. Consequent to a focus on *Saccharomyces boulardii* as the fundamental probiotic yeast, very little is known about hundreds of non-*Saccharomyces* yeasts in terms of their interaction with the human gastrointestinal immune system. The aim of the present study was to evaluate 170 yeast strains representing 75 diverse species for modulation of inflammatory cytokine secretion by human DCs *in vitro*, as compared to cytokine responses induced by a *S. boulardii* reference strain with probiotic properties documented in clinical trials. Furthermore, we investigated whether cytokine inducing interactions between yeasts and human DCs are dependent upon yeast viability or rather a product of membrane interactions regardless of yeast metabolic function. We demonstrate high diversity in yeast induced cytokine profiles and employ multivariate data analysis to reveal distinct clustering of yeasts inducing similar cytokine profiles in DCs, highlighting clear species distinction within specific yeast genera. The observed differences in induced DC cytokine profiles add to the currently very limited knowledge of the cross-talk between yeasts and human immune cells and provide a foundation for selecting yeast strains for further characterization and development toward potentially novel yeast probiotics. Additionally, we present data to support a hypothesis that the interaction between yeasts and human DCs does not solely depend on yeast viability, a concept which may suggest a need for further classifications beyond the current definition of a probiotic.

Editor: Junji Yodoi, Institute for Virus Research, Laboratory of Infection and Prevention, Japan

Funding: The research leading to these results was funded by the EU's Seventh Framework Programme (FP7) under grant agreement PITN-GA-2010-264717, in the form of a research fellowship for IMS. The funders had no role in study design, data collection and analysis, decision to publish, or preparation of the manuscript.

Competing Interests: The authors have read the journal's policy and have the following conflicts to declare. IMS and JEC are employees of Chr. Hansen A/S, a manufacturer of probiotic products, and the described work was carried out at Chr. Hansen A/S facilities in Hørsholm, Denmark.

* E-mail: lj@food.ku.dk

Introduction

The mucosal-associated lymphoid tissues lining the human gastrointestinal tract contain a network of immune cells with the important task of distinguishing potentially dangerous antigens from harmless substances. Dendritic cells (DCs) govern the balance between immunity and tolerance by sampling of intestinal contents and initiating appropriate immune responses to luminal antigens through pattern recognition receptor signaling, cytokine secretion, and their ability to migrate and present antigen to naïve T cells in draining lymph nodes [1,2]. At homeostasis, DCs in the intestinal mucosa are conditioned by commensal microorganisms to promote proliferation of Foxp3[+] regulatory T cells (T$_{regs}$), strong producers of anti-inflammatory IL-10 contributing to intestinal tolerance [1,3,4]. During infection or active inflammation, pathogenic microorganisms bind to pattern recognition receptors expressed by DCs and activate signaling pathways involving MAP kinases and the nuclear transcription factor NFκB resulting in production and secretion of a wide range of chemokines and cytokines with distinct inflammatory effects. In this context, DC secretion of inflammatory cytokines such as TNFα and IL-1β is central for acute, innate inflammatory responses involving

attraction of neutrophils and macrophages to the site of infection. In addition, DCs are central players in the regulation of adaptive immune responses. For example, DC secretion of IL-12 and IL-6 promotes the proliferation of Th1 and Th17 subpopulations, respectively, whereas DC modulation toward an IL-10 secreting phenotype contributes to induction of T$_{reg}$ responses promoting intestinal tolerance [5,6]. Furthermore, efficient antigen presentation relies upon DC maturation, a process involving upregulation of co-stimulatory surface molecules as well as modulation of chemokine receptor expression.

Probiotics are live microorganisms which when administered in adequate amounts confer a health benefit on the host [7]. Based on their role as key regulators of intestinal inflammation, DC involvement in probiotic functionality has been studied extensively [8–11]. The concept of individual commensal microorganisms influencing the makeup of intestinal T cell subsets via interactions with DCs appears to constitute the foundation for immunoregulatory effects of probiotics [4], and several studies have reported probiotic strains resulting in reduction of intestinal inflammation through modulation of DC function [8,9,12–14]. Consequently, modulation of DC cytokine secretion and maturation by various microorganisms has elucidated species and strain specific effects that have guided the selection of novel probiotic strains for further investigation [15–17].

Although the gut microbiota is dominated by bacteria [18], communities of eukaryotic microorganisms are part of the human microbiome [19–21]. In addition, eukaryotes such as food-related yeasts have been utilized for the production of fermented food and beverages and consumed by humans for centuries [22]. Thus, much like for prokaryotes, interactions between eukaryotic microorganisms and the intestinal immune system may influence human health in various ways. While the majority of probiotic microorganisms studied to date are lactic acid producing bacteria, research in yeasts with potentially beneficial influences on human health has mainly revolved around *Saccharomyces boulardii* [23,24], a yeast taxonomically acknowledged as belonging to the *S. cerevisiae* species [25,26] but in the following text referred to as *S. boulardii*.

S. boulardii has been included in numerous randomized controlled trials and strong clinical evidence exists for the use of *S. boulardii* for the prevention of antibiotic associated diarrhea, Traveler's diarrhea, and acute infectious diarrheas [27,28]. In addition, *S. boulardii* has shown a positive impact on disease outcome in clinical studies of inflammatory bowel diseases such as Crohn's disease and ulcerative colitis [27], indicating an ability of *S. boulardii* to influence human immune responses underlying intestinal inflammation. The molecular basis for the beneficial effects of *S. boulardii* has been subject to extensive study, *in vitro* as well as in animal models, and *S. boulardii* has been found to impact inflammatory cytokine production by intestinal epithelial cells [29–34], peripheral blood mononuclear cells (PBMCs) [24], and DCs [23,35–37], reducing inflammatory scores in experimental colitis models in rodents [24,30,32,33,38–40].

Consequent to the intense research focus on *S. boulardii* as the fundamental probiotic yeast, very little is known about hundreds of non-*Saccharomyces* yeasts in terms of their interaction with the human gastrointestinal immune system. Other food-related yeast species typically associated with dairy products such as kefir and traditional cheeses include *Kluyveromyces lactis*, *Kluyveromyces marxianus*, and *Debaryomyces hansenii* [41]. While isolates of all three species have been evaluated for potential probiotic properties in *in vitro* experimental conditions assessing acid and bile survival, and adhesion to and modulation of cytokine secretion from intestinal epithelial cells [41–46], studies of the interactions between these yeasts and specialized immune cells have been far fewer [47,48].

The current definition of probiotics as "live microorganisms which when administered in adequate amounts confer a health benefit on the host" places importance on the viability of probiotic microorganisms at the site of action, presumably the lower small intestines. This has led to numerous studies focusing on the ability of potentially probiotic microorganisms to survive the harsh conditions of the human gastrointestinal tract, i.e. the acidic environment in the gastric sac and the presence of bile salts in the proximal small intestines [41,42,47]. However, while probiotic effects caused by actively secreted molecules will depend on a probiotic microorganism being alive [23,30,34,35], other probiotic effects may depend solely on the interaction of microbial cell wall molecules and surface receptors expressed by host cells without the need for an active metabolic function of the probiotic. Indeed, several studies have found heat killed, UV irradiated, and live bacteria to display equal DC stimulatory patterns *in vitro* [6,10,49–52]. Others have described the failure of nonviable *Saccharomyces* yeasts to prevent pathogen induced cytokine and chemokine expression in cultured epithelial cells [29,43], while a third study reported that *S. boulardii* maintained an inhibitory effect on *Salmonella* induced signaling pathways in epithelial cells even after

being subjected to a membrane disrupting glass bead treatment, thus indicating the likely importance of yeast cell wall structures for the observed inhibition [32].

Multiplexed immunoassays based on the principles of flow cytometry allow for simultaneous determination of numerous soluble proteins in very small sample volumes. The combination of high throughput and impressive accuracy, sensitivity, and reproducibility make these experimental techniques highly relevant for screening purposes where rapid quantification of multiple compounds is critical [53,54].

The aim of the present study was to evaluate a broad spectrum of yeasts (170 strains representing 75 diverse yeast species were included in the study) for modulation of inflammatory cytokine secretion by human DCs, as compared to cytokine responses induced by a *S. boulardii* (Ultra-Levure) reference strain with probiotic properties documented in clinical trials [27]. To our knowledge, this is the first large-scale study of highly diverse yeasts in terms of their modulation of DC function, incorporating secretion levels of several cytokines. Furthermore, we investigated whether cytokine inducing interactions between yeasts and human DCs are dependent upon yeast viability or rather a product of membrane interactions regardless of yeast metabolic function.

Materials and Methods

Yeast strains and growth conditions

Yeast strains included in this study were obtained from CBS (www.cbs.knaw.nl). 170 strains were selected based on a desire to include a broad range of yeast biodiversity (see complete list of included strains in Table 1). Strains were cultured in YPD media (0,5 % yeast extract, 1 % peptone, 1,1 % D-glucose) at 30°C under aerobic conditions. Early stationary growth phase yeast cultures were harvested by centrifugation, washed twice with DC media (RPMI 1640 supplemented with 10 mM HEPES (Sigma-Aldrich, Schnelldorf, Germany) and 50 μM 2-mercaptoethanol (Sigma-Aldrich, Schnelldorf, Germany)), OD adjusted in DC media containing 10 % glycerol, and cryopreserved at −80°C until time of DC stimulation. Viability of frozen yeast cultures was verified by staining with propidium iodide. For some experiments, yeast strains were UV irradiated (70,000 μJ/cm^2 for 5 min) or heat treated (80°C at 650 rpm for 5 min) prior to cryopreservation at −80°C. Yeast cell viability after UV irradiation or heat treatment was assessed by propidium iodide staining (UV <40% intact cells; heat <20% intact cells), and the reproductive ability of UV irradiated and heat treated yeasts was determined by colony counts after 48 h incubation of YPD agar plates at 30°C.

Monocyte-derived DC generation

Immature monocyte-derived DCs were generated *in vitro* by a 6 day procedure as described [50]. Human buffy coats from healthy donors were supplied by Department of Clinical Immunology at Copenhagen University Hospital, Copenhagen, Denmark. Use of human samples with no identifying information was approved by The National Committee on Health Research and the Danish Society for Clinical Immunology, and all donors gave informed written consent upon donation. Briefly, human peripheral blood mononuclear cells were obtained from buffy coats by density gradient centrifugation using Ficoll-Paque PLUS (GE Healthcare, Freiburg, Germany). Monocytes were isolated by positive selection for CD14 using magnetic-activated cell sorting with CD14 microbeads (Miltenyi Biotec, Bergisch Gladbach, Germany) and cultured at a density of 2×10^6 cells/mL in complete DC media (RPMI 1640 supplemented with 10 mM HEPES (Sigma-Aldrich, Schnelldorf, Germany), 50 μM 2-mercaptoethanol (Sigma-Al-

Table 1. Yeast strains included in study.

Genus	Species	Strains
Ambrosiozyma	monospora	CBS2554
Barnettozyma	pratensis	CBS9053, CBS9055
Bensingtonia	yamatoana	CBS9336
Botryozyma	mucatilis	CBS9042, CBS9043
Brettanomyces	custersianus	CBS4806, CBS5207, CBS5208
	naardenensis	CBS7540
Candida	amphixiae	CBS9877
	anneliseae	CBS9837
	atakaporum	CBS9833
	athensensis	CBS9840, CBS9841
	blattae	CBS9871
	bohiensis	CBS9897
	bolitotheri	CBS9832
	bombi	CBS9017
	buenavistaensis	CBS9895
	choctaworum	CBS9831
	chrysomelidarum	CBS9904
	elateridarum	CBS9842
	ghanaensis	CBS8798
	gigantensis	CBS9896
	litsaeae	CBS8799
	michaelii	CBS9878
	palmioleophila	CBS8109
	powellii	CBS8795
	taliae	CBS9838
Citeromyces	siamensis	CBS9152, CBS9153
Cryptococcus	laurentii var. laurentii	CBS8796
	podzolicus	CBS9357, CBS9358
Cryptotrichosporon	anacardii	CBS9549, CBS9551
Debaryomyces	fabryi	CBS10579
	hansenii	CBS116, CBS767, CBS773, CBS1101, CBS1119, CBS1121, CBS1123, CBS1129, CBS1519, CBS1791, CBS1795, CBS1962, CBS2331, CBS2333, CBS4890, CBS5139, CBS5140, CBS6089, CBS6574, CBS7032, CBS7848, CBS8339, CBS9682, CBS9685, CBS9696
	subglobosus	CBS792, CBS1128
Dekkera	anomala	CBS77, CBS4212, CBS4711, CBS7250, CBS8138
	bruxellensis	CBS72, CBS75, CBS96, CBS2547, CBS4459, CBS4482, CBS4601, CBS4602, CBS6055
Geotrichum	cucujoidarum	CBS9893
Hanseniaspora	lachancei	CBS8818, CBS8819
	opuntiae	CBS8820, CBS9791
Kazachstania	exigua	CBS9330
Kluyveromyces	lactis var. drosophilarum	CBS9056
	lactis var. lactis	CBS9057, CBS9058, CBS9059, CBS9060
	marxianus	CBS1553
Kurtzmaniella	cleridarum	CBS8793
Lachancea	fermentati	CBS797
	kluyveri	CBS6545, CBS6546, CBS6547
	thermotolerans	CHCC5756
Lodderomyces	elongisporus	CBS7803
Metschnikowia	arizonensis	CBS9064
	borealis	CBS8431, CBS8432

Table 1. Cont.

Genus	Species	Strains
	gruessii	CBS9029, CBS9030
	koreensis	CBS9066
	kunwiensis	CBS9067, CBS9677, CBS9679, CBS9681
	noctiluminum	CBS9907
	reukaufii	CBS9018, CBS9019, CBS9020, CBS9021, CBS9022
Naumovozyma	castelli	CBS2248, CBS4310, CBS4906
	dairensis	CBS421
Ogataea	dorogensis	CBS9260, CBS9261
Pichia	kluyveri	CHCC11259
	mandshurica	CBS209
	myanmarensis	CBS9786
	sporocuriosa	CBS9200
Rhodosporidium	diobovatum	CBS9081, CBS9084
	sphaerocarpum	CBS9080
	toruloides	CBS14
Rhodotorula	mucilaginosa var. mucilaginosa	CBS9070, CBS9078, CBS9083
Saccharomyces	arboricolus	CBS10644
	bayanus	CBS381, CBS1641, CBS9787
	boulardii	CHCC11905, CHCC11906, 259*, 7103*, 7135*, 7136*, LSB*, Sb.A*, Sb.L*, Sb.P*
	cariocanus	CBS5313, CBS7994, CBS8841
	cerevisiae	CBS1646, CHCCJ4848, CBS6128, CHCC7036, CBS9564
	kudriavzevii	CBS8840
	mikatae	CBS8839, CBS10522, CBS10523
	paradoxus	CBS8442
	pastorianus	CBS1462, CBS1642
Zygosaccharomyces	mellis	CBS711, CBS738
	rouxii	CBS708, CBS733

Strain sources: CBS Centraalbureau voor Schimmelcultures, Utrecht, The Netherlands
CHCC Chr. Hansen Culture Collection, Hørsholm, Denmark
* van der Aa Kühle, A., Jespersen, L., 2003. Systematic and Applied Microbiology 26, 564–571.

drich, Schnelldorf, Germany), 2 mM L-glutamine (Life Technologies Ltd, Paisley, UK), 10 % heat-inactivated fetal bovine serum (Invitrogen, Paisley, UK), 100 U/mL penicillin (Biological Industries, Kibbutz Beit-Haemek, Israel), and 100 µg/mL streptomycin (Biological Industries, Kibbutz Beit-Haemek, Israel)) containing 30 ng/mL human recombinant IL-4 and 20 ng/mL human recombinant GM-CSF (both from Sigma-Aldrich, Saint Louis, USA) at 37°C, 5 % CO_2. Fresh complete DC media containing full doses of IL-4 and GM-CSF was added after three days of culture. At day 6, differentiation to immature DCs was verified by surface marker expression analysis (CD11c >90% expression; CD1a >75% expression).

DC stimulation

Immature DCs were resuspended in fresh complete DC media containing no antibiotics, seeded in 96-well plates at 1×10^5 cells/well, and allowed to acclimate at 37°C, 5 % CO_2, for at least one hour before stimulation. DC stimulation using thawed yeast strains was performed at a yeast:DC ratio of 10:1, and stimulated DCs were incubated for 20 h at 37°C, 5 % CO_2, as time-course experiments had shown a 20 h stimulation time to result in quantifiable levels of all cytokines of interest. After 20 h

stimulation, DC supernatants were sterile filtered through a 0.2 µm AcroPrep Advance 96-well filter plate (Pall Corporation, Ann Arbor, MI, USA) and stored at −80°C until time of cytokine quantification.

DC staining for quantification of co-stimulatory molecules and chemokine receptors

Immediately following 20 h stimulation time, DCs were collected, centrifuged at 200x g for 5 min, and resuspended in cold PBS containing 2 % BSA. Staining was performed using the following monoclonal antibodies: FITC-conjugated anti-human CD80 (clone L307.4), FITC-conjugated anti-human CD86 (clone 2331), APC-conjugated anti-human CCR6 (clone 11A9), FITC-conjugated anti-human CCR7 (clone 150503), and appropriate isotype controls (all from BD Biosciences, Erembodegem, Belgium). DCs were incubated with mAb for 30 min on ice protected from light, followed by repeated wash steps using 1 mL cold PBS 2 % BSA. Finally, DCs were resuspended in PBS 2 % BSA and kept on ice until flow cytometric analysis. Samples were acquired on an LSRFortessa flow cytometer (BD Biosciences, San Jose, CA, USA) using FACSDiva software (BD Biosciences, San Jose, CA, USA).

Cytokine quantification

Secreted levels of IL-12, TNF, IL-10, IL-6, and IL-1β were quantified by the Human Inflammatory Cytokines cytometric bead array (CBA) kit (BD Biosciences, Erembodegem, Belgium) according to the manufacturer's instructions. Briefly, fluorescent beads coated with monoclonal capture antibodies were mixed with PE conjugated detection antibodies and recombinant standards or test samples and allowed to form sandwich complexes during 3 h incubation protected from light. After repeated wash steps, samples were acquired on an LSRFortessa flow cytometer (BD Biosciences, San Jose, CA, USA) and data analysis was performed using the FCAP Array 3 software (BD Biosciences, San Jose, CA, USA). Detection limits for individual cytokines were as follows: 1.9 pg/mL IL-12, 3.7 pg/mL TNF, 3.3 pg/mL IL-10, 2.5 pg/mL IL-6, and 7.2 pg/mL IL-1β.

Multivariate data analysis and statistical analysis

Multivariate data analysis was performed using SIMCA-P+ 13 (Umetrics, Umeå, Sweden). A PCA-class model was generated based on data for induced levels of IL-12, IL-10, IL-6, TNFα, and IL-1β by 12 biological replicates of the *S. boulardii* reference strain, thereby centering the plot point of origin on the cytokine profile induced by the *S. boulardii* reference strain. Induced cytokine data (IL-12, IL-10, IL-6, TNFα, and IL-1β) for the 170 yeast strains included in the screen constituted the prediction data set; i.e. the distance of a given yeast strain from the plot point of origin indicates how closely the induced cytokine profile resembles that of the *S. boulardii* reference strain.

Statistical analysis (one-way ANOVA with Bonferroni's multiple comparison post test) was performed using GraphPad Prism 5 (GraphPad Software, La Jolla, USA).

Results

Yeasts induce highly diverse cytokine profiles in human DCs

Given that modulation of DC cytokine secretion has been linked to probiotic functionality related to intestinal inflammation, we evaluated yeast modulation of DC secretion of five inflammation related cytokines *in vitro*. We exposed human DCs to each yeast strain in duplicate at a 10:1 yeast:DC ratio and quantified secreted levels of IL-12, IL-10, IL-6, TNFα, and IL-1β after 20 h stimulation. Stimulation time was selected based on time-course experiments indicating 20 h as ideal for obtaining quantifiable levels of cytokines produced at a slow rate (IL-12, IL-10, and IL-1β) without reaching saturation conditions for rapidly secreted cytokines (IL-6 and TNFα) (Fig S1). As a point of reference, we included a *S. boulardii* strain with probiotic effects documented in clinical trials [27]. As expected, *S. boulardii* engaged human immune cells, as evidenced by induction of a robust response across all five cytokines (Fig 1). In addition, *S. boulardii* induced high levels of the co-stimulatory molecules CD80 and CD86, indicative of strong activation of the immature DCs (Fig S2), and affected DC chemokine receptor expression, as observed by down-regulation of CCR6 and strong up-regulation of CCR7 (Fig S2), indicating that *S. boulardii* activates immature DCs to a mature phenotype primed for lymph node migration and efficient antigen presentation.

For comparison of the yeast induced DC cytokine profiles, multivariate data analysis was applied as a valuable tool for visualizing and grouping yeast strains based on quantified levels of all five cytokines. A PCA-class model was generated based on data for induced levels of IL-12, IL-10, IL-6, TNFα, and IL-1β by 12 biological replicates of the *S. boulardii* reference strain, thereby

centering the PCA plot point of origin on the cytokine profile induced by the reference strain (Fig 2). Induced cytokine data (IL-12, IL-10, IL-6, TNFα, and IL-1β) for the 170 yeast strains included in the study constituted the prediction data set; i.e. the distance of a yeast strain from the plot point of origin indicates how closely the induced cytokine profile resembles that of the *S. boulardii* reference strain. Visualizing the obtained cytokine profiles in this way revealed the interesting fact that induction of all five cytokines was positively correlated in our study, as shown by the loadings of individual cytokines in the Figure 2 insert.

As displayed in Figure 2, the yeasts included in this study induced highly diverse cytokine profiles in human DCs. Not surprisingly, the majority of *Saccharomyces* yeasts induced cytokine profiles very similar to the *S. boulardii* reference strain, as indicated by their location very close to the plot point of origin in Figure 2. In addition, this overview plot shows *Saccharomyces* yeasts as strong cytokine inducers, with very few non-*Saccharomyces* yeasts inducing cytokine levels higher than the *S. boulardii* reference strain (i.e. not many yeast strains present in the upper right quadrant of the plot). For non-*Saccharomyces* yeasts, we observe a broad range of cytokine inducing properties. For instance, a third of the included *Debaryomyces* strains, several *Dekkera* strains, and all included *Zygosaccharomyces* isolates dominate a distinct cluster of very low cytokine inducing yeasts present at the bottom left corner of the plot (Fig 2).

Distinct differences observed in cytokine inducing properties of individual yeast genera

Next, we focused on the induced DC cytokine profiles of individual yeast genera. Six *Kluyveromyces* strains representing the species *K. marxianus*, *K. lactis* var. *lactis*, and *K. lactis* var. *drosophilarum* were included in our study, and multivariate data analysis of the induced DC cytokine profiles revealed clear species distinctions in immune stimulating capacities (Fig 1A). *K. marxianus* (CBS1553) induced DC cytokine levels statistically indistinguishable (P>0.05) from the profile induced by the *S. boulardii* reference strain for every one of the quantified cytokines (Fig 1B-F). In contrast, the four *K. lactis* var. *lactis* strains (CBS9057, CBS9058, CBS9059, CBS9060) induced much lower levels of cytokines; in particular, induced levels of IL-12, IL-10, and IL-1β were near or below the detection limit of the assay. Strikingly, no significant differences were observed between the DC cytokine profiles induced by the four *K. lactis* var. *lactis* strains (Fig 1B-F, P>0.05 for all quantified cytokines).

For *Debaryomyces* yeasts, 25 of the 28 strains included in our study represented the species *D. hansenii*, and the induced DC cytokine profiles revealed a remarkable diversity in immune stimulating properties (Fig 3A). The strain CBS1121 induced DC cytokine levels similar to the *S. boulardii* reference strain (Fig 3B-F), as indicated by secreted levels of IL-1β, IL-6, and TNFα being statistically indistinguishable from *S. boulardii* induced levels. The *D. hansenii* type strain (CBS767) displayed much poorer cytokine induction capabilities, as seen by significantly lower induction of the pro-inflammatory cytokines IL-12, IL-6, and TNFα. Finally, the *D. hansenii* strain CBS7848 induced a DC cytokine profile characterized by levels of IL-1β, IL-10, IL-6, and TNFα significantly higher than the *S. boulardii* reference strain, yet failed to induce detectable levels of IL-12.

The yeast genus *Metschnikowia* represents a large family of yeasts which, to the best of our knowledge, has not been explored for properties relating to human health. The PCA plot in Figure 4A displays the DC cytokine profiles induced by the 16 isolates representing seven *Metschnikowia* species included in our study. The plot reveals striking species distinctions separating species with

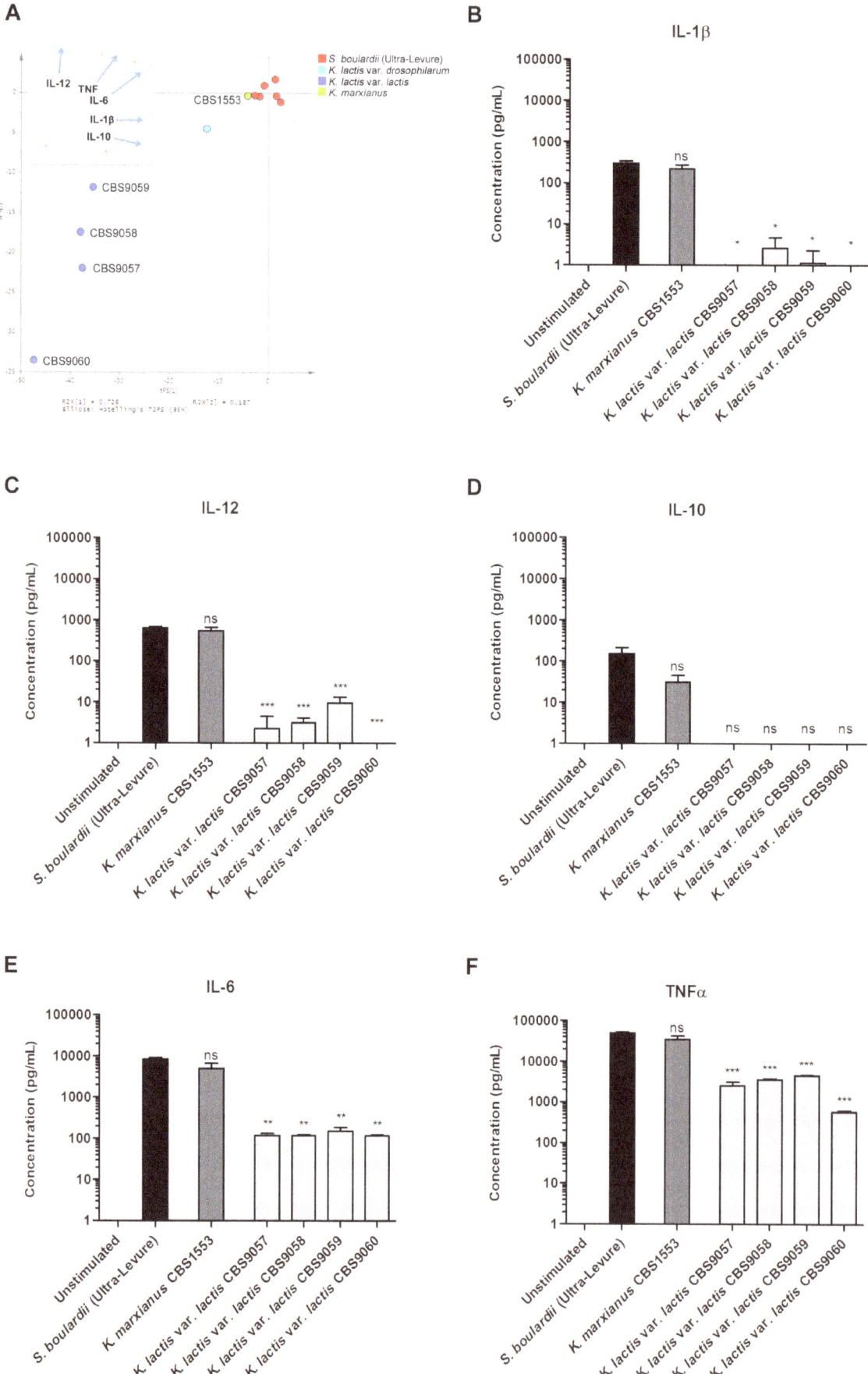

Figure 1. DC cytokine secretion induced by *Kluyveromyces* yeasts. A. Principal component analysis (PCA) scatter plot of DC cytokine profiles induced by *Kluyveromyces* yeast strains and the *S. boulardii* reference strain. Each dot represents induced cytokine data (IL-12, IL-10, IL-6, TNFα, and IL-1β) for one yeast strain, entered as the mean value of two biological replicates, and colored according to yeast species. The six red dots representing biological replicates of the *S. boulardii* included in the PCA model show the deviation in the model. The insert represents loadings of individual cytokines included in the PCA model; for example, high IL-12 inducing yeasts are placed high along the plot Y axis, whereas strong IL-10 inducing yeasts are located to the right along the plot X axis. Levels of **B.** IL-1β, **C.** IL-12, **D.** IL-10, **E.** IL-6, and **F.** TNFα secreted by human monocyte-derived DCs following 20 h stimulation with DC media containing 10 % glycerol (unstimulated) or *S. boulardii* (Ultra-Levure), *K. marxianus* (CBS1553), or *K. lactis* var. *lactis* (CBS9057, CBS9058, CBS9059, CBS9060) at a yeast:DC ratio of 10:1. Data are representative of two independent experiments, error bars represent SEM. One-way ANOVA, Bonferroni's multiple comparison post test, indicating significant differences from cytokine levels induced by *S. boulardii*. ns, not significant; *, $P<0.05$; **, $P<0.01$; ***, $P<0.001$.

highly diverse cytokine inducing properties. While isolates of *M. reukaufii* induced DC cytokine profiles very similar to the *S. boulardii* reference strain across all five cytokines, *M. gruessii* isolates induced robust levels of IL-1β, IL-10, IL-6, and TNFα, yet undetectable levels of IL-12 (Fig 4B-F). In contrast, *M. borealis* displayed poor cytokine inducing properties in general, as seen by an induced DC cytokine profile characterized by significantly lower cytokine levels compared to the *S. boulardii* reference strain.

Yeasts are capable of DC stimulation independently of viability

Next, we investigated whether the observed interactions between yeasts and DCs were dependent upon yeast viability. We hypothesized that yeasts would be able to induce DC activation regardless of metabolic activity and conducted experiments to compare DC stimulation with live, UV irradiated, and heat treated yeast. UV irradiation conditions were designed to generate relatively intact yeast cells unable to reproduce, whereas heat treatment was intended to cause severe yeast cell membrane disruption. Propidium iodide staining confirmed a reduction in the proportion of intact yeast cells to levels below 40% after UV

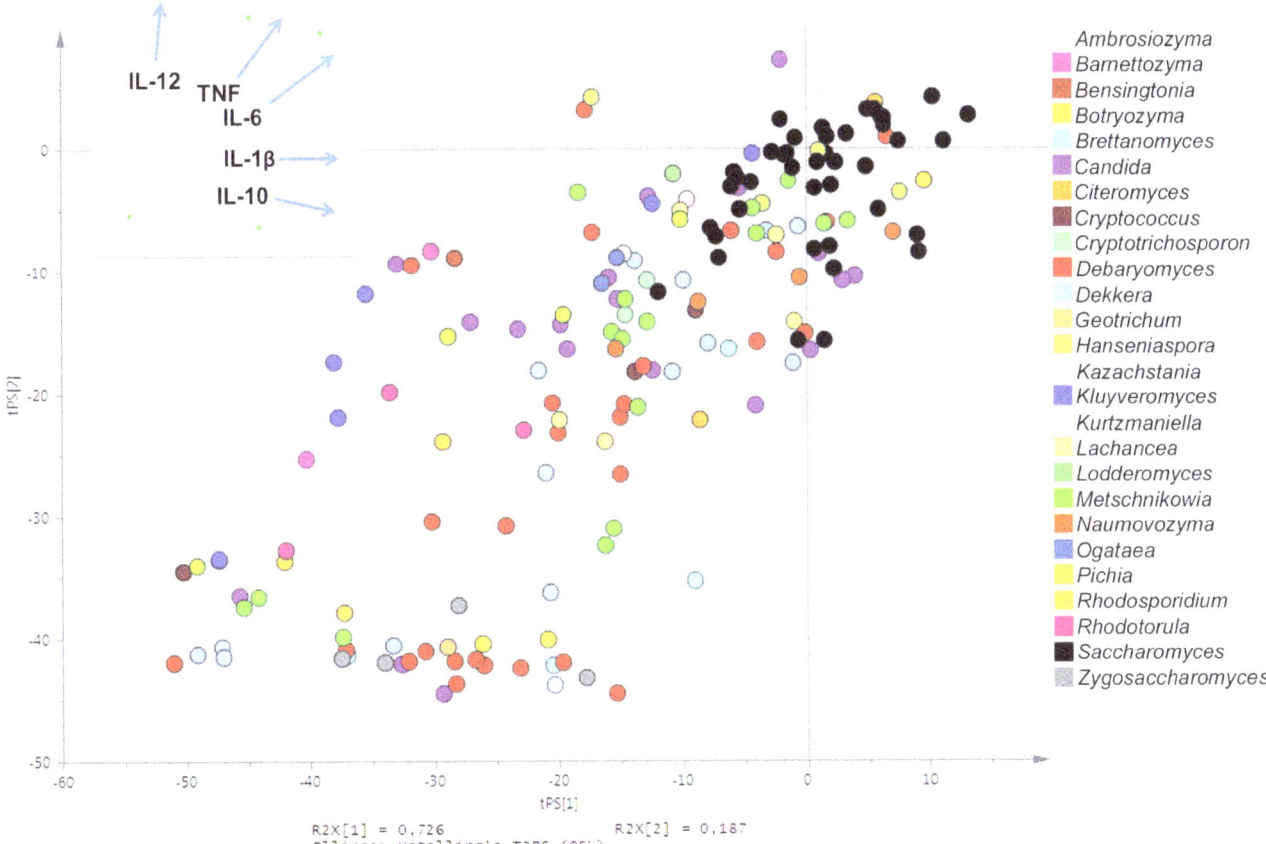

Figure 2. Yeast induced cytokine profiles in human DCs. Principal component analysis (PCA) scatter plot of induced cytokine data for 170 yeast strains included in screen. Each dot represents induced cytokine data (IL-12, IL-10, IL-6, TNFα, and IL-1β) for one yeast strain, entered as the mean value of two biological replicates, and colored according to yeast genera. The plot point of origin is centered on the cytokine profile induced by the *S. boulardii* reference strain, and thus the distance of a given yeast strain from the plot point of origin indicates how closely the induced cytokine profile resembles that of the *S. boulardii* reference strain. The insert represents loadings of individual cytokines included in the PCA model; for example, high IL-12 inducing yeasts are placed high along the plot Y axis, whereas strong IL-10 inducing yeasts are located to the right along the plot X axis.

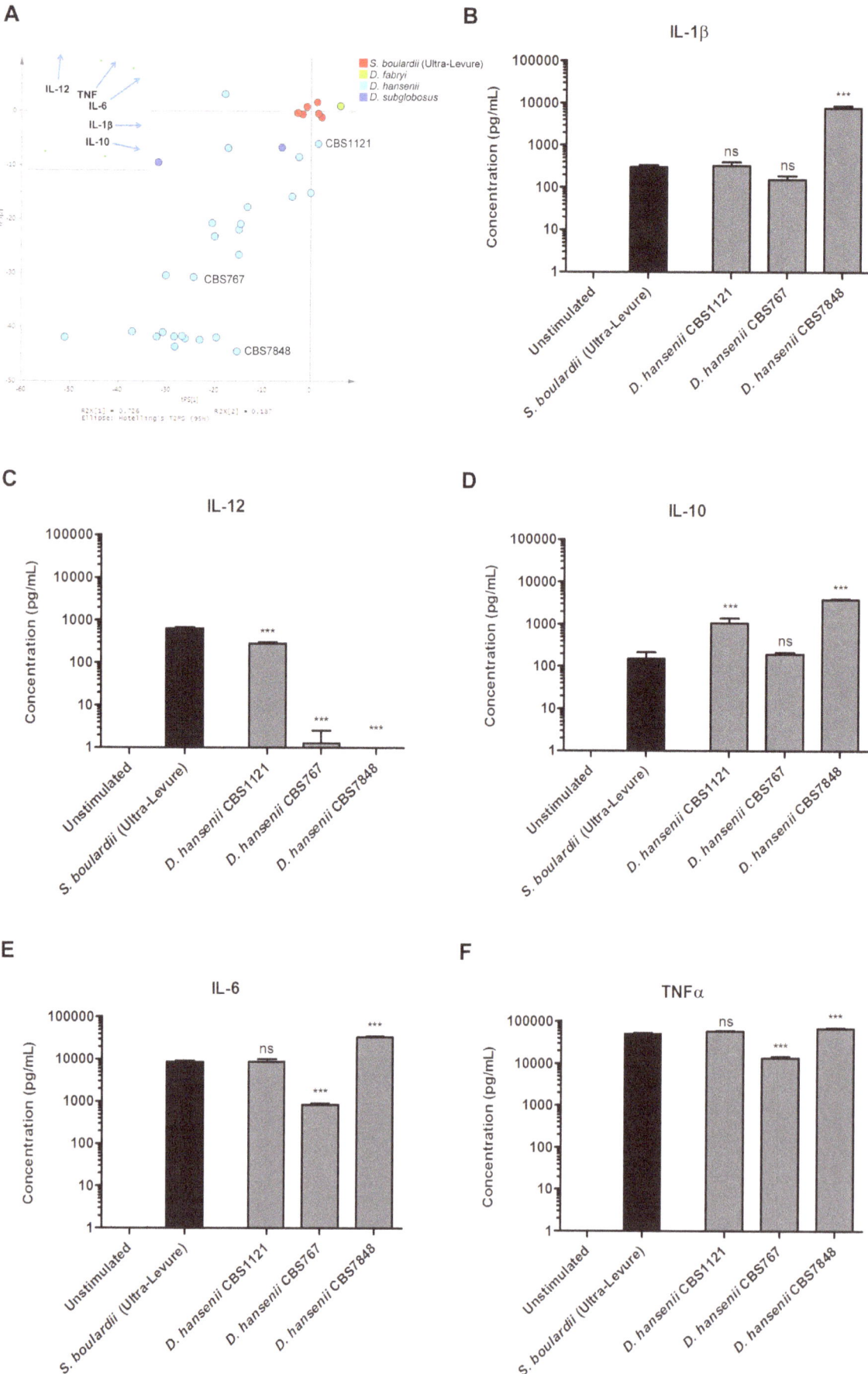

Figure 3. DC cytokine secretion induced by *Debaryomyces* yeasts. A. Principal component analysis (PCA) scatter plot of DC cytokine profiles induced by *Debaryomyces* yeast strains and the *S. boulardii* reference strain. Each dot represents induced cytokine data (IL-12, IL-10, IL-6, TNFα, and IL-1β) for one yeast strain, entered as the mean value of two biological replicates, and colored according to yeast species. The six red dots representing biological replicates of the *S. boulardii* reference strain show the deviation in the model. The insert represents loadings of individual cytokines included in the PCA model; for example, high IL-12 inducing yeasts are placed high along the plot Y axis, whereas strong IL-10 inducing yeasts are located to the right along the plot X axis. Levels of **B.** IL-1β, **C.** IL-12, **D.** IL-10, **E.** IL-6, and **F.** TNFα secreted by human monocyte-derived DCs following 20 h stimulation with DC media containing 10 % glycerol (unstimulated), *S. boulardii* (Ultra-Levure), or *D. hansenii* (CBS1121, CBS767, CBS7848) at a yeast:DC ratio of 10:1. Data are representative of two independent experiments, error bars represent SEM. One-way ANOVA, Bonferroni's multiple comparison post test, indicating significant differences from cytokine levels induced by *S. boulardii*. ns, not significant; *, P< 0.05; **, P<0.01; ***, P<0.001.

irradiation and below 20% after heat treatment, and the reproductive inability of UV irradiated and heat treated yeasts was verified by colony counts (data not shown). As presented in Figure 5, the obtained data show no significant differences in cytokine inducing properties between live, UV irradiated, and heat treated *S. boulardii*. Additionally, UV irradiated and heat treated *S. boulardii* were as effective as live yeast in inducing DC co-stimulatory functions and altering chemokine receptor expression levels (Fig S2). Regardless of UV irradiation or heat treatment, *S. boulardii* induced high levels of the co-stimulatory molecules CD80 and CD86, and modulated DC chemokine receptor expression, as observed by down-regulation of CCR6 and strong up-regulation of CCR7 (Fig S2).

For non-*Saccharomyces* yeasts, we observed a remarkably similar pattern. *K. marxianus* induced DC secretion of IL-12, IL-10, IL-6, TNFα, and IL-1β was unaffected by UV irradiation or heat treatment of the yeast (Fig 5). In contrast, the ability of *K. lactis* var. *lactis* (CBS9057) to induce IL-10 appeared slightly stronger after heat treatment, whereas induced levels of IL-12, IL-6, TNFα, and IL-1β remained unchanged. *D. hansenii* (CBS1121) induced cytokine secretion was unaffected by UV irradiation as well as heat treatment (Fig 5). For *Metschnikowia* yeasts, we observed differential modification of cytokine inducing properties between the three species examined. Whereas *M. gruessii* and *M. borealis* were unaffected by UV irradiation as well as heat treatment, the IL-10 inducing properties of *M. reukaufii* appeared slightly potentiated by the heat treatment (Fig 5).

Discussion

The complex interplay between the gut microbiota and the intestinal immune system as this relates to human health and disease has received increasing attention recently [3,4,19]. Novel understandings of how the composition of resident populations of intestinal microorganisms can significantly impact human health in areas as diverse as obesity, allergy, and inflammatory disorders, is making this an area of intense research [19]. Maintenance of intestinal homeostasis requires a carefully regulated network of immune cells responsible for controlling the delicate balance between populations of potentially pro-inflammatory effector cells such as Th1 and Th17 cells and anti-inflammatory T_{reg} cells [4], and intestinal DCs play a central role in orchestrating both innate and adaptive immune responses to commensal microorganisms as well as invasive pathogens [1,10].

As research in yeasts with a potentially beneficial impact on human health has focused almost exclusively on *S. boulardii*, the aim of the present study was to evaluate immune modulation by a large number of highly diverse yeasts in order to build a data set to serve as a foundation for selecting strains for further mechanistic studies. Choosing an experimental system involving human monocyte-derived DCs resulted in the necessary throughput and by employing a multiplexed immunoassay and multivariate data analysis to incorporate data for yeast induced DC secretion of five

cytokines with distinct inflammatory effects we were able to assess the modulation of DC cytokine secretion by 170 yeast strains representing a broad spectrum of biodiversity. In contrast to studies focusing on modulation of a challenge induced DC cytokine response [23,36], our experimental setup evaluated the ability of yeasts to engage DCs *per se*.

The ability of *S. boulardii* to engage human immune cells and influence cytokine secretion *in vitro* is well established and has been linked to improved inflammatory scores in rodent colitis models. Likewise, commensal bacterial strains have been found to induce strong DC maturation and inflammatory cytokine secretion, yet display tolerogenic properties promoting T cell hyporesponsiveness [10], and strains with known probiotic properties have been shown to induce transient pro-inflammatory host responses in rodents upon initial colonization [55]. Accordingly, we observed robust induction of DC cytokines when human DCs were cultured in the presence of a *S. boulardii* reference strain with probiotic properties documented in clinical trials [27], indicating the expected ability of *S. boulardii* to engage human DCs and impact their secretion of cytokines with distinct inflammatory effects. In addition, we found the *S. boulardii* reference strain capable of inducing co-stimulatory functions and modulating chemokine receptor expression towards an activated DC phenotype primed for lymph node migration and efficient antigen presentation. These findings are supported by reports of *S. boulardii* modulating LPS induced CD80 and CCR7 expression [23,35], and echo reports of commensal bacterial strains inducing DC maturation, as indicated by increased surface expression of CD80, CD86, and CCR7 [8,10]. Our observation that yeasts belonging to the *Saccharomyces* genus are among the strongest cytokine inducing yeasts included in the present study, and that all included *Saccharomyces* yeasts induce DC cytokine profiles very similar to that induced by the *S. boulardii* reference strain, parallels the finding of a recent study where six live yeast strains representing the species *S. bayanus*, *S. cerevisiae*, and *S. pastorianus* induced non-discriminatory cytokine profiles in human PBMCs [24].

The remarkable diversity in cytokine inducing properties observed for non-*Saccharomyces* yeasts reflects the high diversity among the included yeast isolates. As the selection criteria for inclusion in the present study were based on a desire to include a broad representation of yeast biodiversity and thus had no apparent link to properties of immune modulation, it is not surprising that while some non-*Saccharomyces* yeasts exhibit strong cytokine inducing properties, others present as far more immunologically inert.

It is generally accepted that probiotic properties of bacteria are not only species but also strain dependent [56–58]. Interestingly, whereas our data agree with this notion for a number of included yeast species, other yeast genera display distinct species clustering in DC cytokine inducing properties.

The genus *Kluyveromyces* includes food-related yeasts typically isolated from fermented dairy products, and the health-promoting effects associated with the consumption of these products have led

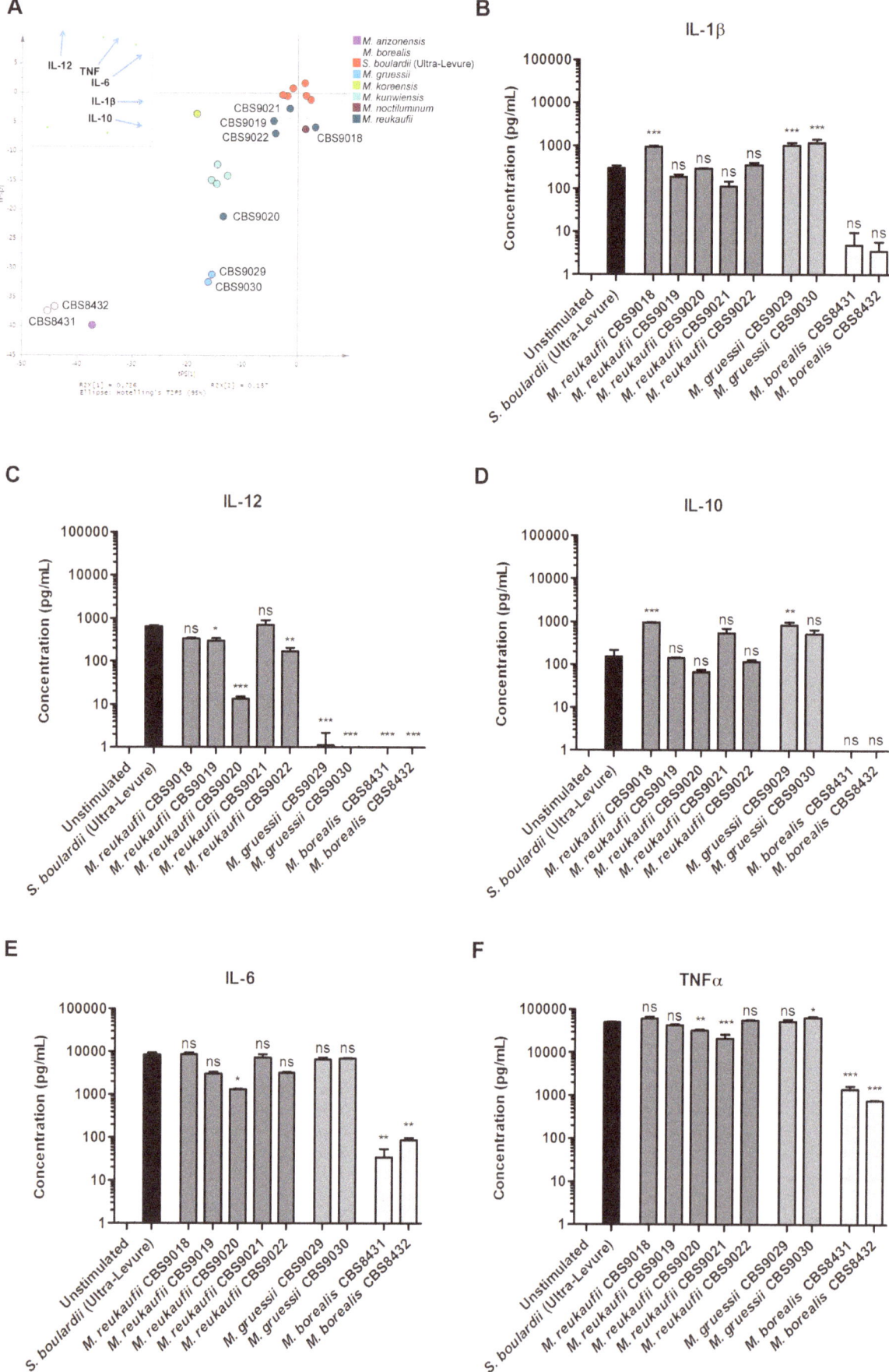

Figure 4. DC cytokine secretion induced by _Metschnikowia_ yeasts. A. Principal component analysis (PCA) scatter plot of DC cytokine profiles induced by _Metschnikowia_ yeast strains and the _S. boulardii_ reference strain. Each dot represents induced cytokine data (IL-12, IL-10, IL-6, TNFα, and IL-1β) for one yeast strain, entered as the mean value of two biological replicates, and colored according to yeast species. The six red dots representing biological replicates of the _S. boulardii_ reference strain show the deviation in the model. The insert represents loadings of individual cytokines included in the PCA model; for example, high IL-12 inducing yeasts are placed high along the plot Y axis, whereas strong IL-10 inducing yeasts are located to the right along the plot X axis. Levels of **B.** IL-1β, **C.** IL-12, **D.** IL-10, **E.** IL-6, and **F.** TNFα secreted by human monocyte-derived DCs following 20 h stimulation with DC media containing 10 % glycerol (unstimulated) or _S. boulardii_ (Ultra-Levure), _M. reukaufii_ (CBS9018, CBS9019, CBS9020, CBS9021, CBS9022), _M. gruessii_ (CBS9029, CBS9030), or _M. borealis_ (CBS8431, CBS8432) at a yeast:DC ratio of 10:1. Data are representative of two independent experiments, error bars represent SEM. One-way ANOVA, Bonferroni's multiple comparison post test, indicating significant differences from cytokine levels induced by _S. boulardii_. ns, not significant; *, P<0.05; **, P<0.01; ***, P<0.001.

to several studies investigating individual isolates for properties related to human health. A recent study evaluating the immune modulatory properties of a _K. marxianus_ strain _in vitro_ found a significant impact on the cytokine secretion of IL-6, TNFα, and IL-1β by human PBMCs [47]. While the _K. marxianus_ isolate included in the present study is not identical, this does support our observation that _K. marxianus_ induces a cytokine profile statistically indistinguishable from that of the _S. boulardii_ reference strain, clearly indicating the ability of _K. marxianus_ to engage specialized immune cells and impact the secretion of cytokines with distinct effects on adaptive immune responses. We observed distinct species differences clearly separating _K. marxianus_ and _K. lactis_ var. _lactis_ in terms of their ability to induce DC cytokine secretion, with all four _K. lactis_ var. _lactis_ strains displaying an apparent inability to induce significant levels of cytokines. This is supported by a study finding _K. lactis_ unable to induce detectable levels of the cytokines IL-6 and TNFα in a system evaluating interactions between yeast and cultured intestinal epithelial cells [44]. As the same study found _K. lactis_ yeasts able to stimulate IL-8 secretion, our data may reflect differences between yeasts mainly interacting with highly abundant epithelial cells triggering innate immune mechanisms such as neutrophil and monocyte recruitment and yeasts capable of engaging specialized immune cells playing an active role in adaptive immune responses.

As discussed above, the species distinction in DC cytokine inducing properties is unmistakable for _Kluyveromyces_ yeasts included in the present study. In contrast, the included isolates representing another genus comprising many food-related yeasts, namely _Debaryomyces_, display highly diverse and strain dependent DC inducing properties. Despite being highly abundant in nature and well known as food-related yeasts typically isolated from dairy, meat, and fermented soy products [59], all of which suggest likely human contact with these microorganisms, _Debaryomyces_ yeasts have not been thoroughly studied for properties relating to human health.

The aim of the present study was to provide a first step in a search for novel yeasts with the ability to impact human health through interactions with the intestinal immune system, and naturally, safety constitutes an important factor when considering novel microorganisms with the potential for development of products aimed at human consumption. Consequently, focusing further studies solely on yeast species approved by the European Food Safety Authority for holding qualified presumption of safety (QPS) becomes an attractive proposition. This would suggest an emphasis on _K. marxianus_, _K. lactis_, and _D. hansenii_, all of which hold QPS status [60] and, in the present study, displayed DC cytokine induction properties worthy of further investigation. However, limiting future efforts to already well-known food-related yeasts would present the inherent risk of missing less-studied yeasts with potentially superior properties. The yeast genus _Metschnikowia_, comprising a large family of yeasts primarily isolated from flowers and bees across the European continent, provides an example. To the best of our knowledge, _Metschnikowia_ yeasts have not been

subject to studies relating to human health, yet the numerous isolates included in our study exhibited a remarkable species distinction in DC cytokine inducing properties warranting further examination.

While the experimental conditions in this study were intended to simulate the _in vivo_ situation where microorganisms encounter mucosal DCs during passage through the human gut; naturally, some limitations need to be kept in mind when interpreting the data. Notably, single strain stimulation of DCs may be a necessity for data analysis when evaluating numerous strains but it fails to account for interactions between different members of the intestinal flora. Given the complexity of the human microbiota, the likely interactions between yeasts and other microorganisms present in the intestinal tract may result in different DC cytokine profiles. This highlights the importance of extending the current knowledge of select strains through additional studies incorporating a more complex mix of microorganisms resembling that of the human gastrointestinal tract.

From an industrial point of view, it is of importance to know whether dead yeast cells can be used in the same manner as viable cells. Additionally, yeast cells present in the human gastrointestinal tract may be either viable or dead, which makes it relevant to investigate whether the immunological responses are the same. Further, as yeast cell wall composition and structure are known to vary depending on yeast growth phase and any treatment, the prospect that cell wall structures of nonviable yeast cells may differ from those of viable yeast cells and in turn potentially trigger a different immunological response, led us to explore whether DC cytokine induction was affected by yeast viability. We observed no significant differences between the DC cytokine inducing properties of live, UV irradiated, and heat treated _S. boulardii_, and in addition, DC co-stimulatory functions were upregulated to an equal extent by live and reproductively unable yeast.

The remarkable consistency in DC cytokine inducing properties observed across very diverse yeast species and genera regardless of viability appears to support our hypothesis that interactions between yeasts and immune cells are likely to rely upon contact between (possibly conserved) yeast cell wall structures and DC surface receptors. Further, our findings suggest that this phenomenon holds true for a number of diverse yeast species beyond the widely studied _Saccharomyces_ genus. In this context, it is interesting to note that a recent study have found yeast cell wall fractions, and particularly β-glucan fractions, prepared from either _S. cerevisiae_ or _C. albicans_, able to modulate intestinal inflammation _in vivo_ [40], potentially through interactions with the C-type lectin receptor Dectin-1 expressed by DCs [61,62]. Thus, while our findings for nonviable yeasts beyond _S. cerevisiae_ and _C. albicans_ appear to extend their observation that yeast cell wall molecules are capable of interacting with immune cells, their findings support our hypothesis that yeast modulation of intestinal inflammation can take place without the need for live yeast cells.

In conclusion, the present study provides the first large-scale study of immune modulation by highly diverse yeasts described in

A

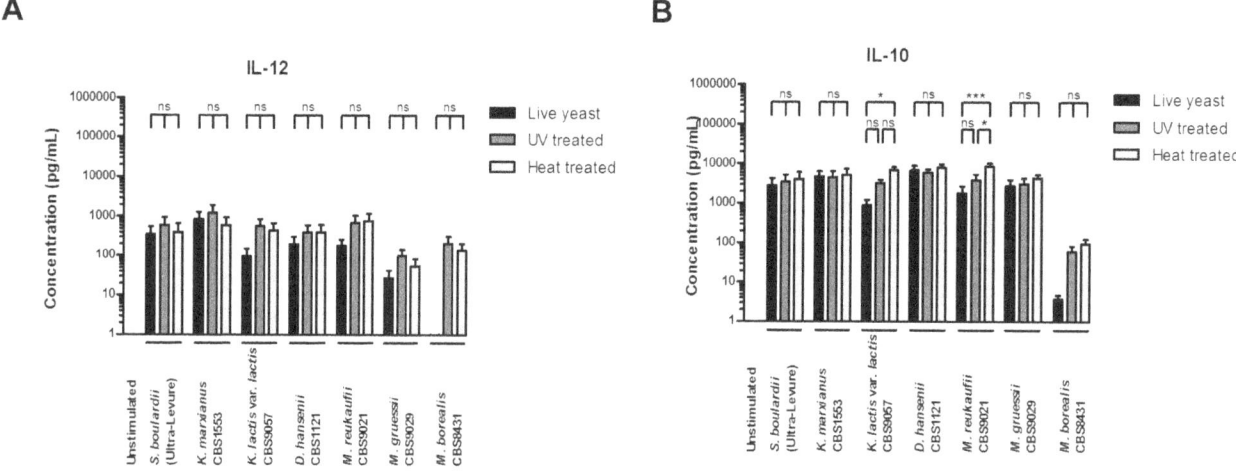

B

C

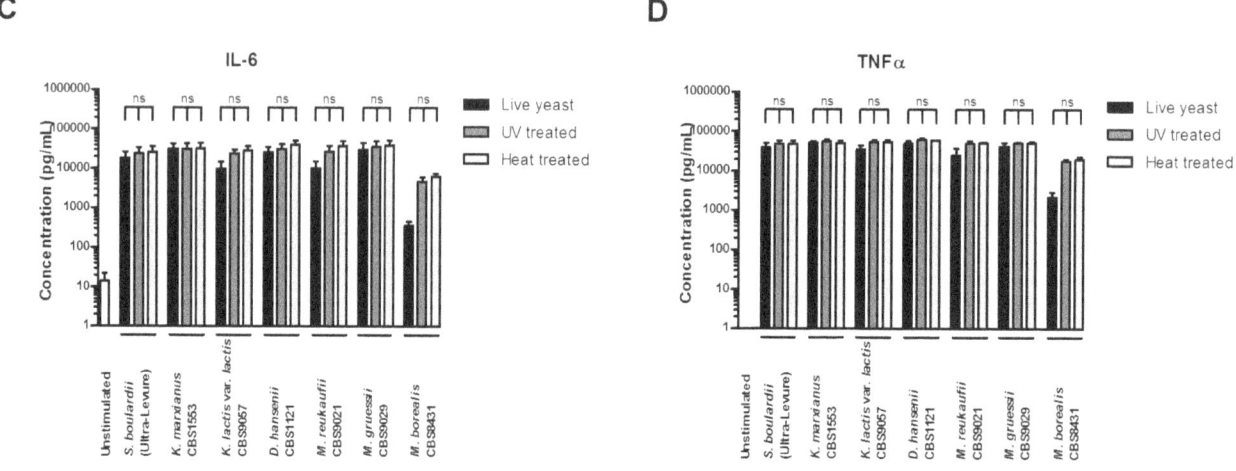

D

E

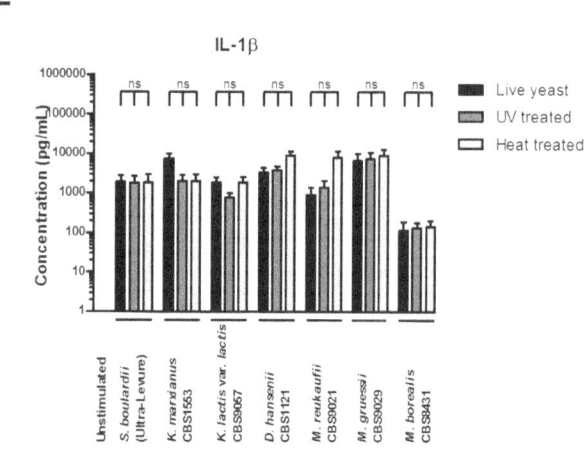

Figure 5. Yeast induced DC cytokine secretion occurs independently of yeast metabolic function. Levels of **A.** IL-12, **B.** IL-10, **C.** IL-6, **D.** TNFα, and **E.** IL-1β secreted by human monocyte-derived DCs following 20 h stimulation with DC media containing 10 % glycerol (unstimulated) or *S. boulardii* (Ultra-Levure), *K. marxianus* (CBS1553), *K. lactis* var. *lactis* (CBS9057), *D. hansenii* (CBS1121), *M. reukaufii* (CBS9021), *M. gruessii* (CBS9029), or *M. borealis* (CBS8431). For each yeast strain, DC stimulation was performed with live yeast, UV irradiated yeast, and heat treated yeast at a yeast:DC ratio of 10:1. Data are expressed as mean±SEM (n = 4). One-way ANOVA, Bonferroni's multiple comparison post test, indicating significant differences between cytokine levels induced by live, UV irradiated, and heat treated yeast for each strain. ns, not significant; *, P<0.05; **, P<0.01; ***, P<0.001.

the scientific literature. Our data clearly demonstrate high diversity in yeast induced cytokine secretion across a broad spectrum of yeasts, and by employing multivariate data analysis we reveal distinct clustering of yeasts inducing similar cytokine profiles in DCs, highlighting clear species distinction within specific yeast genera. The observed differences in induced DC cytokine profiles may indicate distinct modes of interaction between yeasts and human immune cells, and will aid in the selection of strains for further characterization and development toward potentially novel yeast probiotics. Additionally, we present data to support a hypothesis that the interaction between yeasts and human DCs does not solely depend on yeast viability, a concept which may suggest a need for further classifications beyond the current definition of a probiotic.

Supporting Information

Figure S1 Time-course of *S. boulardii* induced DC cytokine secretion supports a 20 h stimulation time. Levels of IL-12, IL-10, IL-1β, IL-6, and TNFα secreted by human monocyte-derived DCs following incubation with *S. boulardii* (Ultra-Levure) at a yeast:DC ratio of 10:1. Data are expressed as mean±SEM (n = 2).

Figure S2 Modulation of DC co-stimulatory functions and chemokine receptor expression occurs independently of yeast metabolic function. DC surface expression of CD80, CD86, CCR6, and CCR7 following 20 h stimulation with DC media containing 10% glycerol (unstimulated) or either live, UV irradiated, or heat killed *S. boulardii* (Ultra-Levure) at a yeast:DC ratio of 10:1. Data are expressed as mean±SEM (n = 4). One-way ANOVA, Bonferroni's multiple comparison post test, indicating significant differences between cytokine levels induced by live, UV treated, and heat killed yeast. ns, not significant; *, P< 0.05; **, P<0.01; ***, P<0.001.

Acknowledgments

We gratefully acknowledge Amparo Gamero Lluna for performing the initial selection of strains included in this study, Stina Rikke Jensen for valuable cell culture advice, Jeanne Olsen for excellent technical assistance, and Jannik Vindeløv for guidance on multivariate data analysis.

Author Contributions

Conceived and designed the experiments: IMS JEC NA IJ. Performed the experiments: IMS JEC. Analyzed the data: IMS JEC. Contributed reagents/materials/analysis tools: IMS JEC. Wrote the paper: IMS.

References

1. Coombes JL, Powrie F (2008) Dendritic cells in intestinal immune regulation. Nat Rev Immunol 8: 435–446.
2. Rescigno M, Urbano M, Valzasina B, Francolini M, Rotta G, et al. (2001) Dendritic cells express tight junction proteins and penetrate gut epithelial monolayers to sample bacteria. Nat Immunol 2: 361–367.
3. Maynard CL, Elson CO, Hatton RD, Weaver CT (2012) Reciprocal interactions of the intestinal microbiota and immune system. Nature 489: 231–241.
4. Hooper LV, Littman DR, Macpherson AJ (2012) Interactions between the microbiota and the immune system. Science 336: 1268–1273.
5. Manicassamy S, Ravindran R, Deng J, Oluoch H, Denning TL, et al. (2009) Toll-like receptor 2-dependent induction of vitamin A-metabolizing enzymes in dendritic cells promotes T regulatory responses and inhibits autoimmunity. Nat Med 15: 401–409.
6. Donkor ON, Ravikumar M, Proudfoot O, Day SL, Apostolopoulos V, et al. (2012) Cytokine profile and induction of T helper type 17 and regulatory T cells by human peripheral mononuclear cells after microbial exposure. Clin Exp Immunol 167: 282–295.
7. Joint FAO/WHO Working Group (2002) Guidelines for the evaluation of probiotics in food.
8. Foligne B, Zoumpopoulou G, Dewulf J, Ben Younes A, Chareyre F, et al. (2007) A key role of dendritic cells in probiotic functionality. PLoS One 2: e313.
9. Kwon HK, Lee CG, So JS, Chae CS, Hwang JS, et al. (2010) Generation of regulatory dendritic cells and CD4+Foxp3+ T cells by probiotics administration suppresses immune disorders. Proc Natl Acad Sci U S A 107: 2159–2164.
10. Baba N, Samson S, Bourdet-Sicard R, Rubio M, Sarfati M (2008) Commensal bacteria trigger a full dendritic cell maturation program that promotes the expansion of non-Tr1 suppressor T cells. J Leukoc Biol 84: 468–476.
11. Konieczna P, Groeger D, Ziegler M, Frei R, Ferstl R, et al. (2012) Bifidobacterium infantis 35624 administration induces Foxp3 T regulatory cells in human peripheral blood: Potential role for myeloid and plasmacytoid dendritic cells. Gut 61: 354–366.
12. Jeon SG, Kayama H, Ueda Y, Takahashi T, Asahara T, et al. (2012) Probiotic bifidobacterium breve induces IL-10-producing Tr1 cells in the colon. PLoS Pathog 8: e1002714.
13. Di Giacinto C, Marinaro M, Sanchez M, Strober W, Boirivant M (2005) Probiotics ameliorate recurrent Th1-mediated murine colitis by inducing IL-10 and IL-10-dependent TGF-beta-bearing regulatory cells. J Immunol 174: 3237–3246.
14. Mann ER, Landy JD, Bernardo D, Peake ST, Hart AL, et al. (2013) Intestinal dendritic cells: Their role in intestinal inflammation, manipulation by the gut microbiota and differences between mice and men. Immunol Lett.
15. Christensen HR, Frokiaer H, Pestka JJ (2002) Lactobacilli differentially modulate expression of cytokines and maturation surface markers in murine dendritic cells. J Immunol 168: 171–178.
16. Weiss G, Christensen HR, Zeuthen LH, Vogensen FK, Jakobsen M, et al. (2011) Lactobacilli and bifidobacteria induce differential interferon-beta profiles in dendritic cells. Cytokine 56: 520–530.
17. Plantinga TS, van Bergenhenegouwen J, Jacobs C, Joosten LA, van't Land B, et al. (2012) Modulation of toll-like receptor ligands and candida albicans-induced cytokine responses by specific probiotics. Cytokine 59: 159–165.
18. Qin J, Li R, Raes J, Arumugam M, Burgdorf KS, et al. (2010) A human gut microbial gene catalogue established by metagenomic sequencing. Nature 464: 59–65.
19. Clemente JC, Ursell LK, Parfrey LW, Knight R (2012) The impact of the gut microbiota on human health: An integrative view. Cell 148: 1258–1270.
20. Scanlan PD, Marchesi JR (2008) Micro-eukaryotic diversity of the human distal gut microbiota: Qualitative assessment using culture-dependent and -independent analysis of faeces. ISME J 2: 1183–1193.
21. Ghannoum MA, Jurevic RJ, Mukherjee PK, Cui F, Sikaroodi M, et al. (2010) Characterization of the oral fungal microbiome (mycobiome) in healthy individuals. PLoS Pathog 6: e1000713.
22. Hatoum R, Labrie S, Fliss I (2012) Antimicrobial and probiotic properties of yeasts: From fundamental to novel applications. Front Microbiol 3: 421.
23. Thomas S, Przesdzing I, Metzke D, Schmitz J, Radbruch A, et al. (2009) Saccharomyces boulardii inhibits lipopolysaccharide-induced activation of human dendritic cells and T cell proliferation. Clin Exp Immunol 156: 78–87.
24. Foligne B, Dewulf J, Vandekerckove P, Pignede G, Pot B (2010) Probiotic yeasts: Anti-inflammatory potential of various non-pathogenic strains in experimental colitis in mice. World J Gastroenterol 16: 2134–2145.
25. van der Aa Kuhle A, Jespersen L (2003) The taxonomic position of saccharomyces boulardii as evaluated by sequence analysis of the D1/D2 domain of 26S rDNA, the ITS1-5.8S rDNA-ITS2 region and the mitochondrial cytochrome-c oxidase II gene. Syst Appl Microbiol 26: 564–571.
26. Vaughan-Martini A, Martini A (2011) *Saccharomyces* Meyen ex reess (1870). In: Kurtzman CP, Fell JW, Boekhout T, editors. The Yeasts: A Taxonomic Study. London, UK: Elsevier. pp. 733.
27. McFarland LV (2010) Systematic review and meta-analysis of saccharomyces boulardii in adult patients. World J Gastroenterol 16: 2202–2222.

28. Dinleyici EC, Eren M, Ozen M, Yargic ZA, Vandenplas Y (2012) Effectiveness and safety of saccharomyces boulardii for acute infectious diarrhea. Expert Opin Biol Ther 12: 395–410.

29. Zanello G, Berri M, Dupont J, Sizaret PY, D'Inca R, et al. (2011) Saccharomyces cerevisiae modulates immune gene expressions and inhibits ETEC-mediated ERK1/2 and p38 signaling pathways in intestinal epithelial cells. PLoS One 6: e18573.

30. Chen X, Kokkotou EG, Mustafa N, Bhaskar KR, Sougioultzis S, et al. (2006) Saccharomyces boulardii inhibits ERK1/2 mitogen-activated protein kinase activation both in vitro and in vivo and protects against clostridium difficile toxin A-induced enteritis. J Biol Chem 281: 24449–24454.

31. Dahan S, Dalmasso G, Imbert V, Peyron JF, Rampal P, et al. (2003) Saccharomyces boulardii interferes with enterohemorrhagic escherichia coli-induced signaling pathways in T84 cells. Infect Immun 71: 766–773.

32. Martins FS, Dalmasso G, Arantes RM, Doye A, Lemichez E, et al. (2010) Interaction of saccharomyces boulardii with salmonella enterica serovar typhimurium protects mice and modifies T84 cell response to the infection. PLoS One 5: e8925.

33. Lee SK, Kim YW, Chi SG, Joo YS, Kim HJ (2009) The effect of saccharomyces boulardii on human colon cells and inflammation in rats with trinitrobenzene sulfonic acid-induced colitis. Dig Dis Sci 54: 255–263.

34. Sougioultzis S, Simeonidis S, Bhaskar KR, Chen X, Anton PM, et al. (2006) Saccharomyces boulardii produces a soluble anti-inflammatory factor that inhibits NF-kappaB-mediated IL-8 gene expression. Biochem Biophys Res Commun 343: 69–76.

35. Thomas S, Metzke D, Schmitz J, Dorffel Y, Baumgart DC (2011) Anti-inflammatory effects of saccharomyces boulardii mediated by myeloid dendritic cells from patients with crohn's disease and ulcerative colitis. Am J Physiol Gastrointest Liver Physiol.

36. Gad M, Ravn P, Soborg DA, Lund-Jensen K, Ouwehand AC, et al. (2011) Regulation of the IL-10/IL-12 axis in human dendritic cells with probiotic bacteria. FEMS Immunol Med Microbiol 63: 93–107.

37. Pothoulakis C (2009) Review article: Anti-inflammatory mechanisms of action of saccharomyces boulardii. Aliment Pharmacol Ther 30: 826–833.

38. Martins FS, Nardi RM, Arantes RM, Rosa CA, Neves MJ, et al. (2005) Screening of yeasts as probiotic based on capacities to colonize the gastrointestinal tract and to protect against enteropathogen challenge in mice. J Gen Appl Microbiol 51: 83–92.

39. Dalmasso G, Cottrez F, Imbert V, Lagadec P, Peyron JF, et al. (2006) Saccharomyces boulardii inhibits inflammatory bowel disease by trapping T cells in mesenteric lymph nodes. Gastroenterology 131: 1812–1825.

40. Jawhara S, Habib K, Maggiotto F, Pignede G, Vandekerckove P, et al. (2012) Modulation of intestinal inflammation by yeasts and cell wall extracts: Strain dependence and unexpected anti-inflammatory role of glucan fractions. PLoS One 7: e40648.

41. Kumura H, Tanoue Y, Tsukahara M, Tanaka T, Shimazaki K (2004) Screening of dairy yeast strains for probiotic applications. J Dairy Sci 87: 4050–4056.

42. Pedersen LL, Owusu-Kwarteng J, Thorsen L, Jespersen L (2012) Biodiversity and probiotic potential of yeasts isolated from fura, a west african spontaneously fermented cereal. Int J Food Microbiol 159: 144–151.

43. Romanin D, Serradell M, Gonzalez Maciel D, Lausada N, Garrote GL, et al. (2010) Down-regulation of intestinal epithelial innate response by probiotic yeasts isolated from kefir. Int J Food Microbiol 140: 102–108.

44. Saegusa S, Totsuka M, Kaminogawa S, Hosoi T (2007) Cytokine responses of intestinal epithelial-like caco-2 cells to non-pathogenic and opportunistic pathogenic yeasts in the presence of butyric acid. Biosci Biotechnol Biochem 71: 2428–2434.

45. Reyes-Becerril M, Salinas I, Cuesta A, Meseguer J, Tovar-Ramirez D, et al. (2008) Oral delivery of live yeast debaryomyces hansenii modulates the main innate immune parameters and the expression of immune-relevant genes in the gilthead seabream (sparus aurata L.). Fish Shellfish Immunol 25: 731–739.

46. Macey BM, Coyne VE (2006) Colonization of the gastrointestinal tract of the farmed south african abalone haliotis midae by the probionts vibrio midae SY9, cryptococcus sp. SS1, and debaryomyces hansenii AY1. Mar Biotechnol (NY) 8: 246–259.

47. Maccaferri S, Klinder A, Brigidi P, Cavina P, Costabile A (2012) Potential probiotic kluyveromyces marxianus B0399 modulates the immune response in caco-2 cells and peripheral blood mononuclear cells and impacts the human gut microbiota in an in vitro colonic model system. Appl Environ Microbiol 78: 956–964.

48. Kourelis A, Kotzamanidis C, Litopoulou-Tzanetaki E, Papaconstantinou J, Tzanetakis N, et al. (2010) Immunostimulatory activity of potential probiotic yeast strains in the dorsal air pouch system and the gut mucosa. J Appl Microbiol 109: 260–271.

49. Weiss G, Rasmussen S, Nielsen Fink L, Jarmer H, Nohr Nielsen B, et al. (2010) Bifidobacterium bifidum actively changes the gene expression profile induced by lactobacillus acidophilus in murine dendritic cells. PLoS One 5: e11065.

50. Zeuthen LH, Fink LN, Frokiaer H (2008) Toll-like receptor 2 and nucleotide-binding oligomerization domain-2 play divergent roles in the recognition of gut-derived lactobacilli and bifidobacteria in dendritic cells. Immunology 124: 489–502.

51. Baba N, Samson S, Bourdet-Sicard R, Rubio M, Sarfati M (2009) Selected commensal-related bacteria and toll-like receptor 3 agonist combinatorial codes synergistically induce interleukin-12 production by dendritic cells to trigger a T helper type 1 polarizing programme. Immunology 128: e523–31.

52. Zeuthen LH, Fink LN, Frokiaer H (2008) Epithelial cells prime the immune response to an array of gut-derived commensals towards a tolerogenic phenotype through distinct actions of thymic stromal lymphopoietin and transforming growth factor-beta. Immunology 123: 197–208.

53. Carson RT, Vignali DA (1999) Simultaneous quantitation of 15 cytokines using a multiplexed flow cytometric assay. J Immunol Methods 227: 41–52.

54. Vignali DA (2000) Multiplexed particle-based flow cytometric assays. J Immunol Methods 243: 243–255.

55. Ruiz PA, Hoffmann M, Szcesny S, Blaut M, Haller D (2005) Innate mechanisms for bifidobacterium lactis to activate transient pro-inflammatory host responses in intestinal epithelial cells after the colonization of germ-free rats. Immunology 115: 441–450.

56. Wall R, Marques TM, O'Sullivan O, Ross RP, Shanahan F, et al. (2012) Contrasting effects of bifidobacterium breve NCIMB 702258 and bifidobacterium breve DPC 6330 on the composition of murine brain fatty acids and gut microbiota. Am J Clin Nutr 95: 1278–1287.

57. Smelt MJ, de Haan BJ, Bron PA, van Swam I, Meijerink M, et al. (2012) L. plantarum, L. salivarius, and L. lactis attenuate Th2 responses and increase treg frequencies in healthy mice in a strain dependent manner. PLoS One 7: e47244.

58. Wells JM (2011) Immunomodulatory mechanisms of lactobacilli. Microb Cell Fact 10 Suppl 1: S17-2859-10-S1-S17. Epub 2011 Aug 30.

59. Suzuki M, Prasad GS, Kurtzman CP (2011) DebaryomycesLodder & kreger-van rij (1952). In: Kurtzman CP, Fell JW, Boekhout T, editors. The Yeasts: A Taxonomic Study. London, UK: Elsevier. pp. 361.

60. EFSA Panel on Biological Hazards (BIOHAZ) (2010) Scientific opinion on the maintenance of the list of QPS biological agents intentionally added to food and feed (2010 update). EFSA Journal 8: 1944.

61. Iliev ID, Funari VA, Taylor KD, Nguyen Q, Reyes CN, et al. (2012) Interactions between commensal fungi and the C-type lectin receptor dectin-1 influence colitis. Science 336: 1314–1317.

62. Brown GD (2006) Dectin-1: A signalling non-TLR pattern-recognition receptor. Nat Rev Immunol 6: 33–43.

16

Probiotics VSL#3 Protect against Development of Visceral Pain in Murine Model of Irritable Bowel Syndrome

Eleonora Distrutti[1]*, **Sabrina Cipriani**[2], **Andrea Mencarelli**[2], **Barbara Renga**[2], **Stefano Fiorucci**[2]

1 S.C. di Gastroenterologia ed Epatologia, Azienda Ospedaliera di Perugia, Perugia, Italy, 2 Dipartimento di Medicina Clinica e Sperimentale, Università degli Studi di Perugia, Perugia, Italy

Abstract

Background and Aims: Irritable bowel syndrome (IBS) is linked to post-inflammatory and stress-correlated factors that cause changes in the perception of visceral events. Probiotic bacteria may be effective in treating IBS symptoms. Here, we have investigated whether early life administration of VSL#3, a mixture of 8 probiotic bacteria strains, protects against development of visceral hypersensitivity driven by neonatal maternal separation (NMS), a rat model of IBS.

Methods: Male NMS pups were treated orally with placebo or VSL#3 from days 3 to 60, while normal, not separated rats were used as controls. After 60 days from birth, perception of painful sensation induced by colorectal distension (CRD) was measured by assessing the abdominal withdrawal reflex (score 0–4). The colonic gene expression was assessed by using the Agilent Whole Rat Genome Oligo Microarrays platform and confirmed by real time PCR.

Results: NMS rats exhibited both hyperalgesia and allodynia when compared to control rats. VSL#3 had a potent analgesic effect on CRD-induced pain without changing the colorectal compliance. The microarray analysis demonstrated that NMS induces a robust change in the expression of subsets of genes (CCL2, NOS3, THP1, NTRK1, CCR2, BDRKRB1, IL-10, TNFRSF1B, TRPV4, CNR1 and OPRL1) involved in pain transmission and inflammation. TPH1, tryptophan hydroxylase 1, a validated target gene in IBS treatment, was markedly upregulated by NMS and this effect was reversed by VSL#3 intervention.

Conclusions: Early life administration of VSL#3 reduces visceral pain perception in a model of IBS and resets colonic expression of subsets of genes mediating pain and inflammation.

Transcript profiling: Accession number of repository for expression microarray data is GSE38942 (http://www.ncbi.nlm.nih.gov/geo/query/acc.cgi?acc=GSE38942).

Editor: Yvette Tache, University of California, Los Angeles, United States of America

Funding: This study was supported in part by CDS investments (Italy). The funders had no role in study design, data collection and analysis, decision to publish, or preparation of the manuscript.

Competing Interests: The authors have declared that no competing interests exist.

* E-mail: eleonoradistrutti@katamail.com

Introduction

Irritable bowel syndrome (IBS) is a disorder characterized by chronic abdominal pain and discomfort associated with alterations in bowel habits in the absence of a demonstrable pathology [1]. Alterations in bowel habits are likely related to dysregulation of autonomic system in the gut, whereas symptoms of abdominal pain and discomfort are thought to involve additional changes in the perception of visceral events, in the form of hyperalgesia or allodynia [2–3]. Evidence is growing to support the notion that IBS might be a post-inflammatory and stress-correlated condition [4–5] and chronic gut inflammatory processes are thought to play a role in its pathogenesis.

The neonatal maternal separation model (NMS) is an early life stress experience that resets the expression of neurotransmitters, receptors and neurotransporters in the central nervous system (CNS) and predisposes adult rats to develop hyperalgesia to nociceptive visceral stimuli [6–9]. The altered physiological responses and visceral hyperalgesia of NMS rats are consistent with changes observed in IBS patients [6] making the NMS model a useful tool to investigate the pathophysiological mechanism of visceral hypersensitivity in functional gastroenterological disorders [10].

The intestinal microbiota plays essential roles in nutrient absorption and metabolism, immune stimulation, satiety and pain. An altered composition of intestinal microbiota has been reported in IBS patients [11–12] while its modification by probiotic diet reduces visceral hypersensitivity in experimental models of abdominal pain by modulating neural functions [13–16]. The probiotic VSL#3 is a mixture of 8 Gram-positive bacteria strains [17]. VSL#3 improves the outcome of patients with chronic intestinal inflammation [18–22], ameliorates abdominal bloating [23], reduces flatulence scores and delays colonic transit without altering bowel function in IBS patients [24] and children [25]. Nevertheless, the mechanism that underlies the beneficial effects of VSL#3 in these settings is still poorly defined.

In the present study we have made an attempt to identify mechanisms involved in beneficial effects exerted by VSL#3 intervention in a model of visceral hypersensitivity induced in rats by NMS by using a global gene expression analysis. Our results indicates that NMS causes both allodynia and hyperalgesia and influences the expression of a wide array of genes, including genes known to mediates inflammation and pain. Our results also demonstrate that VSL#3 intervention was effective in both reverting NMS-induced visceral hypersensitivity and resetting the complex network of genes involved in inflammation and pain.

Materials and Methods

Animals

Male, Wistar rats (200–250 g, Charles River, Monza, Italy) were housed in plastic cages and maintained under controlled conditions with 12-hour light/dark cycles (lights on at 07.00). Tap water and standard laboratory chow were freely available. Food was withheld for 12 hours before CRD recordings. All the animals were individually trained by spending 2–3 hours per day in a Plexiglas cage for 2–3 days. This allowed them to adjust to a movement-restriction environment similar to that adopted during the distending procedure. All experimental procedures described below were approved by the institutional animal research committees of University of Perugia (Permit Number: 98/2010-B) and were in accordance with nationally approved guidelines for the treatment of laboratory animals. All experiments were performed in conscious rats and conducted in a blind manner in that the observers were not aware of the identity of drugs administered to each animal.

The Neonatal Maternal Separation Model

A neonatal maternal separation (NMS)-induced visceral hyperalgesia rat model has been previously established [6]. Because its characteristics mimic the symptoms of IBS patients, it is often used to study the mechanism of visceral hyperalgesia and to evaluate the pharmacological effects of potential IBS therapies [10], [26]. Briefly, pups in the NMS group were separated from their mothers and placed into individual cages in another room 180 min daily from postnatal day 2 to day 14, whereas normally-handled (NH) pups remained undisturbed in their home cage with the dam. All pups were weaned on postnatal day 22, and only male pups were used in the present study to avoid hormonal cycle induced variations. Male rats on postnatal day 60 were used in a series of CRD experiments.

Table 2. Sense and antisense probes for genes related to "Pain" annotation.

Oligo Name	Sense	Antisense
rCCL2	atgcagttaatgccccactc	ttccttattggggtcagcac
rNOS3	caatcttcgttcagccatca	gggtccagccatgttgaata
rNTRK1	gtctggtgggtcagggacta	cacacatcactctcggtgct
rTPH1	gacatctttcccctgctgaa	tctttgaagccaggatggtc
rCCR2	ctgcccctacttgtcatggt	ggcctggtctaagtgcatgt
rBDKRB1	ccccgtgactgctatcatct	agaccaggaaggaggctacc
rIL10	ggagtgaagaccagcaaagg	ggcaacccaagtaaccctta
rTNFRSF1B	ggctcagatgtgctgtgcta	atgcagatggttccagacct
rTRPV4	cgatatgaggcgacaggact	gggagcacttgagaagcaac
rCNR1	agagcatcatcatccacacg	tcaacaccaccaggatcaga
rOPLR1	aagagatcgagtgcctggtg	agcacagggatgatgaagga

Experimental Design - Effects of VSL#3 on Colonic Nociception and Compliance

Rats were divided in 7 groups of 5 animals each. All rats, except those of Group 1 (healthy, intact) and Group 3 (NMS, intact), were treated daily by gavage with placebo or VSL#3 at the dose of 17 billions in 100 µL saline according to the experimental design (Table 1) [27]. Placebo was administered only to animals of Group 2 that corresponds to healthy animals in which CRD was performed and Group 4 that corresponds to NMS rats in which CRD was performed. Group 5, Group 6 and Group 7 included NMS rats treated with VSL#3 from day 3 to day 60, from 3 to day 15 and from day 45 to day 60 respectively (Table 1). Microarray studies were conducted only in Group 2 (healthy, CRD, named Group C), Group 4 (NMS, CRD, named Group M) and Group 5 (NMS, VSL#3 administered from day 3 to day 60, CRD, named Group D). At the end of the studies, animals were sacrified and colon, blood and spinal cord collected for further determinations.

CRD and Behavioral Testing

The distending protocol was performed as previously described [28–32]. The night before CRD experiments, the balloons (7–8 mm diameter) were inflated and left overnight so that the latex stretched and the balloons became compliant. On the testing day, each rat was sedated with ether inhalation and the latex balloon (1.5 cm long) and the probe catheter (0.5 cm) was inserted

Table 1. Experimental design.

Name of the group	Abbreviation	Treatment
Normal group	Group 1	Healthy, intact
Control group	Group 2, named Group C*	Healthy, placebo, CRD
NMS group	Group 3	NMS, intact
Neonatal-Maternal separation, NMS group	Group 4, named Group M*	NMS, placebo, CRD
Probiotic Diet, NMS+VSL#3 (day 3–60) group	Group 5, named Group D*	NMS, VSL#3 from day 3 to day 60, CRD
NMS+VSL#3 (day 3–15) group	Group 6	NMS, VSL#3 from day 3 to day 15, CRD
NMS+VSL#3 (day 45–60) group	Group 7	NMS, VSL#3 from day 45 to day 60, CRD

*Groups in which microarray analysis was performed.

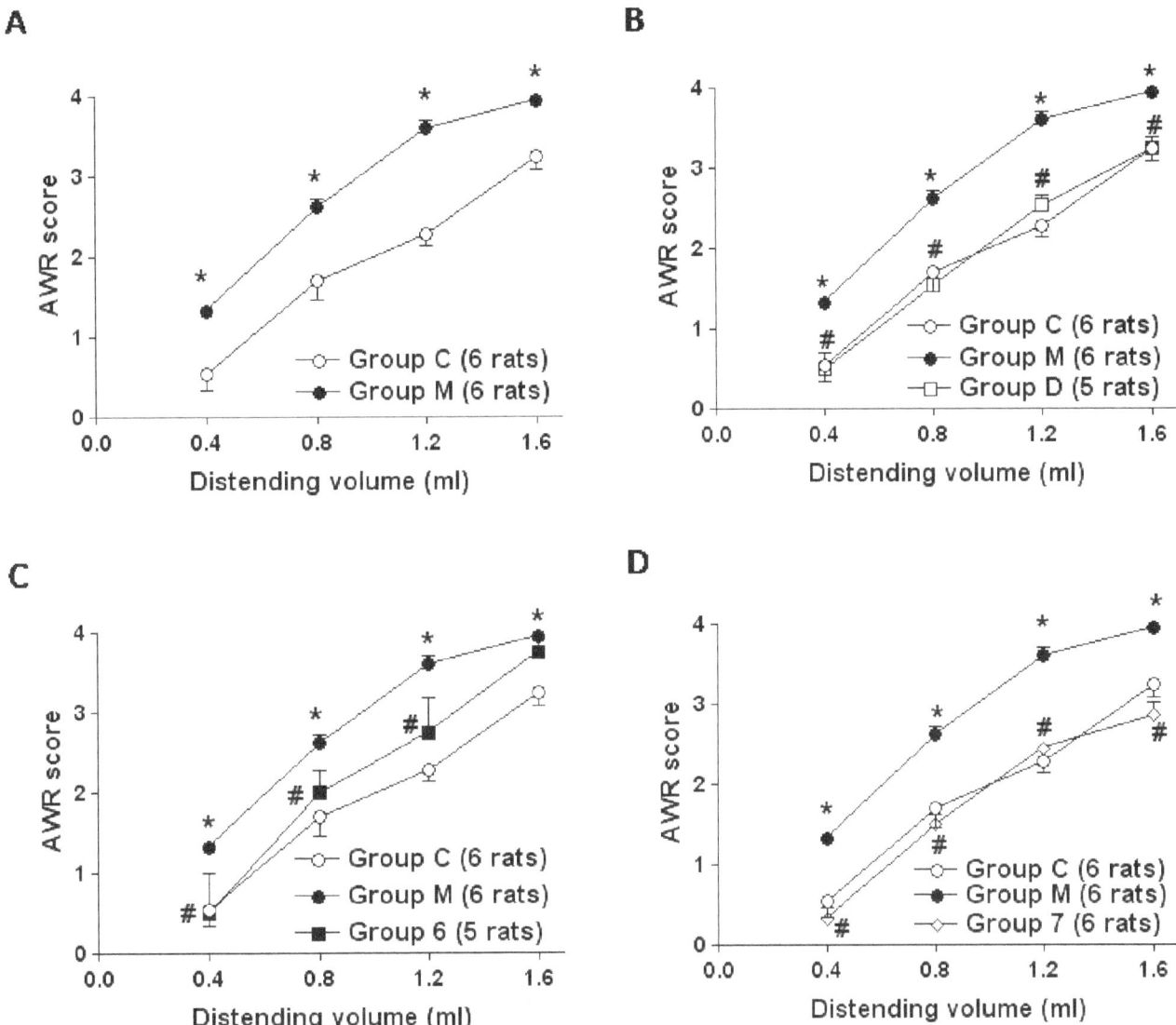

Figure 1. VSL#3 reverses hypersensitivity induced by NMS. (A) Group C and Group M are the same in all comparison experiments. In healthy animals (Group C, 6 rats) CRD (0.4–1.6 ml water) elicited a volume-dependent increase of the AWR scores while caused allodynia and hyperalgesia in NMS animals (Group M, 6 rats). (B, D) VSL#3 administered from day 3 to day 60 (Group D, 5 rats) and from day 45 to day 60 (Group 7, 6 rats) restored normal sensitivity. (C) Probiotic intervention from day 3 to day 15 (Group 6, 5 rats) was only partially effective in visceral sensitivity as hyperalgesia persisted. Statistical comparisons were performed by the Mann-Whitney test for unpaired data and by the Wilcoxon signed rank test for paired data when two group of data were analyzed, and by the ANOVA for non parametric data, Kruskal-Wallis followed by Dunns comparison of selected pairs of column, when more than two groups of data were analyzed. *$p < 0.05$ vs Group C; #$p < 0.05$ vs Group M.

intrarectally and fixed at the base of the tail. The balloon was connected via a double barreled cannula to a pressure transducer to continuously monitoring the colorectal pressure by a computer (PowerLab PC, A.D. Instruments, Milford, MA, USA) and to a syringe for inflation/deflation of the balloon. The rats were then housed in a small Plexiglas cage ($20 \times 8 \times 8$ cm) on an elevated platform and allowed to regain consciousness and adapted for 1 hour. After recovery from sedation, the rats underwent the CRD procedure and behavioral response was tested in all groups except groups 1 and 3 in which no CRD was performed. Infusion of water was performed by hands. CRD of 20 seconds performed every 5 minutes was applied in increment of 0.4 ml starting from 0.4 ml and increasing to 1.6 ml water. To achieve an accurate measurement of the colonic parameters and perception, each distension was repeated twice and data were averaged for analysis.

Behavioral responses and colonic parameters collected during the first and the second sets of CRD were assessed and compared among all groups [28–32]. The behavioral response to CRD was assessed by measuring the abdominal withdrawal reflex (AWR) using a semiquantitative scoring system [33–34]. The AWR is an involuntary motor reflex similar to the visceromotor reflex, but it has the great advantage that the latter requires abdominal surgery to implant recording electrodes and wires in the abdominal muscle wall, which may cause additional sensitization [34]. Measurement of the AWR consists of visual observation of the rat's response to graded CRD by a blinded observer and assignment of an AWR score according with the behavioral scale previously described [33], in which grade 0 corresponds to no behavioral response to CRD, grade 1 corresponds to brief head movement at the onset of the stimulus followed by immobility, grade 2 corresponds to a mild

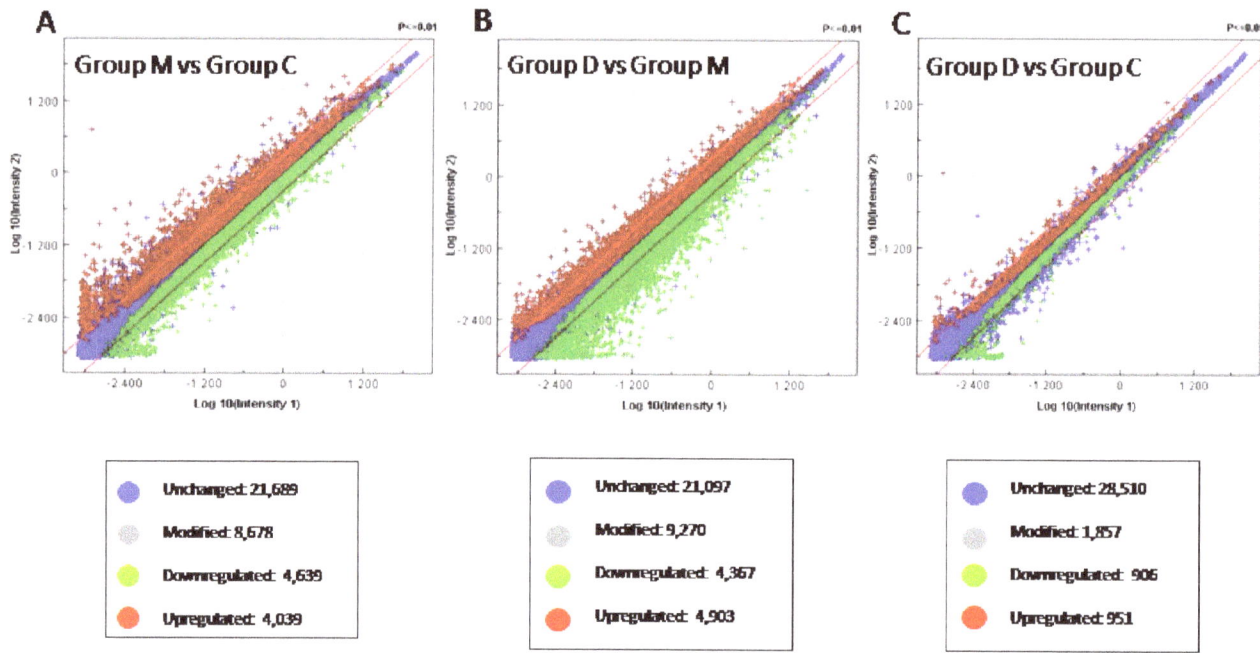

Figure 2. VSL#3 reverses the NMS-induced alteration of gene expression: global microarray analysis. (A) Expression of 8,678 (28.57% of total) genes was globally modified in NMS animals compared with normal rats (Group M vs Group C): 4,039 (13.29%) were upregulated and 4,639 (15.27%) were downregulated. (B) VSL#3 effectively resected gene expression in NMS rats (Group D vs Group M): 9,270 (30.5% of total) genes were globally modified, 4,903 (16.14 of total%) were upregulated, while 4,367 (14.38% of total) were downregulated. (C) VSL#3 administration to NMS rats caused a pattern of gene expression that was similar to that of control rats (Group D vs Group C). Microarray data from 3–4 replicates.

contraction of abdominal muscles although the rat does not lift the abdomen off the platform, grade 3 corresponds to a strong contraction of the abdominal muscles with the lifting of the abdomen off the platform, and grade 4 corresponds to a severe contraction of the abdominal muscles manifested by body arching and the lifting of the abdomen and of the pelvic structures and scrotum. The rats that did not show any behavioral response (i.e. score 0) were excluded. To determine the effect of placebo or VSL#3 on colonic smooth muscle, the compliance of the colon during CRD was obtained from colorectal volume and pressure and expressed as ml/mmHg [28–32].

Microarray Analysis

Microarray analysis was performed on colonic samples from Group C, Group M and Group D. The same segment was taken in each animal starting approximaly 2 cm from the anus. The data discussed in this publication have been deposited in NCBI's Gene Expression Omnibus (GEO) and are accessible through GEO Series accession number GSE38942 (http://www.ncbi.nlm.nih.gov/geo/query/acc.cgi?acc=GSE38942) [35–36].

All microarray analysis were performed by Miltenyi Biotec, GmbH Bioinformatics, German.

RNA extraction. The RNA was isolated from rat tissue samples by using standard RNA extraction protocols (Trizol), the RNA quality-checked via the Agilent 2100 Bioanalyzer platform (Agilent Technologies) from the following treatment groups:

Group C = healthy rats+CRD

Group M = NMS rats+CRD

Group D = NMS rats+probiotic Diet from day 3 to day 60+CRD

For each condition, four biological replicates exist.

Linear T7-based amplification of RNA. For the linear T7-based amplification step, 100 ng of each total RNA sample was used. To produce Cy3-labeled cRNA, the RNA samples were amplified and labeled using the Agilent Low Input Quick Amp Labeling Kit (Agilent Technologies) following the manufacturer's protocol. Yields of cRNA and the dye-incorporation rate were measured with the ND-1000 Spectrophotometer (NanoDrop Technologies).

Hybridization of agilent whole genome oligo microarrays. The hybridization procedure was performed according to the Agilent 60-mer oligo microarray processing protocol using the Agilent Gene Expression Hybridization Kit (Agilent Technologies). Briefly, 0.6 μg Cy3-labeled fragmented cRNA in hybridization buffer was hybridized overnight (17 hours, 65°C) to Agilent Whole Rat Genome Oligo Microarrays 8×60 K using Agilent's recommended hybridization chamber and oven.

Scanning results. Fluorescence signals of the hybridized Agilent Microarrays were detected using Agilent's Microarray Scanner System (Agilent Technologies).

Image and data analysis. The Agilent Feature Extraction Software (FES) was used to read out and process the microarray image files. The software determines feature intensities (including background subtraction), rejects outliers and calculates statistical confidences. For determination of differential gene expression FES derived output data files were further analyzed using the Rosetta Resolver gene expression data analysis system (Rosetta Biosoftware). All samples were labelled with Cy3, here, the ratio experiments are designated as control versus (vs) sample experiments (automated data output of the Resolver system).

The ratios are always calculated by dividing sample signal intensity through control signal intensity.

The bioinformatics data analysis of eleven microarray datasets obtained from one-color hybridization of rat RNAs on Agilent Whole Rat Genome Oligo Microarrays 8×60 K was performed.

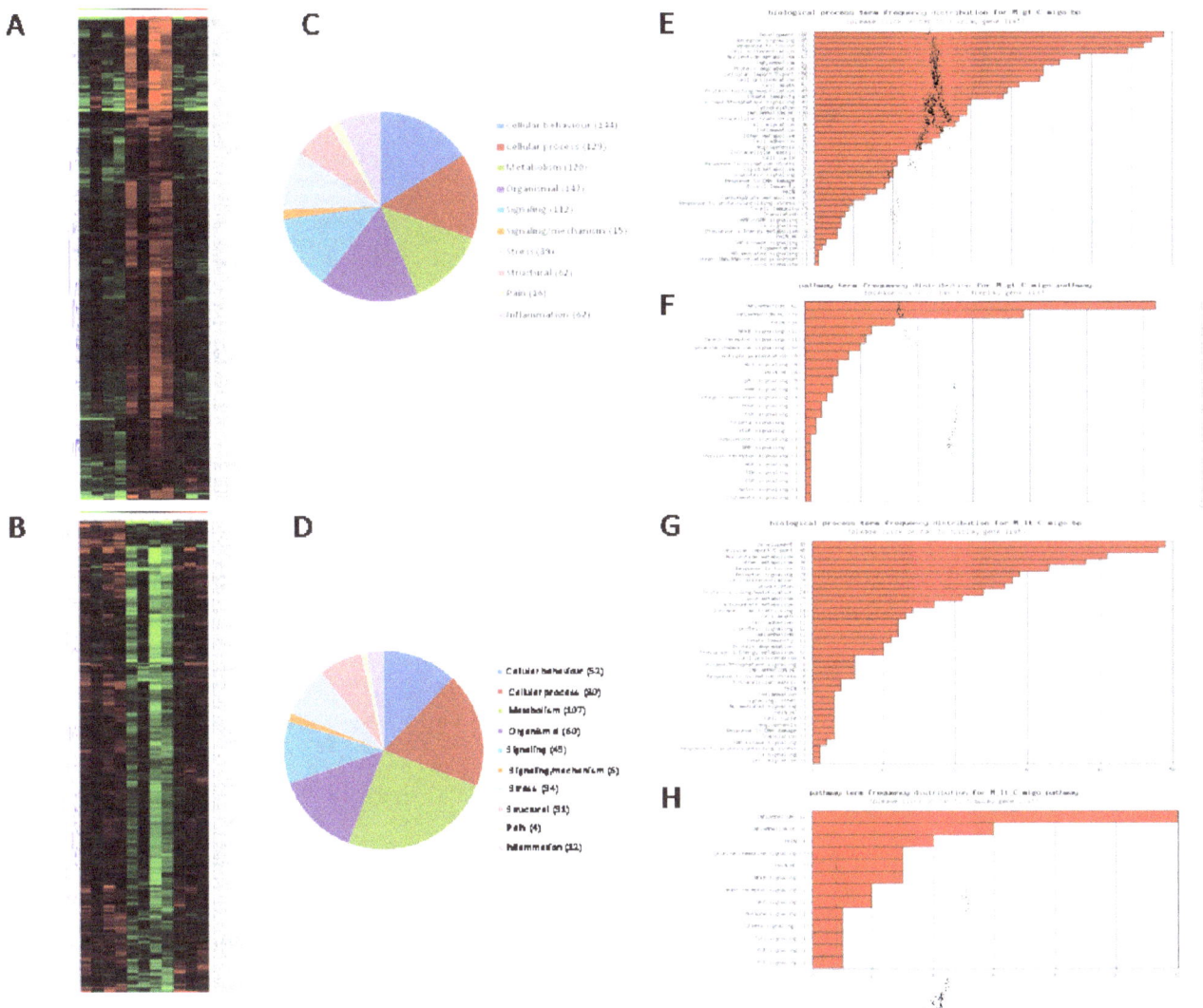

Figure 3. DGA and functional analysis of the effect of NMS on gene expression. (A–B) Heat maps of the significant genes (p<0.05 and 2-fold up- or downregulation) in NMS rats (Group M, lanes M5–M8) in comparison to VSL#3 treated NMS rats (Group C, lanes C1–C4). In comparison to control rats, NMS modulated the colonic expression of 665 genes (2% of total), 353 were upregulated and 312 downregulated. The expression of the majority of these genes returned to normal value after VSL#3 administration (Group D, lanes D9–D11). (C–D) Pie charts representing biological processes, molecular functions and cellular components among differentially expressed genes (up- or downregulation respectively). (E–G) Frequency distribution of the annotations for upregulated and downregulated genes respectively. (F–H) Frequency distribution of the signaling pathways for upregulated and downregulated genes respectively. gt: greater than; lt: lower than; migo_bp: gene ontology (GO) curated by Miltenyi Bioinformatics biological processs/function; migo_pathways: gene ontology (GO) curated by Miltenyi Bioinformatics pathways; Inflammation_HC: list of genes associated with inflammation from GO data but with more stringent selection criteria; Pain_HC: list of genes associated with pain from GO data but with more stringent selection criteria. Microarray data from 3–4 replicates.

As the aim of the study was to identify differentially expressed genes in the comparisons between all conditions, the differentially expressed genes were further filtered for functional associations related to neurotransmitters/mediators of pain, as well as for associations with cytokines/immunity and inflammatory responses.

Table 3. Number of genes that were modified ±2 folds (up- or downregulation) in different experimental groups.

Groups	Number of significantly modified genes (% of total)	Number of significanly upregulated genes (% of total)	Number of significantly downregulated genes (% of total)
Group M vs Group C	665/30,367 (2.18)	353/30,367 (1.16)	312/30,367 (1.02)
Group D vs Group M	779/30,367 (2.56)	411/30,367 (1.35)	368/30,367 (1.21)
Group D vs Group C	108/30,367 (0.35)	64/30,367 (0.21)	44/30,367 (0.14)

A

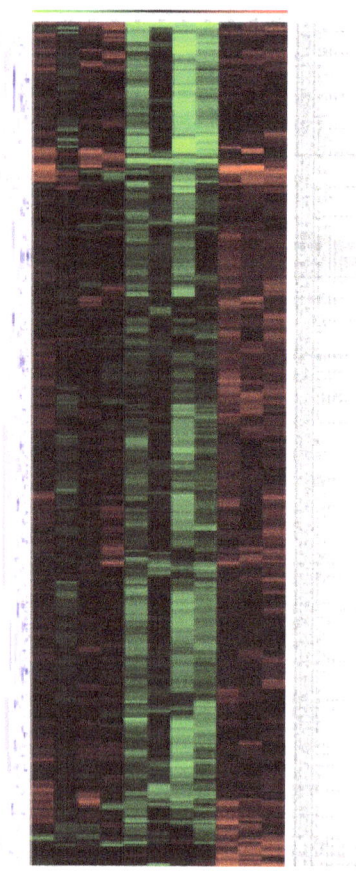

C

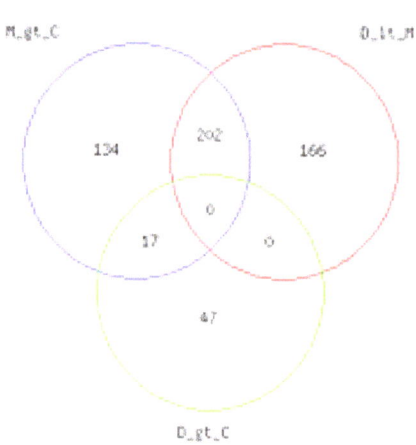

B

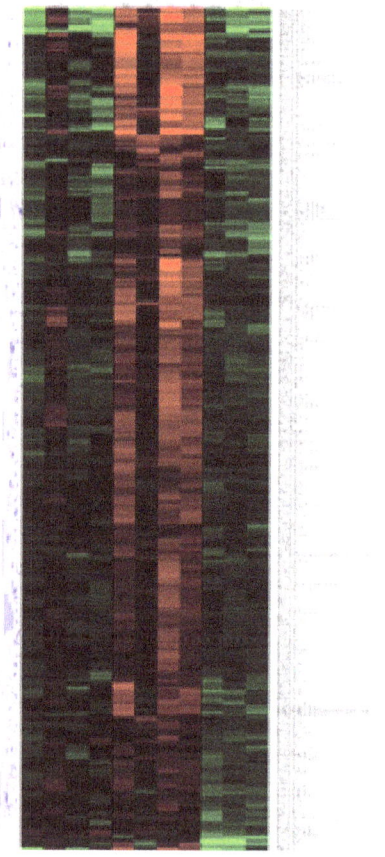

D

Figure 4. DGA and Venn diagrams of the effect of VSL#3 on NMS-induced alteration of gene expression. (A–B) Heat maps analysis of changes caused by VSL#3 administration (Group D, lanes D9–D11) to NMS rats (group M, lanes M5–M8) on genes whose expression was modulated >2-folds (up- or downregulation, p<0.05). (C–D) Venn diagrams of modulated genes. NMS induced an upregulation of 353 genes, 202 of which are downregulated by the probiotic treatment while 312 genes were downwregulated, 139 of which are upregulated by VSL#3 administration. Microarray data from 3–4 replicates. gt: greater than; lt: lower than.

The data was processed as follows:

1. Preprocessing of the data, including normalization and correlation analysis
2. Differential gene expression analysis (DGA) for the following groups:

   – Group M versus Group C
   – Group D versus Group M
   – Group D versus Group C

The analyses aim at distinguishing expression changes between all groups of samples so that six discriminatory gene sets (for each group up- and downregulated genes) were analyzed. A combination of statistical methods and the magnitude of expression difference (fold change) were applied in order to identify genes with differential expression between two sample groups. For the detection of discriminatory expression, genes had been selected that show a statistically significant deviation in the test compared to the reference group (ANOVA p-value ≤0.05, tukey p≤0.05). At the same time, it was required that the average expression value was at least 2-fold higher or lower than the reference average. For allowing a visually appealing display as a red/green heatmap, the

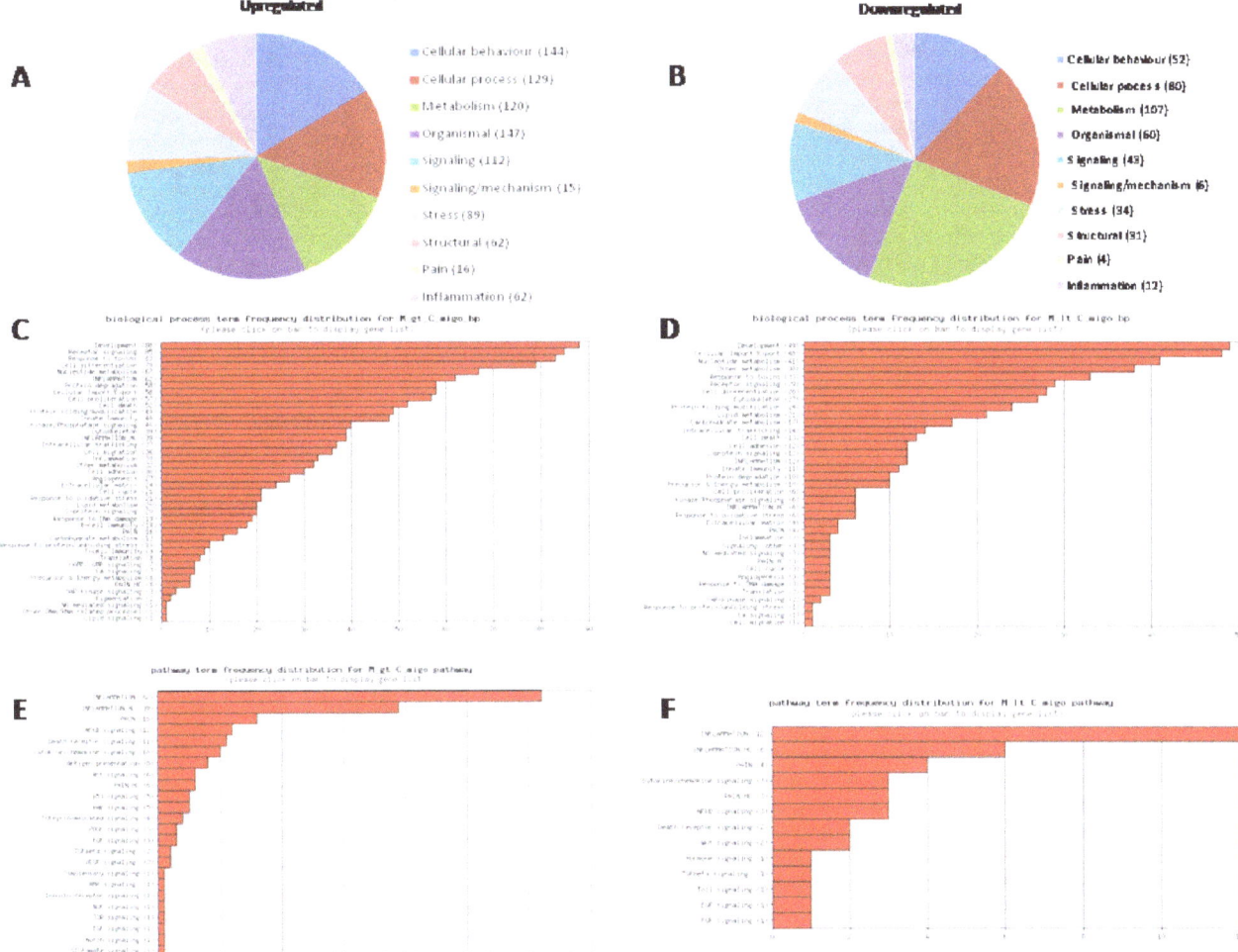

Figure 5. Functional analysis of the effect of VSL#3 on NMS-induced alteration of gene expression. (A–B) Pie charts representing biological processes, molecular functions and cellular components among differentially expressed genes. (C–D) Frequency distribution of the annotations for upregulated and downregulated genes respectively. Note that in animals treated with probiotic diet there were more genes belonging to the annotations "Pain" and "Inflammation" that were downregulated than those upregulated. (E–F) Frequency distribution of the signaling pathways for upregulated and downregulated genes respectively. gt: greater than; lt: lower than; migo_bp: gene ontology (GO) curated by Miltenyi Bioinformatics biological processs/function; migo_pathways: gene ontology (GO) curated by Miltenyi Bioinformatics pathways; Inflammation_HC: list of genes associated with inflammation from GO data but with more stringent selection criteria; Pain_HC: list of genes associated with pain from GO data but with more stringent selection criteria.

Table 4. Upregulated genes specifically involved in pain transmission in NMS rats in comparison with normal animals.

Identifier	Short name	Long name
A_42_P47339	CXCL1	chemokine (C-X-C motif) ligand 1 (melanoma growth stimulating activity, alpha)
A_42_P695401	CCL2	chemokine (C-C motif) ligand 2
A_43_P12508	PTGER2	prostaglandin E receptor 2 (subtype EP2)
A_44_P198620	NOS3	nitric oxide synthase 3, endothelial cell
A_44_P306204	TPH1	tryptophan hydroxylase 1*
A_44_P371339	IL6	interleukin 6
A_44_P430547	NTRK1	neurotrophic tyrosine kinase, receptor, type 1
A_64_P048210	TPH1	tryptophan hydroxylase 1*
A_64_P057941	MRGPRG	MAS-related GPR, member G
A_64_P100793	CCR2	chemokine (C-C motif) receptor 2
A_64_P118628	BDKRB1	bradykinin receptor B1
A_64_P122382	IL10	Interleukin 10
A_64_P125973	TNFRSF1B	tumor necrosis factor receptor superfamily, member 1b
A_64_P129316	PTGS2	prostaglandin-endoperoxide synthase 2
A_64_P130174	TRPV4	transient receptor potential cation channel, subfamily V, member 4
A_64_P130184	TRPV2	transient receptor potential cation channel, subfamily V, member 2

*Two different probe sequences are used for the same gene on the array.

expression values were converted to "virtual ratios" by referencing each individual intensity signal to median of all intensities. The base-2 logarithms of these virtual ratios were used for heatmap display. In each heat map the lanes from C1 to C4 corresponded to 4 rats of Group C, the lanes from M5 to M8 corresponded to 4 rats of Group M, and the lanes from D9 to D11 corresponded to 3 rats of Group D. A comparison of the discriminatory gene sets among different groups was showed by using the Venn diagrams.

3. Functional grouping analysis, including custom bioinformatics to identify associations of differentially expressed genes obtained in any of the DGAs with:

  – neurotransmitters and/or mediators of pain
  – cytokines, immunity, and inflammatory responses.

The functional grouping and annotation analysis provides an overview of the different biological processes and pathways, which are modulated in the discriminatory analyses. Here, the reporters were annotated with information from various databases in order to find common features among the genes sharing similar expression characteristics. The annotations used were derived from Gene Ontology (GO), which provides information on molecular function, as well as various pathway resources for information on involvement in biological signalling pathways. In addition, the gene lists obtained were also compared to known targets of certain signaling pathways. The significantly modified genes were grouped into 10 principal pathways, namely Cellular behavior, Cellular process, Metabolism, Organismal, Signaling, Signaling/mechanism, Stress, Structural, Pain, Inflammation. GO classification separated the genes involved in different biological processes, molecular functions and cellular components taking into account that the same gene could be present in more than one pathway. The annotation analysis was then completed by representative bar charts that give an overview of the biological categories found most frequently among the genes of the input reporter set. As the number of genes in the categories varies considerably, the size of the bars does not indicate a particular biological importance or over-representation.

4. Interaction network: core inflammatory pathway

The observations made in previous analyses were confirmed in the interaction graphs of key regulatory molecules of the core inflammatory pathways generated with Cytoscape (www.cytoscape.org).

Table 5. Downregulated genes specifically involved in pain transmission in NMS rats in comparison with normal animals.

Identifier	Short name	Long name
A_44_P228891	FAAH	fatty acid amide hydrolase
A_44_P472874	TRPA1	transient receptor potential cation channel, subfamily A, member 1
A_44_P484909	P2RX4	purinergic receptor P2X, ligand-gated ion channel 4
A_64_P109894	P2RX4	angiotensin II receptor, type 1b

Table 6. Upregulated genes specifically involved in pain transmission in VSL#3-treated NMS rats in comparison with not treated NMS animals.

Identifier	Short Name	Long Name
A_43_P13083	TACR2	tachykinin receptor 2
A_43_P13418	KCNK2	potassium channel, subfamily K, member 2
A_44_P1030258	CNR1	cannabinoid receptor 1 (brain)
A_44_P166967	NPY1R	neuropeptide Y receptor Y1
A_44_P408520	ENSRNOT00000022509	opiate receptor-like 1
A_44_P472874	TRPA1	transient receptor potential cation channel, subfamily A, member 1
A_64_P003854	IAPP	islet amyloid polypeptide
A_64_P058336	UCN	urocortin

PCR Analysis

Quantification of the expression of selected genes was performed by quantitative real-time PCR (qRT-PCR). 1 µl of the remaining RNA from colon samples that were used for gene array was incubated with DNase I and reverse-transcribed with Superscript II (Invitrogen) according to manufacturer specifications. For real-time PCR, 1 µl of template was used in a 25-µl reaction containing a 0.2 µM concentration of each primer and 12.5 µl of 2× SYBR Green PCR Master Mix (Bio-Rad Laboratories, Hercules, CA). All reactions were performed in duplicate using the following cycling conditions: 2 min at 95°C, followed by 50 cycles of 95°C for 10 s and 60°C for 30 s using an iCycler iQ instrument (Bio-Rad Laboratories). The mean value of the duplicates for each sample was calculated and expressed as cycle threshold (C_T). The amount of gene expression was then calculated as the difference (ΔC_T) between the C_T value of the sample for the target gene and the mean C_T value of that sample for the endogenous control (GAPDH). Relative expression was calculated as the difference ($\Delta\Delta C_T$) between the ΔC_T values of the test and control samples for each target gene. The relative level of expression was measured as $2^{-\Delta\Delta C_T}$. All PCR primers were designed using the software PRIMER3-OUTPUT using published sequence data obtained from the NCBI database (Table 2).

Statistical Analysis

Behavioral data are presented as mean ± SE, with sample sizes of at least 5 rats per group. Statistical comparisons were performed by the Mann-Whitney test for unpaired data and by the Wilcoxon signed rank test for paired data when two group of data were analyzed, and by the ANOVA for non parametric data, Kruskal-Wallis followed by Dunns comparison of selected pairs of column, when more than two groups of data were analyzed. An associated probability (p value) of less that 5% was considered significant.

For microarray analysis a combination of statistical methods (ANOVA p-value ≤0.05, tukey p≤0.05) and the magnitude of expression difference (2-fold change higher or lower than the reference average) was applied. For allowing a visually appealing display as a red/green heatmap, the expression values were converted to "virtual ratios" by referencing each individual intensity signal to median of all intensities. The base-2 logarithms of these virtual ratios were used for heatmap display in which lanes from 1 to 4 correspond to control rats, lanes from 5 to 8 correspond to NMS animals and lanes from 9 to 11 correspond to NMS rats treated with probiotic diet (VSL#3 3 60 day). A comparison of the discriminatory gene sets among different groups was performed by using the Venn diagrams.

Table 7. Downregulated genes specifically involved in pain transmission in VSL#3-treated NMS rats in comparison with not treated NMS animals.

Identifier	Short Name	Long Name
A_42_P695401	CCL2	chemokine (C-C motif) ligand 2
A_42_P714311	CCL3	chemokine (C-C motif) ligand 3
A_44_P198620	NOS3	nitric oxide synthase 3, endothelial cell
A_44_P430547	NTRK1	neurotrophic tyrosine kinase, receptor, type 1
A_64_P033800	SLC1A2	solute carrier family 1 (glial high affinity glutamate transporter), member 2
A_64_P048210	TPH1	tryptophan hydroxylase 1
A_64_P100793	CCR2	chemokine (C-C motif) receptor 2
A_64_P118628	BDKRB1	bradykinin receptor B1
A_64_P122382	IL10	interleukin 10
A_64_P125973	TNFRSF1B	tumor necrosis factor receptor superfamily, member 1b*
A_64_P130174	TRPV4	transient receptor potential cation channel, subfamily V, member 4
A_64_P165297		tumor necrosis factor receptor superfamily, member 1b*

*Two different probe sequences are used for the same gene on the array.

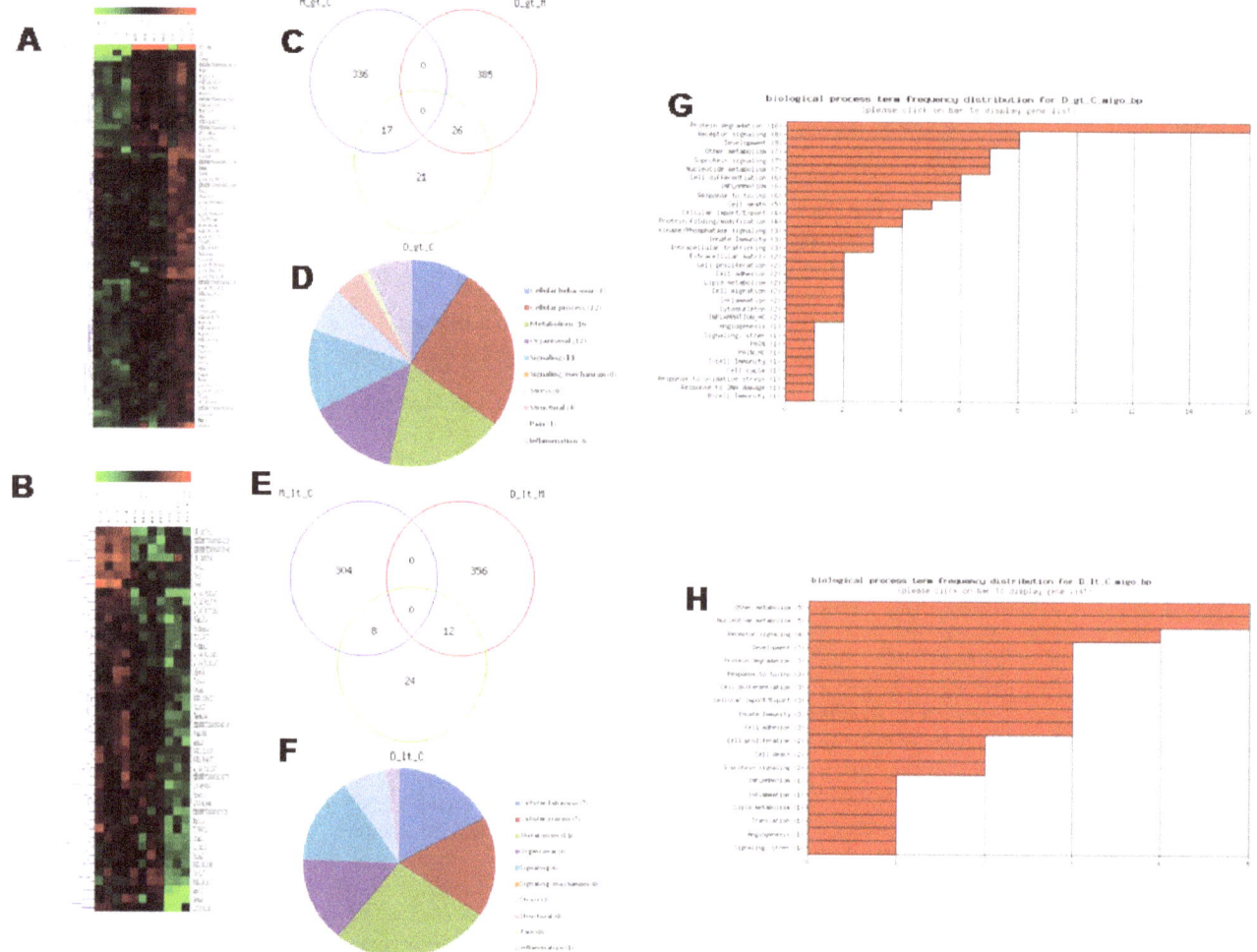

Figure 6. DGA and functional analysis of gene expression of VSL#3-treated NMS rats in comparison with controls. (A–B) Heat maps analysis of changes caused by VSL#3 administration (Group D, lanes D9–D11) to NMS rats on genes whose expression was modulated >2-folds (up- or downregulation, p<0.05) in comparison with control group (Group C, lanes C1–C4). NMS rats treated with VSL#3 showed upregulation and downregulation of 64 and 44 genes respectively, indicating that the probiotic diet resets the colonic expression of genes to values observed in control rats. (C–E) Venn diagrams of regulated genes in Group D in comparison with Group C. In NMS rat, VSL#3 administration induced an upregulation of 411 genes, 26 of which are also upregulated in Group C, and on a total of 368 downregulated genes in Group M, 12 were also dowregulated in Group C. The functional analysis in Group D vs Group C was illustrated by pie charts representing biological processes, molecular functions and cellular components among differentially expressed genes (D and F for up- and downregulated genes respectively). (G–H) Frequency distribution of the annotations for up- and downregulated genes respectively. Note that in VSL#3-treated NMS rats only 1 gene belonging to the category "Pain" (the opiate receptor-like 1 gene) was upregulated and, in generally, a lower number of gene were significantly modified in group D in comparison with group C. gt: greater than; lt: lower than; migo_bp: gene ontology (GO) curated by Miltenyi Bioinformatics biological process/ function; Inflammation_HC: list of genes associated with inflammation from GO data but with more stringent selection criteria; Pain_HC: list of genes associated with pain from GO data but with more stringent selection criteria. Microarray data from 3–4 replicates.

Results

VSL#3 Reverses Both Allodynia and Hyperalgesia in NMS Rats

In all experimental settings, animals were awake and no changes in the consciousness state were produced by CRD and VSL#3 intervention. For each CRD, two sequential distension-effect curves were constructed and the two obtained scores were averaged. In control animals (Group C), CRD (0.4–1.6 ml water) elicited a volume-dependent increase of the AWR scores which was rapid in onset, persisted for the duration of the distension period (Figure 1A) and returned to the baseline immediately after the distension was stopped. In the NMS animals (Group M), CRD induced both allodynia and hyperalgesia (Figure 1A). In NMS

Table 8. Upregulated genes specifically involved in pain transmission in VSL#3-treated NMS rats in comparison with normal animals.

Identifier	Short Name	Long Name
A_44_P408520	ENSRNOT00000022509	opiate receptor-like 1

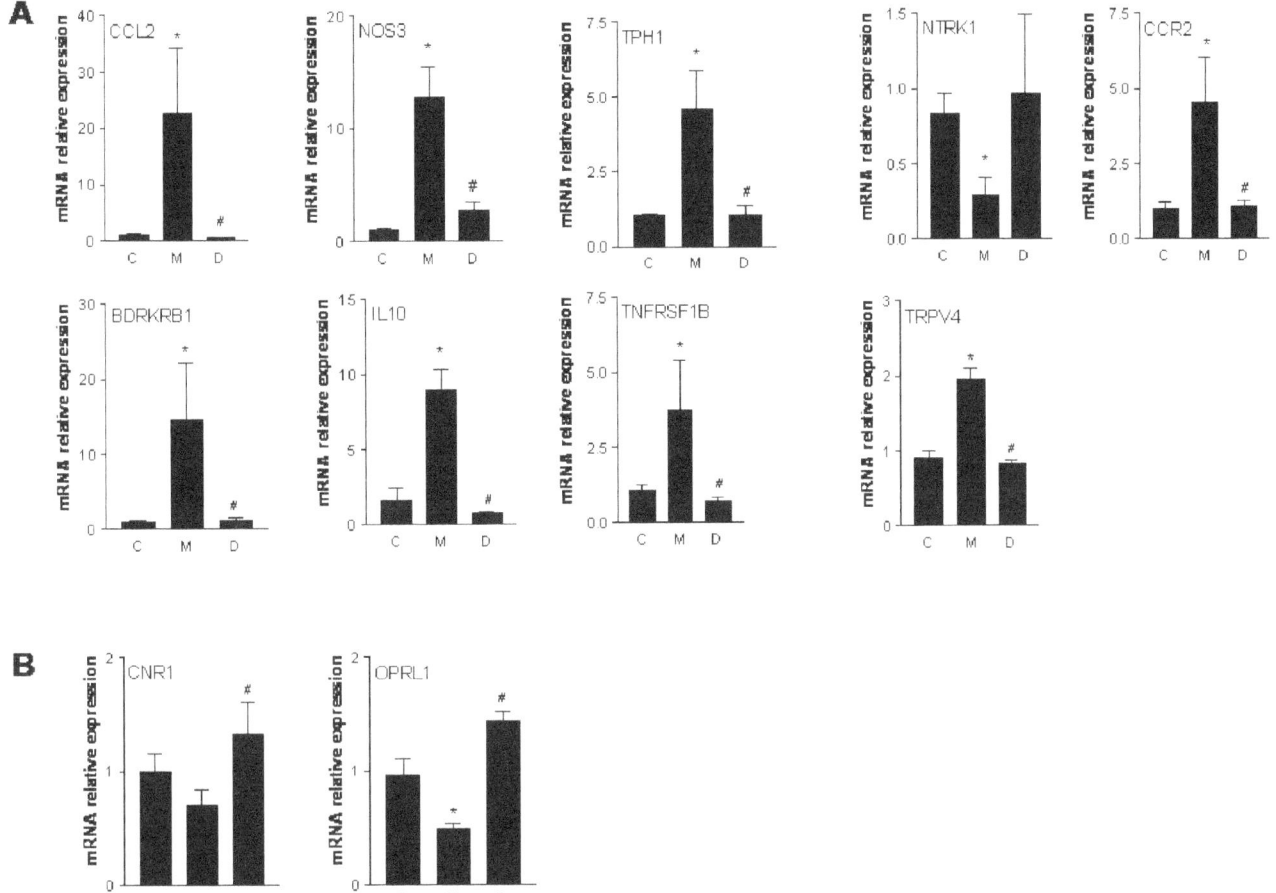

Figure 7. Confirmation of microarray data by qRT-PCR analysis. (A) Colon expression of the "Pain" annotation-related genes that were upregulated in Group M and downregulated in Group D. (B) Gene expression of the "Pain" annotation-related genes that were upregulated in Group D in comparison with Group M. *$p < 0.05$ vs Group C; #$p < 0.05$ vs Group M.

animals, VSL#3 administered from day 3 to day 60 (Group D) or from day 45 to day 60 (Group 7) caused the complete restoration of normal sensitivity (Figure 1B and D respectively), while probiotic intervention from day 3 to day 15 (Group 6) was only partially effective in reducing the visceral hypersensitivity (Figure 1C), as hyperalgesia persisted. *$p < 0.05$ versus Group C; #$p < 0.05$ versus Group M. Neither maternal deprivation and VSL#3 intervention had any effect on colorectal compliance (Figure S1).

Effect of NMS and VSL#3 Treatment on Gene Expression

Global microarray analysis. Global microarray analysis, performed on colonic tissue from groups C, M and D, detected several differentially expressed genes (data are available at the website http://www.ncbi.nlm.nih.gov/geo/query/acc.cgi?acc = GSE38942). On a total of 30,367 genes, expression of 8,678 (28.57%) genes was modified in NMS in comparison with healthy rats, 4,039 of which were upregulated and 4,639 downregulated (Figure 2A). Treating NMS rats with VSL#3 caused changes in the expression of 9,270 genes (30.5% of total): 4,903 genes were upregulated and 4,367 downregulated (Figure 2B). VSL#3 almost completely restored the initial pattern of gene expression in NMS rats. Thus, only 1,857 genes were modified in group D when compared with control rats, corresponding to 6.11% of total genes, with slight differentiation

between up- and downregulation (51.22% and 48.78% respectively) (Figure 2C).

DGA and functional analysis. When the combination of statistical methods (p-value < 0.05) and the magnitude of gene expression difference (fold change at least ± 2) was applied, 665 gene (2.18% of total) were significantly modulated in NMS animals compared to naïve rats (Table 3). The hierarchical clustering of these genes provided good separation based on the expression of the genes in the three groups of animals. In Figure 3A and B, red and green colours indicate upregulated and downregulated genes respectively, while black colour indicates no significant changes in gene expression. When NMS animals (lanes M5–M9) were compared with naive rats (lanes C1–C4), we detected a significant up- and downregulation of 353 and 312 genes respectively (Figure 3A and B). Moreover, the expression of these genes in group M was almost completely reverted by VSL#3 intervention (lanes D9–D11) (Figure 3A and B). Genes differentially expressed in NMS animals were classified according to their putative Gene Ontology (GO) based on significant similarity with known genes recorded in public databases (Figure 3C and D for up- and downregulated genes respectively). The frequency distribution of significantly modified genes in representative GO-derivative biological processes/functions (migo_bp) (Figure 3E and G for up- and downregulated genes respectively) and GO-derivative biological pathways (migo_pathways) (Figure 3F and

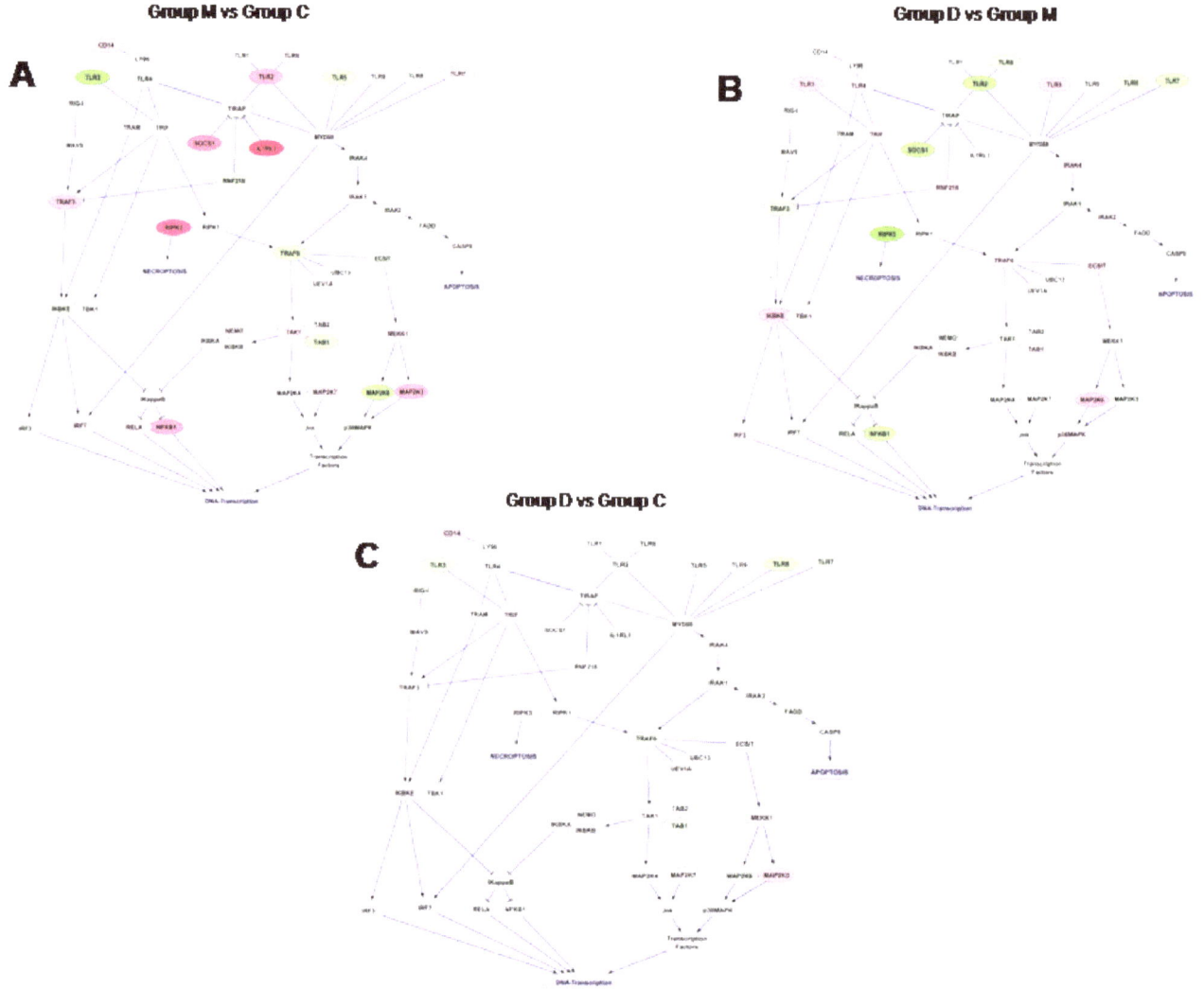

Figure 8. Interaction network: core inflammatory pathway. (A–B) The expression of several genes of the pro-inflammatory pathways which were upregulated (red color) or downregulated (green color) in group M compared with group C was neutralized upon treatment with the probiotic diet. This effect was observed for both receptors expressed on the cell surface and components of various signal transduction cascades including NF kB, IRFs and MAPK pathways. (C) VSL#3 almost completely restored the initial pattern of inflammation-related genes.

H for up- and downregulated genes respectively) for group M vs group C is also illustrated. The up- and downregulated genes specifically involved in pain transmission are illustrated in Table 4 and Table 5 respectively, while the lists of genes involved in inflammation are available in the GEO system at the website http://www.ncbi.nlm.nih.gov/geo/query/acc. cgi?acc=GSE38942. Interestingly the number of upregulated genes was higher than that of downregulated genes in "Pain" and "Inflammation" categories, as only 4 and 12 genes were downregulated, while 15 (tryptophan hydroxylase 1 gene was identified by two different primers) and 62 genes were upregulated in the two functional annotations, respectively. Most notable among those genes known for their role in mediating pain in both humans and rodents, were tryptophan hydroxylase 1 (TPH1), chemokine [C–C motif] ligand 2 (CCL2), chemokine [C–C motif] receptor 2 (CCR2), bradykinin receptor B1 (BDKRB1), tumor necrosis factor receptor superfamily member 1b (TNFRSF1B), prostaglandin E receptor 2 (subtype EP2) (PTGER2), nitric oxide synthase 3 endothelial cell (NOS3), neurotrophic tyrosine kinase

receptor type 1 (NTRK1), interleukin 6 (IL6) and transient receptor potential cation channel subfamily V members 4 and 2 (TRPV4 and TRPV2) (Table 4). Additionally, migo_bp and migo_pathways identified several upregulated genes also associated with immune and inflammatory responses (http://www.ncbi.nlm.nih.gov/geo/query/acc.cgi?acc=GSE38942).

In NMS rats, probiotic intervention reset the abnormal gene regulation caused by maternal deprivation. The hierarchical clustering and the heat map (Figure 4A and B for up- and downregulated genes respectively) demonstrated that, when compared with NMS (lanes M5–M8), VSL#3 treated rats (lanes D9–D11) showed a robust change in the expression of 779 genes with 411 and 368 up- or downregulated genes (Table 3). Moreover, VSL#3 administration attenuated changes in the expression of majority of genes: mostly of them were normalized to the level of Group C (lanes C1–C4) (Figure 4A and B). The comparison of regulated gene sets between group M compared to group C and group D compared to group M showed that the probiotic diet reversed or significantly attenuated changes occurring in NMS model. Panel C of Figure 4 shows a VENN diagram of the genes upregulated by maternal separation relatively to healthy animals, in comparison to the set of genes downregulated in group D compared to group M. The number of affected genes was comparable, but the majority of genes (202) was found at the intersection – these gene were upregulated in group M but downregulated in group D. The overlap between gene sets downregulated in samples of NMS group and those upregulated by probiotic diet compared to mother deprivation alone is also very high. Indeed, 139 genes were found at the intersection, indicating that the probiotic intervention also reversed the effects of mother separation on these genes (Figure 4D). The functional analysis conducted in group D in comparison with group M showed an overexpression of 20 genes correlated with "Pain" and "Inflammation" categories, while 79 genes were globally downregulated, demonstrating that VSL#3 treatment was counterregulatory on gene expression caused by NMS (Figure 5A and B respectively). The frequency distribution of significantly modified genes in representative GO-derivative biological processes/functions (migo_bp) and GO-derivative biological pathways (migo_pathways) for group D in comparison to group M is also illustrated in Figure 5 that shows the signaling pathways in which upregulated (panels C and E) and downregulated (panels D and F) genes were involved. Interestingly, VSL#3 induced downregulation of several gene related to antigen presentation and to immune and inflammatory response. VSL#3 intervention up- or downregulates 8 and 11 genes (TNFRSF1B was identified by two different primers) correlated with "Pain" (Tables 6 and 7 respectively). Among the formers were genes that play a significant role in antinociception, including cannabinoid receptor 1 and opiate receptor-like 1 (Table 6). Moreover, 9 of the 15 genes correlated to "Pain" that were upregulated in NMS rats in comparison with healthy animals were normalized after treatment with VSL#3 (Table 7), indicating that a possible mechanism of action of the probiotic diet was the regulation of genes encoding for protein mediating painful signals. Among others, TPH1, CCL2, CCR2, NOS3, NTRK1, BDKRB1, IL10, TNFRSF1B and TRPV4 genes were significantly downregulated by VSL#3.

When the gene expression driven by administering NMS rats with VSL#3 was compared to that of naïve rats, only a minority of genes were still modified by NMS (Figure 6). Indeed, group D (lanes D9–D11) showed up- and downregulation of 64 and 44 genes respectively in comparison with group C (lanes C1–C4) (Figure 6A and B respectively). Moreover, Venn diagrams demonstrated an overlap of genes that were significantly up- and downregulated among the three groups (Figure 6C and E respectively). The functional GO analysis revealed that only 1 gene belonging to pain annotation was still significantly upregulated in group D in comparison with Group C (Table 8), while no gene was downregulated, demonstrating that VSL#3 treatment in NMS rats induced an almost complete normalization of gene expression. Finally, the functional analysis showed an overexpression of a total of 6 genes correlated with "Inflammation" category, while only 1 gene was globally downregulated (Figure 6D and F respectively). The frequency distribution of significantly modified genes in representative GO-derivative biological processes/functions (migo_bp) conducted in group D in comparison with group C is also reported (Figure 6G and H).

PCR Analysis

The results of PCR analysis confirmed the microarray data for nociceptive (except for the NTRK1 gene) (Figure 7A) and antinociceptive genes (CNR1 and OLR-1; Figure 7B). *p<0.05 versus Group C; #p<0.05 versus Group M.

Interaction Network: Core Inflammatory Pathway

The observations made in the previous analyses were confirmed by analysis of the interaction graphs. Many genes of the proinflammatory pathways which are upregulated (in red) or downregulated (in green) in the NMS group compared to healthy animals (Figure 8A) were normalized upon additional treatment with probiotic diet (Figure 8B). This effect is observed for both receptors expressed on the cell surface and components of various signal transduction cascades including NFκB, IRFs and MAPK pathways. In contrast, NMS rats administered VSL#3 gained an almost complete normalization of the inflammation-related genes pattern when compared with naïve rats (Figure 8C).

Discussion

In the present study we have shown that the neonatal stress caused by NMS alters the expression of large cohort of genes in the colonic tissue and that these changes might be mechanistically linked to development of pain. Additionally, we have shown that this pattern could be corrected by administering NMS rats with a probiotic diet.

Early stress life events in the form of maternal separation induce permanent alterations in the rat and predispose to develop increased colonic motility and permeability [6] and hypersensitivity to mucosal inflammation in adulthood [37]. Further, maternal deprivation results in permanent changes in the central nervous system including unrestrained secretion of corticotrophin-releasing factor [38], alterations in serotoninergic, noradrenergic and dopaminergic systems [39], decreased expression of benzodiazepine and γ-aminobutyric acid type A receptors [40]. All pathways are potentially involved in development and maintenance of visceral pain.

In the present study we have shown that NMS rats exhibit visceral hyperalgesia and allodynia in response to CRD, indicating that stress in early life induces profound and long-standing changes in the development of the central nervous system and a modification of the neural pathways that result in an increase of pain perception. These changes were attenuated by administering rats with a probiotic mixture VSL#3. The timing of VSL#3 administration, however, had relevance. Indeed, VSL#3 reduces visceral pain not only when administered for the duration of the study, but also when it was administered in the last 2 weeks before CRD. In contrast, when VSL#3 intervention was performed for the first 2 weeks, it reverses allodynia but not hyperalgesia [41].

The effects of VSL#3 on NMS-induced hypersensitivity is not due to a change of the state of consciousness or modification of colorectal tone since VSL#3 administration had no effect on colorectal compliance (see Figure S1).

To elucidate the mechanisms of visceral hypersensitivity, we adopted a microarray approach. Analysis of colonic genetic pattern of NMS rats demonstrated that exposure to a stressful stimulus in the early life altered the expression of approximately one third (28%) of whole genome. However, the combination of a restricted DGA with the functional analysis revealed that less than 3% of genes, belonging to several annotation categories, are significantly up- or downregulated in NMS rats. The larger number of genes was grouped in the "Metabolism" category. This group includes carbohydrate and protein metabolisms. Our data agree with those of Lopes et al. who have shown that NMS affects the neuromuscular protein profile of the rat colon, in particular the expression of proteins whose functions are related to purine metabolism, protein folding and carbohydrate metabolism [42].

Although we have shown that NMS influences expression of many genes, we have focused our attention on the "Pain" annotation category, finding 15 upregulated and 4 downregulated genes that may have relevance in the development and maintenance of visceral pain in this model. Interestingly, the GO-based functional analyses has revealed also a correlation between "Inflammation" and "Pain" categories, as many genes are included in both annotations.

In the present study, data converge onto the hypothesis that VSL#3 intervention inhibits CRD-induced pain by reversing the alterations of genetic pattern caused by early maternal deprivation, specifically by influencing those pathways involved in pain and inflammation. First, the global genic analysis has demonstrated that VSL#3 induces the change of approximately 30% of the genes in NMS rats in comparison with non treated NMS animals and, more interestingly, the global genic pattern of NMS rats exposed to VSL#3 was similar to that of healthy animals, as only 6% of genes were modified. In other words, the genic pattern of NMS rats treated with probiotic diet was similar to that of naïve animals, as the expression of great majoriry of genes (94%) was similar in the two groups of animals. Second, DGA analysis confirms that VSL#3 reversed the alterations induced by maternal deprivation. Indeed, VSL#3 caused the significant modification of 779 genes in comparison with NMS, not treated rats and the great majority of genes that are significantly up- or downregulated in NMS rats are then counter-regulated by VSL#3 administration. In contrast, only 108 genes are significantly modified in VSL#3-treated NMS animals in comparison with naïve rats. Third, both the functional analysis and the qRT-PCR studies have demonstrated that the expression of several genes related to "Pain" and "Inflammation" annotation categories was counter-regulated by VSL#3 intervention. In particular, TPH1, CCL2, CCR2, NOS3, NTRK1, BDKRB1, IL10, TNFRSF1B and TRPV4, all genes encoding for proteins involved in nociception, are upregulated in NMS, hypersensitive rats and downregulated in NMS animals treated with VSL#3. Among them, TPH1 has a central role in nociception. This gene encodes for the tryptophan hydroxylase 1, an enzyme that catalyzes the first and rate limiting step in the biosynthesis of serotonin. Tryptophan hydroxylase 1 is involved in both neuropathic and visceral pain in animal models of nociception and in humans disorders. Thus, Crohn's disease patients who experience IBS-like symptoms are characterized by increased expression of tryptophan hydroxylase 1 in the colon [43]. Further, inhibition of tryptophan hydroxylase 1 in the gastrointestinal tract in IBS patients reduces mucosal production of serotonin and ameliorates symptoms [44-45]. A phase 2 clinical trial with the tryptophan hydroxylase 1 inhibitor LX1031 in patients with non-constipating IBS has been recently reported [46]. Fourth, exposure to VSL#3 upregulates the expression of several genes related to induction of analgesia, as cannabinoid receptor 1 (brain) (CNR1) [16], [47] and opiate receptor-like 1 (ENSRNOT0000002250), although the role of the latter gene in both nociception and analgesia is still controversial [48-49]. These findings agree with previous observations indicating that probiotics induce the expression of δ-opioid and cannabinoid receptors in intestinal epithelial cells, mediating analgesic functions in the gut in a way that mirrors the effect of morphine [16]. Fifth, It has been demonstrated recently that VSL#3 up- or downregulates the expression of several genes related to immunomodulation and inflammation [50], including the expression of the proinflammatory chemokine MCP-1/CCL2 in macrophages of children affected by Crohn disease [51]. Moreover, VSL#3 corrects the inflammation-driven dysregulation of PPARγ, FXR and leptin in a rodent model of inflammation [52]. Present findings are consistent with these observation indicating that VSL#3 counter-regulates genes involved in the inflammatory cascade, including CCL2, NOS3, IL10 and TNFRSF1B and genes that encode for factors that regulate the innate and adaptive immune response, as TLRs, NFκB and MAPKs, thus inhibiting inflammatory and, indirectly, nociceptive processes.

In conclusion, this report illustrates a novel approach to the analysis of the mechanisms underlying the pathogenesis of pain in experimental model of IBS. This approach allowed the identification of novel regulatory mechanisms and specific patterns of genic expression caused by exposure to probiotic that might have clinical readouts in conditions of visceral pain and stress-correlated intestinal pathologies.

Supporting Information

Figure S1 (A) Group C and Group M are the same in all comparison experiments. (A) Maternal deprivation and (B, C, D) VSL#3 intervention performed in different periods had any effect on colorectal compliance.

Acknowledgments

We thank Jutta Kollett for technical assistance to interpret the microarray data.

Author Contributions

Conceived and designed the experiments: ED SF. Performed the experiments: SC AM BR. Analyzed the data: SC. Wrote the paper: ED BR SF.

References

1. Thompson WG, Longstreth GF, Drossman DA, Heaton KW, Irvine EJ, et al. (1999) Functional bowel disorders and functional abdominal pain. Gut 45: 1143–1147.

2. Mayer EA, Gebhart GF (1994) Basic and clinical aspects of visceral hyperalgesia. Gastroenterology 107: 271–293.

3. Distrutti E, Salvioli B, Azpiroz F, Malagelada JR (2004) Rectal function and bowel habit in irritable bowel syndrome. Am J Gastroenterol 99: 131–137.

4. De Giorgio R, Barbara G (2008) Is irritable bowel syndrome an inflammatory disorder? Curr Gastroenterol 10: 385–390.

5. Kiank C, Taché Y, Larauchehttp://www.sciencedirect.com/science?_ob = ArticleURL&_udi = B6WC1-4X1SB98-1&_user = 3782037&_coverDate = 01%2F31%2F2010&_rdoc = 1&_fmt = high&_orig = search&_sort = d&_docanchor = &view = c&_acct = C000061366&_version = 1&_urlVersion = 0 &_userid = 3782037&md5 = 9e1206c20c8fa2c8d54d2add66976cb2 - implicit0 M (2010) Stress-related modulation of inflammation in experimental models of bowel disease and post-infectious irritable bowel syndrome: Role of corticotropin-releasing factor receptors. Brain Behav Immun 1: 41–48.

6. Coutinho SV, Plotsky PM, Sablad M, Miller JC, Zhou H et al. (2002) Neonatal maternal separation alters stress-induced responses to viscerosomatic nociceptive stimuli in rat. Am J Physiol Gastrointest Liver Physiol 282: G307–316.

7. Gareau MG, Jury J, Yang PC, MacQeen G, Perdue MH (2006) Neonatal maternal separation causes colonic dysfunction in rat pups including impaired host resistance. Pediatric Research 59: 83–88.

8. Pihoker C, Owens MJ, Kuhn CM, Schanberg SM, Nemeroff CB (1993) Maternal separation in neonatal rats elicits activation of the hypothalamic-pituitary-adrenocortical axis - a putative role for corticotropin-releasing factor. Psychoneuroendocrinology 18: 485–493.

9. Vicentic A, Francis D, Moffett M, Lakatos A, Rogge G et al. (2006) Maternal separation alters serotonergic transporter densities and serotonergic 1A receptors in rat brain. Neuroscience 140: 355–365.

10. Ren TH, Wu J, Yew D, Ziea E, Lao L, et al. (2007) Effects of neonatal maternal separation on neurochemical and sensory response to colonic distension in a rat model of irritable bowel syndrome. American Journal of Physiology 292: G849–856.

11. Fuller R (1989) Probiotics in man and animals. J Appl Bacteriol 66: 365–378.

12. Kassinen A, Krogius-Kurikka L, Makivuokko H, Rinttilä T, Paulin L, et al. (2007) The fecal microbiota of irritable bowel syndrome patients differs significantly from that of healthy subjects. Gastroenterology 133: 23–33.

13. McKernan DP, Fitzgerald P, Dinan TG, Cryan JF. (2010) The probiotic *Bifidobacterium infantis 35624* displays visceral antinociceptive effects in rats. Neurogastroenterol Mot 22: 1026–e268.

14. Desbonnet L, Garrett L, Clarke G, Kiely B, Cryan JF et al. (2010) Effects of the probiotic *bifidobacterium infantis* in the maternal separation model of depression. Neuroscience 170: 1178–1189.

15. Ma X, Mao YK, Wang B, Huizinga JD, Bienenstock J, et al. (2009) Lactobacillus reuteri ingestion prevents hyperexcitability of colonic DRG neurons induced by noxious stimuli. Am J Physiol Gastrointest Liver Physiol 296: G868–875.

16. Rousseaux C, Thuru X, Gelot A, Barnich N, Neut C, et al. (2007) Lactobacillus acidophilus modulates intestinal pain and induces opioid and cannabinoid receptors Nat Med 13: 35–37.

17. Venturi A, Gionchetti P, Rizzello F, Johansson R, Zucconi E et al. (1999) Impact on the composition of the faecal flora by a new probiotic preparation: preliminary data on maintenance treatment of patients with ulcerative colitis. Aliment Pharmacol Ther 8: 1103–1108.

18. Gionchetti P, Rizzello F, Venturi A, Brigidi P, Matteuzzi D, et al. (2000) Oral bacteriotherapy as maintenance treatment in patients with chronic pouchitis: a double-blind placebo-controlled trial. Gastroenterology 119: 305–309.

19. Gionchetti P, Rizzello F, Helwig U, Venturi A, Lammers KM, et al. (2003) Prophylaxis of pouchitis onset with probiotic therapy: a double-blind, placebo-controlled trial. Gastroenterology 124: 1202–1209.

20. Mimura T, Rizzello F, Helwig U, Poggioli G, Schreiber S et al. (2004) Once daily high dose probiotic therapy (VSL#3) for maintaining remission in recurrent or refractory pouchitis Gut 53: 108–114.

21. Bibiloni R, Fedorak RN, Tannock GW, Madsen KL, Gionchetti P et al. (2005) VSL#3 probiotic-mixture induces remission in patients with active ulcerative colitis. Am J Gastroenterol 100: 1539–1546.

22. Wall GC, Schirmer LL, Anliker LE, Tigges AE (2011) Pharmacotherapy for Acute Pouchitis. Ann Pharmacother 45: 1127–1137.

23. Kim HJ, Camilleri M, McKinzie S, Lempke MB, Burton DD, et al. (2003) A randomized controlled trial of a probiotic, VSL#3, on gut transit and symptoms in diarrhoea-predominant irritable bowel syndrome. Aliment Pharmacol Ther 17: 895–904.

24. Kim HJ, Vazquez Roque MI, Camilleri M, Stephens D, Burton DD, et al. (2005) A randomized controlled trial of a probiotic combination VSL#3 and placebo in irritable bowel syndrome with bloating. Neurogastroenterol Motil 17: 687–696.

25. Guandalini S, Magazzù G, Chiaro A, La Balestra V, Di Nardo G, et al. (2010) VSL#3 improves symptoms in children with irritable bowel syndrome: a multicenter, randomized, placebo-controlled, double-blind, crossover study. J Pediatr Gastroenterol Nutr 51: 24–30.

26. Chung EK, Zhang X, Li Z, Zhang H, Xu H, et al. (2007) Neonatal maternal separation enhances central sensitivity to noxious colorectal distention in rat. Brain Res 1153: 68–77.

27. Mencarelli A, Cipriani S, Renga B, Bruno A, D'Amore C, et al. (2012) VSL#3 resets insulin signaling and protects against NASH and atherosclerosis in a model of genetic dyslipidemia and intestinal inflammation. PLoS One 7(9): e45425.

28. Fiorucci S, Distrutti E, Cirino G, Wallace JL (2006) The emerging roles of hydrogen sulfide in the gastrointestinal tract and liver. Gastroenterology 131: 259–271.

29. Distrutti E, Sediari L, Mencarelli A, Renga B, Orlandi S et al. (2006) 5-Amino-2-hydroxybenzoic acid 4-(5-thioxo-5H-[1,2]dithiol-3yl)-phenyl ester (ATB-429), a hydrogen sulfide-releasing derivative of mesalamine, exerts antinociceptive effects in a model of postinflammatory hypersensitivity. J Pharmacol Exp Ther 319: 447–458.

30. Distrutti E, Sediari L, Mencarelli A, Renga B, Orlandi S, et al. (2006) Evidence that hydrogen sulfide exerts antinociceptive effects in the gastrointestinal tract by activating KATP channels. J Pharmacol Exp Ther 316: 325–335.

31. Distrutti E, Mencarelli A, Renga B, Caliendo G, Santagada V, et al. (2009) A nitro-arginine derivative of trimebutine (NO₂-Arg-Trim) attenuates pain induced by colorectal distension in conscious rats. Pharmacol Res 5: 319–329.

32. Distrutti E, Cipriani S, Renga B, Mencarelli A, Migliorati M, et al. (2010) Hydrogen sulphide induces μ opioid receptor-dependent analgesia in a rodent model of visceral pain. Mol Pain 6: 36.

33. Al-Chaer ED, Kawasaki M, Pasricha PJ (2000) A new model of chronic visceral hypersensitivity in adult rats induced by colon irritation during postnatal development. Gastroenterology 119: 1276–1285.

34. Ness TJ, Gebhart GF (1990) Visceral pain: a review of experimental studies. Pain 41: 167–234.

35. Edgar R, Domrachev M, Lash AE (2002) Gene Expression Omnibus: NCBI gene expression and hybridization array data repository. Nucleic Acids Res 30: 207–210.

36. Barrett T, Troup DB, Wilhite SE Ledoux P, Evangelista C, et al. (2011) NCBI GEO: archive for functional genomics data sets–10 years on. Nucleic Acids Res 39: D1005–1010.

37. Barreau F, Ferrier L, FioramontiJ, Bueno L (2004) Neonatal maternal deprivation triggers long term alterations in colonic ephitelial barrier and mucosal immunity in rats. GUT 53: 501–506.

38. O'Malley D, Dinan TG, Cryan JF (2011) Neonatal maternal separation in the rat impacts on the stress responsivity of central corticotropin-releasing factor receptors in adulthood. Psychopharmacology 214: 221–229.

39. Arborelius L, Eklund MB (2007) Both long and brief maternal separation produces persistent changes in tissue levels of brain monoamine in middle-aged female rats. Neurosci 145: 738–750.

40. Caldji C, Francis D, Sharma S, Plotsky PM, Meaney MJ. (2000) The effects of early rearing environment on the development of GABAA and central benzodiazepine receptors levels and novelty-induced fearfulness in the rat. Neuropsychopharmacology 22: 219–229.

41. Ohtsuka Y, Ikegami T, Izumi H, Namura M, Ikeda T, et al. (2012) Effects of Bifidobacterium breve on inflammatory gene expression in neonatal and weaning rat intestine. Pediatr Res 71: 46–53.

42. Lopes LV, Marvin-Guy LF, Fuerholz A, Affolter M, Ramadan Z et al. (2008) Maternal deprivation affects the neuromuscular protein profile of the rat colon in response to an acute stressor later in life. J Proteomics 71: 80–88.

43. Minderhoud IM, Oldenburg B, Schipper ME, ter Linde JJ, Samsom M. (2007) Serotonin synthesis and uptake in symptomatic patients with Crohn's disease in remission. Clin Gastroenterol Hepatol 75: 714–720.

44. Bian ZX, Li Z, Huang ZX, Chen HL, Xu HX, et al. (2009) Unbalanced expression of protease-activated receptors-1 and -2 in the colon of diarrhea-predominant irritable bowel syndrome patients. J Gastroenterol 44: 666–674.

45. Faure C, Patey N, Gauthier C, Brooks EM, Mawe GM (2010) Serotonin signaling is altered in irritable bowel syndrome with diarrhea but not in functional dyspepsia in pediatric age patients. Gastroenterology 139: 249–258.

46. Brown PM, Drossman DA, Wood AJ, Cline GA, Frazier KS et al. (2011) The tryptophan hydroxylase inhibitor LX1031 shows clinical benefit in patients with nonconstipating irritable bowel syndrome. Gastroenterology 141: 507–516.

47. Talwar R, Potluri VK (2011) Cannabinoid 1 (CB1) receptor-pharmacology, role in pain and recent developments in emerging CB1 agonists. CNS Neurol Disord Drug Targets 10: 536–544.

48. Largent-Milnes TM, Vanderah TW (2010) Recently patented and promising ORL-1 ligands: where have we been and where are we going? Expert Opin Ther Pat 20: 291–305.

49. Mustazza C, Bastanzio G. (2011) Development of nociceptin receptor (NOP) agonists and antagonists. Med Rec Rev 31: 605–648.

50. Evrard B, Coudeyras S, Dosgilbert A, Charbonnel N, Alamé J et al. (2011) Dose-dependent immunomodulation of human dendritic cells by the probiotic Lactobacillus rhamnosus Lcr35. PLoS One 6: e18735.

51. Lin YP, Thibodeaux CH, Peña JA, Ferry GD, Versalovic J. (2008) Probiotic Lactobacillus reuteri suppress proinflammatory cytokines via c-Jun. Inflamm Bowel Dis 14: 1068–1083.

52. Mencarelli A, Distrutti E, Renga B, D'Amore C, Cipriani S, et al. (2011) Probiotics modulate intestinal expression of nuclear receptor and provide counter-regulatory signals to inflammation-driven adipose tissue activation. PLoS One 6: e22978.

17

Screening and Characterization of Purine Nucleoside Degrading Lactic Acid Bacteria Isolated from Chinese Sauerkraut and Evaluation of the Serum Uric Acid Lowering Effect in Hyperuricemic Rats

Ming Li[1❾], **Dianbin Yang**[1❾], **Lu Mei**[1,2], **Lin Yuan**[3], **Ao Xie**[1], **Jieli Yuan**[1]*

1 Department of Microecology, School of Basic Medical Science, Dalian Medical University, Dalian, Liaoning, China, **2** Department of Gastroenterology, the Second Affiliated Hospital of Zhengzhou University, Zhengzhou, Henan, China, **3** Faculty of Agricultural, Life and Environmental Sciences, University of Alberta, Edmonton, Alberta, Canada

Abstract

Hyperuricemia is well known as the cause of gout. In recent years, it has also been recognized as a risk factor for arteriosclerosis, cerebrovascular and cardiovascular diseases, and nephropathy in diabetic patients. Foods high in purine compounds are more potent in exacerbating hyperuricemia. Therefore, the development of probiotics that efficiently degrade purine compounds is a promising potential therapy for the prevention of hyperuricemia. In this study, fifty-five lactic acid bacteria isolated from Chinese sauerkraut were evaluated for the ability to degrade inosine and guanosine, the two key intermediates in purine metabolism. After a preliminary screening based on HPLC, three candidate strains with the highest nucleoside degrading rates were selected for further characterization. The tested biological characteristics of candidate strains included acid tolerance, bile tolerance, anti-pathogenic bacteria activity, cell adhesion ability, resistance to antibiotics and the ability to produce hydrogen peroxide. Among the selected strains, DM9218 showed the best probiotic potential compared with other strains despite its poor bile resistance. Analysis of 16S rRNA sequences showed that DM9218 has the highest similarity (99%) to *Lactobacillus plantarum* WCFS1. The acclimated strain DM9218-A showed better resistance to 0.3% bile salt, and its survival in gastrointestinal tract of rats was proven by PCR-DGGE. Furthermore, the effects of DM9218-A in a hyperuricemia rat model were evaluated. The level of serum uric acid in hyperuricemic rat can be efficiently reduced by the intragastric administration of DM9218-A ($P<0.05$). The preventive treatment of DM9218-A caused a greater reduction in serum uric acid concentration in hyperuricemic rats than the later treatment ($P<0.05$). Our results suggest that DM9218-A may be a promising candidate as an adjunctive treatment in patients with hyperuricemia during the onset period of disease. DM9218-A also has potential as a probiotic in the prevention of hyperuricemia in the normal population.

Editor: Ralf Andreas Linker, Friedrich-Alexander University Erlangen, Germany

Funding: This work was supported by the Key Project of the National Twelfth-Five Year Research Program of China (2012BAI35B02), the National Program on Key Basic Research Project (973 Program, 2013CB531405), and the Research Fund for the Doctoral Program of Higher Education, China (RFDP, 20132105120012). The funders had no role in study design, data collection and analysis, decision to publish, or preparation of the manuscript.

Competing Interests: The authors have declared that no competing interests exist.

* Email: jieli_yuan_dmu@163.com

❾ These authors contributed equally to this work.

Introduction

Hyperuricemia is caused by abnormally high level of uric acid in the blood [1]. It is known to cause the formation of solid uric acid crystals within joints, which results in a painful condition commonly known as gout [2]. In recent years, hyperuricemia has also been recognized as a risk factor for arteriosclerosis, cerebrovascular and cardiovascular diseases [3], chronic kidney disease [4] and nephropathy in diabetic patients [5,6]. Several studies have reported that experimentally induced hyperuricemia in rodents can cause metabolic syndrome, hypertension and renal disease [7–9]. Lowering uric acid was proven to improve renal function, proteinuria and tubulointerstitial damage in type 2 diabetic *db/db* mice. The mechanism is likely caused by blocking uric acid-induced intrarenal inflammation [10].

Factors contributing to hyperuricemia vary from genetics, insulin resistance, hypertension and renal insufficiency, to obesity, diet and the consumption of alcoholic beverages [11]. Foods high in the purines adenine and hypoxanthine are more potent in exacerbating hyperuricemia [12]. Therefore patients with hyperuricemia have to strictly control their diet. However, accurate information on which food products and which nutrients affect plasma uric acid concentration are limited, and thus the dietary recommendations are currently unclear [13]. In addition, the purine compounds disodium 5'-guanylate and disodium 5'-inosinate are the main components of many flavor enhancers that are widely used in modern food production; thus, a strict diet means the changing both food flavors and eating habits. Drugs used to treat hyperuricemia are effective but with many side-effects. For example, allopurinol can induce hypersensitivity syndrome (AHS), which may lead to the death of patients [14]. In comparison to restricting the diet, the use of purine compound degrading probiotics is a promising alternative for the prevention of hyperuricemia.

Lactic acid bacteria (LAB) are Gram positive, acid-tolerant, fermenting rods or cocci that produce lactic acid as the major metabolic end-product of carbohydrate fermentation [15,16]. They are used in the manufacture of dairy products such as acidophilus milk, yogurt, cheeses and pickled vegetables. Numerous LAB strains have been continuously screened for desirable characteristics, such as stimulation of immune system [17], antitumor activity [18], stabilization of gut microbiota [19] and inhibition of pathogenic species. These beneficial properties make LAB strains valuable as probiotics, and many of them are used as starter microorganisms in yogurt fermentation [20]. However, no serum uric acid-lowering LAB and systemic evaluation of the probiotic characteristics have been reported.

In this study, we report for the first time the screening of purine nucleoside degrading LAB strains isolated from Chinese sauerkraut, and evaluate the probiotic characters of selected candidate strains. The effects of the optimal strain on a hyperuricemia rat model were also presented. We believe that our results will provide a reference for the development of hyperuricemia-preventing probiotics.

Materials and Methods

Isolation of LAB

To isolate LAB, different fresh Chinese sauerkrauts were purchased from local markets in Dalian, China. Each sample (10 g) was blended with 50 mL of 0.85% NaCl solution and further diluted in a 10-fold dilution series with 0.85% NaCl solution ($10^{-1} \sim 10^{-8}$). Each diluted solution was spread-plated onto de Man, Rogosa and Sharpe (MRS) agar (Difco, USA). The plates were incubated in anaerobic chest with AnaeroPack (Beijing B-Y Tech Co., Inc., China) at 37°C for 48 h. The bacterial colonies appeared on the plates were picked and streaked on fresh MRS agar plates. Single colonies were again picked and stored in 20% glycerol at −80°C. The isolates were screened for catalase activity and Gram staining, and only those that were catalase-negative and Gram-positive were selected for further studies. Prior to experiments, the stocks were propagated twice in MRS broth at 37°C for 24 h.

Screening of guanosine and inosine degrading LAB strains

A HPLC solution system was used to detect both inosine and guanosine simultaneously. The procedure is as follow: 33.7 mg of inosine and 35.7 mg of guanosine was dissolved in 100 mL K_3PO_4 solution (100 mmol/L, pH = 7.0) to make inosine-guanosine solution. After filtration (0.22 µm), 5, 10, 15, and 20 µL inosine-guanosine solutions were injected into a HPLC device (LC-20A, Shinadzu Corporation, Japan) equipped with a variable wavelength detector and a Cosmosil-5C$_{18}$-AR-II column (4.6×250 mm, Cosmosil, Japan). The isocratic elution was performed with a NaClO$_4$-H$_3$PO$_4$ solution (0.1 µmol/L NaClO$_4$, and 0.187 mol/L H$_3$PO$_4$ in dH$_2$O), at flow rate of 1 mL/min. Contents of inosine and guanosine were identified at 254 nm by retention time of 14.906 and 10.889 min, respectively, and quantified by interpolation of calibration curves. The deduced

standard curves were $A_{ino} = 5 \times 10^7 C_{ino} + 22844$, R = 0.9999, and $A_{gua} = 5 \times 10^7 C_{gua} - 10822$, R = 0.9999. A: the peak area, C: concentration (g/L).

To evaluate the inosine and guanosine assimilating ability, LAB strain was inoculated in MRS and cultured for 48 h at 37°C under anaerobic conditions. 2 mL of the culture broth was centrifuged at 4,000×g, 4°C for 10 min. The cells were then washed twice with 1 mL 0.85% NaCl, resuspended with 750 µL of inosine-guanosine solution and incubated at 37°C for 60 min, with shaking (120 rpm). After that, the solution was centrifuged at 4,000×g, 4°C for 10 min. 270 µL of the supernatant was removed. 30 µL HClO$_4$ (0.1 mol/L) was added into the supernatant, mixed thoroughly to prevent further degradation. 20 µL of the mixture was injected into the HPLC device after filtration. The remaining inosine and guanosine contents were calculated by the formula deduced above. The degrading speed and rate of inosine or guanosine by different LAB strains were calculated according to the following formula: $V = (0.9C\text{-}X)/60$, $\alpha = [(0.9C\text{-}X)/0.9C] \cdot 100\%$. V: degrading speed (g/L/min), X: the remaining content of inosine or guanosine (g/L).

To evaluate the degradation of purine compounds by cell-free extracts of LAB, 2 mL of the culture broth was centrifuged at 4,000×g, 4°C for 10 min. The cells were then washed twice with 1 mL 0.85% NaCl, resuspended with 750 µL of inosine-guanosine solution and were sonicated as described by Feliu et al [21]. After that, the extracts were incubated at 37°C for 60 min and 120 min, with shaking (120 rpm). The contents of inosine and guanosine were analyzed by HPLC as mentioned above.

Characterization of candidate strains as potential probiotics

Acid tolerance. Resistance to acidic conditions was tested by inoculating 10^8 CFU/mL (final bacterial concentration) of LAB in MRS broth with different pH values [22]. The pH of MRS was adjusted to 2.0, 3.0 and 4.0 using 0.1 N HCl. After cultivation for 4 h at 37°C, serial dilutions were performed and the growth of the bacterial strains was recorded by viable plate counting. This assay was performed in triplicate.

Bile tolerance. The bile salt solutions were prepared using ox gall powder (Sigma, St. Louis, Mo. USA) at final concentrations of 0.1%, 0.2% and 0.3% [23]. 10^8 CFU/mL (final bacterial concentration) of freshly prepared LAB were inoculated into 10 mL of the autoclaved solutions and incubated at 37°C for 4 h before viable plate counting. This assay was performed in triplicate.

Tolerance to pepsin and trypsin. 10^8 CFU/mL (final bacterial concentration) of freshly prepared LAB cultures were inoculated into MRS broth with different concentrations of pepsin (0.59 µg/mL, 0.72 µg/mL, and 1.48 µg/mL), and trypsin (0.336 µg/mL, 0.592 µg/mL, and 0.723 µg/mL). Growth of strains was enumerated 4 h after incubated at 37°C. This assay was performed in triplicate.

Test of antimicrobial activity. To evaluate the antimicrobial ability of the isolates, agar spot test was performed [22]. The pathogenic bacteria used in this study are *Escherichia coli* ATCC 8739, *Staphyloccocus aureus* ATCC 6538P, *Micrococcus luteus* MTCC 2470, *Salmonella typhi* ATCC 786, *Pseudominas aeruginosa* ATCC 25619, and *Staphylococcus epidermidis* ATCC12228. Briefly, 2.5 µL of a fresh LAB culture was spotted on MRS agar plates and incubated at 37°C for 24 h. After that the agar surface was covered with 15 mL of LB agar inoculated with the pathogenic strains at concentration of 10^6 CFU/mL. The plates were then incubated at 37°C. Inhibition of growth was

Table 1. The abilities of candidate LAB strains to assimilate inosine and guanosine.

Strain	V_{ino} (g/L/min)	α_{ino} (%)	Vgua (g/L/min)	α_{gua} (%)	Strain	V_{ino} (g/L/min)	α_{ino} (%)	Vgua (g/L/min)	α_{gua} (%)
DM9010	1.14E-04	1.96	2.63E-04	4.53	DM9234	2.68E-04	4.60	1.10E-04	1.89
DM9013	3.95E-04	6.78	6.02E-04	10.38	DM9237	1.36E-04	2.33	3.67E-04	6.33
DM9014	6.77E-04	11.61	6.95E-04	11.98	DM9241	4.82E-04	8.27	6.92E-04	11.93
DM9017	7.88E-04	13.52	8.63E-04	14.88	DM9242	1.31E-03	22.47	1.58E-03	27.18
DM9019	4.43E-04	7.60	3.96E-04	6.83	DM9247	5.18E-04	8.89	1.00E-03	17.31
DM9023	4.82E-04	8.27	2.23E-04	3.84	DM9250	4.62E-04	7.92	9.82E-04	16.93
DM9037	2.44E-04	4.19	3.54E-04	6.10	DM9258	3.64E-04	6.24	4.95E-04	8.54
DM9040	9.04E-04	15.51	8.62E-04	14.86	DM9500	9.03E-04	15.49	9.03E-05	1.55
DM9059	2.41E-04	3.89	3.05E-04	5.26	DM9501	7.53E-04	12.92	9.63E-05	1.72
DM9065	1.87E-04	3.21	9.32E-05	1.61	DM9502	7.97E-04	13.66	2.79E-04	4.81
DM9077	5.97E-04	10.25	8.22E-05	1.42	DM9503	6.53E-04	11.20	1.21E-04	2.01
DM9091	7.76E-04	13.31	7.31E-04	12.59	DM9504	9.18E-04	15.75	7.71E-04	13.30
DM9125	3.36E-04	5.77	7.45E-04	12.85	DM9505	1.74E-03	29.85	1.76E-03	30.37
DM9155	4.87E-04	8.35	7.31E-04	12.60	DM9506	9.70E-04	16.64	3.15E-04	5.43
DM9157	3.31E-04	5.67	2.50E-04	4.31	DM9507	3.29E-04	5.63	1.63E-04	2.82
DM9166	2.71E-04	4.64	6.13E-04	10.58	DM9508	6.65E-04	11.41	2.36E-04	4.07
DM9168	3.88E-04	6.64	6.95E-04	11.99	DM9509	7.91E-04	13.58	3.37E-04	5.77
DM9172	3.22E-04	5.52	2.50E-04	4.31	DM9510	1.57E-04	2.69	5.67E-05	0.99
DM9176	2.69E-04	4.61	4.13E-04	7.12	DM9514	1.20E-03	20.58	8.02E-04	13.82
DM9183	3.95E-04	6.77	7.76E-04	13.38	DM9515	7.44E-04	12.75	9.87E-05	1.74
DM9185	4.53E-04	7.77	6.53E-04	11.26	DM9517	7.08E-04	12.14	2.59E-04	4.46
DM9194	3.67E-04	6.29	5.61E-04	9.67	DM9518	9.83E-04	16.86	6.02E-04	10.38
DM9200	6.53E-04	11.20	6.34E-04	10.93	DM9519	5.99E-04	10.27	8.12E-05	1.39
DM9206	4.47E-04	7.66	3.21E-04	5.53	DM9520	7.76E-04	13.31	9.20E-05	1.61
DM9207	2.45E-04	4.21	5.18E-04	8.93	DM9521	3.27E-04	5.61	6.17E-04	10.63
DM9213	5.48E-04	9.40	9.30E-04	16.03	DM9529	5.93E-04	10.18	1.52E-04	2.62
DM9218	5.79E-03	99.31	5.78E-03	99.64	DM9530	6.84E-04	6.39	6.67E-07	0.01
DM9233	6.97E-04	11.96	9.77E-04	16.85	DM9218-A	5.79E-03	99.48	5.81E-03	99.73

Values are Means, n = 3. V_{ino}, the assimilating speed of inosine by LAB strain; α_{ino}, the assimilating rate of inosine by LAB strain; V_{gua}, the assimilating speed of inosine by LAB strain; α_{gua}, the assimilating rate of inosine by LAB strain. Underlines indicate strains with assimilating speed higher than 1.00×10^{-3} g/L/min. DM9218-A, the acclimated DM9218 strain.

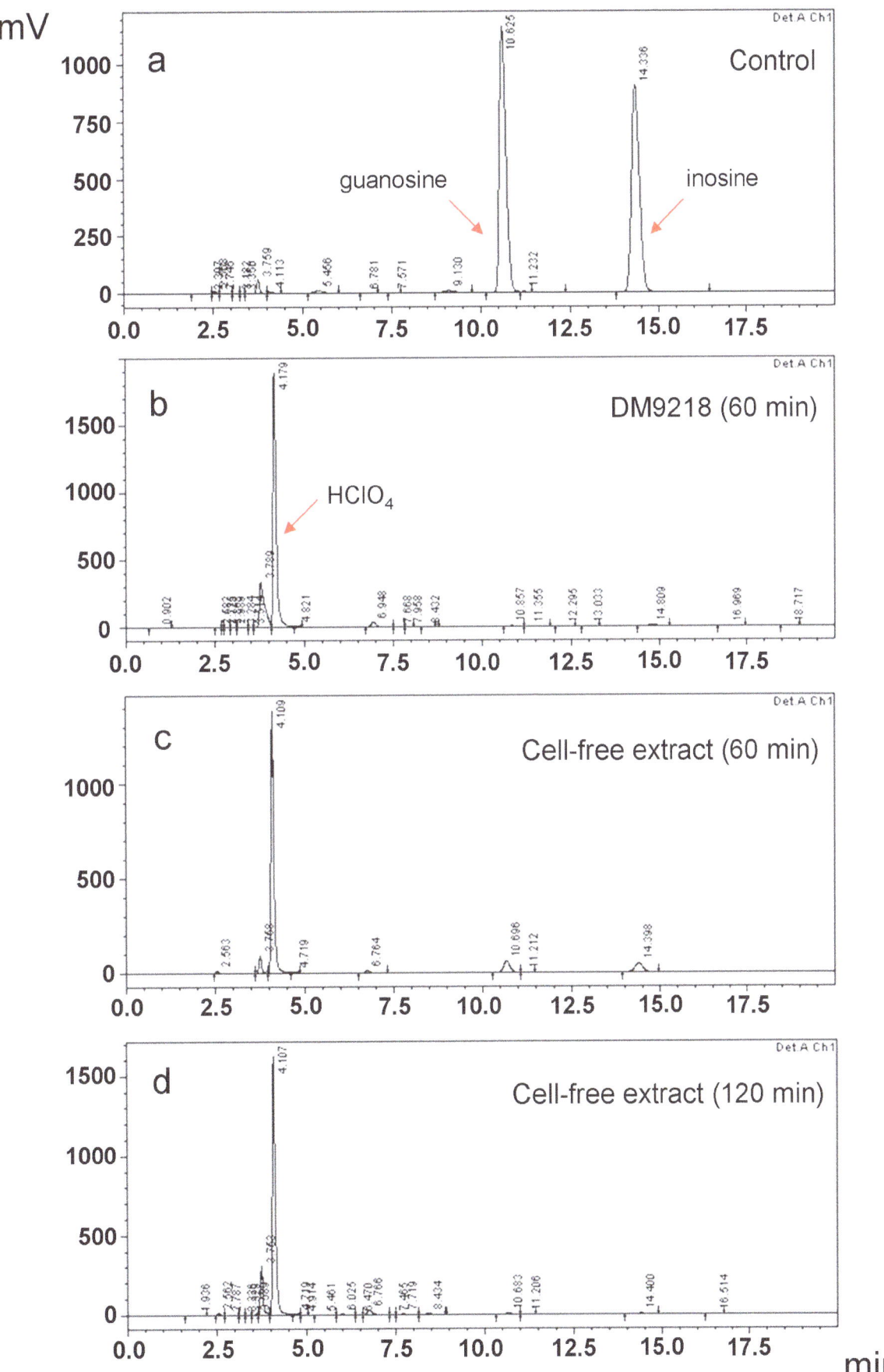

Figure 1. The degradation of inosine and guanosine by DM9218. (a) Control: inosine and guanosine solution without inoculation of bacteria (270 μL of the inosine and guanosine solution was incubated at 37°C for 60 min, after that 30 μL HClO₄ was added, 20 μL of the mixture was analyzed by HPLC). (b) DM9218 living cells were incubated with inosine and guanosine solution for 60 min. (c) Cell-free extracts of DM9218 were incubated with inosine and guanosine solution for 60 min. (d) Cell-free extracts of DM9218 were incubated with inosine and guanosine solution for 120 min. See details in methods.

detected after 24 h by measuring the diameters of inhibition zones. The test was performed in triplicate.

Measurement of H₂O₂ production. The freshly cultured cells of LAB strains were harvested by centrifugation at 10,000 rpm for 10 min at 0°C. 2 g of the cell pellet was resuspended in 20 mL of cold, sterile phosphate buffer (adjusted to pH 6.5). The cells were then incubated at 5°C for a period of 5 days under anaerobic condition. Measurements of H₂O₂ were done according to the methods of Villegas and Gilliland [24].

Cell adhesion ability. The adhesion ability of candidate strains to the human enterocyte-like Caco-2 cells was evaluated using the method described previously [29]. Briefly, the cells were grown in Dulbecco's minimal essential medium (DMEM) (Invitrogen, Germany) with 100 U/mL penicillin and 100 mg/mL streptomycin. Before the adherence assays, Caco-2 cells were cultured in 2 mL of the medium without antibiotics for 10 days. After standardization of conditions and preparation of monolayer (1×10^5 cells/well) of cell lines, 1 mL of the medium was replaced with 1 mL of LAB suspension (10^8 CFU/mL in DMEM). The inoculated cultures were then incubated for 3 h at 37°C in 5% CO₂. The infected cells were washed 3 times with sterile PBS (pH 7.8), fixed for 2 h with 10% formaldehyde, Gram-stained, and observed microscopically (1,500×magnification, with oil immersion). The adherent bacteria from 20 randomly selected microscopic fields were counted. All samples were analyzed in triplicate.

The sensibility of the strains to the antibiotics. Disk diffusion susceptibility tests were performed according to Clinical and Laboratory Standards Institute (CLSI) standard procedure [25]. Lab strains were inoculated in MRS and incubated at 37°C for 24 h. The culture broth was then diluted to a concentration of 6×10^8 CFU/mL, and spreaded onto the entire surface of a dried MRS agar plate using a sterile cotton swab. Antibiotic discs (see contents of antibiotics in Tab. 3) were placed on the surface of each MRS plate. After incubation for 48 h at 37°C, the diameter (in mm) of the inhibition zone around each disk was measured to classify the antibiotic sensitivity of each isolate. All samples were analyzed in triplicate.

Identification of strain DM9218

The 16S rDNA sequence was amplified using the primers described by Dubernet *et al* [26]. The sequence data was submitted to GenBank of NCBI, under accession number KF753248 and compared with the similar reference species using BLAST program, which is available at http://www.ncbi.nlm.nih.gov/. Phylogenetic tree was constructed based on the 16S rRNA gene sequences using Neighbor-joining method by MEGA 6.0 program.

DGGE analysis and 16S rDNA sequencing

The metagenomic DNA was extracted from the frozen feces of randomly selected rats (7/group) by the QIAamp DNA stool mini kit (Qiagen, Germany). PCR was conducted using universal primers F338+GC clamp and R518 targeting the hyper variable V3 region of 16S rRNA gene [27]. The resulting 16S rDNA amplicons were analyzed using the DCode system (Bio-Rad, USA) according to descriptions of Joossens *et al* [27]. Fragments of

interest were excised from the gel and macerated, and the suspension was incubated for 10 min at 98°C. The supernatant was used in PCR with F338 and R518 primers. The obtained PCR products were purified using the QIAquick PCR purification kit (Qiagen, Germany) and send for sequencing (Takara, Japan).

Animal test

The study protocol was approved by the Animal Care Committee of the Dalian Medical University, China (SCXK-2008-0002). 40 male Wistar rats, aged 30 days, were obtained from the SPF animal center of Dalian Medical University. They were divided randomly into 5 experimental groups (8/group). The control group was given food and water *ad libitum*. The hyperuricemia group, the allopurinol group, and the DM9218-After group were first treated with high purine diet for 7 days. The high purine diet (per 100 g) contains 87 g yeast extract (Sigma, USA), and 1.5 g ribonucleic acid from torula yeast (R6625, Sigma, USA). The DM9218-Before group was treated with high purine diet and DM9218-A cell solution (1.2×10^9 CFU/mL in 0.85% NaCl, 1 mL per day) for 7 days. On the 8th day, except the control group, other groups were injected intraperitoneally by potassium oxonate (0.35 mg/100 g body weight per day)-carboxymethylcellulose sodium (CMC-Na) solution (3 g/L) for 7 days. Meanwhile, the allopurinol group was administrated with 4.2 mg/100 g (body weight) of allopurinol (Shenyang No. 1 Pharmaceutical Co. LTD, China) per day, and the DM9218-After group was treated with DM9218-A cell solution (1.2×10^9 CFU/mL in 0.85% NaCl, 1 mL per day) for 7 days. The rat body weight was recorded randomly during the 14 days.

Blood samples were collected at 0, 7th, and 14th d. Blood was taken from the tail vein, and centrifuged immediately at 1,500×g for 15 min at 4°C to obtain serum. Serum samples were stored at −80°C until analyses. The levels of serum UA, UN, and Cr were detected according to the methods described previously [28].

Statistical analysis

Statistical analysis was performed by using SAS 9.1 system (SAS Institute Inc, USA) and GraphPad Prism 5 (Graph Pad Software, La Jolla, CA, USA). Data are presented as Means±S, $P < 0.05$ was considered as significant.

Results

Screening of inosine and guanosine degrading LAB strains

A total of fifty-five LAB strains were isolated from different Chinese sauerkraut samples, and they were all tested for inosine and guanosine assimilating abilities based on HPLC detection. Results are summarized in Table 1. The majority of the tested fifty-five LAB strains exhibited purine assimilation abilities, with the average assimilation rate of 11.30% for inosine, and 10.20% for guanosine. Among them, strains DM9218, DM9242, and DM9505 were found to assimilate purine nucleosides at greater than 1.00×10^{-3} g/L/min, and assimilation rates over 25.00%. In particular, DM9218 showed the highest rates of 99.31% and 99.64%, for inosine and guanosine respectively. These three stains were selected for further study.

To clarify whether the purine compounds were degraded by LAB, cell-free extracts of DM9218 were incubated with the inosine and guanosine solution for 60 and 120 min, and the concentration of purine compounds analyzed by HPLC. Results in Fig. 1 show that after 60 min incubation, the concentration of inosine and guanosine was decreased by 85.76% and 86.20%, respectively (Fig. 1c). After 120 min incubation, the two purine compounds were degraded by 98.81% and 98.98% respectively (Fig. 1d), which was comparable with purine degradation by living DM9218 cells (Fig. 1b). Cell-free extracts of DM9242 and DM9505 were also incubated with inosine and guanosine for 120 min. 20.28% and 25.16% degradation of inosine and 24.25% and 28.97% degradation of guanosine were observed for DM9242 and DM9505 respectively.

Evaluation of the probiotic potential of DM9218, DM9242 and DM9505

The tolerance of the three candidate strains to acid and bile salt are summarized in Table 2, which shows there was little variation among the candidate LAB strains. The acid tolerance test revealed that strains DM9505 and DM9218 could survive at pH values as low as 2, but DM9242 could only survive at a pH over 3. The 3 strains were not tolerant to bile salt. When the medium contained 0.2% bile salt, DM9218 and DM9242 could grow weakly, but DM9505 could not grow at all. When the concentration of bile salt was increased to 0.3%, no strains grew. All the three strains could grow well in media supplemented by 0.59~1.48 µg/mL pepsin or 33.60~72.32 U/g trypsin.

The antimicrobial activities of strain DM9218, DM9242 and DM9505 are shown in Fig. 2. All strains exhibited specific inhibitory activities against some of the pathogenic microorganisms tested. Strain DM9242 showed strong inhibitory activities especially to *Salmonella typhi* and *Staphylococcus aureus*. Strain DM9505 strongly inhibited *Pseudomonas aeruginosa*. Strain DM9218 showed high inhibitory activities to *Escherichia coli* and *Staphylococcus aureus*.

The H_2O_2 content produced by candidate strains in cold, sterile phosphate buffer (pH 6.5) was determined (Fig. 3a). DM9505 and DM9218 were found to produce 14.17 µg/mL and 5.32 µg/mL of H_2O_2, respectively, while minimal H_2O_2 was detected in the phosphate buffer inoculated with DM9242.

The cell adhesion ability of the candidate strains to human enterocyte-like Caco-2 cells was tested. Results are shown in Fig. 3b. DM9242 showed the highest adhesion value of 50.1±8.35 cells/Caco-2 cell. This was followed by DM9218, with an adhesion value of 45.5±5.67, and DM9505 with an adhesion value of 22.7±7.86.

The sensitivity of the three strains to antibiotics (Table 3) was tested by disc diffusion assay against 16 different antibiotics from 9 categories. The results demonstrated that the strains were susceptible to antibiotics belonging to the penicillins (including penicillin, piperacillin, and amoxicillin), cephalosporins (cephalothin), macrolides (erythromycin), and ansamycins (rifampicin). They were found to be resistant to the aminoglycosides (gentamicin), quinolones (levofloxacin, nalidixic acid, ciprofloxacin), and glycopeptides (vancomycin). The three strains showed intermediate sensitivity to the other antibiotics tested.

Identification of DM9218 by 16S rDNA sequencing

Owing to its superior performance in degrading inosine and guanosine, and its superior probiotic characteristics compared with the two other candidates, DM9218 was selected for further study. It was first identified by 16S rDNA sequencing. The 16S rDNA sequence of DM9218 had the highest identity (99%) with

Table 2. The tolerance of candidate LAB strains to biological barriers (Log CFU/mL).

Strains	Initial counts[a]	pH			Bile salt (%)		
		2	3	4	0.1	0.2	0.3
DM9218	8.59±0.18	4.36±0.10	6.25±0.13	9.05±0.16	2.86±0.13	-	-
DM9242	9.02±0.11	-[b]	2.56±0.24	3.12±0.14	1.36±0.18	-	-
DM9505	8.98±0.23	3.11±0.23	5.87±0.18	8.15±0.24	-	-	-

Strains	Initial counts[a]	Pepsin (µg/mL)			Trypsin (U/g)		
		0.59	0.72	1.48	33.60	52.96	72.32
DM9218	8.59±0.18	9.24±0.47	9.44±0.52	8.99±0.46	8.98±0.61	9.13±0.41	9.04±0.48
DM9242	9.02±0.11	9.63±0.33	9.21±0.18	9.02±0.84	9.07±0.19	9.16±0.11	8.99±0.71
DM9505	8.98±0.23	9.01±0.13	9.11±0.87	9.11±0.77	9.04±0.27	9.00±0.53	8.99±0.37

[a]Each value represents Mean±S, n=3.
[b]No growth.

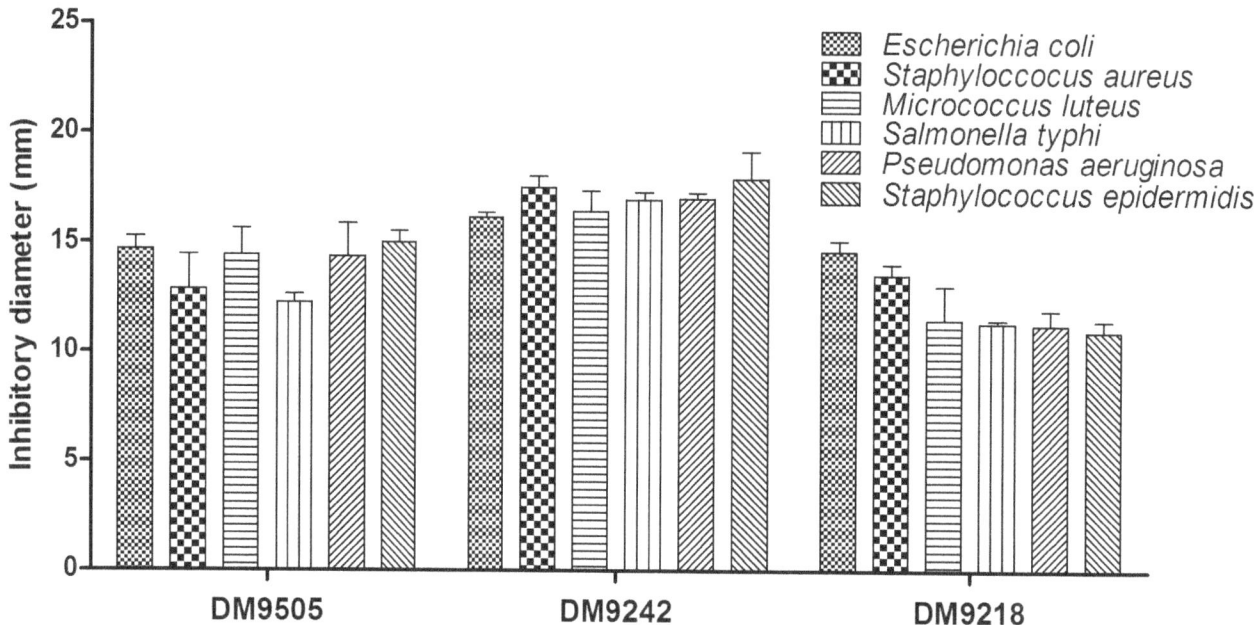

Figure 2. The antimicrobial activities of DM9218, DM9242, and DM9505. All values are Means ± S, n = 3.

the 16S rDNA of *Lactobacillus plantarum* WCFS1 [29] (Fig. 4). It also showed high similarity with the 16S rDNA of *Lactobacillus brevis* ATCC 367 (95%) [30] and *Lactobacillus rhamnosus* GG (93%) [31].

Acclimation of DM9218 to bile salt

To overcome the poor tolerance to bile salt, we acclimatized DM9218 to bile salt by cultivating it on MRS media containing increasing concentrations (0.1–0.3%) of bile salt. After ten generations, the clones growing on 0.3% bile salt containing MRS agar media were picked, and their growth pattern in liquid media with different concentrations of bile salt tested. Compared with DM9218, the acclimated strain (DM9218-A) showed no difference when grown in MRS medium without adding bile salt (Fig. 5a and b). However, it showed tolerance to 0.2% and 0.3% bile salt, the percentage survival increasing to 75.10% and 48.97%

respectively after 5 hours incubation. Furthermore, the inosine and guanosine degradation abilities of DM9218-A were not affected (See Table 1).

Detection of DM9218-A in rats

To investigate whether DM9218-A can survive in the gastro-intestinal (GI) tract of rats, feces taken from rats before and after intragastric administration were analyzed as an indicator of gut microbiota composition. The DGGE profile of 16S rRNA gene amplicons obtained from a pure culture of DM9218-A strain revealed the existence of several copies of the 16S rRNA gene, with one copy staining most intensively on the gel. DGGE analysis of 16S rRNA products obtained from fecal contents of DM9218-A treated rats (3 to 14 days samples after administration) revealed the band corresponding to the dominant band of 16S rDNA from DM9218-A (Fig. 6). Suspected bands, as well as the dominant

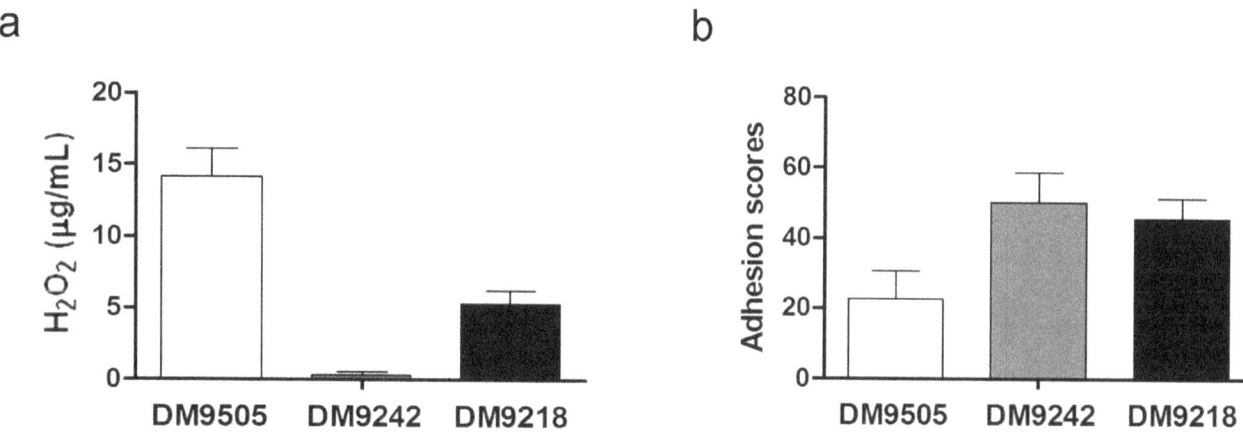

Figure 3. Production of H$_2$O$_2$ by DM9218, DM9242 and DM9505, and their adhesion abilities to Caco-2 cells. (a) Production of H$_2$O$_2$ by strains in cold, sterile phosphate buffer (pH 6.5) incubated at 5°C for 5 days. (b) Adhesion abilities to Caco-2 cells. The adhesion scores indicate values of LAB cells adhered to one Caco-2 cell. All values are Means ± S, n = 3.

Table 3. The Susceptibility of candidate LAB strains to different antibiotics.

Category	Antibiotics	Dose (µg/disc)	Susceptibility		
			DM9218	DM9242	DM9505
Penicillins	Penicillin (P)	10 IU	I	S	S
	Ampicillin (AMP)	10	I	S	I
	Piperacillin (PRL)	100	S	S	S
	Amoxicillin (AMX)	10	S	S	S
Cephalosporins	Cefazolin (CZ)	30	I	S	I
	Cefalotin (KF)	30	S	S	S
Aminoglycosides	Gentamycin (CN)	10	R	R	R
Quinolones	Levofloxacin (LE)	5	R	R	R
	nalidixic acid (NA)	30	R	R	R
	Ciprofloxacin (CIP)	5	R	R	R
Macrolides	Azithromycin (AZM)	15	I	S	I
	Erythromycin (E)	15	S	S	S
Tetracyclines	Tetracycline (TE)	30	I	S	I
Ansamycins	Rifampicin (RD)	5	S	S	S
Glycopeptides	Vancomycin (VA)	30	R	R	R
Combination	Sulfamethoxazolum-Trimethoprimum (SXT)	23.75–1.25	I	I	S

S, susceptible, the diameter of inhibition zone ≥17 mm; I, intermediate, the diameter of inhibition zone between 12–17 mm; R, resistant, the diameter of inhibition zone ≤12 mm resistance.

pure culture band, were sequenced queried against the NCBI genome collection. The results confirmed that all eluted bands belonged to *Lactobacillus plantarum*. No band corresponding to the dominant pure DM9218-A culture band was detected in control rats or fecal content of rats before treatment. Along with the demonstration of the ability of DM9218-A to survive in the GI tract of the rats, the results obtained by DGGE analysis showed no significant shift in the intestinal microbial community of the treated rats.

Effects of DM9218-A on hyperuricemia in rats

To examine the effect of DM9218-A on the serum uric acid (UA) level of hyperuricemic rats, we induced hyperuricemia by intraperitoneal injection of potassium oxonate into rats, supplemented with a high purine diet. Fig. 7a shows the outline of the animal study: one group of rats was intragastrically administrated DM9218-A from the very beginning of hyperuricemia induction, they were considered the preventive DM9218-Before group. After 7-days, 4 groups (all except the control group) were injected with potassium oxonate for 7 days, 2 groups were treated with DM9218-A or allopurinol, and they were defined as the DM9218-After and allopurinol groups respectively. No differences in food consumption (data not shown) and body weight (Fig. 7b) were detected among the experimental groups during the trial ($P > 0.05$). Blood samples were taken on days 0, 7, and 14. Fig. 7c (top) shows that after 7 days' high purine diet, the blood UA level of hyperuricemic rats increased to 153.25 µmol/L, which is significantly higher than control rats (79.00 µmol/L, $P = 0.0016$), 14 days after high purine diet combined with 7 days potassium oxonate injection, the blood UA level of hyperuricemic rats reached to 311.75 µmol/L, which is over 3.45 fold that of control rats ($P = 0.0001$). It can therefore be considered that under these experimental conditions, hyperuricemia was successfully established. The administration of DM9218-A to rats from the first day

(DM9218-Before group) resulted in a dramatically lower level of blood UA compared with the control group at the 14th day ($P = 0.0015$); however this level was still higher than that of healthy controls ($P = 0.0113$). From the 8th day, we treated the hyperuricemic rats with allopurinol. Results of the 14th day samples indicated that allopurinol significantly reduced blood UA to a healthy level ($P > 0.05$ compared with healthy control). The blood UA level in rats administrated DM9218-A from the 8th day were lower than the hyperuricemic group ($P = 0.0211$), but still higher than the healthy group ($P = 0.0454$). The preventive treatment of DM9218-A caused a greater reduction in serum UA concentration as compared with the DM9218-After group ($P = 0.0487$). Effects of DM9218-A on blood urea nitrogen (UN) and creatinine (Cr) contents of rats were also evaluated. Fig. 7c (middle and bottom) shows that no significant differences were detected among the control group, hyperuricemia model group and the rats treated with DM9218-A or allopurinol ($P > 0.05$).

Discussion

Purine rich foods such as meat and seafood, as well as alcoholic beverages potently exacerbate hyperuricemia, the major factor causing gout. Over the past ten years, more and more studies have documented the biological interference of uric acid with other diseases, such as hypertension, to initiate endothelial dysfunction, vascular damage and renal disease [32,33]. Normalizing serum uric acid level is thus absolutely essential to reduce the risk of complications. However, prescribing uric acid-lowering drugs to individuals with asymptomatic hyperuricemia is not recommended by current policy because of insufficient evidence [32]. Therefore, probiotics with the ability to degrade purine compounds in food deserve to be investigated exhaustively for the prevention of hyperuricemia.

LAB have been investigated intensively over the past few decades as they were found to play important roles in gastroin-

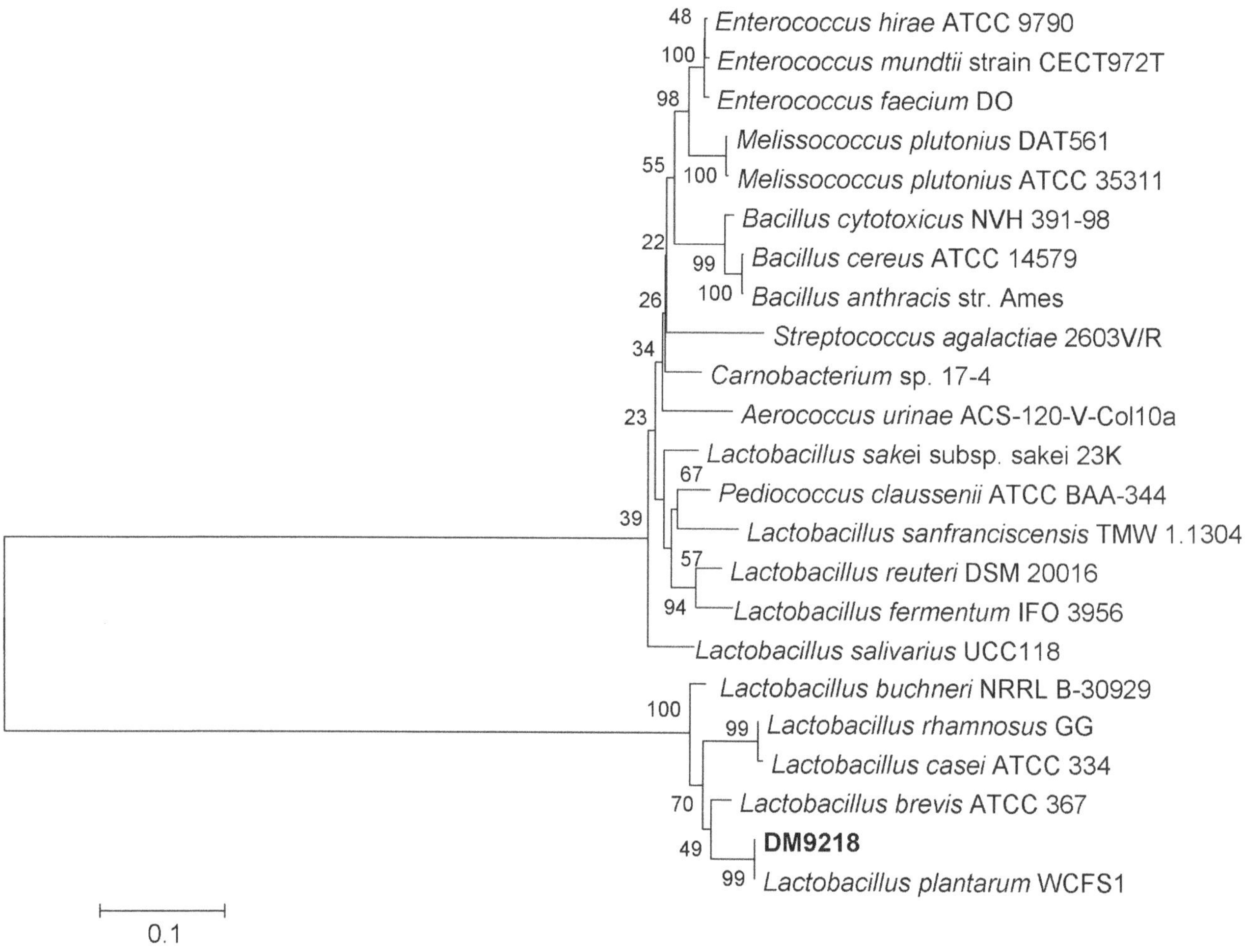

48 ┌ *Enterococcus hirae* ATCC 9790
100 └ *Enterococcus mundtii* strain CECT972T
98 └ *Enterococcus faecium* DO
┌ *Melissococcus plutonius* DAT561
55 ┌ 100 └ *Melissococcus plutonius* ATCC 35311
┌ *Bacillus cytotoxicus* NVH 391-98
22 99 ┌ *Bacillus cereus* ATCC 14579
26 100 └ *Bacillus anthracis* str. Ames
34 └ *Streptococcus agalactiae* 2603V/R
└ *Carnobacterium* sp. 17-4
23 └ *Aerococcus urinae* ACS-120-V-Col10a
┌ *Lactobacillus sakei* subsp. sakei 23K
67 ┌ *Pediococcus claussenii* ATCC BAA-344
└ *Lactobacillus sanfranciscensis* TMW 1.1304
39 57 ┌ *Lactobacillus reuteri* DSM 20016
94 └ *Lactobacillus fermentum* IFO 3956
└ *Lactobacillus salivarius* UCC118
┌ *Lactobacillus buchneri* NRRL B-30929
100 99 ┌ *Lactobacillus rhamnosus* GG
└ *Lactobacillus casei* ATCC 334
70 ┌ *Lactobacillus brevis* ATCC 367
49 **DM9218**
99 └ *Lactobacillus plantarum* WCFS1

0.1

Figure 4. Phylogenetic Neighbor-Joining tree of DM9218 based on 16S rDNA sequences. The phylogenetic tree was constructed using the neighbor-joining method by MEGA 6.0 based on 1372 bp of the 16S rDNA sequences. The numbers at the nodes are bootstrap confidence levels (percentage) from 1,000 replicates. The scale bar represents 0.1 substitutions per nucleotide position. Reference sequences were obtained from the GenBank nucleotide sequence database (NCBI).

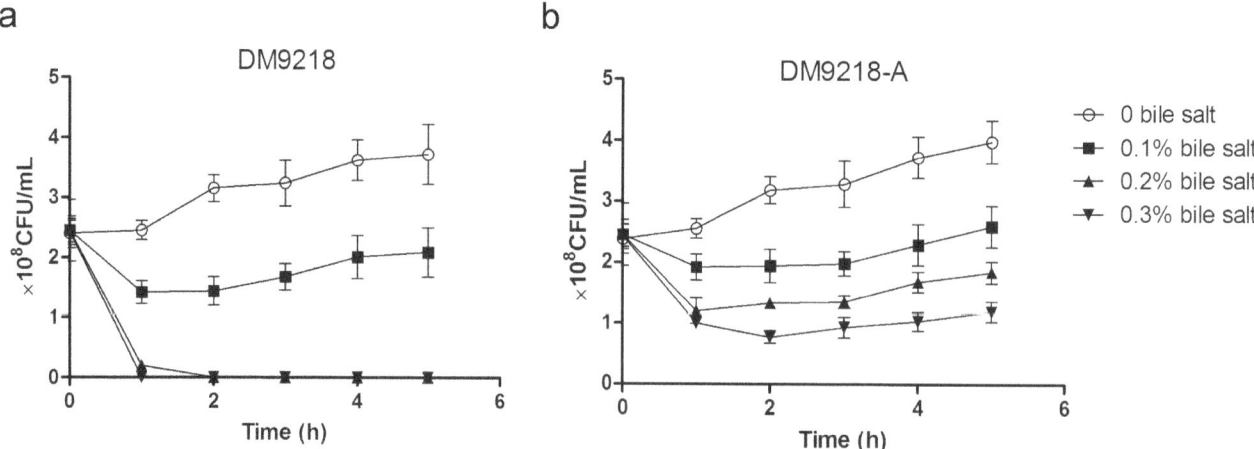

Figure 5. The growth of DM9218 and DM9218-A in bile salt containing MRS media. (a) The growth of DM9218 in bile salt-containing media. (b) The growth of DM9218-A (acclimated DM9218) in bile salt containing media. 10^8 CFU/mL of freshly prepared LAB were inoculated into 10 mL of the autoclaved solutions and incubated at 37°C for 1~5 h before viable plate counting. All values are Means ± S, n = 3.

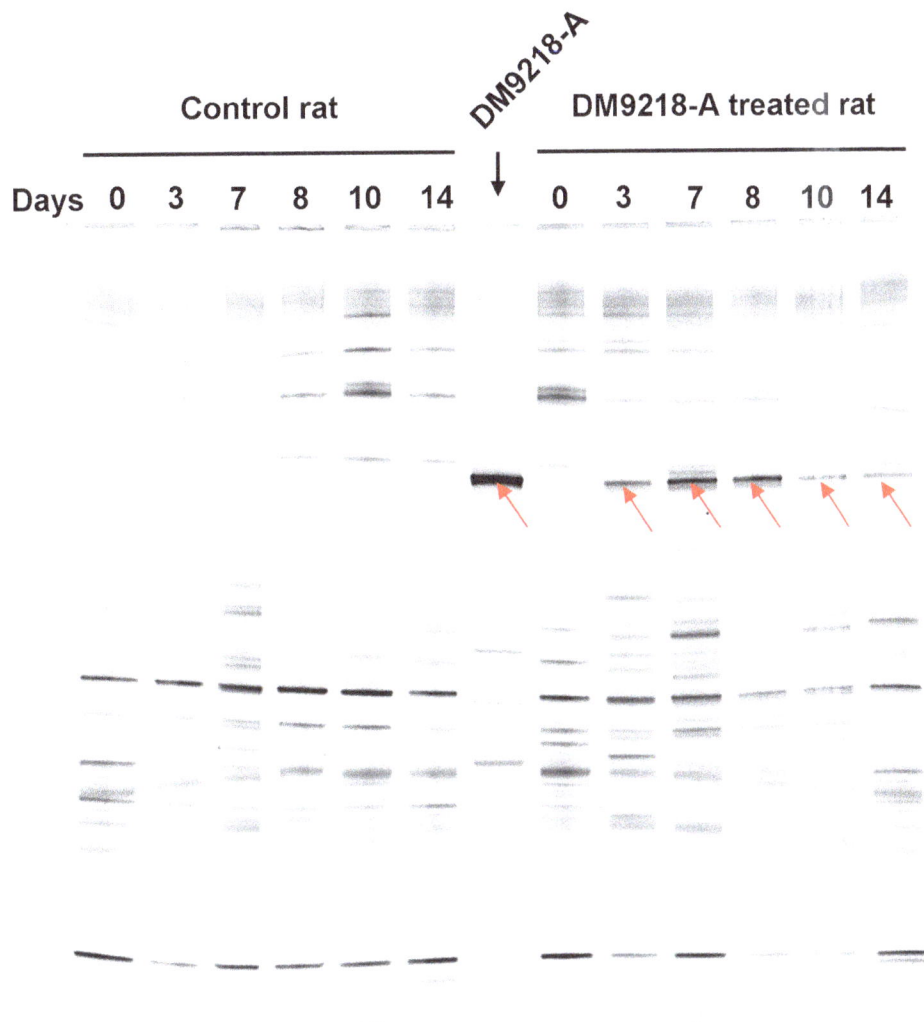

Figure 6. Detection of DM9218-A in rats. The control rats (n = 7) was given food and water *ad libitum*. The DM9218-A treated rats (n = 7) were first intragastrically administrated DM9218-A cell solution (1.2×10^9 CFU/mL in 0.85% NaCl, 1 mL per day) for 7 days, and then given food and water *ad libitum* for 7 days. DGGE profiles of V3 region of 16S rRNA gene amplicons obtained in PCRs with F338/R518 set of primers. Red arrows indicate eluted and sequenced bands. Lanes indicate the DM9218-A pure culture, and the faecal contents of DM9218-A treated or untreated rats taken at indicated sampling days.

testinal transit and food processing [34]. Many of them were proven to have characteristics beneficial to human health [20]. Chinese sauerkraut, the typical representative of Chinese traditional fermented foods, is a rich natural source of LAB. Many species of LAB have been isolated from Chinese sauerkraut and used as probiotics [35]. The objective of this study was to isolate a collection of probiotic LAB from Chinese sauerkraut, screen for purine compound degrading strains and evaluate the potential of the candidate strains for the prevention and treatment of hyperuricemia.

Most of the fifty-five isolates were found to assimilate purine nucleosides at different speeds, with three of them possessing high assimilation capabilities. However it was still unclear whether the purine nucleosides were degraded or incorporated into cells. To clarify this, the degradation of inosine and guanosine by cell-free extracts of DM9218, DM9242 and DM9505 were analyzed. The results showed that the cell-free extracts were able to degrade the purine compounds, and the degradation rates (during 120 min incubation) were comparable to that of living LAB cells.

Microorganisms degrade nucleosides mainly through the biosynthesis of nucleoside hydrolases. These nucleosidases are widely found in plants and microorganisms but have not yet been detected in mammals [36]. Nucleoside hydrolases break the β-glycoside bonds of nucleosides and release nitrogenous bases and pentose. The correlation between inosine and guanosine assimilating abilities of the candidate strains were tested by SAS system and showed that there is a strong positive correlation between the assimilation of inosine and guanosine (Fig. S1). This suggests the existence of nucleoside hydrolases in the tested LAB strains.

To determine whether the tested strains can survive in the gastrointestinal environment and colonize successfully, the biological characteristics of the three candidate strains were evaluated,

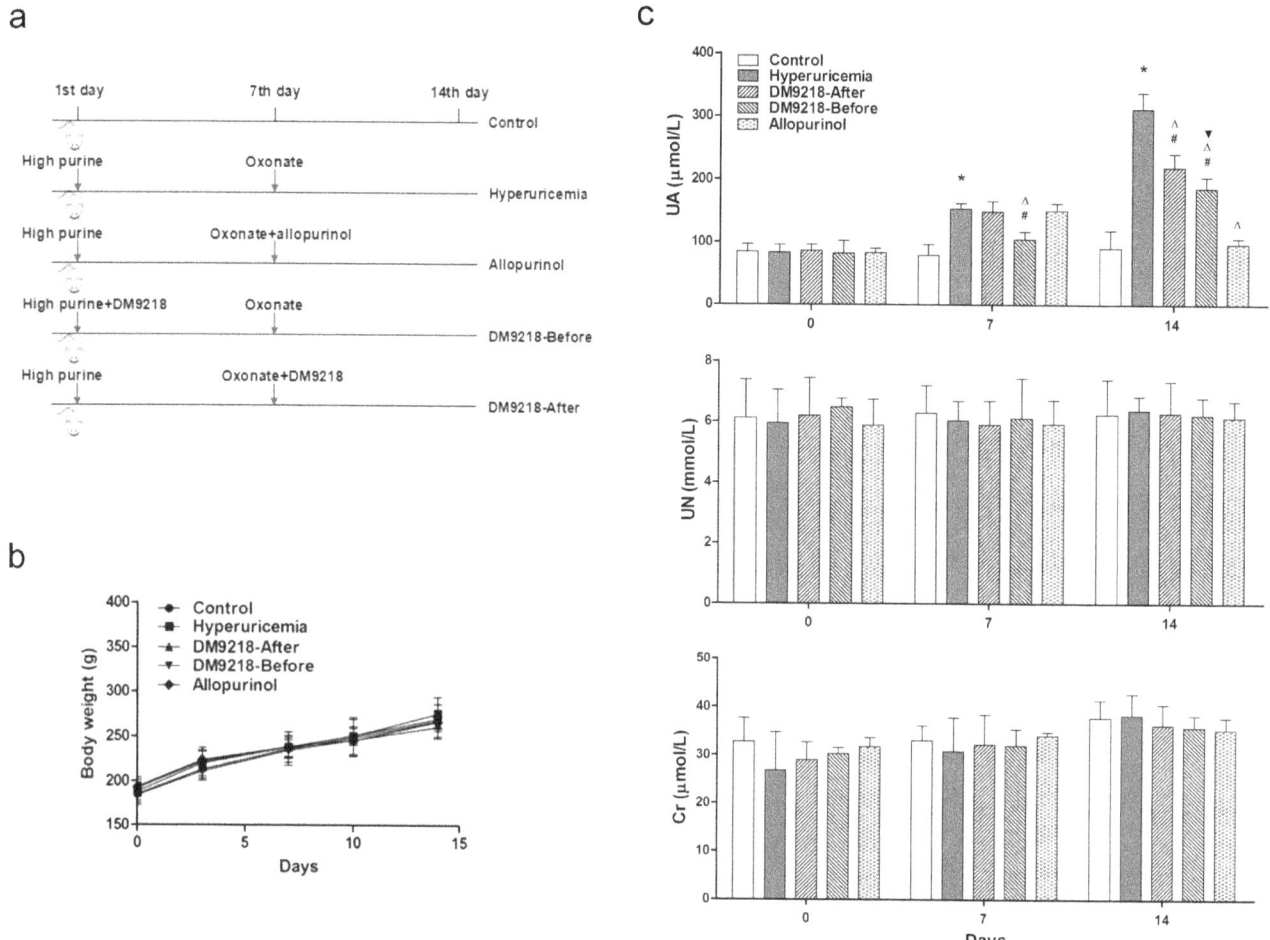

Figure 7. The animal test. (a) The outline of animal test. (b) The body weights of rats were measured randomly at day 0, 3, 7, 10, and 14. Data are Means ± S, n = 8. No differences in body weight gain were detected among the different experimental groups. (c) The level of serum UA (uric acid), * means $P < 0.05$ compare with control rats, △ means $P < 0.05$ compare with hyperuricemic rats, # means $P < 0.05$ compare with allopurinol treated rats, ▼ means $P < 0.05$ compare with DM9218-After group. (d) The level of serum UN (urea nitrogen). (e) The level of serum Cr (creatinine). All values are Means ± S, n = 8.

including tolerance to acid and bile salt, antimicrobial activity, sensitivity to drugs and cell adhesion ability. Results showed that all strains possess certain inhibitory activities against the tested pathogenic bacteria. In general, the antibacterial ability of LAB is derived from the main characteristics: the production of lactic acid by fermentation of sugars, lowering the pH of the environment, making it unsuitable for the growth of other bacteria; and the production of antibacterial substances, such as H_2O_2 and bacteriocins. Testing of the H_2O_2-producing ability of the candidate strains suggested that there is no clear relationship between their antibacterial activities and the production of H_2O_2, it is likely that the inhibition is due to the accumulation of organic acids. The three candidate strains all have moderate tolerance to acid, but poor tolerance to bile salt. DM9218 and DM9242 can survive in 0.1% bile salt containing medium, but DM9505 could not even grow in 0.1% bile salt containing medium. Adhesion to the intestinal epithelium is an important requisite for allowing probiotics to modulate the immune system. Therefore the cell adhesive abilities of the three strains to Caco-2 cells were also evaluated. According to Maccaferri and colleagues [37], more than 40 bacterial cells adhered to one Caco-2 cell is defined as strongly adhesive; thus, DM9242 and DM9218 can be classified as

strongly adhesive strains, while the adhesive ability of DM9505 is relatively weak. Considering the superior purine nucleosides degrading ability and moderate biological characteristics of strain DM9218, it was chosen as the optimal strain for further study.

Bile is secreted by the liver cells and aids the digestion of lipids in the small intestine. It has a strong inhibitory effect on intestinal microbiota, especially on Gram positive bacteria. Therefore, the ability to survive in a bile salt containing environment is a prerequisite condition for DM9218 to exert its biological activity *in vivo*. Because of its poor tolerance to bile salt, we decided to acclimatize DM9218 to bile salt. According to the industrial standard for the attenuation of LAB strains, we gradually acclimatized DM9218 from 0 to 0.3% bile salt containing media. After acclimation, the DM9218-A strain showed increased tolerance to 0.2% and 0.3% bile salt, while its purine degrading ability was not affected.

The survival of DM9218-A in the GI tract of rats was determined by PCR-DGGE. The positive PCR-DGGE result provided evidence that DM9218-A successfully survived in the digestive tract of rats during the process of intragastric administration, as the DNA of dead bacteria is degraded by nucleases released in the GI tract [38]. After the administration of DM9218-

A stopped, the intestinal population of DM9218-A decreased (14th day), although it was still detectable. Moreover, DM9218-A treated rats displayed a microbial profile that was similar to untreated rats, which suggests that no perturbation of the host gut microbiota will be caused by intragastric administration of DM9218-A.

It is currently unclear whether the intake of purine compounds can cause an elevation of serum uric acid. However, it has been difficult to establish an animal hyperuricemia model to test this [39], because commonly used laboratory animals such as rats, mice and rabbits, all express urate oxidase. This enzyme oxidizes the poorly soluble uric acid to water soluble allantoin, thus it is very difficult to induce hyperuricemia in these animals. As for humans and apes, mutation of the liver uricase gene resulted in inactivation of uric acid oxidase [40], therefore the occurrence of hyperuricemia in human is much higher than other mammalian. The methods of inducing hyperuricemia in rats include gene knockout, high purine diet, injection of potassium oxonate and the combination of high purine diet with potassium oxonate injection. The combination method can efficiently elevate serum uric acid in rats, and compared with potassium oxonate or high purine diet alone, the effect is greater and more stable and the duration of hyperuricemia longer. We therefore adopted this method to induce hyperuricemia in rats.

After injection of potassium oxonate and high purine diet, the level of serum uric acid in treated rats was elevated more than 2.45 fold compared with controls. Application of DM9218-A to the hyperuricemic rats significantly decreased the uric acid level, but it was still higher than the allopurinol treated group. This is because allopurinol directly inhibits xanthine oxidase, a key enzyme in the purine metabolite pathway, which blocks the production of uric acid. In contrast, DM9218-A competes with the intestinal epithelium for the absorption of nucleosides in food. Therefore

there are still free purine bases that can be absorbed by the epithelium and eventually metabolized into uric acid. The experimental results also suggest that, once applied to humans, the preventive effect of DM9218-A against hyperuricemia may be stronger than the effect of treatment after hyperuricemia has been induced. No significant differences in the level of serum creatinine and urea nitrogen were observed during the two week study period, which suggests that, because of the short modeling time, no serious renal injury has been induced in the rats.

Conclusions

Although further systemic safety assessment is necessary, our results suggest that DM9218-A is a promising candidate as an adjunctive treatment in patients with hyperuricemia, especially during the onset period of disease. DM9218-A also has potential as a probiotic in the prevention of hyperuricemia in the normal population.

Supporting Information

Figure S1   The correlation between abilities of tested strains to assimilate inosine and guanosine. The assimilation abilities of candidate strains were tested by SAS 9.1 System for the Pearson correlation coefficients.

Author Contributions

Conceived and designed the experiments: JY. Performed the experiments: ML DY LM. Analyzed the data: ML DY LY LM. Contributed reagents/materials/analysis tools: JY. Contributed to the writing of the manuscript: ML LY AX JY.

References

1. Chizynski K, Rozycka M (2005) [Hyperuricemia]. Pol Merkur Lekarski 19(113): 693–696.
2. Doghramji PP, Wortmann RL (2012) Hyperuricemia and gout: new concepts in diagnosis and management. Postgrad Med 124(6): 98–109.
3. Kim SM, Choi YW, Seok HY, Jeong KH, Lee SH, et al. (2012) Reducing serum uric acid attenuates TGF-beta1-induced Profibrogenic progression in type 2 diabetic nephropathy. Nephron Exp Nephrol 121(3–4): e109–121.
4. Ndrepepa G, Cassese S, Braun S, Fusaro M, King L, et al. (2013) A gender-specific analysis of association between hyperuricaemia and cardiovascular events in patients with coronary artery disease. Nutr Metab Cardiovasc Dis 23(12): 1195–1201.
5. Lv Q, Meng XF, He FF, Chen S, Su H, et al. (2013) High serum uric acid and increased risk of type 2 diabetes: a systemic review and meta-analysis of prospective cohort studies. PloS One 8(2): e56864.
6. Shcherbak AV, Kozlovskaia LV, Bobkova IN, Balkarov IM, Lebedeva MV, et al. (2013) [Hyperuricemia and the problem of chronic kidney disease]. Ter Arkh 85(6): 100–104.
7. Mazzali M, Hughes J, Kim YG, Jefferson JA, Kang DH, et al. (2001) Elevated uric acid increases blood pressure in the rat by a novel crystal-independent mechanism. Hypertension 38(5): 1101–1106.
8. Kang DH, Nakagawa T, Feng L, Watanabe S, Han L, et al. (2002) A role for uric acid in the progression of renal disease. J Am Soc Nephrol 13(12): 2888–2897.
9. Nakagawa T, Mazzali M, Kang DH, Kanellis J, Watanabe S, et al. (2003) Hyperuricemia causes glomerular hypertrophy in the rat. Am J Nnephrol 23(1): 2–7.
10. Kosugi T, Nakayama T, Heinig M, Zhang L, Yuzawa Y, et al. (2009) Effect of lowering uric acid on renal disease in the type 2 diabetic db/db mice. Am J Physiol Renal Physiol 297(2): F481–488.
11. Sun SZ, Flickinger BD, Williamson-Hughes PS, Empie MW (2010) Lack of association between dietary fructose and hyperuricemia risk in adults. Nutr Metab 7 (16): 16.
12. Brule D, Sarwar G, Savoie L (1992) Changes in serum and urinary uric acid levels in normal human subjects fed purine-rich foods containing different amounts of adenine and hypoxanthine. J Am Coll Nutr 11(3): 353–358.
13. Zgaga L, Theodoratou E, Kyle J, Farrington SM, Agakov F, et al. (2012) The association of dietary intake of purine-rich vegetables, sugar-sweetened beverages and dairy with plasma urate, in a cross-sectional study. PloS One 7(6): e38123.
14. Yun J, Mattsson J, Schnyder K, Fontana S, Largiader CR, et al. (2013) Allopurinol hypersensitivity is primarily mediated by dose-dependent oxypurinol-specific T cell response. Clin Exp Allergy 43(11): 1246–1255.
15. Klaenhammer TR, Barrangou R, Buck BL, Azcarate-Peril MA, Altermann E (2005) Genomic features of lactic acid bacteria effecting bioprocessing and health. FEMS microbiol Rev 29(3): 393–409.
16. Kanmani P, Satish Kumar R, Yuvaraj N, Paari KA, Pattukumar V, et al. (2013) Probiotics and its functionally valuable products-a review. Crit Rev Food Sci 53(6): 641–658.
17. Isolauri E, Sutas Y, Kankaanpaa P, Arvilommi H, Salminen S (2001) Probiotics: effects on immunity. Am J Clin Nutr 73(2 Suppl): 444S–450S.
18. Ostlie HM, Helland MH, Narvhus JA (2003) Growth and metabolism of selected strains of probiotic bacteria in milk. Int J Food Microbiol 87(1–2): 17–27.
19. Fuller R, Gibson GR (1997) Modification of the intestinal microflora using probiotics and prebiotics. Scan J Gastroenterol 222: 28–31.
20. Chang JH, Shim YY, Cha SK, Chee KM (2010) Probiotic characteristics of lactic acid bacteria isolated from kimchi. J Appl Microbiol 109(1): 220–230.
21. Feliu JX, Cubarsi R, Villaverde A (1998) Optimized release of recombinant proteins by ultrasonication of E. coli cells. Biotechno Bioengeneer 58: 536–540.
22. Jacosben C, Rosenfeldt N, Hayford A, Moller P, Michalsen K, et al. (1999) Screening of probiotic activities of forty seven strains of Lactobacillus spp. by in vitro techniques and evaluation of the colonization ability of five selected strains in humans. Appl Environ Microbiol 65: 4949–4956.
23. Perelmuter K, Fraga M, Zunino P (2008) In vitro activity of potential probiotic Lactobacillus murinus isolated from the dog. J Appl Microbiol 104: 1718–1725.
24. Villegas E, Gilliland SE (1998) Hydrogen peroxide production by Lactobacillus delbrueckii subsp. lactis at 5°C. J Food Sci 63: 1070–1074.
25. Wikler M (2010) National Committee for Clinical Laboratory Standards, Performance standards for antimicrobial susceptibility testing, sixth Information supplement (M100-S20). National Committee for Clinical Laboratory Standards 2010.
26. Dubernet S, Desmasures N, Gueguen M (2002) A PCR-based method for identification of lactobacilli at the genus level. FEMS microbiol Lett 214(2): 271–275.

27. Joossens M, Huys G, De Peter V, Verbeke K, Rutqeerts P, et al. (2011) Dysbiosis of the faecal microbiota in patients with Crohn's disease and their unaffected relatives. Gut 60: 631–637.

28. Ahmadzadeh A, Jalali A, Assar S, Khalilian H, Zandian K, et al. (2011) Renal tubular dysfunction in pediatric patients with beta-thalassemia major. Saudi J Kidney Dis Transpl 22(3): 497–500.

29. Siezen RJ, Francke C, Renckens B, Boekhorst J, Wels M, et al. (2012) Complete resequencing and reannotation of the Lactobacillus plantarum WCFS1 genome. J Bacteriol 194(1): 195–196.

30. Makarova K, Slesarev A, Wolf Y, Sorokin A, Mirkin B, et al. (2006) Comparative genomics of the lactic acid bacteria. Proc Natl Acad Sci U S A 103(42): 15611–15616.

31. Kankainen M, Paulin L, Tynkkynen S, von Ossowski I, Reunanen J, et al. (2009) Comparative genomic analysis of Lactobacillus rhamnosus GG reveals pili containing a human- mucus binding protein. Proc Natl Acad Sci USA 106(40): 17193–17198.

32. Soltani Z, Rasheed K, Kapusta DR, Reisin E (2013) Potential role of uric acid in metabolic syndrome, hypertension, kidney injury, and cardiovascular diseases: is it time for reappraisal? Curr Hypertens Rep 15(3): 175–181.

33. Zoccali C, Mallamaci F (2013) Uric Acid, Hypertension, and Cardiovascular and Renal Complications. Curr Hypertens Rep 15(6): 531–537.

34. Wang CY, Lin PR, Ng CC, Shyu YT (2010) Probiotic properties of Lactobacillus strains isolated from the feces of breast-fed infants and Taiwanese pickled cabbage. Anaerobe 16(6): 578–585.

35. Tao X, Guan Q, Song S, Hao M, Xie M (2012) Dynamic changes of lactic acid bacteria flora during Chinese sauerkraut fermentation. Food Control 26: 178–181.

36. Guimaraes AP, Oliveira AA, da Cunha EF, Ramalho TC, Franca TC (2011) Analysis of Bacillus anthracis nucleoside hydrolase via in silico docking with inhibitors and molecular dynamics simulation. J Mol Model 17(11): 2939–2951.

37. Maccaferri S, Klinder A, Brigidi P, Cavina P, Costabile A (2012) Potential probiotic Kluyveromyces marxianus B0399 modulates the immune response in Caco-2 cells and peripheral blood mononuclear cells and impacts the human gut microbiota in an in vitro colonic model system. Appl Enviro Microb 78(4): 956–964.

38. Bertazzoni-Minelli E, Benini A, Marzotto M, Sbarbati A, Ruzzenente O, et al. (2004) Assessment of novel probiotic Lactobacillus casei strains for the production of functional dairy foods. Int Dairy J 14: 723–736.

39. Sanchez-Lozada LG, Soto V, Tapia E, Avila-Casado C, Sautin YY, et al. (2008) Role of oxidative stress in the renal abnormalities induced by experimental hyperuricemia. Am J Physiol Renal Physiol 295(4): F1134–1141.

40. Johnson RJ, Gaucher EA, Sautin YY, Henderson GN, Angerhofer AJ, et al. (2008) The planetary biology of ascorbate and uric acid and their relationship with the epidemic of obesity and cardiovascular disease. Med Hypotheses 71(1): 22–31.

18

Modulation of Intestinal Microbiota by the Probiotic VSL#3 Resets Brain Gene Expression and Ameliorates the Age-Related Deficit in LTP

Eleonora Distrutti[1]*, **Julie-Ann O'Reilly**[2], **Claire McDonald**[2], **Sabrina Cipriani**[3], **Barbara Renga**[4], **Marina A. Lynch**[2], **Stefano Fiorucci**[4]

1 S.C. di Gastroenterologia ed Epatologia, Azienda Ospedaliera di Perugia, Perugia, Italy, 2 Trinity College Institute for Neuroscience, Department of Physiology, Trinity College, Dublin, Ireland, 3 Dipartimento di Medicina, Università degli Studi di Perugia, Perugia, Italy, 4 Dipartimento di Scienze Chirurgiche e Biomediche, Università degli Studi di Perugia, Perugia, Italy

Abstract

The intestinal microbiota is increasingly recognized as a complex signaling network that impacts on many systems beyond the enteric system modulating, among others, cognitive functions including learning, memory and decision-making processes. This has led to the concept of a microbiota-driven gut–brain axis, reflecting a bidirectional interaction between the central nervous system and the intestine. A deficit in synaptic plasticity is one of the many changes that occurs with age. Specifically, the archetypal model of plasticity, long-term potentiation (LTP), is reduced in hippocampus of middle-aged and aged rats. Because the intestinal microbiota might change with age, we have investigated whether the age-related deficit in LTP might be attenuated by changing the composition of intestinal microbiota with VSL#3, a probiotic mixture comprising 8 Gram-positive bacterial strains. Here, we report that treatment of aged rats with VSL#3 induced a robust change in the composition of intestinal microbiota with an increase in the abundance of *Actinobacteria* and *Bacterioidetes*, which was reduced in control-treated aged rats. VSL#3 administration modulated the expression of a large group of genes in brain tissue as assessed by whole gene expression, with evidence of a change in genes that impact on inflammatory and neuronal plasticity processes. The age-related deficit in LTP was attenuated in VSL#3-treated aged rats and this was accompanied by a modest decrease in markers of microglial activation and an increase in expression of BDNF and synapsin. The data support the notion that intestinal microbiota can be manipulated to positively impact on neuronal function.

Editor: Manabu Sakakibara, Tokai University, Japan

Funding: VSL#3 was supplied by CD Investments s.r.l, Rome, Italy. The funders had no role in study design, data collection and analysis, decision to publish, or preparation of the manuscript.

Competing Interests: The authors declare that they received funding from a commercial source (CD Investments s.r.l, Rome, Italy).

* Email: eleonoradistrutti@katamail.com

Introduction

Age-related changes in the brain contribute to the time-related deterioration in cognitive function. These include neuroinflammatory and oxidative changes with the associated glial activation, as well as loss of synaptic connections and perhaps neurons, and reduced neurogenesis [1]. The deterioration in cognitive function is manifest by poorer spatial learning [2], while the alterations at the level of the synapse are reflected by loss of plasticity, for example a poorer ability of animals to sustain long-term potentiation (LTP). Significantly, when these age-related neuroinflammatory changes are attenuated by treating aged rats with polyunsaturated fatty acids [3], statins [4] or a combination of vitamin D and dexamethasone [5,6], LTP is partially restored. Similarly, the expression of brain-derived neurotrophic factor (BDNF), which induces neurogenesis [7] and enhances the ability of rats to sustain LTP, positively correlates with LTP [8]. Additionally, an increased expression of hippocampal BDNF has been associated with restoration of LTP in middle-aged rats [9].

Whereas the brain is protected to a significant degree by the existence of the blood brain barrier, it has become clear that it is not the immune-privileged organ that was considered in the past [10]. Indeed peripheral infections have been known for many years to impact on neuronal function and the CNS effects of sickness behaviour have been well rehearsed [10]. Peripheral administration of lipopolysaccharide increases inflammatory and oxidative stress in brain, and glial activation in the hippocampus, to a greater extent in aged, compared with young, rats [11]. Consistently, age is associated with increased vulnerability to infections, while infections have been shown to accelerate progression of diseases such as Alzheimer's disease and multiple sclerosis [12,13]. These findings support the notion that a communication network between the CNS and the periphery exists. This communication network is consolidated by the association between psychiatric conditions, like anxiety, and the inflammatory changes in the gastrointestinal tract that typify inflammatory bowel disease [14].

A reciprocal interaction exists between the gut microbiota and CNS function [15]. For example, it has been shown that the stress associated with neonatal maternal separation induces cognitive dysfunction and impacts on composition of gut microbiota [16,17], although the effects change with the time and duration of separation [18,19]. Similarly, mice kept in germ-free conditions, in which development of the hypothalamic-pituitary-adrenal axis is impaired, exhibit deficits in cognitive function [20]. Interestingly, stress negatively impacts on cognitive function in mice infected with *Citrobacter rodentium*, but this deficit is normalized if animals are pretreated with probiotics [20], which effectively attenuated inflammatory changes in the colon and infection-induced decrease in hippocampal BDNF.

Anti-inflammatory effects of the probiotic strains, *Lactobacillus rhamnosus* [21] and *Lactobacillus reuteri* [22] and VSL#3 [23] have been reported. VSL#3 is a mixture of 8 different strains of bacteria, namely *Streptococcus thermophilus* DSM24731, *Bifidobacterium breve* DSM24732, *Bifidobacterium longum* DSM24736, *Bifidobacterium infantis* DSM24737, *Lactobacillus acidophilus* DSM24735, *Lactobacillus plantarum* DSM24730, *Lactobacillus paracasei* DSM24733, *Lactobacillus delbrueckii subspecies Bulgaricus* DSM24734. VSL#3 administration to mice rendered colitic by exposure to the barrier-braking agent sodium dextran sulphate, attenuated changes in COX2, iNOS, TNFα and IL-6 in the colon [24]. This effect has been linked to the ability of the probiotic mixture to increase IL-10 [25]. Further, VSL#3 has been shown to reduce pain in an animal model of visceral hypersensitivity (i.e. the hyperalgesia induced by neonatal maternal separation) and this was associated with treatment-induced changes in expression of genes encoding for proteins involved in nociception and inflammation [26], further emphasizing a modulatory role for VSL#3 in stressful conditions. Clinical studies have confirmed the VSL#3 is effective in reducing inflammation and symptoms in clinical settings [27,28].

In the present study, we set out to assess whether treatment with VSL#3 modulates neuronal functions and LTP in young and aged rats. We found that the age-related deficit in LTP was markedly attenuated in rats receiving VSL#3, while gene array analysis revealed that the probiotic mixture resets the expression of several genes in the brain.

Materials and Methods

Animals

Young (3 months; 250–350 g) and aged (20–22 months; 550–600 g) male Wistar rats (Bantham and Kingman, UK) were housed in a controlled environment (temperature: 20–22°C; 12:12 h light/dark cycle) in the BioResources Unit, Trinity College, Dublin. Young and aged rats were subdivided into 2 groups: those which were given VSL#3 (CD Investments s.r.l, Rome, Italy) at the dose of 12.86 bn living bacteria/kg/day [29] in maple syrup (90–160 μl) for 6 weeks [young VSL#3-treated (YV) and aged VSL#3-treated (AV)] and control rats which received only maple syrup [young control-treated (YC) and aged control-treated (AC)]. Experiments started at day 0 and ended after 6 weeks (day 42). Rats had free access to food and water and were maintained under veterinary supervision for the duration of the experiment. All experiments were carried out under licence from the Department of Health and Children (Ireland) and with ethical approval from Trinity College Ethical Committee.

Analysis of intestinal microbiota: DNA extraction, amplification, digestion and fragment sizing

To collect stool samples at the beginning (day 0) and end of the treatment period (day 42), rats were individually placed in a clean cage separated with paper and faecal pellets were placed in individual tubes and stored at −80°C. Samples were shipped from Trinity College Dublin to the University of Perugia on dry ice. DNA was extracted from 200 mg of frozen stool samples using the QIAamp DNA Stool mini Kit (Qiagen) according to manufacturer's instructions. Primers 8f (AGAGTTTGATCCTGGCTCAG) and 536r (GWATTACCGCGGCKGCTG) were applied to 200 ng DNA in order to amplify a part of the 16S rRNA using the PCR protocol for Phusion High Fidelity DNA Polymerase (New England Biolabs). Forward primers were fluorescently labeled (WellRED D4dye, Sigma-Proligo, St. Louis, MO) to allow detection of the fragments by capillary electrophoresis. The polymerase chain reaction (PCR) was as follows: 98°C for 30''; 40 cycles at 98°C for 10''; 61°C for 30''; 72°C for 30''; and a final extension at 72°C for 10 minutes. The PCR product (~528 base pairs) was purified using the QIAmp PCR purification kit (Qiagen). To produce terminal restriction fragments (T-RF) 10 μl of PCR product was digested using the restriction enzyme Hha I (New England Biolabs). The mix was adjusted to a final volume of 20 μl with DNase/RNase-free water and the DNA was digested at 37°C for 4 hours. The precise length of T-RF amplicons was determined by performing capillary gel electrophoresis with a CEQ 8000 Genetic Analysis System (Beckman Coulter). Four μl of fluorescently labeled fragments, 35.5 μl of sample loading solution (SLS) (Beckman Coulter) and 0.5 μl of 600 bp DNA standard size (Beckman Coulter) were mixed and separated using the frag4 protocol. An electropherogram with peaks of different size was obtained for each stool sample. Fragment analysis was performed using CEQ software version 9.0.

Bioinformatic analysis of T-RFLP data

MiCA (Microbial Community Analysis III) on-line software (http://mica.ibest.uidaho.edu/about.php) was used to build a putative reference database of T-RF of the gut. The analysis was performed using as reference the H.Q. database. Primers 8f and 536r, Hha I restriction enzyme and T-RFLP data obtained from CEQ software (Fragment sizes, migration time and peak area) were applied to PAT (Phylogenetic Assignment tool of MiCA) and a library of probable species was obtained for each sample. Almost all species found in this study belong to the four most populated bacterial phyla, namely *Bacteroidetes, Proteobacteria, Firmicutes*, and *Actinobacteria*. Thus, values reported in this analysis are expressed as percentage of these four phyla.

Microarray analysis

According to Data Availability Statement of PlosOne, the data discussed in this publication have been deposited in NCBI's Gene Expression Omnibus (GEO) [30] and are accessible through GEO Series accession number **GSE51381**. All microarray analysis were performed by MiltenyiBiotec, GmbH Bioinformatics, German. Microarray analysis was performed on cortical samples from young control-treated rats (YC), young VSL#3-treated rats (YV), aged control-treated rats (AC), aged VSL#3-treated rats (AV).

Preparation, amplification and hybridization of RNA

RNA was isolated from rat tissue samples by using standard RNA extraction protocols (Trizol) and the RNA was quality-checked via the Agilent 2100 Bioanalyzer platform (Agilent Technologies). Four replicates were assessed per sample. For the linear T7-based amplification step, 100 ng of each RNA sample was used. To produce Cy3-labeled cRNA, the RNA samples were amplified and labelled using the Agilent Low Input Quick Amp Labeling Kit (Agilent Technologies) following the manufacturer's protocol. Yields of cRNA and the dye-incorporation rate were measured with the ND-1000 Spectrophotometer (NanoDrop Technologies).

The hybridization procedure was performed according to the Agilent 60-mer oligo microarray processing protocol using the Agilent Gene Expression Hybridization Kit (Agilent Technologies). Briefly, 0.6 μg Cy3-labeled fragmented cRNA in hybridization buffer was hybridized overnight (17 hours, 65°C) to Agilent Whole Rat Genome Oligo Microarrays 8×60K using Agilent's recommended hybridization chamber and oven. Fluorescence signals of the hybridized Agilent Microarrays were detected using Agilent's Microarray Scanner System (Agilent Technologies).

Image and data analysis

The Agilent Feature Extraction Software (FES) was used to read out and process the microarray image files. The software determines feature intensities (including background subtraction), rejects outliers and calculates statistical confidences. For determination of differential gene expression FES derived output data files were further analyzed using the Rosetta Resolver gene expression data analysis system (Rosetta Biosoftware). All samples were labelled with Cy3. The ratio experiments are designated as control versus (vs) sample experiments (automated data output of the Resolver system) with the ratios calculated by dividing sample signal intensity through control signal intensity.

The bioinformatics data analysis of eleven microarray datasets obtained from one-colour hybridization of rat RNAs on Agilent Whole Rat Genome Oligo Microarrays 8×60K was performed. Pre-processing of the data, including normalization and correlation analysis, was followed by differential gene expression analysis (DGA) for AC versus YC, AC versus AV, YC versus YV and YC versus AV. These analyses aimed to distinguish changes in expression among the four groups of samples so that eight discriminatory gene sets (for each group up- and downregulated genes) were analyzed. A combination of statistical methods and the magnitude of expression difference (fold change) were applied in order to identify genes with differential expression between two sample groups. For the detection of discriminatory expression, genes were selected that show a statistically significant deviation in the test compared with the reference group (ANOVA p-value ≤ 0.05, Tukey p≤0.05). Average expression value was at least 1.5-fold higher or lower than the reference average. To enable construction of a red/green heatmap, the expression values were converted to "virtual ratios" by referencing each individual intensity signal to median of all intensities. The base-2 logarithms of these virtual ratios were used to prepare the heatmap display. A comparison of the discriminatory gene sets among different groups is shown using Venn diagrams.

Microarray validation by PCR

Quantification of the expression of selected genes was performed by quantitative real-time PCR (qRT-PCR). 1 μl of the remaining RNA from cortex samples that were used for gene array was incubated with DNase I and reverse-transcribed with Superscript II (Invitrogen) according to manufacturer specifications. For real-time PCR, 10 ng of template was used in a 20μl reaction containing 0.2 μM of each primer and 10 μl of KAPA SYBR FAST (KapaBiosystem). PCR primers were designed using the software PRIMER3-OUTPUT using published sequence data obtained from the NCBI database.

rPLA2G3 s: gcaccaacgaaggagaagag
rPLA2G3 as: gcaagggtgagatggtttgt
rAlox15s: tacctgtggttggttggaca
rAlox15as: ggcgtcatccgtgagataat
rNid2s:gccttcagagccagatgttc
rNid2as:ggtcctccagtcctaccaca

Analysis of LTP in vivo

At the end of the 6 week period during which animals received VSL#3 in maple syrup or maple syrup alone, rats were anaesthetized with urethane (1.5 g/kg urethane intraperitoneally) and assessed for their ability to sustain LTP as previously described [31]. Briefly, when rats reached a state of deep anaesthesia, verified by the absence of a pedal reflex, they were placed in a stereotaxic frame and a bipolar stimulating electrode was slowly lowered into the perforant path (4.4 mm lateral and 8.0 mm posterior to Bregma) and a unipolar recording electrode was slowly lowered into the dorsal cell body region of the dentate gyrus (2.5 mm lateral and 3.9 mm posterior to Bregma). The depth of the electrodes was adjusted in the cell body region to obtain potentials with maximum amplitude, and the stimulus strength was chosen to ensure that a population spike of approximately 1 mV was evident. Test shocks were delivered at 30 s intervals for up to 1 hour to establish a stable baseline and, after this time, recordings were collected for 15 minutes prior to delivery of 3 trains of high-frequency stimuli (250 Hz for 200 ms; 30 s inter-train interval). Recording at test shock frequency resumed for the remaining 45 minutes of the experiment. The slope of the excitatory post-synaptic potential (epsp; mV/ms) was used as a measure of excitatory synaptic transmission in the dentate gyrus. At the end of the experiment, rats were killed by cervical dislocation, the hippocampus and cortex dissected free and snap frozen for later analysis.

RT-PCR and Western blot analysis from hippocampal tissue

Total RNA was extracted from snap-frozen hippocampal tissue using a NucleoSpin RNAII isolation kit (Macherey-Nagel Inc., Germany) as described by the manufacturer. RNA integrity and total RNA concentration were assessed, and cDNA synthesis was performed as described previously [32]. Real-time PCR was performed using Taqman Gene Expression Assays (Applied Biosystems, Germany) which contain forward and reverse primers, and a FAM-labeled MGB Taqman probe for each gene of interest. The assay IDs for the genes examined in this study were as follows: CD11b (Mm001271265_m1), CD68 (Rn01495631_g1), GFAP (DETAILS), BDNF (DETAILS and NGF (DETAILS). Each well contained 20 μl comprising diluted cDNA (9 μl), primer (1 μl) and Taqman Universal PCR Master Mix (10μl) and samples were assayed in duplicate. Each run (40 cycles) consisted of 3 stages, 95°C for 10 min, 95°C for 15 sec for each cycle (denaturation) and finally the transcription step at 60°C for 1 min. The endogenous control, used to normalize gene expression data, was β-actin and β-actin expression was conducted using a gene expression assay containing forward and reverse primers (primer limited) and a VIC-labeled MGB Taqmanprobe (Applied Biosystems, Germany; Assay ID: 4352341E). Gene expression was calculated relative to the endogenous control samples and to the control sample giving an RQ value (2^{-DDCt}, where CT is the threshold cycle).

For Western blot analysis, hippocampal homogenates were prepared in lysis buffer as described [33]. Lysates were centrifuged (20000 x g, 12 min) and supernatant samples (10 μg) were added to NuPAGE LDL sample buffer, heated at 70°C for 10 min and

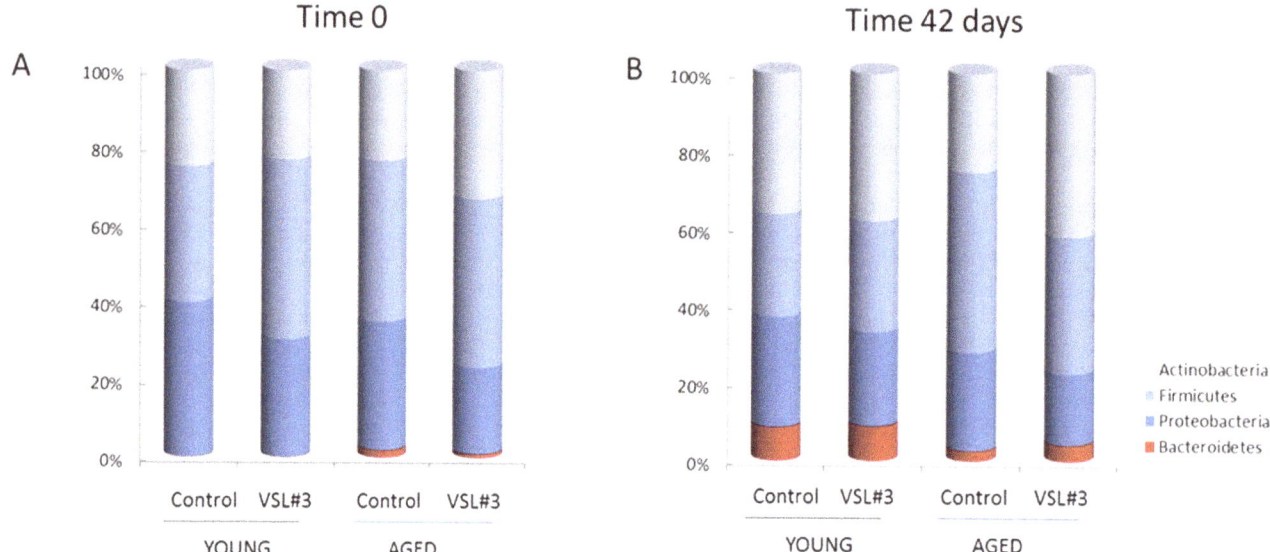

Figure 1. Phylogenetic profiles of gut microbiote in young and aged rats treated with a control diet or VSL#3. (A-B). Terminal restriction fragments (T-RFs) of microbial communities from feces of young control rats (YC), young rats administered VSL#3 (YV), aged control rats (AC) and aged rats administered VSL#3 (AV) from day 0 (panel A) and 42 (panel B)were assigned to hierarchical taxonomic groups using the on-line software MiCA. Values of taxonomic ranks are expressed in percentage as a proportion of the reference library used for analysis (H.Q. database).

separated on 4–12% gradient gels (Invitrogen, UK). Proteins were transferred to nitrocellulose membrane (Sigma, UK) and blocked for 1 hour in Tris-buffered saline-0.05% Tween-20 (TBS-T) and 5% bovine serum albumin (BSA). Membranes were incubated overnight at 4°C with anti-synaptophysin (1:5,000; Sigma, UK) in TBS-T/1% BSA, washed and incubated with secondary antibody (1:5000 in 5% BSA/TBS-T; Sigma, UK) for 2 h. Immunoreactive bands were detected using enhanced chemiluminescence (Amersham, UK) and blots were stripped (Re-blot Plus; Chemicon International, Temecula, CA) and re-probed using anti-β-actin (1:4000 in 5% BSA/TBS-T; Sigma, UK) and a peroxidase conjugated secondary antibody (1:1000 in 5% BSA/TBS-T; Sigma, UK). Bands were quantified by densitometry (Labworks v4.5, MediaCybernetics, Bethesda, MD). Values were normalized for protein loading using the actin protein expression values.

Statistical analysis

Data were analyzed using either Student's t-test for independent means, or analysis of variance (ANOVA) followed by post hoc Student Newman–Keuls test or Tukey test to determine which conditions were significantly different from each other. Data are expressed as means with standard errors.

For microarray analysis a combination of statistical methods (ANOVA p-value ≤0.05, Tukey p≤0.05) and the magnitude of expression difference (1,5-fold change higher or lower than the reference average) was applied.

Results

Analysis of intestinal microbiota composition

At day 0, young control-treated rats (YC) and young rats treated with VLS#3 (YV) showed only 3 of the 4 most populated bacterial phyla, as *Bacteroidetes* were not detected (Figure 1A). In YC at day 0, *Proteobacteria* was the most representative bacterial family ($39.8\pm3.3\%$), followed by *Firmicutes*, ($35.1\pm2.0\%$) and *Actinobacteria* ($25\pm3.9\%$). Noteworthy, at day 0 there was no significant difference in bacterial composition of YV in comparison with YC,

as *Bacteroidetes* were not detected, while *Firmicutes* represented $47\pm4.6\%$ of the total, followed by *Actinobacteria* ($30\pm6.8\%$) and *Proteobacteria* ($23\pm2.3\%$) (Figure 1A). In samples from YC at day 42, the bacterial composition spontaneously changed, as *Bacteroidetes* appeared ($9\pm1.8\%$ of the total; p<0.05 versus day 0), while a trend towards an increased percentage of *Actinobacteria* ($36\pm8.8\%$), and a decreased percentage of both *Firmicutes* and *Proteobacteria* was observed ($27\pm3.2\%$ and $28\pm8\%$ respectively) (Figure 1B). At day 42, samples from YV showed the same microbiota composition as that of YC, as *Bacteroidetes and Actinobacteria* phyla increased ($9\pm2.7\%$ and $38\pm11\%$ respectively) while both *Firmicutes and Proteobacteria* decreased ($29\pm1.7\%$ and $24\pm9.1\%$ respectively) (Figure 1B). These results indicate that microbiota spontaneously changed during the 6 week period of the experiment, and this effect was independent of VSL#3 administration.

At day 0, the percentage of the 4 main bacterial families was similar in the aged control-treated rats (AC) and aged VSL#3-treated rats (AV), but the *Bacteroidetes* phylum was significantly greater in AC in comparison with YC (p<0.05; Figure 1A). Microbiota composition did not significantly change between day 0 and day 42 in AC rats, while in AV there was a significant increase in *Bacteroidetes* phylum at day 42 in comparison with day 0 (p<0.05; Figure 1A and B). Furthermore, in AV there was a trend in the increase of *Actinobacteria* and a reduction of *Firmicutes* between day 0 and day 42 (Figure 1A and B).

Global microarray analysis of brain tissue

Global microarray analysis, performed on cortical tissue obtained from the rats in each of the 4 groups, AC, YC, AV, and YV, screened 30,367 genes and detected several differentially-expressed genes (data are available at website http://www.ncbi. nlm.nih.gov/geo/query/acc.cgi?token = kjirwsoqzlapbit&acc = GS E51381). When the combination of statistical methods (p-value < 0.05) and the magnitude of gene expression difference (fold change at least ±1.5) was applied, 333 genes (1.1% of total) were found to be significantly modulated in aged rats compared with young rats

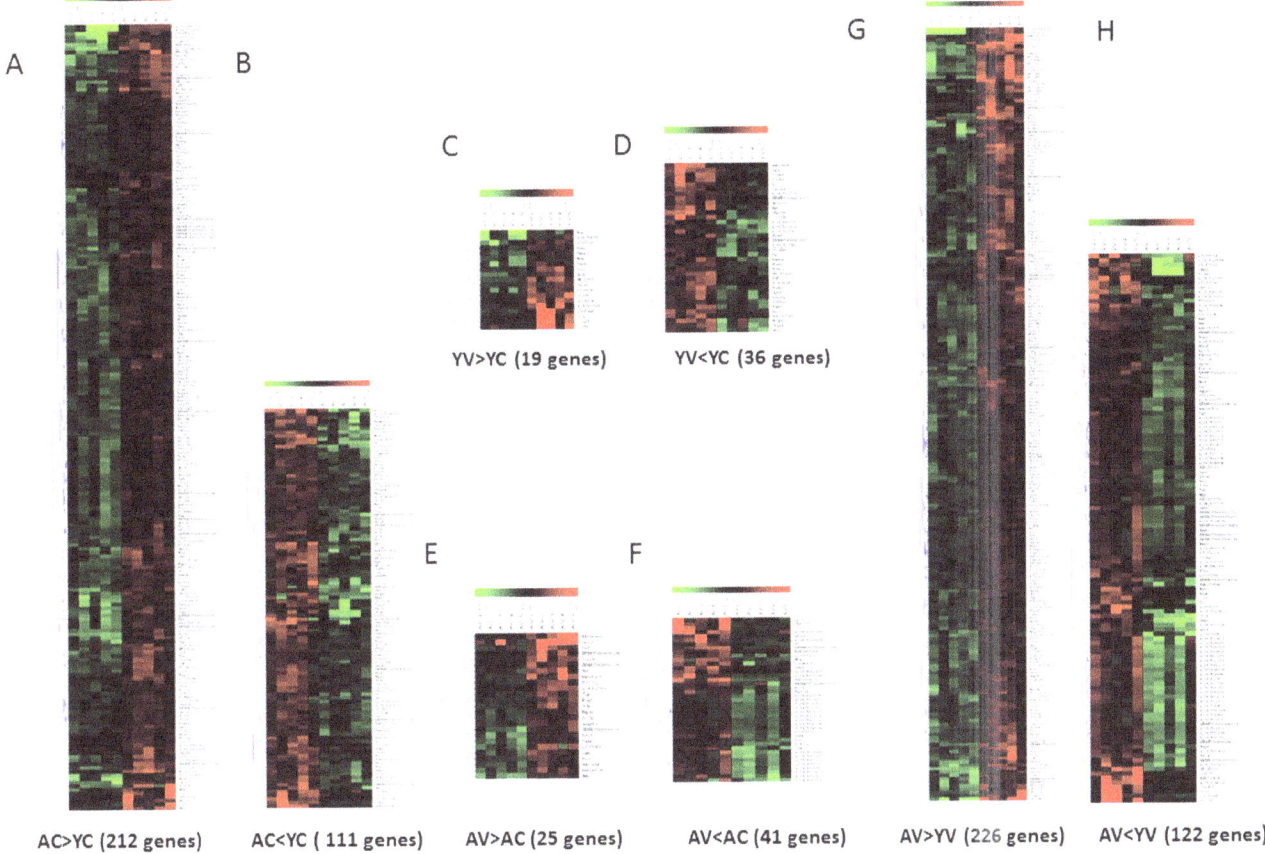

AC>YC (212 genes)   AC<YC (111 genes)   AV>AC (25 genes)   AV<AC (41 genes)   AV>YV (226 genes)   AV<YV (122 genes)

Figure 2. Microarray analysis of genes in cortical tissue of young and aged rats shows modulation by age and VSL#3. (A–B) Heat maps of the significantly regulated genes (p<0.05 and 1.5-fold up- or downregulation) in cortical tissue of control-treated aged rats (lanes AC1–AC5) in comparison with control-treated young rats (lanes YC1-YC5). We detected changes in the expression of 323 genes (about 1% of total). Thus, 212 were upregulated and 111 downregulated. VSL#3 treatment effectively modulates the expression of genes in cortical tissue of both young (panel C and D for upregulated and downregulated genes respectively) and aged (panel E and F for upregulated and downregulated genes respectively) rats. Comparison between YV and AV demonstrates that more that 300 genes are significantly modulated (panel G and H for upregulated and downregulated genes respectively). Microarray data from 5 rats are presented.

(Figures 2–4). The hierarchical clustering of these genes provided good separation based on the expression of the genes in the four groups of rats. In Figures 2–4, red and green colours indicate upregulated and downregulated genes respectively, while black colour indicates those genes whose expression was not changed. As shown in Figures 2A and 2B, when the pattern of gene expression in the brain cortex of aged control-treated rats (lanes AC1-AC5) was compared with that of young control-treated rats (lanes YC1-YC5), we detected a significant up- and downregulation of 212 and 111 genes, respectively. The complete list of these genes is available at the above mentioned website. We have then investigated whether VSL#3 could modulate these age-associated changes. To this end, the pattern of gene expression in the brain cortex of aged rats treated with VLS#3 (lanes AV1-AV5) was compared with the pattern of gene expression in tissues obtained from control-treated young rats (lanes YC1-YC5). The results of this comparison are shown in Figure 3 and 4. We found that the two animal groups differ for expression of 339 genes, 226 of which were significantly upregulated (Figure 3A) while 113 were downregulated (Figure 4A). The effect of VSL#3 was further analysed by using Venn diagrams which demonstrate that 8 genes which were upregulated in AC in comparison with YC, were then downregulated by VLS#3 treatment (AV, Figure 3B), while 3 downregulated by VLS#3 treatment (AV, Figure 4B),

genes that were downregulated in AC in comparison with YC were then upregulated by VLS#3 treatment (AV, Figure 4B), indicating that the probiotic intervention at least partially reversed the effects of aging on these genes. The list of these genes is shown in Figure 3C and 4C.

Further analysis of 226 genes that were upregulated in AV in comparison with YC, demonstrated that 102 new genes were induced by VSL#3 treatment *per se* (Figure 3B). Similarly, among the 113 genes that were downregulated in AV in comparison with YC, 69 were new genes specifically modified by administering aged rats with VSL#3 (Figure 4B).

The effect of probiotic treatment was further investigated by comparing the pattern of gene expression in the cortical tissues obtained from both young and aged rats treated or not treated with the VSL#3. Of relevance, we found that the probiotic diet *per se* effectively modulated the expression of several genes in both groups. This effect was downregulatory in nature as demonstrated by the fact that, while VSL#3 administration induced a significant modification in expression of 55 genes in YV in comparison with YC, 19 genes were upregulated and 36 genes were downregulated (Figure 2C and D respectively). Consistently, among the 66 genes whose expression was modulated by administering aged rats with

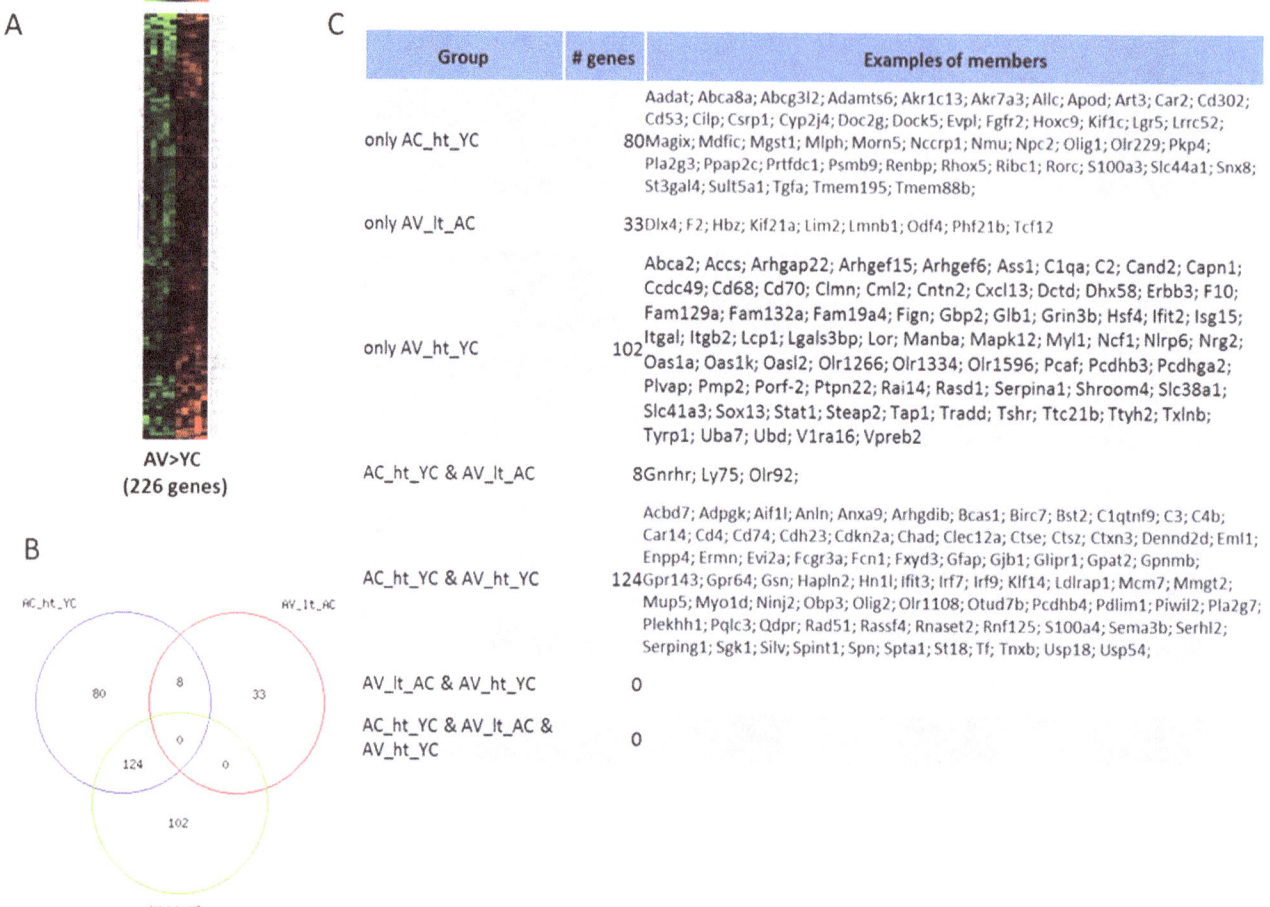

Figure 3. Microarray analysis of genes in cortical tissue of young and aged rats shows modulation by age and VSL#3. (A) Heat maps analysis of changes caused by VSL#3 administration in VSL#3-treated aged rats (lanes AV1-AV5) in comparison with control-treated young rats (lanes YC1-YC5) demonstrate that expression of 226 genes was modulated >1.5-fold (p<0.05). (B) Venn diagram of modulated genes indicates that age was associated with an upregulation of 226 genes;8 of these genes were downregulated by the probiotic treatment. (C) The table presents the complete list of all genes that were significantly modulated in control-treated aged rats in comparison with control-and VSL#3-treated young rats and VSL#3-treated aged rats. Microarray data from 5 rats are presented. ht: higher than; lt: lower than.

VSL#3, 25 were upregulated while 41 were downregulated (AV vs AC; Figure 2E and F and Figure 3B and 4B respectively).

Finally, the effect of VSL#3 administration was further analysed by comparing the effect of VSL#3 administration to aged and young rats. This comparison demonstrate that young and aged rats administered VSL#3 differ in the expression of 338 genes, 226 of which were upregulated and 122 genes were significantly downregulated (Figure 2G and H respectively).

PCR analysis performed on a subset of genes strongly involved in neurodegenerative processes confirmed the microarray data for Alox15 gene that was downregulated in AC in comparison with YC (p<0.05 versus YC; Figure S1B) and PLA2G3 gene that was overexpressed in AC and return to levels of YC with probiotic diet (p<0.05 versus YC; p<0.05 versus AC; Figure S1C).

Analysis of LTP and markers of inflammation in the hippocampus

To evaluate whether VLS#3 treatment exerted any effect on synaptic plasticity, we assessed the effect of delivery of a high frequency train of stimuli to the perforant path on changes in the dentate gyrus. The high frequency train of stimuli induced an immediate and sustained increase in epsp slope in YC and YV; the

mean percentage increases in epsp slope in the last 10 minutes of the experiment in these 2 groups of rats were 122.2 (±0.92, SEM) and 136.3 (±0.37, SEM). The change in epsp slope following high frequency stimulation in AC was significantly reduced and the mean percentage change in the last 10 minutes of the experiment was 98.0 (±0.56, SEM). In contrast, AV sustained LTP in a manner similar to the YC and the mean percentage change in epsp slope in the last 10 minutes of the experiment in AV was 132.2 (±0.69, SEM; Figure 5A).

Previous data have indicated that the age-related deficit in LTP is associated with microglial activation and therefore we assessed expression of 2 markers of activation, CD68 mRNA and CD11b mRNA. The data show a significant effect of age in each marker (p<0.05 AC vs YC; Figure 5B and C). CD68 mRNA and CD11b mRNA were reduced in AV compared with AC but this difference reached significance only in the case of CD68 mRNA (p<0.05 AV vs AC; Figure 5B and C).

Impaired LTP has also been associated with decreased BDNF expression and with decreased expression of synaptic proteins and therefore we assessed BDNF mRNA in tissue prepared from these rats showing a significant VSL#3 treatment effect in both young and aged rats (p<0.05 YV vs YC; p<0.05 AV vs AC; Figure 6A).

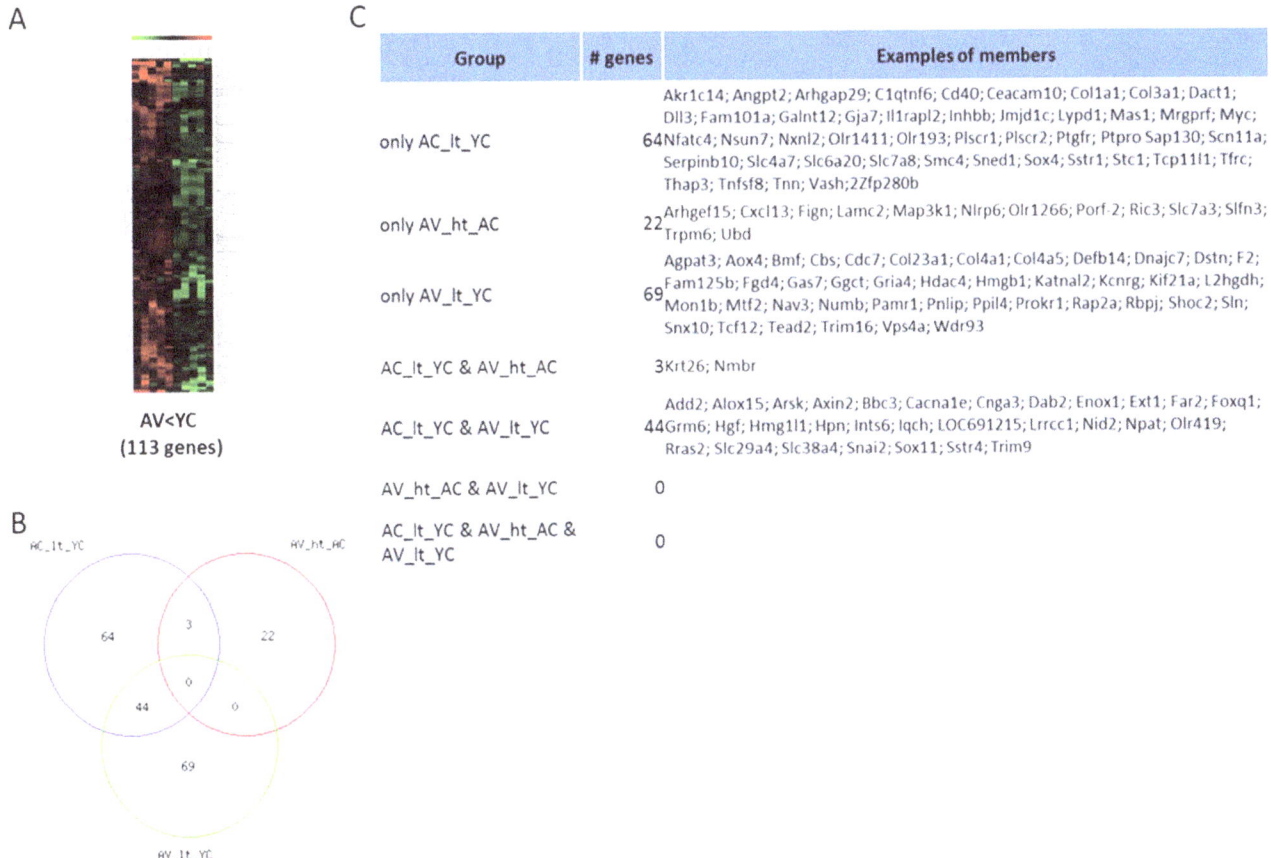

Figure 4. Microarray analysis of genes in cortical tissue of young and aged rats shows modulation by age and VSL#3. (A) Heat maps analysis of changes caused by VSL#3 administration in VSL#3-treatedagedrats (lanes AV1-AV5) in comparison with control-treated young rats (lanes YC1-YC5) demonstrate that expression of 113 genes was modulated <1.5-folds (p<0.05). (B) Venn diagram of modulated genes shows that age was associated with a downregulation of 113 genes;3 of these genes were upregulated by the probiotic treatment. (C) The table presents the complete list of all genes that were significantly modulated in control-treated aged rats in comparison with control- and VSL#3-treated young rats and VSL#3-treated aged rats. Microarray data from 5 rats are presented. ht: higher than; lt: lower than.

In addition, a significant VSL#3 treatment effect on synapsin was observed in aged rats (p<0.05 AV vs AC; Figure 6B), whereas a significant treatment effect was observed in the postsynaptic protein drebrin in young rats (p<0.05 YV vs YC; Figure 6C). We also observed a significant age effect in syntaxin (p<0.05 AC vs YC; Figure 6D), but no change in PSD95 was observed (Figure 6E). These data suggest that VSL#3 may have a synaptotrophic effect driven by BDNF.

Discussion

A growing body of evidence supports the suggestion that the gut microbiota might have an effect on normal human behavior and that alterations in its composition or metabolism can play a role in the pathophysiology of psychiatric and neurologic diseases and on neuronal function which is impaired with age.

In the present study comparison between young and aged animal allows us to investigate whether age influenced intestinal microbiota composition, brain functions and brain gene expression and whether treatment of aged rats with the probiotic VSL#3 attenuated these alterations. By bioinformatic approach we have made a wide array of comparisons among all groups of animals including adult control rats and adult rats administered with VSL#3 (i.e. AV vs AC), but also we have arnessed to investigate whether the probiotic treatment reversed the age-induced

alterations of brain functions by comparing this group with young control rats (i.e. AV versus YC). Although the latter comparison might be less circumstantial, the data shown in Figures 3-6 strongly support the notion that age related changes might be attenuated by the probiotic intervention, which also ameliorates the age-related deficit in LTP, a commonly-used method for assessing synaptic plasticity.

The composition of gut microbiota changes with age. We detected both spontaneous and VSL#3-induced changes in the composition of the rat gut microbiota. Four main bacterial phyla were identified and an age-dependent fluctuation in their relative abundance was documented throughout the study period. At day 0, young and aged rats differed for *Bacteroidetes* composition; this phylum was significantly more abundant in aged rats, while the other bacterial phyla were similar. Interestingly, at day 42, the relative percentage of *Bacteroidetes* increased also in young rats while *Firmicutes* decreased, indicating that bacterial population spontaneously changes during life. These observations support the notion that age represents an important regulatory factor for the gut microbiota both in rats [34] and humans [35,36].

The data indicate that long-term VSL#3 administration reshapes the intestinal microbiota only in aged rats. The most significant change was an increase in the *Bacteroidetes* and *Actinobacteria* amount in the aged rats treated with VSL#3. This

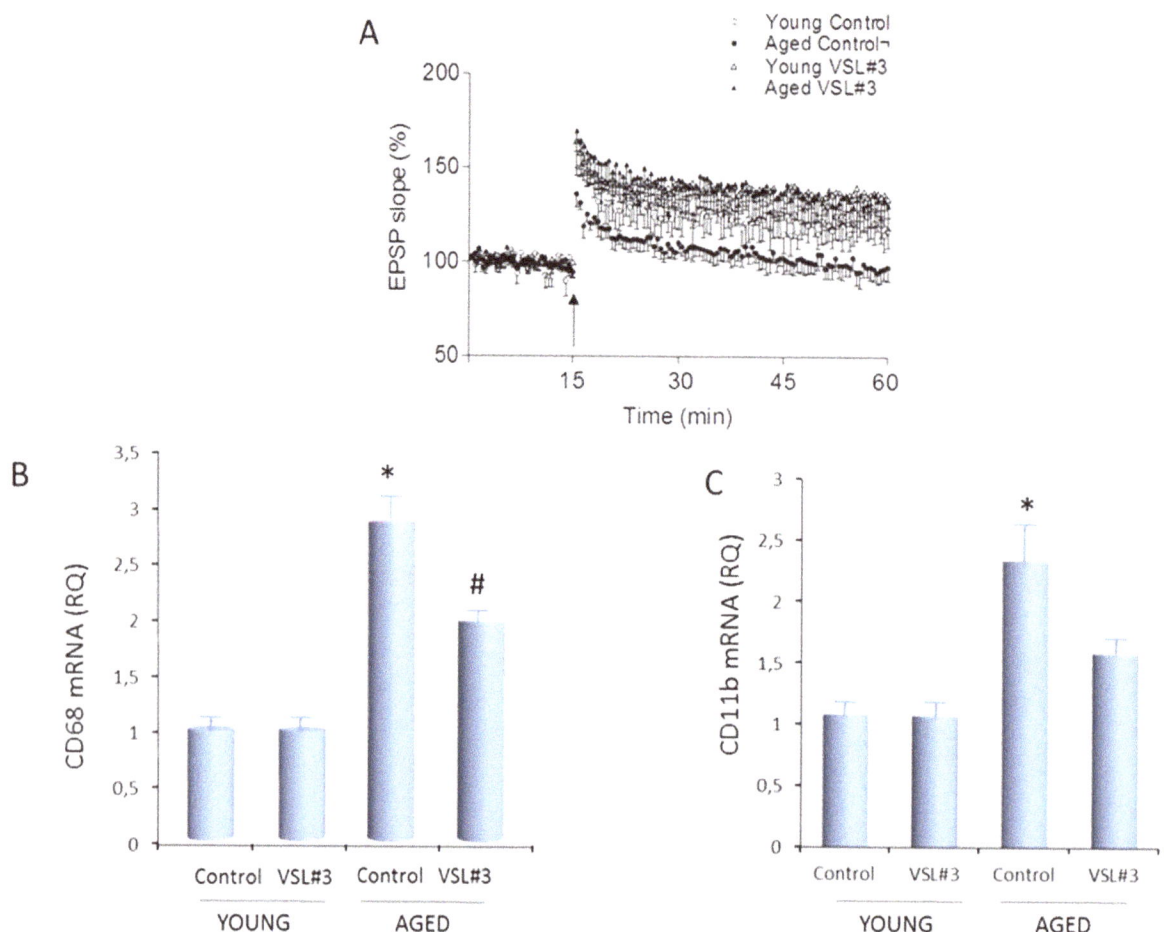

Figure 5. VLS#3 treatment attenuates the age-related decrease in LTP and the age-related increase in microglial activation. (A) Delivery of a high frequency train of stimuli to the perforant path (arrow) induced an immediate and sustained increase in EPSP slope in control-treated young rats and this effect was markedly decreased in control-treated aged rats. Aged rats treated with VLS#3 sustained LTP in a manner similar to young rats. (B, C) Expression of CD68 mRNA and CD11b mRNA was significantly increased in hippocampal tissue prepared from control aged rats compared with control young rats (*p<0.05 AC vs YC). Expression of both markers was reduced in VSL#3-treated aged rats compared with control aged rats, and the difference was statistically significant onlyin the case of CD68 mRNA (#p<0.05 AV vs AC).

change was comparable to that which occurred spontaneously in young rats whether or not they received VSL#3. Because the *Actinobacteria* phylum includes the *Bifidobacteria* species, one of the main components of VSL#3, these changes suggest that VSL#3 treatment effectively modulates the composition of the intestinal microbiota. Previous studies have shown that VSL#3 promotes a variety of local/intestinal effects including regulation of intestinal permeability and host innate immune response [37–42]. In addition to these well-characterized local effects on the gut epithelium, gut immune function and enteric nervous system, long-distance effects of the microbiota on metabolism, liver, adipose tissue and brain have been reported [23].

In the present study, we report that age *per se* resulted in a significantly reshaping of the expression of ~1% of the entire genoma of rat brain cortex. While the interpretation of modulation in gene expression in the context of complex physiological changes such age is difficult, the bioinformatic approach used in this study to characterize the whole brain rat genome allowed the detection of a well defined signal-to-noise effect on the expression of over 300 genes. Thus, in comparison with young rats, the cortical tissue of aged rats was characterized by upregulation in the expression of 212 genes and downregula-

tion in expression of 111 genes. Among the genes that were upregulated were several that are key in inflammation including Interferon regulatory factor 7 (Irf7), a gene that is significantly downregulated in prefrontal cortex of patients with major depressive disorders [43]. Changes were also observed in GFAP which codes for glial fibrillary acidic protein expressed in astrocytes of central nervous system [44], and phospholipase A2, group III (PLA2G3) that is the highest expressed gene in a neuronal model of oxidative stress induced by the free radical-generating system xanthine/xanthine oxidase [45]. Changes in the expression of PLA2G3 were confirmed by RT-PCR (Figure S1) suggesting that reshaping of this gene is an adaptive change that occurs with age. Importantly PLA2G3 drives apoptotic cell death and its overexpression has been associated with Alzheimer's disease [45]. Changes in the expression of other genes associated with increased risk of developing Alzheimer's disease were also observed including sorting nexin 8 (SNX8) [46] and apolipopro-tein D (Apod), which is up-regulated in pathological and stress condition including Alzheimer's disease [47]. Upregulation of oligodendrocyte transcription factor 1 (Olig 1) which controls differentiation and myelin production during inflammation [48] and interferon regulatory factor 9 (Irf9) which is upregulated by

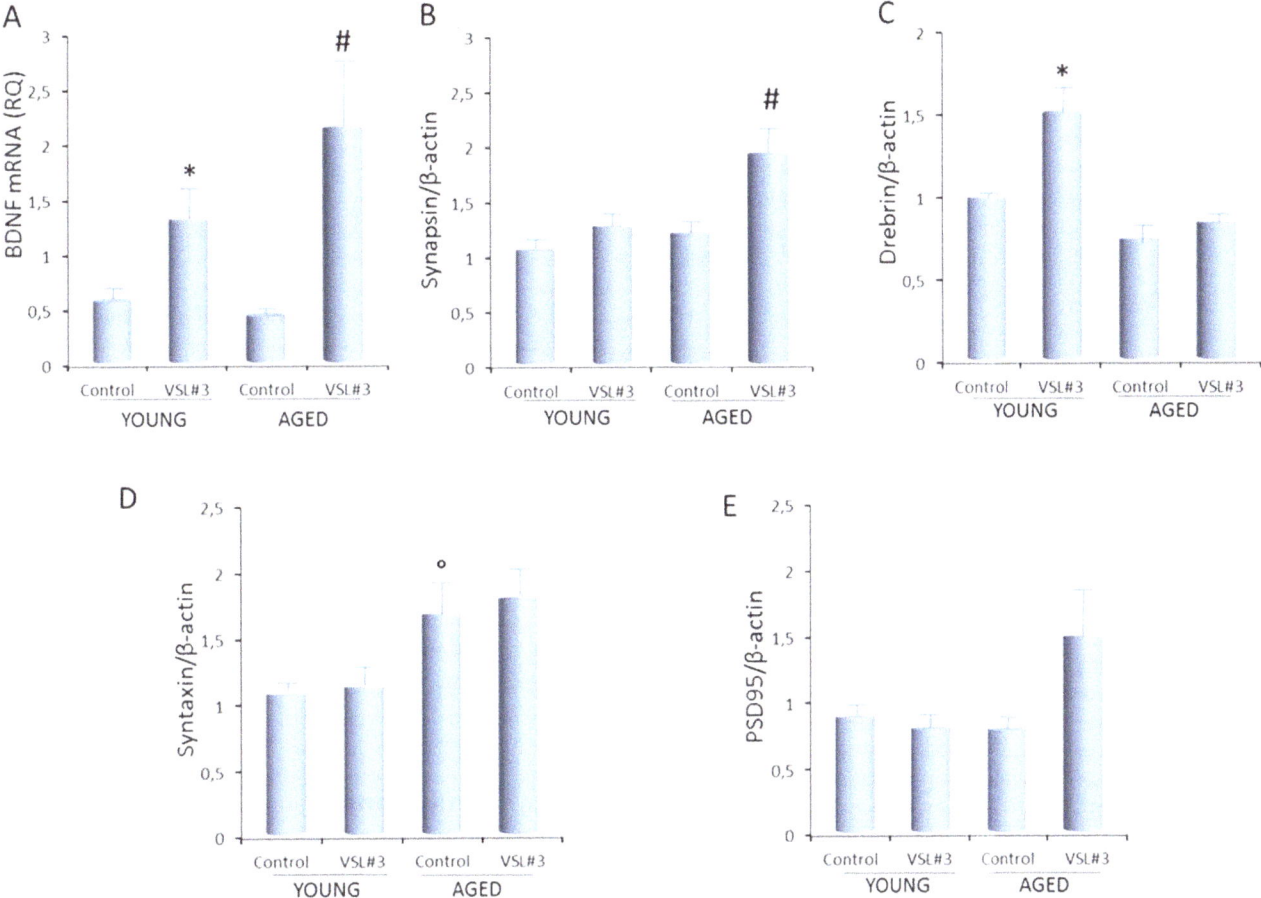

Figure 6. VLS#3 treatment increases BDNF and modulates the age-related changes in synaptic proteins. (A) VLS#3 treatment significantly increased expression of BDNF mRNA in hippocampal tissue(*p<0.05 YV vs YC; #p<0.05 AV vs AC). (B) A significant VSL#3 treatment effect on synapsin was observed in aged rats (#p<0.05 AV vs AC). (C) A significant treatment effect was observed in drebrin in young rats (*p<0.05 YV vs YC). (C) A significant age effect was observed in syntaxin (°p<0.05 AC vs YC). (E) No effect in PSD95 was observed.

neuronal injury [49], suggests that age is associated with resetting of genes that are commonly associated with inflammation [1].

Among the genes that were downregulated in cortical tissue of aged, compared with young, rats were several encoding for proteins related with neuronal development and plasticity. These include nidogen 2 (osteonidogen) (Nid2), a membrane protein expressed in cortex and striatum, whose expression has been reported to decrease with age [50]. This age-related decline in the expression of Nid2 was confirmed in this study by RT-PCR (Figure S1). Interestingly Nid2 prevents aggregation of β amyloid and destabilizes preformed fibrils of β amyloid [50]. Other genes whose expression was negatively modulated by age were the neuromedin B receptor (NMBR) which is highly expressed in area of the brain involved in memory and emotional processing [51] and arachidonate 15-lipoxygenase (Alox15) which is reported to be involved in development of Alzheimer's disease [52].

One important finding of the present study was the observation that VSL#3 treatment resets the expression of a number of genes in the cortex. Thus treatment of aged rats with VSL#3 resulted in the resetting of at least 66 genes of which 25 were upregulated and 41 were downregulated. Significantly, VSL#3 treatment attenuated the age-related changes in 3 genes that impact on inflammation, PLA2G3, Nid2 and Alox15.

Our data agree with previous studies which demonstrated that intestinal microbiota modulates brain gene expression and alters the profiles of canonical signaling pathways, neurotransmitter turnover and synaptic-related proteins which, in turn, influence brain development and function [53]. While changes in the expression of genes in the brain might be functionally relevant, the mechanism involved in their regulation by VSL#3 remains to be determined. One possible mechanism mediating the gut-brain communication may be via established neuronal circuits. Recent data have shown that an impact of probiotics on the brain requires the integrity of the vagus nerve and gut microbiota can elicit signals via the vagal nerve to the brain and *vice versa* [54]. Moreover modulation of transmitters (e.g., serotonin, melatonin, gamma-aminobutyric acid, histamines, and acetylcholine) within the gut is yet another possible mechanism of action that could mediate the effects of the gut microbiota. Alternatively, metagenomics or metatranscriptome studies have demonstrated that ingestion of probiotics impacts on bacterial metabolic activities in the gut redirecting the host metabolism; according to this proposal, changes in microbiota-produced signaling molecules (including amino acid metabolites, short chain fatty acids and neuroactive substances) might be involved [55]. In our study, VSL#3 induced not only a change on gene expression, but also altered the expression of several proteins involved in aging and inflammation, indicating that modulation of these molecules may play a key role in the expression of brain genes.

To evaluate whether the VSL#3-induced changes in gene expression impacted on neuronal function, we assessed LTP in the hippocampus, which has been shown to be adversely affected by age [32,56–61]. LTP is widely used as an indicator of healthy brain function and, accordingly, it is impaired in a number of neurodegenerative disease models which are associated with inflammatory changes. Thus deficits in LTP, accompanied by cognitive dysfunction, have been reported in models of Alzheimer's disease. The present data show that LTP was robustly decreased in aged rats providing further evidence of the negative impact of age on synaptic plasticity. An age-related decrease in cognitive function associated with loss of synaptic plasticity, specifically a deficit in LTP, has been consistently reported [56] and has been attributed variously to dysregulation in calcium handling by cells, altered receptor expression and receptor-mediated signaling, loss of synapses and decreased neurotrophic support [57]. Oxidative and neuroinflammatory changes as a consequence of microglial activation have also been shown to negatively impact on LTP [1]. In the context of neuroinflammatory changes, agents which decrease microglial activation in the brain of aged rats, for example minocycline, atorvastatin, rosiglitazone and polyunsaturated fatty acids, like eicosapentaenoic acid, attenuate the age-related deficit in LTP [32,58–60]. Here we show that VSL#3 attenuates the age-related decrease in LTP. The mechanisms involved in regulation of LTP by VSL#3 are likely to be many; the most parsimonious explanation is that the anti-inflammatory effects of the treatment observed in the gut [61,62] extend to the brain [63–64]. In this context, we provide evidence that VSL#3 modulates hippocampal expression of two markers of microglial activation (and therefore inflammation), CD68 and CD11b, confirming the previously-described inverse correlation between LTP and inflammatory changes. An important observation made in this study is that VSL#3 regulates the expression of specific mediators of synaptic plasticity including BDNF, synapsin and syntaxin in the hippocampus of aged rats. BDNF is essential for maintaining LTP and its role has been long recognized; it's specific function still remains to be clarified [65] though it induces neurogenesis [66] and synaptogenesis [67] and the increased hippocampal expression of BDNF by exercise, has

been shown to enhance the ability of aged rats to sustain LTP [68,69]. BDNF was increased in hippocampus of VSL#3-treated rats and that this was associated with an increase in synapsin, suggesting that VSL#3 exerted a synaptotrophic effect, though the mechanism involved remains to be elucidated.

In conclusion, we have shown that age regulates the expression of several genes in cortical tissue and adversely affected synaptic function. Altering the intestinal microbiota of aged rats by treatment with VSL#3 modulated the expression of a cohort of genes in the cortex, some of which impact on inflammatory processes. We suggest that this effect, together with its neurotrophic/synaptotrophic effect, contributes to the ability of VSL#3 in attenuating the age-related impairment of LTP.

Supporting Information

Figure S1 Confirmation of microarray data by qRT-PCR analysis (A) Cortical expression of the Nid2 detected by PCR did not change in the four groups of rats. (B-C) PCR confirmed the gene array data for Alox15 and PLA2G3 respectively. *p<0.05 vs Group YC; #p<0.05 vs Group AC.

Acknowledgments

JuttaKollet, Ph.D., Dipl. Biol. Senior Scientist Bioinformatics, Miltenyi-Biotec GmbH Bioinformatics, Friedrich-Ebert-Strasse 68 51429 Bergisch-Gladbach, Germany.

Author Contributions

Conceived and designed the experiments: ED MAL SF. Performed the experiments: J-AO CM SC BR. Analyzed the data: J-AO CM SC BR ED. Contributed reagents/materials/analysis tools: SF MAL. Contributed to the writing of the manuscript: ED SF MAL. LTP and markers of inflammation in the hippocampus: J-AO CM. PCR on brain genes: SC. Intestinal microbiota composition: BR. Analysis of LTP and markers of inflammation in the hippocampus: J-AO CM. Analysis of global microarray of brain tissue: ED SC. Analysis of PCR on brain genes: SC. Analysis of intestinal microbiota composition: BR.

References

1. Lynch MA (2010) Age-related neuroinflammatory changes negatively impact on neuronal function. Front Aging Neurosci 1: 6.
2. Lynch MA (2004) Long-term potentiation and memory. Physiological Reviews 84: 87–136.
3. Kelly L, Grehan B, Chiesa AD, O'Mara SM, Downer E, et al. (2011) The polyunsaturated fatty acids, EPA and DPA exert a protective effect in the hippocampus of the aged rat. Neurobiol Aging 32: 2318 e2311–2315.
4. Clarke RM, Lyons A, O'Connell F, Deighan BF, Barry CE, et al. (2008) A pivotal role for interleukin-4 in atorvastatin-associated neuroprotection in rat brain. J Biol Chem 283: 1808–1817.
5. Moore M, Piazza A, Nolan Y, Lynch MA (2007) Treatment with dexamethasone and vitamin D3 attenuates neuroinflammatory age-related changes in rat hippocampus. Synapse 61: 851–861.
6. Moore ME, Piazza A, McCartney Y, Lynch MA (2005) Evidence that vitamin D3 reverses age-related inflammatory changes in the rat hippocampus. Biochem Soc Trans 33: 573–577.
7. Taliaz D, Stall N, Dai DE, Zangen A (2010) Knockdown of brain-derived neurotrophic factor in specific brain sites precipitates behaviors associated with depression and reduces neurogenesis. Mol Psychiatry 15: 80–92.
8. O'Callaghan RM, Griffin EW, Kelly AM (2009) Long-term treadmill exposure protects against age-related neurodegenerative change in the rat hippocampus. Hippocampus 19: 1019–1029.
9. Rex CS, Lauterborn JC, Lin CY, Kramar EA, Rogers GA, et al. (2006) Restoration of long-term potentiation in middle-aged hippocampus after induction of brain-derived neurotrophic factor. J Neurophysiol 96: 677–685.
10. Dantzer R, Kelley KW (2007) Twenty years of research on cytokine-induced sickness behavior. Brain Behav Immun 21: 153–160.

11. Godbout JP, Chen J, Abraham J, Richwine AF, Berg BM, et al. (2005) Exaggerated neuroinflammation and sickness behavior in aged mice following activation of the peripheral innate immune system. Faseb J 19: 1329–1331.
12. Honjo K, van Reekum R, Verhoeff NP (2009) Alzheimer's disease and infection: do infectious agents contribute to progression of Alzheimer's disease? Alzheimers Dement 5: 348–360.
13. Simpson S Jr, Taylor B, Dwyer DE, Taylor J, Blizzard L, et al. (2012) Anti-HHV-6 IgG titer significantly predicts subsequent relapse risk in multiple sclerosis. Mult Scler 18: 799–806.
14. Reber SO (2012) Stress and animal models of inflammatory bowel disease—an update on the role of the hypothalamo-pituitary-adrenal axis. Psychoneuroendocrinology 37: 1–19.
15. Cryan JF, Dinan TG (2012) Mind-altering microorganisms: the impact of the gut microbiota on brain and behaviour. Nature Reviews Neuroscience 13: 701–712.
16. O'Mahony SM, Marchesi JR, Scully P, Codling C, Ceolho AM, et al. (2009) Early life stress alters behavior, immunity, and microbiota in rats: implications for irritable bowel syndrome and psychiatric illnesses. Biol Psychiatry 65: 263–267.
17. Aisa B, Tordera R, Lasheras B, Del Rio J, Ramirez MJ (2007) Cognitive impairment associated to HPA axis hyperactivity after maternal separation in rats. Psychoneuroendocrinology 32: 256–266.
18. Suri D, Veenit V, Sarkar A, Thiagarajan D, Kumar A, et al. (2013) Early stress evokes age-dependent biphasic changes in hippocampal neurogenesis, BDNF expression, and cognition. Biol Psychiatry 73: 658–666.
19. Wang L, Jiao J, Dulawa SC (2011) Infant maternal separation impairs adult cognitive performance in BALB/cJ mice. Psychopharmacology 216: 207–218.

20. Gareau MG, Wine E, Rodrigues DM, Cho JH, Whary MT, et al. (2011) Bacterial infection causes stress-induced memory dysfunction in mice. Gut 60: 307–317.

21. Evrard B, Coudeyras S, Dosgilbert A, Charbonnel N, Alame J, et al. (2011) Dose-dependent immunomodulation of human dendritic cells by the probiotic Lactobacillus rhamnosus Lcr35. PLoS One 6:e18735.

22. Lin YP, Thibodeaux CH, Pena JA, Ferry GD, Versalovic J (2008) Probiotic Lactobacillus reuteri suppress proinflammatory cytokines via c-Jun. Inflammatory bowel diseases 14: 1068–1083.

23. Mencarelli A, Distrutti E, Renga B, D'Amore C, Cipriani S, et al. (2011) Probiotics modulate intestinal expression of nuclear receptor and provide counter-regulatory signals to inflammation-driven adipose tissue activation. PLoS One 6:e22978.

24. Dai C, Zheng CQ, Meng FJ, Zhou Z, Sang LX, et al. (2013) VSL#3 probiotics exerts the anti-inflammatory activity via PI3k/Akt and NF-kappaB pathway in rat model of DSS-induced colitis. Mol Cell Biochem 374: 1–11.

25. Reiff C, Delday M, Rucklidge G, Reid M, Duncan G, et al. (2009) Balancing inflammatory, lipid, and xenobiotic signaling pathways by VSL#3, a biotherapeutic agent, in the treatment of inflammatory bowel disease. Inflammatory Bowel Diseases 15: 1721–1736.

26. Distrutti E, Cipriani S, Mencarelli A, Renga B, Fiorucci S (2013) Probiotics VSL#3 Protect against Development of Visceral Pain in Murine Model of Irritable Bowel Syndrome. PLoS One 8:e63893.

27. Mimura T, Rizzello F, Helwig U, Poggioli G, Schreiber S, et al. (2004) Once daily high dose probiotic therapy (VSL#3) for maintaining remission in recurrent or refractory pouchitis. Gut 53: 108–114.

28. Kim HJ, Camilleri M, McKinzie S, Lempke MB, Burton DD, et al. (2003) A randomized controlled trial of a probiotic, VSL#3, on gut transit and symptoms in diarrhoea-predominant irritable bowel syndrome. Alimentary Pharmacology &Therapeutics 17: 895–904.

29. Dharmani P, De Simone C, Chadee K (2013) The probiotic mixture VSL#3 accelerates gastric ulcer healing by stimulating vascular endothelial growth factor. PLoS One 8:e58671.

30. Edgar R, Domrachev M, Lash AE (2002) Gene Expression Omnibus: NCBI gene expression and hybridization array data repository. Nucleic Acids Res 30: 207–210.

31. Costello DA, Watson MB, Cowley TR, Murphy N, Murphy Royal C, et al. (2011) Interleukin-1alpha and HMGB1 mediate hippocampal dysfunction in SIGIRR-deficient mice. J Neurosci 31: 3871–3879.

32. Lyons A, Murphy KJ, Clarke R, Lynch MA (2011) Atorvastatin prevents age-related and amyloid-beta-induced microglial activation by blocking interferon-gamma release from natural killer cells in the brain. J Neuroinflammation 8: 27.

33. Lyons A, Griffin RJ, Costelloe CE, Clarke RM, Lynch MA (2007) IL-4 attenuates the neuroinflammation induced by amyloid-betain vivo and in vitro. J Neurochem 101: 771–781.

34. Benno Y, Nakao H, Uchida K, Mitsuoka T (1992) Impact of the advances in age on the gastrointestinal microflora of Beagle dogs. J Vet Med Sci 54: 703–706.

35. Hopkins MJ, Sharp R, Macfarlane GT (2001) Age and disease related changes in intestinal bacterial populations assessed by cell culture, 16S rRNA abundance, and community cellular fatty acid profiles. Gut 48: 198–205.

36. Yatsunenko T, Rey FE, Manary MJ, Trehan I, Dominguez-Bello MG, et al. (2012) Human gut microbiome viewed across age and geography. Nature 486: 222–227.

37. Ewaschuk J, Endersby R, Thiel D, Diaz H, Backer J, et al. (2007) Probiotic bacteria prevent hepatic damage and maintain colonic barrier function in a mouse model of sepsis. Hepatology 46: 841–850.

38. Jobin C (2010) Probiotics and ileitis: could augmentation of TNF/NFκB activity be the answer? Gut Microbes 1: 196–9.

39. Claes IJ, De Keersmaecker SC, Vanderleyden J, Lebeer S (2011) Lessons from probiotic-host interaction studies in murine models of experimental colitis. Mol Nutr Food Res 55: 1441–53.

40. Macho Fernandez E, Valenti V, Rockel C, Hermann C, Pot B, et al. (2011) Anti-inflammatory capacity of selected lactobacilli in experimental colitis is driven by NOD2-mediated recognition of a specific peptidoglycan-derived muropeptide. Gut 60: 1050–9.

41. Viladomiu M, Hontecillas R, Yuan L, Lu P, Bassaganya-Riera J (2013) Nutritional protective mechanisms against gut inflammation. J Nutr Biochem 24: 929–39.

42. Qiu X, Zhang M, Yang X, Hong N, Yu C (2013) Faecali bacterium prausnitzii upregulates regulatory T cells and anti-inflammatory cytokines in treating TNBS-induced colitis. J Crohn Colitis 7:e558–68.

43. Mamdani F, Berlim MT, Beaulieu MM, Labbe A, Merette C, et al. (2011) Gene expression biomarkers of response to citalopram treatment in major depressive disorder. Transl Psychiatry 1:e13.

44. Baba H, Nakahira K, Morita N, Tanaka F, Akita H, et al. (1997) GFAP gene expression during development of astrocyte. Dev Neurosci 19: 49–57.

45. Martínez-García A, Sastre I, Recuero M, Aldudo J, Vilella E, et al. (2010) PLA2G3, a gene involved in oxidative stress induced death, is associated with Alzheimer's disease. J Alzheimers Dis 22: 1181–1187.

46. Rosenthal SL, Wang X, Demirci FY, Barmada MM, Ganguli M, et al. (2012) Beta-amyloid toxicity modifier genes and the risk of Alzheimer's disease. Am J Neurodegener Dis 1: 191–198.

47. Bhatia S, Knoch B, Wong J, Kim WS, Else PL, et al. (2012) Selective reduction of hydroperoxyeicosatetraenoic acids to their hydroxy derivatives by apolipoprotein D: implications for lipid antioxidant activity and Alzheimer's disease. Biochem J 442: 713–721.

48. Glezer I, Lapointe A, Rivest S (2006) Innate immunity triggers oligodendrocyte progenitor reactivity and confines damages to brain injuries. FASEB J 20: 750–752.

49. Khorooshi R, Owens T (2010) Injury-induced type I IFN signaling regulates inflammatory responses in the central nervous system. Immunol 185: 1258–1264.

50. Hawkes CA, Gatherer M, Sharp MM, Dorr A, Yuen HM, et al. (2013) Regional differences in the morphological and functional effects of aging on cerebral basement membranes and perivascular drainage of amyloid-β from the mouse brain. Aging Cell 12: 224–236.

51. Yamada K, Santo-Yamada Y, Wada K (2003) Stress-induced impairment of inhibitory avoidance learning in female neuromedin B receptor-deficient mice. Physiol Behav 78: 303–309.

52. Zhao Y, Bhattacharjee S, Jones BM, Hill J, Dua P, et al. (2013) Regulation of Neurotropic Signaling by the Inducible, NF-kB-Sensitive miRNA-125b in Alzheimer's Disease (AD) and in Primary Human Neuronal-Glial (HNG). Cells Mol Neurobiol: Nov 29.

53. Diaz Heijtz R, Wang S, Anuar F, QianY, Björkholm B, et al. (2011) Normal gut microbiota modulates brain development and behavior. PNAS 108: 3047–3052.

54. BorovikovaLV, Ivanova S, Zhang M, Yang H, Botchkina GI, et al. (2000) Vagus nerve stimulation attenuates the systemic inflammatory response to endotoxin. Nature 405: 458–462.

55. Holmes E, Li JV, Marchesi JR, Nicholson JK (2012) Gut microbiota composition and activity in relation to host metabolic phenotype and disease risk. Cell Metab 16: 559–64.

56. Kumar A (2011) Long-Term Potentiation at CA3-CA1 Hippocampal Synapses with Special Emphasis on Aging, Disease, and Stress. Front Aging Neurosci 3: 7.

57. Burke SN, Barnes CA (2006) Neural plasticity in the ageing brain. Nat Rev Neurosci 7: 30–40.

58. Griffin R, Nally R, Nolan Y, McCartney Y, Linden J, et al. (2006) The age-related attenuation in long-term potentiation is associated with microglial activation. J Neurochem 99: 1263–1272.

59. Lynch AM, Loane DJ, Minogue AM, Clarke RM, Kilroy D, et al. (2007) Eicosapentaenoic acid confers neuroprotection in the amyloid-beta challenged aged hippocampus. Neurobiol Aging 28: 845–855.

60. Loane DJ, Deighan BF, Clarke RM, Griffin RJ, Lynch AM, et al. (2009) Interleukin-4 mediates the neuroprotective effects of rosiglitazone in the aged brain. Neurobiol Aging 30: 920–931.

61. Reiff C, Delday M, Rucklidge G, Reid M, Duncan G, et al. (2009) Balancing inflammatory, lipid, and xenobiotic signaling pathways by VSL#3, a biotherapeutic agent, in the treatment of inflammatory bowel disease. Inflamm Bowel Dis 15: 1721–36.

62. Dai C, Zheng CQ, Meng FJ, Zhou Z, Sang LX, et al. (2013) VSL#3 probiotics exerts the anti-inflammatory activity via PI3k/Akt and NF-κB pathway in rat model of DSS-induced colitis. Mol Cell Biochem 374: 1–11

63. Bajaj JS, Ridlon JM, Hylemon PB, Thacker LR, Heuman DM, et al. (2012) Linkage of gut microbiome with cognition in hepatic encephalopathy. Am J Physiol Gastrointest Liver Physiol 302: G168–175

64. Bajaj JS, Hylemon PB, Ridlon JM, Heuman DM, Daita K, et al. (2012) Colonic mucosal microbiome differs from stool microbiome in cirrhosis and hepatic encephalopathy and is linked to cognition and inflammation. Am J Physiol Gastrointest Liver Physiol 303: G675–685.

65. Edelmann E, Lessmann V, Brigadski T (2014) Pre- and postsynaptic twists in BDNF secretion and action in synaptic plasticity. Neuropharmacology 76: 610–627.

66. Taliaz D, Stall N, Dar DE, Zangen A (2010) Knockdown of brain-derived neurotrophic factor in specific brain sites precipitates behaviors associated with depression and reduces neurogenesis. Mol Psychiatry 15: 80–92.

67. Cunha C, Brambilla R, Thomas KL (2010) A simple role for BDNF in learning and memory? Front Molecular Neurosci 3: 1.

68. O'Callaghan RM, Griffin EW, Kelly AM (2009) Long-term treadmill exposure protects against age-related neurodegenerative change in the rat hippocampus. Hippocampus 19: 1019–1029.

69. Rex CS, Lauterborn JC, Lin CY, Kramar EA, Rogers GA, et al. (2006) Restoration of long-term potentiation in middle-aged hippocampus after induction of brain-derived neurotrophic factor. J Neurophysiol 96: 677–685.

19

Lactobacillus gasseri SBT2055 Reduces Infection by and Colonization of *Campylobacter jejuni*

Keita Nishiyama[1], Yasuyuki Seto[2], Kazuki Yoshioka[3], Tsutomu Kakuda[3], Shinji Takai[3], Yuji Yamamoto[1], Takao Mukai[1]*

1 Department of Animal Science, School of Veterinary Medicine, Kitasato University, Towada, Aomori, Japan, 2 Milk Science Research Institute, Megmilk Snow Brand Co. Ltd., Kawagoe, Saitama, Japan, 3 Faculty of Veterinary Medicine, School of Veterinary Medicine, Kitasato University, Towada, Aomori, Japan

Abstract

Campylobacter is a normal inhabitant of the chicken gut. Pathogenic infection with this organism in humans is accompanied by severe inflammation of the intestinal mucosal surface. The aim of this study was to evaluate the ability of *Lactobacillus gasseri* SBT2055 (LG2055) to inhibit the adhesion and invasion of *Campylobacter jejuni in vitro* and to suppress *C. jejuni* colonization of chicks *in vivo*. Pretreatment with LG2055 significantly reduced adhesion to and invasion of a human epithelial cell line, Intestine 407, by *C. jejuni* 81–176. Methanol (MeOH)-fixed LG2055 also reduced infection by *C. jejuni* 81–176. However, proteinase K (ProK)-treated LG2055 eliminated the inhibitory effects. Moreover, LG2055 co-aggregated with *C. jejuni* 81–176. ProK treatment prevented this co-aggregation, indicating that the co-aggregation phenotype mediated by the proteinaceous cell-surface components of LG2055 is important for reducing *C. jejuni* 81–176 adhesion and invasion. In an *in vivo* assay, oral doses of LG2055 were administered to chicks daily for 14 days after oral inoculation with *C. jejuni* 81–176. At 14 days post-inoculation, chicks treated with LG2055 had significantly reduced cecum colonization by *C. jejuni*. Reduction in the number of *C. jejuni* 81–176 cells adhering to and internalized by human epithelial cells demonstrated that LG2055 is an organism that effectively and competitively excludes *C. jejuni* 81–176. In addition, the results of the chick colonization assay suggest that treatment with LG2055 could be useful in suppressing *C. jejuni* colonization of the chicks at early growth stages.

Editor: Stefan Bereswill, Charité-University Medicine Berlin, Germany

Funding: This study was supported in part by a Grant-in-Aid for Scientific Research C (21580334) from the Japan Society for the Promotion of Science (JSPS). This work was also partially supported by Megmilk Snow Brand Co. Ltd. The funders had no role in study design, data collection and analysis, decision to publish, or preparation of the manuscript.

Competing Interests: This study was partially financially supported by Megmilk Snow Brand Co. Ltd. Yasuyuki Seto is an employee of Megmilk Snow Brand Co. Ltd. There are no patents, products in development or marketed products to declare.

* Email: mukai@vmas.kitasato-u.ac.jp

Introduction

Campylobacter jejuni infection (campylobacteriosis), which leads to gastroenteritis in humans, is one of the leading causes of enteric bacterial infections [1–3]. *C. jejuni* colonization in the ileum and colon can interfere with normal functions of the human gastrointestinal (GI) tract. *C. jejuni* infection is characterized by fever, abdominal cramps, and diarrhea [1], [4], [5]. Epithelial cell adhesion and invasion of the GI tract by the pathogen play important roles in the pathogenesis of the disease and are associated with other well defined disease traits, including induction of cell death and disruption of mucosal barrier function [2], [4–8]. Therefore, preventing *C. jejuni* adhesion to and invasion of human epithelial cells is critical for reducing campylobacteriosis.

Chickens are a natural host for *Campylobacter* species, in particular, *C. jejuni*. One study detected considerable *Campylobacter* contamination in chicken carcasses, with an average prevalence of over 60% [9]. Chicken meat products are the main source of campylobacteriosis in humans [10], [11]. Thus, reducing

C. jejuni colonization of the chick intestinal tract has been proposed as a strategy to reduce the disease burden in humans [4], [12].

Probiotic bacteria, such as *Lactobacillus* strains, can competitively inhibit *C. jejuni* colonization and infection [13]. Competitive inhibition of *C. jejuni* by *Lactobacillus* may occur through several key mechanisms, including competition for attachment sites [14], co-aggregation with the pathogen [15], [16], antimicrobial compound production [17–19] hydrogen peroxide production [20], and lactic acid production [21]. *Lactobacillus gasseri* SBT2055 (LG2055), a strain isolated from human feces, exhibits good fermentative properties and strong resistance to artificial gastric and bowel fluid [22]; furthermore, this strain can colonize the GI tract of humans and of mice [22], [23]. Moreover, LG2055 in the human intestine significantly decreases the number of fecal staphylococci [22]. LG2055 administration is effective for protection against influenza A virus infection in mice [24]. However, despite these findings, the inhibitory effect of LG2055 on *C. jejuni* infection is unknown at present.

The *C. jejuni* 81–176 strain, originally isolated during an outbreak of campylobacteriosis [25], can persistently colonize the GI tract of chicks [25], [26] and invade human epithelial cells [27]. In the present work, we evaluated the ability of LG2055 to inhibit adhesion and invasion of *C. jejuni* 81–176 *in vitro* as well as its ability to inhibit *C. jejuni* 81–176 colonization of chicks *in vivo*. We also sought to clarify the inhibitory mechanisms of LG2055.

Materials and Methods

Bacterial strains and growth conditions

C. jejuni 81–176 was cultured at 37°C under microaerophilic conditions on Mueller–Hinton (MH) agar plates (BD Difco, NJ, USA) for 48 h. The bacteria were transferred to a biphasic medium in 25-cm^2 tissue culture flasks (Techno Plastic Products, Switzerland), containing MH agar and 5 mL of MH broth, and then cultured to an OD_{600} of 0.5. The *L. gasseri* SBT2055 strain (LG2055) in this study was provided by Megmilk Snow Brand Co., Ltd. (Saitama, Japan) and was cultured on de Man-Rogosa-Sharpe (MRS) agar plates (BD Difco) at 37°C for 24 h under anaerobic conditions. The bacteria were transferred to MRS broth and cultured to an OD_{600} of 2.0.

Epithelial cell line and culture conditions

The epithelial cell line Intestine 407 (Int407) was obtained from the American Type Culture Collection (VA, USA). The cells were cultured in RPMI 1640 medium (Gibco, NY, USA) with 10% (v/v) fetal bovine serum (FBS), 10 U/mL penicillin (Gibco), and 10 µg/mL streptomycin (Gibco). Cells were maintained in 25-cm^2 flasks and then seeded onto 12-well plates (37°C, 5% CO_2). Cells were grown for a minimum of 10 days, and the medium was changed every 3 days. Antibiotics and FBS were removed from the cells at least 24 h before the infection assays.

Anti-*C. jejuni* activity

Anti-*Campylobacter* activity was evaluated by the agar well diffusion method [21] with several modifications. Overnight culture supernatants of LG2055 were collected, and either boiled for 6 min, neutralized to pH 7 with 6 N NaOH, or left untreated. Supernatants were subsequently filter sterilized (0.22-µm filter). MH agar plates were overlaid with 15 mL of molten MH soft agar (0.75%) inoculated with 300 µL of *C. jejuni* 81–176 cultures, standardized to an OD_{600} of 1.0 in MH broth. Wells of 5-mm diameter were cut into agar plates, and 25 µL of the LG2055 supernatant was added to each well. After 24 h incubation under microaerophilic conditions at 37°C, the diameter of the zone of inhibition around each well was measured.

Tissue culture adhesion and invasion assays

Adhesion and invasion assays were performed as described previously, with some modifications [14]. Confluent monolayers of Int407 cells (approx. 2×10^5 cells/well) were washed twice with phosphate-buffered saline (PBS; pH 7.4). Bacterial cells were resuspended in 1 mL of RPMI 1640 medium to a multiplicity of infection (MOI) of 0.1:1 to 1000:1 for *C. jejuni* 81–176 and a ratio of interaction (ROI) of 0.1:1 to 1000:1 for LG2055. For the adhesion assay, bacterial cells were then added to each well and incubated at 37°C in 5% CO_2 for 4 h. Subsequently, the wells were washed three times with PBS to remove non-adherent bacteria and incubated with 500 µL of 0.1% (v/v) Triton X-100 in PBS. Serial dilutions of the bacteria were plated onto MH or MRS agar plates and cultured under microaerophilic or anaerobic

conditions at 37°C for 24 h, respectively. After incubation, colony-forming units (CFU) were counted.

For the invasion assay, bacterial cells were added to each well with Int407 monolayers, and the plates were incubated at 37°C in 5% CO_2 for 4 h. The wells were then washed three times with PBS, fresh RPMI 1640 medium containing 100 µg/mL gentamicin (Wako, Osaka, Japan) was added, because gentamicin (100 µg/mL) effectively kills all extracellular bacteria [28], and then the samples were incubated further for 2 h. The cells were then washed three times with PBS. The count for the invasive bacteria was performed, as described above, by using MH agar plates.

In the assay for inhibition of adhesion and invasion, LG2055 cells (ROI of 10:1, 100:1, and 500:1) were added to the Int407 monolayers 1 h before inoculation with *C. jejuni* 81–176 (MOI of 100:1 for invasion, 500:1 for adhesion). The *C. jejuni* 81–176 suspension was added to the Int407 monolayers without washing off the LG2055 cells, and the mixture was incubated at 37°C in 5% CO_2 for 4 h. Subsequently, the wells were washed three times with PBS to remove non-adherent bacteria. The assay to quantify adhesion and invasion was the same as described above.

Methanol (MeOH) fixation and proteinase K (ProK) treatment of LG2055

MeOH fixation and ProK treatment of LG2055 were performed as described previously [29], with some modifications. Briefly, LG2055 cells were resuspended in PBS to a concentration of 1×10^8 CFU/mL (ROI of 500:1). The bacterial cell suspensions were treated with an equal volume of ice-cold MeOH for 10 min or with 0.2 mL of RPMI 1640 containing 1 mg of ProK for 2 h at 37°C. Bacterial suspensions were centrifuged ($6,000 \times g$, 5 min, for 4°C), and the bacterial pellet was resuspended in 1 mL of RPMI 1640. Microscopic visualization of the gram-stained samples confirmed that the lactobacilli were intact after MeOH fixation and ProK treatment. The adhesion and invasion tests were performed as described above.

Self- and co-aggregation assays

The ability of bacteria to self- and co-aggregate was assessed according to a previously described method [16] with some modifications. Bacterial cells were suspended in 3 mL of PBS at a concentration of 1×10^8 CFU/mL. To determine self-aggregation, bacterial suspensions were statically incubated in aliquots at 25°C for 3 h. To determine the ability of bacteria to co-aggregate, equal volumes of LG2055 or ProK-treated LG2055 and *C. jejuni* 81–176 suspensions were mixed, and the mixtures were statically incubated at 25°C for 3 h.

To further characterize the self- and co-aggregation, surface plasmon resonance (SPR) studies using a Biacore X instrument (GE Healthcare, NJ, USA) were performed. LG2055 or ProK-treated LG2055 were immobilized on a CM5 dextran sensor chip with 5650 and 5820 resonance units (RUs), respectively, by using amine-coupling reagents (GE Healthcare). The analytes, which were bacterial cells, were suspended in PBS containing 0.005% surfactant P20 (pH 7.4) to a concentration of 1×10^8 CFU/mL. Multicycle experiments were performed as described [30], with an analyte flow rate of 3 µL/min for 10 min at 25°C per cycle.

Chick colonization assay

Chicken colonization studies were performed as described previously [31]. White leghorn chicken eggs were supplied by a commercial farm (Koiwai Farm, Ltd., Iwate, Japan) and maintained in an egg incubator until the chicks hatched. Approx. 24 h after hatching, chicks were randomly assigned to two groups. Bacterial cells were washed and resuspended in ice-cold PBS prior to inoculation. All birds were administered 1×10^6 CFU of *C. jejuni* 81–176 in a 100-µL suspension by oral gavage. Twenty-four hours after oral gavage, LG2055 (1×10^8 CFU in 100 µL) was orally administered daily to one group of *C. jejuni*-inoculated birds. PBS was administered to the other group of birds as a control. Chicks were sacrificed at 14 days post-inoculation, and the cecal contents were diluted in ice-cold PBS to 0.1 g/mL. Ten-fold serial dilutions of each sample were prepared and then plated on MH agar containing trimethoprim and cefoperazone to select for *C. jejuni*. MH agar plates were incubated under microaerophilic conditions at 37°C for 48 h. Cecal pH was measured with a compact pH meter (Model B-212; Horiba, Kyoto, Japan). The chicken experiments were performed at the Department of Animal Science, School of Veterinary Medicine, Kitasato University. All animal experiments were performed using protocols approved by the Institutional Animal Care and Use Committee at Kitasato University.

Quantification of bacterial cells by real-time PCR

Quantification of bacterial cells by real-time PCR using 16S rDNA gene-specific primers was performed as described previously [32]. DNA was extracted from Int407 cells infected with bacterial cells and from chick cecal contents using an InstaGene Matrix (Bio-Rad Laboratories, CA, USA) or QIAamp DNA Stool Mini Kit (Qiagen, Hilden, Germany), respectively. Quantitative PCR was performed on a StepOnePlus Real-Time PCR System (Applied Biosystems, CA, USA). Duplicate 10 µL PCR reactions were carried out using Power SYBR Green Master Mix (Applied Biosystems) according to the manufacturer's instructions. Bacterial primers (F_eub: 5'-TCCTACGGGAGGCAGCAGT-3', R_eub: 5'-GGACTACCAGGGTATCTAATCCTGTT-3') [33], *Lactobacillus* spp. primers (LactoF: 5'-TGGAAACAGRTGCTAATACCG-3', LactoR: 5'-GTCCATTGTGGAAGATTCCC-3') [34], and *L. gasseri*-specific primers (LactoF, Lgass R: 5'-CAGTTACTACCTCTATCTTTCTTCACTAC-3') [34] were used at final concentrations of 300, 200, and 400 nM, respectively. The reaction conditions were as follows: 95°C for 10 min followed by 40 cycles at 95°C for 15 s, annealing at the optimal temperature [33], [34] for 15 s, and elongation at 72°C for 20 s. The DNA samples extracted from *Escherichia coli* K12 (for total bacteria), *L. reuteri* JCM1112^T (for *Lactobacillus* spp.), and LG2055 (for *L. gasseri* strains) were used as real-time PCR standards. Standard bacterial DNA was prepared from a known number of bacterial cells (i.e., the number of CFU) and was found to be linear over the range of DNA concentrations from approx. 10^2–10^8 cells per PCR mixture. Bacteria were quantified using the standard-curve method, and all reactions were performed in triplicate in three independent experiments.

Histochemical analysis

Cecal tissues of chicks (14 days old, 4 birds/group) were fixed without washing in 4% paraformaldehyde phosphate buffer solution (Wako) for 4 h at 25°C. Paraffin-embedded tissues were cut into 3-µm sections. After de-waxing in xylene and rinsing in alcohol, antigen retrieval was performed by placing the slides in 50 mM citrate buffer (pH 6.0) in an autoclave for 10 min at 110°C. Sections were then incubated with 5% (w/v) bovine serum albumin (BSA)–PBS solution for 3 h. Biotin-conjugated anti-*C. jejuni* antibody (1:300 dilution; Fitzgerald Industries, MA, USA) was resuspended in 2% BSA–PBS solution and incubated for 12 h at 4°C. After washing, Cy3-conjugated streptavidin (1:1000; GE Healthcare) was added and incubated for 40 min. Counterstaining

was performed with 4',6-diamidino-2-phenylindole (DAPI; Wako). The sections were briefly washed with PBS and mounted. Pictures were obtained with an Olympus BX53 Microscope (Tokyo, Japan).

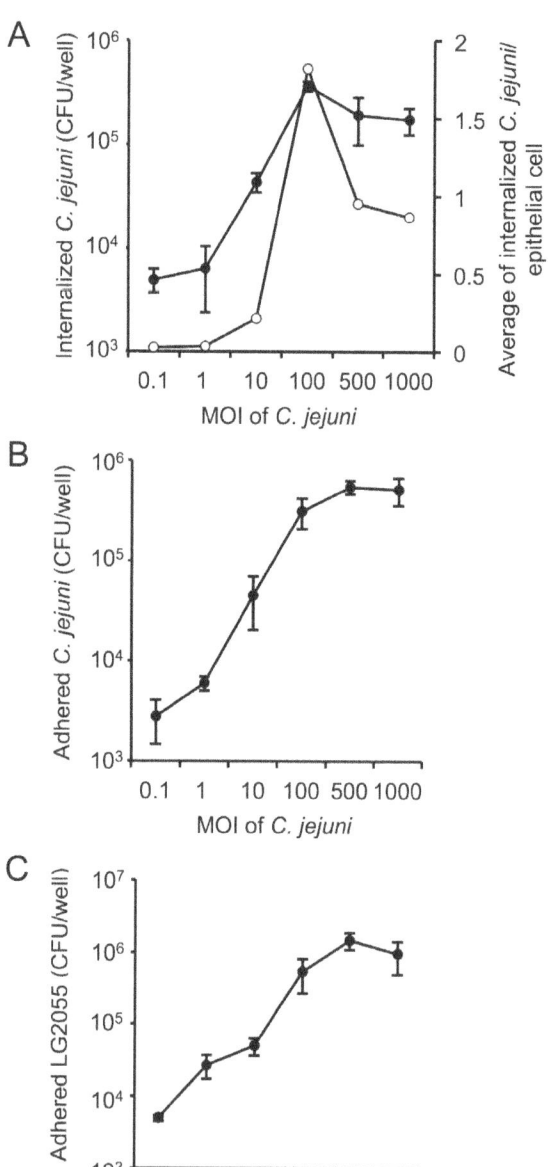

Figure 1. Invasion and adhesion by *C. jejuni* 81–176 or adhesion by LG2055 to Int407 cells. Effect of varying MOI on efficiency of *C. jejuni* 81–176 invasion (**A**) and adhesion (**B**). Int407 cells were infected with *C. jejuni* 81–176. The invasion and adhesion assays were conducted as described in materials and methods. Results are presented as the mean CFU per well (●), and the resulting average number of internalized *C. jejuni* 81–176 per epithelial cell (○). (**C**) Effect of varying ROI on LG2055 adhesion efficiency. Int407 cells were treated with LG2055. The adhesion assay was conducted as described in materials and methods. Results are presented as the mean CFU per well. All experimental error bars indicate standard deviations (n = 3).

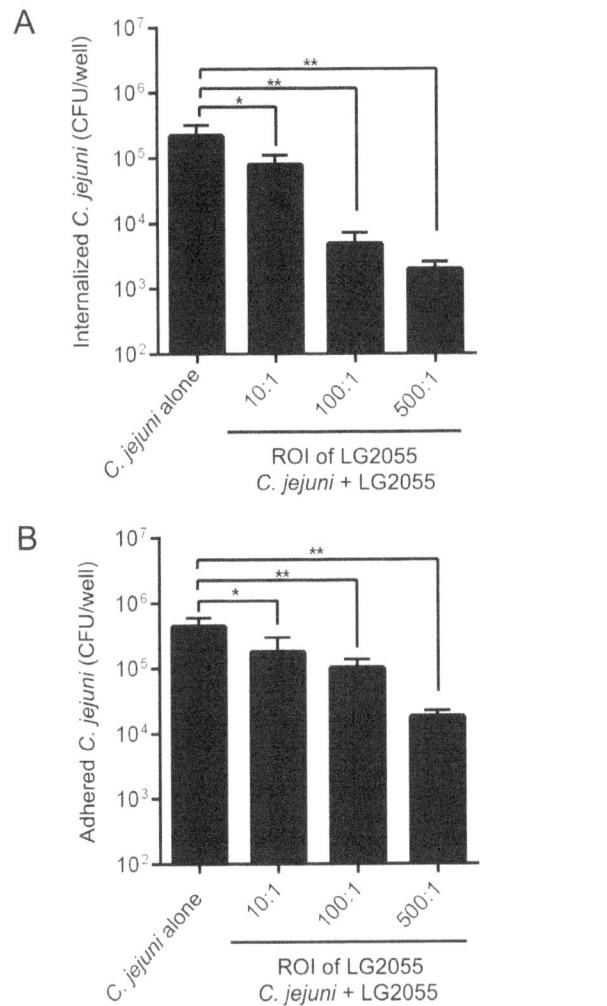

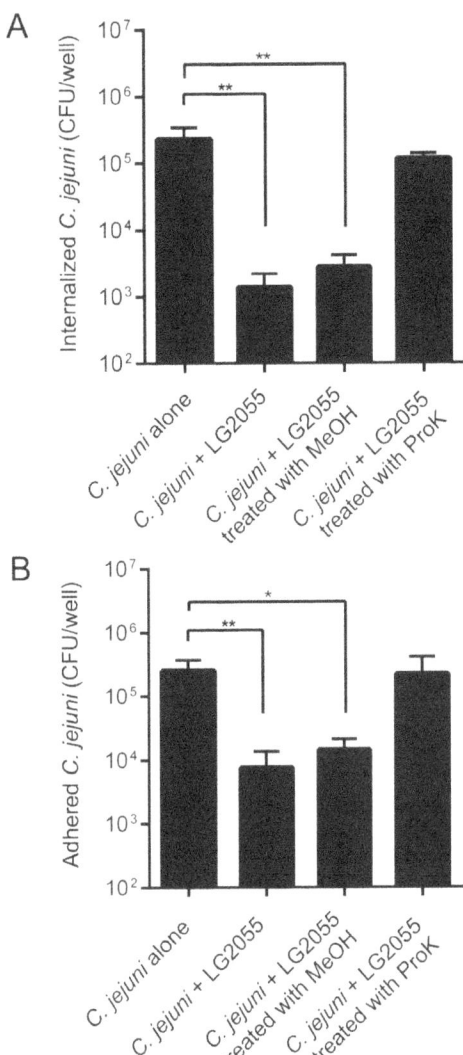

Figure 2. Inhibition of *C. jejuni* **81–176 invasion and adhesion to Int407 cells by LG2055.** Assays for *L. gasseri*-mediated inhibition of invasion (**A**) and adhesion (**B**) were performed by pre-incubation of Int407 cells with LG2055 at the ROI of 10:1, 100:1, or 500:1 followed by the addition of *C. jejuni* 81–176 at the MOI of 100:1 (for invasion), or 500:1 (for adhesion). Results are presented as the mean CFU per well. The asterisks indicate that the number of *C. jejuni* 81–176 was statistically different (*$p<0.05$, **$p<0.01$) than that of the untreated groups, as determined by one-way ANOVA with the *post hoc* Dunnett's test. Error bars indicate standard deviations (n = 5).

Figure 3. Inhibition of *C. jejuni* **81–176 invasion and adhesion to Int407 cells by MeOH-fixed or ProK-treated LG2055.** LG2055 cells were treated with MeOH or ProK prior to infection of Int407 cells. Assays for LG2055-mediated inhibition of invasion (**A**) and adhesion (**B**) were performed by pre-incubation of Int407 cells with treated or untreated LG2055 at the ROI of 500:1 followed by the addition of *C. jejuni* 81–176 at the MOI of 100:1 (for invasion), or 500:1 (for adhesion). Results are presented as the mean CFU per well. The asterisks indicate that the number of *C. jejuni* 81–176 was statistically different (*$p<0.05$, **$p<0.01$) than that of the untreated groups, as determined by one-way ANOVA with the *post hoc* Dunnett's test. Error bars indicate standard deviations (n = 5).

Statistical analyses

Prism 6 (GraphPad Software) was used for all statistical analyses. Significant differences were determined by using one-way analysis of variance (ANOVA) with *post hoc* Dunnett's test, or Mann–Whitney U test. "n" represents the number of individual experiments. Differences with *p* values less than 0.05 were considered significant.

Results

Capacity for invasion of and adhesion to Int407 cells

We tested the invasive properties of *C. jejuni* 81–176 *in vitro*. *C. jejuni* 81–176 was co-cultured with Int407 cells at MOIs ranging from 0.1:1 to 1000:1. *C. jejuni* 81–176 was internalized by Int407 cells in a dose-dependent manner (MOI$\leq$100:1) (Figure 1A). We quantified the average number of internalized bacterial cells per

Int407 cell by dividing the number of internalized bacteria by the total number of epithelial cells per well at each MOI tested. The highest possible number of internalized bacterial cells, which we observed at an MOI of 100:1, was approx. 1.8 bacterial cells per epithelial cell (Figure 1A). This level of internalization is similar to the results of previous studies [35]. For comparison, the adhesive properties of *C. jejuni* 81–176 were also assessed. *C. jejuni* 81–176 adhered dose-dependently to Int407 cells, with the highest number of adherent bacteria at an MOI of 500:1 (Figure 1B).

We next tested the adhesion abilities of LG2055 *in vitro*. LG2055 was co-cultured with Int407 cells at ROIs ranging from 0.1:1 to 1000:1. LG2055 also adhered to Int407 cells in a dose-

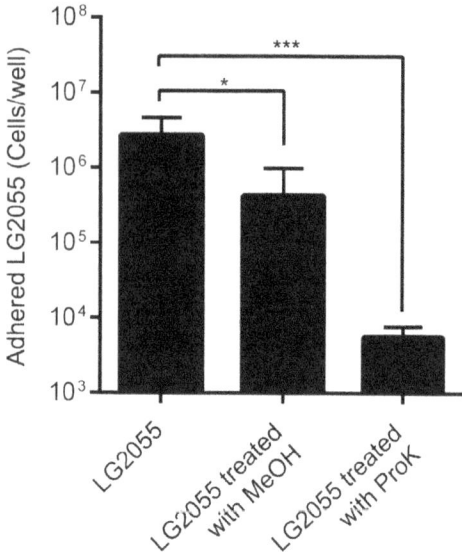

Figure 4. Quantification of adhesion of MeOH-fixed or ProK-treated LG2055 strains to Int407 cells in an adhesion inhibition assay. The numbers of *L. gasseri* strains were quantified by real-time PCR using *L. gasseri*-specific primers. The asterisks indicate that the number of LG2055 was statistically different (*$p<0.05$, ***$p<0.001$) than that of the untreated LG2055, as determined by one-way ANOVA with the *post hoc* Dunnett's test. Error bars indicate standard deviations (n = 5).

dependent manner (Figure 1C), with the highest number of adherent bacteria at an ROI of 500:1.

Inhibition of *C. jejuni* 81–176 adhesion and invasion by LG2055

To examine whether LG2055 competitively inhibited the adhesion and invasion of *C. jejuni* 81–176 *in vitro*, Int407 cells were pre-incubated with LG2055 at an ROI of 10:1, 100:1, or 500:1, and then Int407 cells were incubated with *C. jejuni* 81–176 at an MOI of 100:1 (for invasion), or 500:1 (for adhesion), respectively. Pre-incubation with LG2055 resulted in approx. 3- to 100-fold decrease in the number of internalized bacteria (Figure 2A) and approx. 2.5- to 25-fold decrease in adherent bacteria (Figure 2B).

Characterization of *C. jejuni* 81–176 inhibition effects by LG2055

Surface components of probiotic bacteria have been shown as potential reactants with pathogens [13], [15], [16], [29], [36]. To evaluate *C. jejuni* 81–176 inhibition caused by LG2055, we focused on the proteinaceous cell surface components of LG2055 that were involved in inhibition by examining the effects of ProK-treated LG2055 on *C. jejuni* 81–176 invasion and adhesion. ProK treatment eliminated LG2055-mediated inhibition of *C. jejuni* 81–176 invasion (Figure 3A) and adhesion (Figure 3B). By contrast, in an agar diffusion assay, untreated and heat-treated LG2055 supernatants greatly inhibited *C. jejuni* 81–176 growth (Figure S1). Therefore, we next examined whether the metabolically active components of LG2055 were involved in *C. jejuni* 81–176 inhibition. MeOH-fixed LG2055 significantly decreased invasion by approx. 83-fold (Figure 3A) and significantly reduced adhesion by 17-fold (Figure 3B), similar to the levels of inhibition observed with the intact cells.

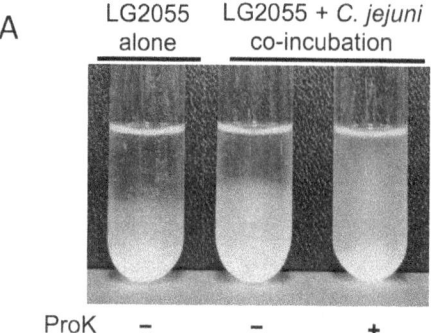

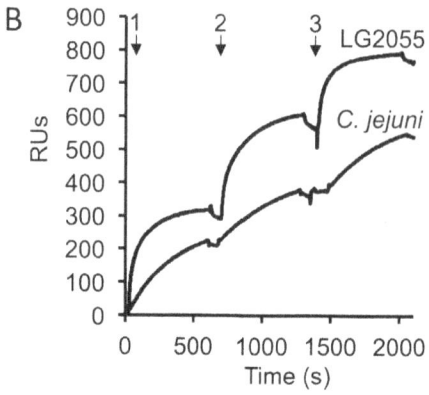

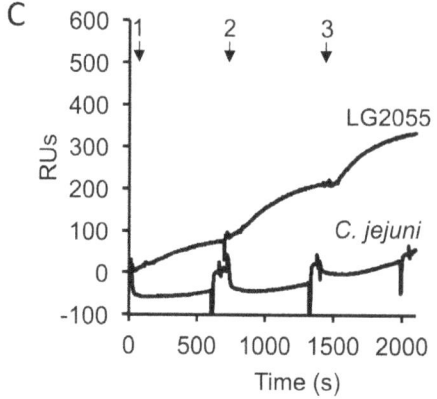

Figure 5. Co-aggregation of LG2055 and *C. jejuni* 81–176. (A) The left tube shows self-aggregation of LG2055, the center tube shows co-aggregation of LG2055 with *C. jejuni* 81–176, and the right tube shows co-aggregation of ProK-treated LG2055 with *C. jejuni* 81–176. **(B, C)** Interaction of bacterial cells by SPR analysis. The SPR biosensors presented the following amounts of immobilized ligands: 5650 RUs, LG2055 **(B)**; and 5820 RUs, ProK-treated LG2055 **(C)**. The live bacterial cells were suspended in running buffer and injected as an analyte during each cycle. The SPR analysis was conducted as described in materials and methods. Arrows indicate sample inject point.

The adhesion property of LG2055 in an adhesion inhibition assay was assessed using quantitative real-time PCR. The adhesion capacity was significantly reduced by treating the cells with MeOH ($p<0.05$) and ProK ($p<0.001$), compared with that of untreated LG2055 cells (Figure 4). In particular, the adhesion of ProK-treated LG2055 cells was greatly reduced by approx. 470-fold. These results showed that the inhibitory factor is a protein and/or cell-surface component.

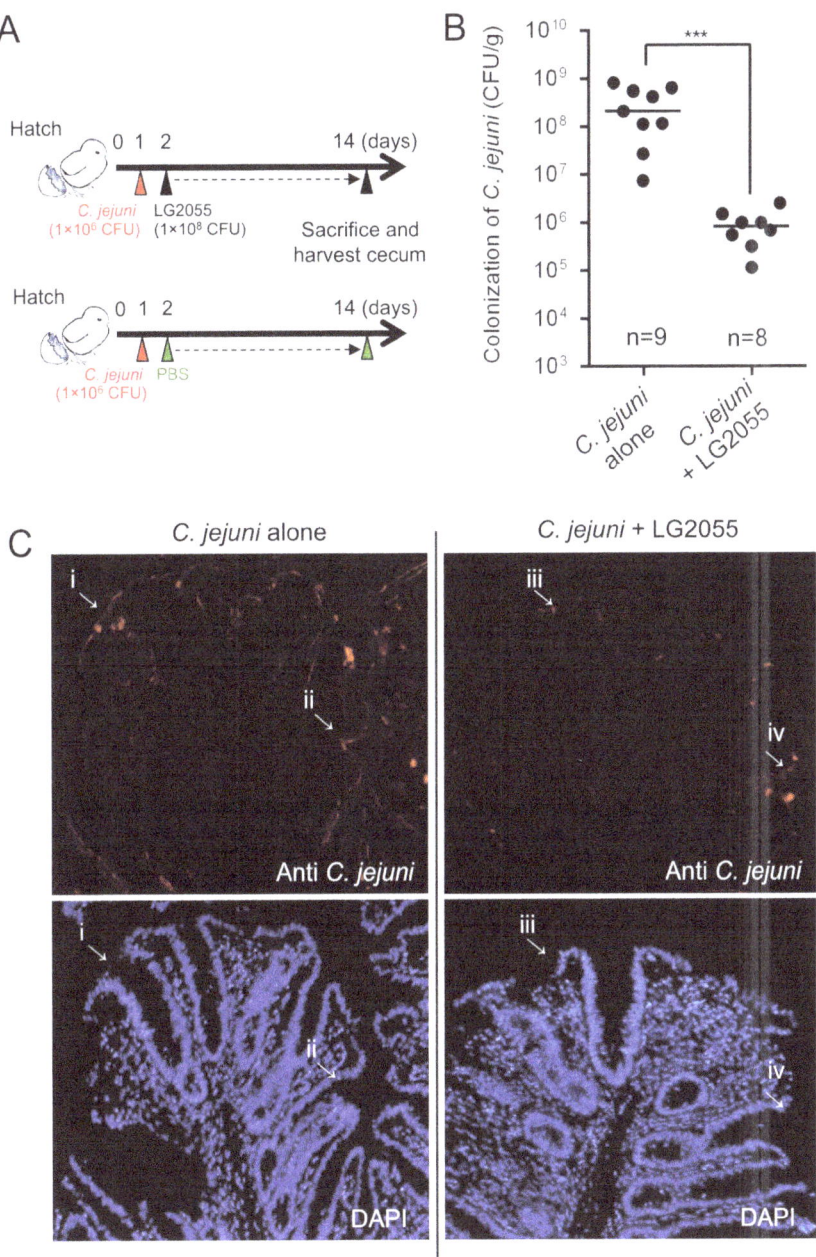

Figure 6. Colonization of *C. jejuni* in the chick cecum. (**A**) Twenty-four hours after hatching, chicks were inoculated orally. LG2055 was administered orally to one group of *C. jejuni* 81–176-inoculated birds daily. PBS was administered to the other group of birds as a control. (**B**) Chicks were sacrificed at days 14 post-inoculation, and colonization with *C. jejuni* was quantified by counting CFU using MH agar plates. The bar indicates the median CFU for each group, which was determined using all birds within the group. The LG2055-administered group of birds exhibited significant suppression of *C. jejuni* colonization (***$p<0.001$), as determined by the Mann-Whitney U test. "n" indicates the number of chicks in each group of 9 that were colonized with *C. jejuni* (limit of detection, 10^3 CFU/g of cecal contents). (**C**) Immunohistochemical staining of chick cecal tissue at 14 days post-inoculation. Cecal sections were probed with anti-*Campylobacter* antiserum or stained with DAPI. Fluorescence microscopy shows *C. jejuni* colonization (arrows: i–iv) over the mucosal surface of the cecal tissue at 100× magnification.

Characterization of LG2055 co-aggregation with *C. jejuni* 81–176

Co-aggregation, a probiotic interaction with pathogenic microorganisms, may lead to the formation of a barrier that prevents colonization by the pathogens [36–38], which is usually facilitated by surface-exposed proteinaceous components [15], [16]. Aggregation of LG2055 was examined by statically incubating bacterial suspensions over a period of 3 h. The strain exhibited a strong self-

aggregation phenotype (Figure 5A, left tube). Co-incubation of LG2055 with *C. jejuni* 81–176 also resulted in a similar aggregation phenotype (Figure 5A, middle tube), while ProK pretreatment of LG2055 abolished co-aggregation (Figure 5A, right tube).

To further characterize the aggregation, we examined the interaction between bacterial cells by SPR analysis. We measured the ability of live LG2055 and *C. jejuni* 81–176 cells to bind

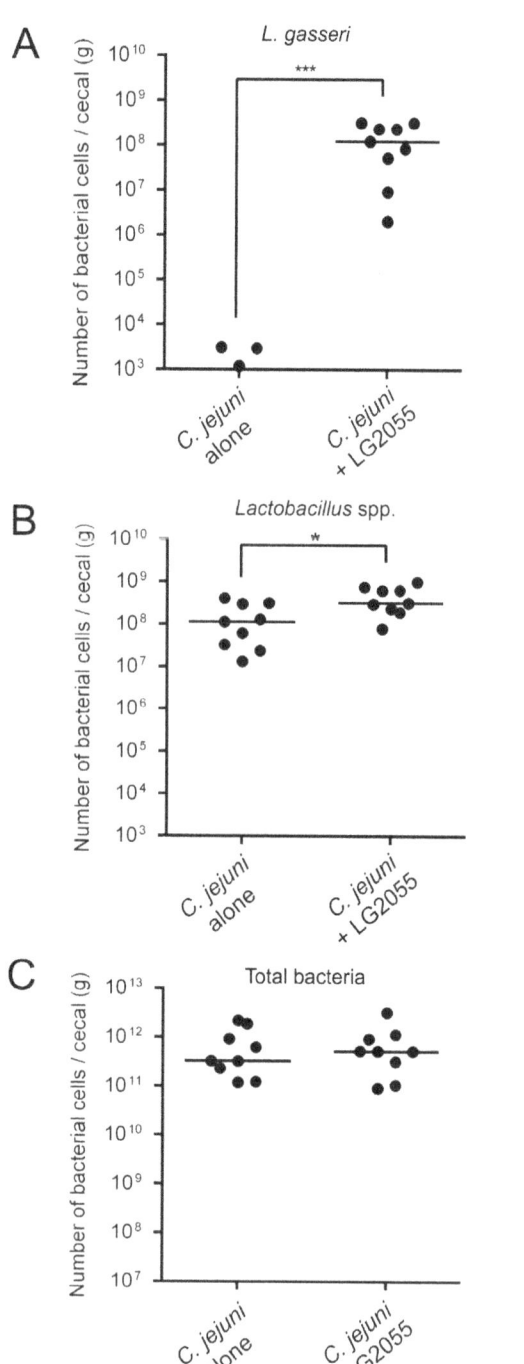

not bind to the bacterial ligand (Figure 5C). In contrast, live LG2055 cells showed moderate binding to the ligand.

Inhibition of *C. jejuni* colonization in chicks by LG2055

The inhibitory effect of LG2055 on *C. jejuni* colonization in chicks was examined. Approx. 24 h after hatching, chicks were pre-inoculated orally with *C. jejuni* 81–176, and then LG2055 was administered daily (Figure 6A). *C. jejuni* cells were counted in the untreated control group at 14 days post-inoculation; the median number of *C. jejuni* in the ceca increased to 2.0×10^8 CFU/g of cecal content at 14 days post-inoculation (Figure 6B). In contrast, the median level of *C. jejuni* colonization in chicks treated with LG2055 significantly decreased by approx. 250-fold, compared with the levels of the untreated control group. In addition, immunohistochemical staining of chick cecal tissue and fluorescence microscopy showed *C. jejuni* colonization over the mucosal surface of the cecal tissue (Figure 6C, arrows i–iv). The fluorescence signal of *C. jejuni* was lower in chicks treated with LG2055 (Figure 6C, arrows iii and iv) than in the chicks from the untreated control group (Figure 6C, arrows i and ii).

Quantitative real-time PCR analysis of 16S rDNA with species- or genus-specific primers was used to quantify cecal bacteria at 14 days post-inoculation. Colonization by *L. gasseri* and *Lactobacillus* spp. in the group treated with LG2055 was significantly higher than in the untreated control group, but colonization did not alter the total bacteria in the ceca (Figure 7A–C). The number of *L. gasseri* cells in the ceca of the untreated group was below the limit of detection by real-time PCR ($<10^3$ cells/g) (Figure 7A). There were no significant differences in cecal pH values (Figure S2) and body weight. In addition, both groups of birds displayed similar appetites during the experiment.

Discussion

LG2055 possesses a variety of probiotic activities and roles, including production of bacteriocin [39], inducing suppression of the number of fecal staphylococci [22], protection against influenza A virus infection [24], regulating adipose tissue growth in rats [40], and exerting an anti-obesity effect in humans [41]. Campylobacteriosis, the most commonly reported form of food-borne gastroenteritis, is mainly caused by the zoonotic pathogen *C. jejuni* [3], [5], [10], [12]. We have now demonstrated that LG2055 can also reduce the adhesion and invasion of *C. jejuni in vitro* and can inhibit pathogen colonization of chicks *in vivo*. In this study, we characterized the effective competitive ability of LG2055 to exclude this pathogen, which may be useful for the prevention of campylobacteriosis.

Intestinal epithelial cells are the initial site for host interactions with *Campylobacter*. Int407 cell lines of human intestinal epithelium origin are considered one of the most appropriate models for studying *Campylobacter* infection [42]. In this study, we established the ability of LG2055 to reduce *C. jejuni* 81–176 adhesion to Int407 cells. Treatment with LG2055 resulted in greater than 2-log-fold reduction in *C. jejuni* 81–176 invasion of the epithelial cells. Previous papers reported a protective effect of probiotic strains on epithelial cells infected with *Campylobacter*. For example, *L. helveticus* R0052 adheres to human colonic T84 and Int407 cells, reducing *C. jejuni* invasion into these cells by 35%–55% following co-culture with probiotics [14]. However, comparison of previous results with our current results shows that LG2055 is more effective at inhibiting *C. jejuni* infection.

An anti-*Campylobacter* analysis showed that LG2055 inhibits *C. jejuni* 81–176 growth by means of an acidic component instead of a heat-labile component. Lactic acid produced by *L. crispatus*

Figure 7. Quantification of gut bacteria at 14 days post-inoculation. The numbers of *L. gasseri* (**A**), *Lactobacillus* spp. (**B**), and total bacteria (**C**) in cecal samples were quantified by real-time PCR using 16S rDNA gene-specific primers. The LG2055-administered group of birds showed significantly increased levels of *L. gasseri* and *Lactobacillus* spp. (*$p<0.05$, ***$p<0.001$), as determined by the Mann-Whitney U test.

LG2055 cells immobilized on a biosensor. Both bacteria bound to the immobilized cells during the three-cycle experiment (Figure 5B). Moreover, in a three-cycle binding experiment using immobilized ProK-treated LG2055, live *C. jejuni* 81–176 cells did

JCM5810 inhibits *C. jejuni* growth and colonization *in vitro* and *in vivo* [21]. Although acidity induced by LG2055 might affect *C. jejuni* 81–176 colonization, pretreatment with MeOH-fixed LG2055 reduced pathogen adhesion and invasion. In addition, ProK treatment of LG2055 eliminated the inhibitory effect. Taken together, these data suggest that inhibition of *C. jejuni* 81–176 invasion of and adhesion to epithelial cells might involve constitutive proteinaceous components on the exposed surface of LG2055, rather than soluble metabolically active components.

Furthermore, we showed that LG2055, but not ProK-treated LG2055, co-aggregated with *C. jejuni* 81–176. These results support our observations of the *in vitro* inhibition effects. Previous reports also demonstrated effective *in vitro* co-aggregation of *Campylobacter* strains with a surface component of *Lactobacillus* [15], [16]. Thus, we presumed that co-aggregation properties of LG2055 with *C. jejuni* 81–176 mediated by proteinaceous, cell-surface component(s) of LG2055 play a key play in preventing this pathogen infection. On the other hand, ProK treatment of LG2055 dramatically reduced the ability to adhere to Int407 cells. Competition with pathogens for adhesion and colonization of the mucosal surface are possible protective mechanisms employed by probiotics [14], [29], [36]. This competition suggests that the inhibition mechanisms against infection and colonization involve other distinct processes in addition to the co-aggregation phenotype.

In our *in vivo* model of *Campylobacter* colonization, we demonstrated that LG2055 was able to significantly reduce colonization in chicks at 14 days post-inoculation. Histochemical analysis also showed reduced adhesion of *C. jejuni* on the cecal mucosa of chicks treated with LG2055. A previous study demonstrated that most *Campylobacter* colonization occurs at an age of two to four weeks [43], which is probably delayed because of the presence of maternally derived antibodies inhibiting *Campylobacter* colonization in young chicks [44]. Moreover, once *Campylobacter* colonization is detected in chickens, the prevalence of *Campylobacter* colonization increases from 5%–95% within several days [45]. Therefore, our results imply that, although LG2055 cannot completely inhibit *C. jejuni* colonization in chicks, treatment with LG2055 could be useful in suppressing pathogen colonization of the chicks at early growth stages, thereby helping to prevent pathogen infection. On the other hand, a previous study indicated that there is a high risk of *Campylobacter* colonization in chicken throughout the feeding period [43]. Therefore, we will need perform more long-term chicken colonization experiments to further evaluate the *Campylobacter* inhibition effects caused by LG2055.

Inoculation with LG2055 increased the number of *Lactobacillus* spp. and *L. gasseri* cells, as detected by quantitative real-time PCR, and the LG2055 strain exhibited a strong self-aggregation phenotype *in vitro*. Self-aggregation of probiotic strains correlates with adhesion to mucous components and biofilm formation, which is a prerequisite for colonization of the GI tract [36], [46–50]. Taken together, our results show that LG2055 may be able to establish persistent colonization and may become a part of the dominant microbiota. Moreover, the effects of acid production alone do not explain the inhibitory effects of LG2055 on *C. jejuni* colonization, as cecal pH values changed minimally. Some probiotic treatments reduce colonization by *Campylobacter*. Specifically, certain *Lactobacillus* strains, which are employed by the poultry industry and produce bacteriocins against *Campylobacter*, efficiently reduce the colonization of this pathogen in chickens during commercial processing [17], [19]. The anti-colonization effect by LG2055 is not yet applicable for use in the poultry industry; however, the co-aggregation phenotype and (or) adhesion capacity are important factors that should be considered in designing competitive exclusion strategies to reduce pathogen loads in livestock.

In conclusion, we have demonstrated that LG2055 is an organism that effectively and competitively excludes *C. jejuni*. This exclusion is evidenced by the reduction in the number of *C. jejuni* cells adhering to and internalized by epithelial cells of humans and chicks. Data from our *in vitro* experiments indicate that the co-aggregation phenotype and/or adhesion mediated by proteinaceous surface component(s) of LG2055 might be responsible for the reduction in *C. jejuni* infection and colonization. Extracellular aggregation-promoting factors (APFs) and genes encoding them for several *Lactobacillus* species, including *L. gasseri*, have been characterized previously [51]. APFs are secreted proteins that are associated with a diverse number of functional roles, including self-aggregation [52], maintenance of cell shape [53], and adhesion [47], [49]. Thus, further analysis is required to confirm the role and involvement of proteinaceous cell-surface components such as APFs in the inhibitory properties of LG2055.

Supporting Information

Figure S1 Inhibition of *C. jejuni* 81–176 by LG2055. Anti-*Campylobacter* activity was assessed using spotted lactobacilli cell-free culture supernatant. Overnight LG2055 culture supernatants were collected and either heat-treated (boiled), neutralized with NaOH, or left untreated. The supernatants were added to MH agar plates seeded with *C. jejuni* and incubated for 24 h at 37°C.

Figure S2 The effect of cecal pH on administered LG2055. The plot shows the mean pH for each group, which was determined using all birds within the group. Error bars indicate standard deviations. "n" indicates the number of chicks in each group.

Acknowledgments

The authors would like to thank Aya Takamine, Ayuha Niwano, and Ryo Ito (Kitasato University) for their technical assistance.

Author Contributions

Conceived and designed the experiments: KN YS YY TM. Performed the experiments: KN KY. Analyzed the data: KN. Contributed reagents/materials/analysis tools: TK ST TM. Wrote the paper: KN YY TM.

References

1. Allos BM (2001) *Campylobacter jejuni* Infections: update on emerging issues and trends. Clin Infect Dis 32: 1201–1206.

2. Zilbauer M, Dorrell N, Wren BW, Bajaj-Elliott M (2008) *Campylobacter jejuni*-mediated disease pathogenesis: an update. Trans R Soc Trop Med Hyg 102: 123–129.

3. Kubota K, Kasuga F, Iwasaki E, Inagaki S, Sakurai Y, et al. (2011) Estimating the burden of acute gastroenteritis and foodborne illness caused by *Campylobacter*, *Salmonella*, and *Vibrio parahaemolyticus* by using population-based telephone survey data, Miyagi Prefecture, Japan, 2005 to 2006. J Food Prot 74: 1592–1598.

4. Young KT, Davis LM, Dirita VJ (2007) *Campylobacter jejuni*: molecular biology and pathogenesis. Nat Rev Microbiol 5: 665–679.

5. Dasti JI, Tareen AM, Lugert R, Zautner AE, Gross U (2010) *Campylobacter jejuni*: a brief overview on pathogenicity-associated factors and disease-mediating mechanisms. Int J Med Microbiol 300: 205–211.

6. Kalischuk LD, Inglis GD, Buret AG (2007) Strain-dependent induction of epithelial cell oncosis by *Campylobacter jejuni* is correlated with invasion ability and is independent of cytolethal distending toxin. Microbiology 153: 2952–2963.

7. Beltinger J, Del Buono J, Skelly MM, Thornley J, Spiller RC, et al. (2008) Disruption of colonic barrier function and induction of mediator release by strains of *Campylobacter jejuni* that invade epithelial cells. World J Gastroenterol 14: 7345–7352.

8. Boehm M, Krause-Gruszczynska M, Rohde M, Tegtmeyer N, Takahashi S, et al. (2011) Major host factors involved in epithelial cell invasion of *Campylobacter jejuni*: role of fibronectin, integrin beta1, FAK, Tiam-1, and DOCK180 in activating Rho GTPase Rac1. Front Cell Infect Microbiol 1: 17.

9. Suzuki H, Yamamoto S (2009) *Campylobacter* contamination in retail poultry meats and by-products in the world: a literature survey. J Vet Med Sci 71:255–261.

10. Altekruse SF, Stern NJ, Fields PI, Swerdlow DL (1999) *Campylobacter jejuni*–an emerging foodborne pathogen. Emerg Infect Dis 5: 28–35.

11. Hermans D, Van Deun K, Martel A, Van Immerseel F, Messens W, et al. (2011) Colonization factors of *Campylobacter jejuni* in the chicken gut. Vet Res 42: 82.

12. Rosenquist H, Nielsen NL, Sommer HM, Nørrung B, Christensen BB (2003) Quantitative risk assessment of human campylobacteriosis associated with thermophilic *Campylobacter* species in chickens. Int J Food Microbiol 83: 87–103.

13. Gaggia F, Di Gioia D, Baffoni L, Biavati B (2011) The role of protective and probiotic cultures in food and feed and their impact in food safety. Trends Food Sci Technol 22: 58–66.

14. Wine E, Gareau MG, Johnson-Henry K, Sherman PM (2009) Strain-specific probiotic (*Lactobacillus helveticus*) inhibition of *Campylobacter jejuni* invasion of human intestinal epithelial cells. FEMS Microbiol Lett 300: 146–152.

15. Schachtsiek M, Hammes WP, Hertel C (2004) Characterization of *Lactobacillus coryniformis* DSM 20001T surface protein Cpf mediating coaggregation with and aggregation among pathogens. Appl Environ Microbiol 70: 7078–7085.

16. Tareb R, Bernardeau M, Gueguen M, Vernoux J-P (2013) *In vitro* characterization of aggregation and adhesion properties of viable and heat-killed forms of two probiotic *Lactobacillus* strains and interaction with foodborne zoonotic bacteria, especially *Campylobacter jejuni*. J Med Microbiol 62: 637–649.

17. Stern NJ, Svetoch EA, Eruslanov BV, Perelygin VV, Mitsevich EV, et al. (2006) Isolation of a *Lactobacillus salivarius* strain and purification of its bacteriocin, which is inhibitory to *Campylobacter jejuni* in the chicken gastrointestinal system. Antimicrob Agents Chemother 50: 3111–3116.

18. Campana R, Federici S, Ciandrini E, Baffone W (2012) Antagonistic activity of *Lactobacillus acidophilus* ATCC 4356 on the growth and adhesion/invasion characteristics of human *Campylobacter jejuni*. Curr Microbiol 64: 371–378.

19. Svetoch EA, Eruslanov BV, Levchuk VP, Perelygin VV, Mitsevich EV, et al. (2011) Isolation of *Lactobacillus salivarius* 1077 (NRRL B-50053) and characterization of its bacteriocin, including the antimicrobial activity spectrum. Appl Environ Microbiol 77: 2749–2754.

20. Zhao T, Doyle MP (2006) Reduction of *Campylobacter jejuni* on chicken wings by chemical treatments. J Food Prot 69: 762–767.

21. Neal-McKinney JM, Lu X, Duong T, Larson CL, Call DR, et al. (2012) Production of organic acids by probiotic lactobacilli can be used to reduce pathogen load in poultry. PLoS One 7: e43928.

22. Fujiwara S, Seto Y, Kimura A, Hashiba H (2001) Establishment of orally-administered *Lactobacillus gasseri* SBT2055SR in the gastrointestinal tract of humans and its influence on intestinal microflora and metabolism. J Appl Microbiol 90: 343–352.

23. Takahashi H, Fujita T, Suzuki Y, Benno Y (2006) Monitoring and survival of *Lactobacillus gasseri* SBT2055 in the human intestinal tract. Microbiol Immunol 50: 867–870.

24. Nakayama Y, Moriya T, Sakai F, Ikeda N, Shiozaki T, et al. (2014) Oral administration of *Lactobacillus gasseri* SBT2055 is effective for preventing influenza in mice. Sci Rep 4:4638.

25. Korlath JA, Osterholm MT, Judy LA, Forfang JC, Robinson RA (1985) A point-source outbreak of campylobacteriosis associated with consumption of raw milk. J Infect Dis 152: 592–596.

26. Hendrixson DR, DiRita VJ (2004) Identification of *Campylobacter jejuni* genes involved in commensal colonization of the chick gastrointestinal tract. Mol Microbiol 52: 471–484.

27. Oelschlaeger TA, Guerry P, Kopecko DJ (1993) Unusual microtubule-dependent endocytosis mechanisms triggered by *Campylobacter jejuni* and *Citrobacter freundii*. Proc Natl Acad Sci U S A 90: 6884–6888.

28. Watson RO, Galán JE (2008) *Campylobacter jejuni* survives within epithelial cells by avoiding delivery to lysosomes. PLoS Pathog 1:e14.

29. Spurbeck RR, Arvidson CG (2010) *Lactobacillus jensenii* surface-associated proteins inhibit *Neisseria gonorrhoeae* adherence to epithelial cells. Infect Immun 78: 3103–3111.

30. Chu H, Pazgier M, Jung G, Nuccio S-P, Castillo PA, et al. (2012) Human α-defensin 6 promotes mucosal innate immunity through self-assembled peptide nanonets. Science 337: 477–481.

31. Kakuda T, DiRita VJ (2006) Cj1496c encodes a *Campylobacter jejuni* glycoprotein that influences invasion of human epithelial cells and colonization of the chick gastrointestinal tract. Infect Immun 74: 4715–4723.

32. Matsumoto M, Sakamoto M, Benno Y (2009) Dynamics of fecal microbiota in hospitalized elderly fed probiotic LKM512 yogurt. Microbiol Immunol 53: 421–432.

33. Nadkarni MA, Martin FE, Jacques NA, Hunter N (2002) Determination of bacterial load by real-time PCR using a broad-range (universal) probe and primers set. Microbiology 148: 257–266.

34. Byun R, Nadkarni MA, Chhour K, Martin FE, Jacques NA, et al. (2004) Quantitative analysis of diverse *Lactobacillus* species present in advanced dental caries. J Clin Microbiol 42: 3128–3136.

35. Hu L, Kopecko DJ (1999) *Campylobacter jejuni* 81–176 associates with microtubules and dynein during invasion of human intestinal cells. Infect Immun 67: 4171–4182.

36. Collado MC, Meriluoto J, Salminen S (2008) Adhesion and aggregation properties of probiotic and pathogen strains. Eur Food Res Technol 226: 1065–1073.

37. Collado MC, Meriluoto J, Salminen S (2007) Measurement of aggregation properties between probiotics and pathogens: *in vitro* evaluation of different methods. J Microbiol Methods 71: 71–74.

38. Twetman L, Larsen U, Fiehn N-E, Stecksén-Blicks C, Twetman S (2009) Coaggregation between probiotic bacteria and caries-associated strains: an *in vitro* study. Acta Odontol Scand 67: 284–288.

39. Kawai Y, Saitoh B, Takahashi O, Kitazawa H, Saito T, et al. (2000) Primary amino acid and DNA sequences of gassericin T, a lactacin F-family bacteriocin produced by *Lactobacillus gasseri* SBT2055. Biosci Biotechnol Biochem 64: 2201–2208.

40. Sato M, Uzu K, Yoshida T, Hamad EM, Kawakami H, et al. (2008) Effects of milk fermented by *Lactobacillus gasseri* SBT2055 on adipocyte size in rats. Br J Nutr 99: 1013–1017.

41. Kadooka Y, Sato M, Imaizumi K, Ogawa A, Ikuyama K, et al. (2010) Regulation of abdominal adiposity by probiotics (*Lactobacillus gasseri* SBT2055) in adults with obese tendencies in a randomized controlled trial. Eur J Clin Nutr 64: 636–643.

42. Konkel ME, Mead DJ, Hayes SF, Cieplak W (1992) Translocation of *Campylobacter jejuni* across human polarized epithelial cell monolayer cultures. J Infect Dis 166: 308–315.

43. van Gerwe T, Miflin JK, Templeton JM, Bouma A, Wagenaar JA, et al. (2009) Quantifying transmission of *Campylobacter jejuni* in commercial broiler flocks. Appl Environ Microbiol 75: 625–628.

44. Sahin O, Luo N, Huang S, Zhang Q (2003) Effect of *Campylobacter*-specific maternal antibodies on *Campylobacter jejuni* colonization in young chickens. Appl Environ Microbiol 69:5372–5379.

45. Van Gerwe TJWM, Bouma A, Jacobs-Reitsma WF, Van den Broek J, Klinkenberg D, et al. (2005) Quantifying transmission of *Campylobacter* spp. among broilers. Appl Environ Microbiol 71:5765–5770.

46. Kos B, Suskovic J, Vukovic S, Simpraga M, Frece J, et al. (2003) Adhesion and aggregation ability of probiotic strain *Lactobacillus acidophilus* M92. J Appl Microbiol 94: 981–987.

47. Goh YJ, Klaenhammer TR (2010) Functional roles of aggregation-promoting-like factor in stress tolerance and adherence of *Lactobacillus acidophilus* NCFM. Appl Environ Microbiol 76: 5005–5012.

48. Mackenzie DA, Jeffers F, Parker ML, Vibert-Vallet A, Bongaerts RJ, et al. (2010) Strain-specific diversity of mucus-binding proteins in the adhesion and aggregation properties of *Lactobacillus reuteri*. Microbiology 156: 3368–3378.

49. Hevia A, Martínez N, Ladero V, Alvarez MA, Margolles A, et al. (2013) An extracellular Serine/Threonine-rich protein from *Lactobacillus plantarum* NCIMB 8826 is a novel aggregation-promoting factor with affinity to mucin. Appl Environ Microbiol 79: 6059–6066.

50. Frese SA, Mackenzie DA, Peterson DA, Schmaltz R, Fangman T, et al. (2013) Molecular characterization of host-specific biofilm formation in a vertebrate gut symbiont. PLoS Genet 9: e1004057.

51. Ventura M, Jankovic I, Walker DC, Pridmore RD, Zink R (2002) Identification and characterization of novel surface proteins in *Lactobacillus johnsonii* and *Lactobacillus gasseri*. Appl Environ Microbiol 68: 6172–6181.

52. Reniero R, Cocconcelli P, Bottazzi V, Morelli L (1992) High frequency of conjugation in *Lactobacillus* mediated by an aggregation-promoting factor. J Gen Microbiol 138: 763–768.

53. Jankovic I, Ventura M, Meylan V, Rouvet M, Elli M, et al. (2003) Contribution of aggregation-promoting factor to maintenance of cell shape in *Lactobacillus gasseri* 4B2. J Bacteriol 185: 3288–3296.

20

The Effect of Probiotics Supplementation on *Helicobacter pylori* Eradication Rates and Side Effects during Eradication Therapy

Yini Dang[1], Jan D. Reinhardt[2,3,4], Xiaoying Zhou[1], Guoxin Zhang[1]*

1 Department of Gastroenterology, The First Affiliated Hospital of Nanjing Medical University, Nanjing, China, and First Clinical Medical College of Nanjing Medical University, Nanjing, China, **2** Institute for Disaster Management and Reconstruction, Sichuan University, Chengdu, China, and Hong Kong Polytechnical University, Hung Hom, Hong Kong, China, **3** Department of Health Sciences, University of Lucerne, Lucerne, Switzerland, **4** Swiss Paraplegic Research, Nottwil, Switzerland

Abstract

Background: Previous meta-analyses reported that probiotics improve the effectiveness of *Helicobacter pylori* (*H. pylori*) eradication during antibiotic therapy, while results regarding a possible reduction of side effects remained inconclusive. Moreover, the effectiveness of different strains of probiotics has not been studied so far. It is further conceivable that probiotics will produce additional effects only if antibiotics are relatively ineffective.

Methods: This meta-analysis includes eligible randomized controlled trials examining effects of probiotics supplementation on eradication rates (ER) and side effects, published up to May 2014. Sub-group analysis was performed to compare different probiotic strains and antibiotic therapies with different effectiveness in controls (ER <80% vs.>80%). Publication bias was assessed with funnel plots and Harbord's test. The quality of the trials was assessed with the Cochrane risk of bias tool.

Results: Thirty-three RCTs involving a total of 4459 patients met the inclusion criteria in case of eradication rates of which 20 assessed total side effects in addition. Overall, the pooled eradication rate in probiotics supplementation groups was significantly higher than in controls (ITT analysis: RR 1.122, 95% CI 1.086–1.159, PP analysis: RR 1.114, 95% CI 1.070–1.159). Sub group-analysis could, however, confirm this finding only for four individual strains (*Lactobacillus acidophilus*, *Lactobacillus casei* DN-114001, *Lactobacillus gasseri*, and *Bifidobacterium infantis* 2036) and for relatively ineffective antibiotic therapies. There was a significant difference between groups in the overall incidence of side effects (RR 0.735, 95% CI 0.598–0.902). This result was, however, only confirmed for non-blinded trials.

Conclusions: The pooled data suggest that supplementation with specific strains of probiotics compared with eradication therapy may be considered an option for increasing eradication rates, particularly when antibiotic therapies are relatively ineffective. The impact on side effects remains unclear and more high quality trials on specific probiotic strains and side effects are thus needed.

Editor: Pere-Joan Cardona, Fundació Institut d'Investigació en Ciències de la Salut Germans Trias i Pujol. Universitat Autònoma de Barcelona. CIBERES, Spain

Funding: This work was supported by grants from the National Natural Science Foundation of China (No. 81072032 and No. 81270476). The funders had no role in study design, data collection and analysis, decision to publish, or preparation of the manuscript.

Competing Interests: The authors have declared that no competing interests exist.

* Email: guoxinz@njmu.edu.cn

Introduction

'*Helicobacter pylori* (*H. pylori*) is a common bacterium infecting about half of the world's population. It is causally associated with a diverse spectrum of gastrointestinal disorders' [1]. Eradication of *H. pylori* is necessary for the management of *H. pylori*-related complications. The recommended first approach for *H. pylori* eradication is standard triple antibiotic therapy. Other choices include sequential therapy and quadruple therapy [2]. However, due to antibiotic resistance and patient non compliance, several studies showed widespread failure of antibiotic therapy [3,4]. Driven by the growing necessity for alternative solutions to

eradication regimens, some studies have started to focus on probiotics [5], i.e. 'live microorganisms which when administered in adequate amounts confer a health benefit on the host' [6]. The most commonly used probiotic bacteria belong to the genera *Lactobacillus* and *Bifidobacterium*, and these also include several yeasts such as *Saccharomyces boulardii* [7]. It has been hypothesized that probiotics could improve *H. pylori* eradication and reduce side effects during therapy.

Although seven meta-analyses [8-14] on this topic have been conducted over the past seven years, we perform an additional meta-analysis based on the following considerations: 1) The number of articles included in six of the seven previously

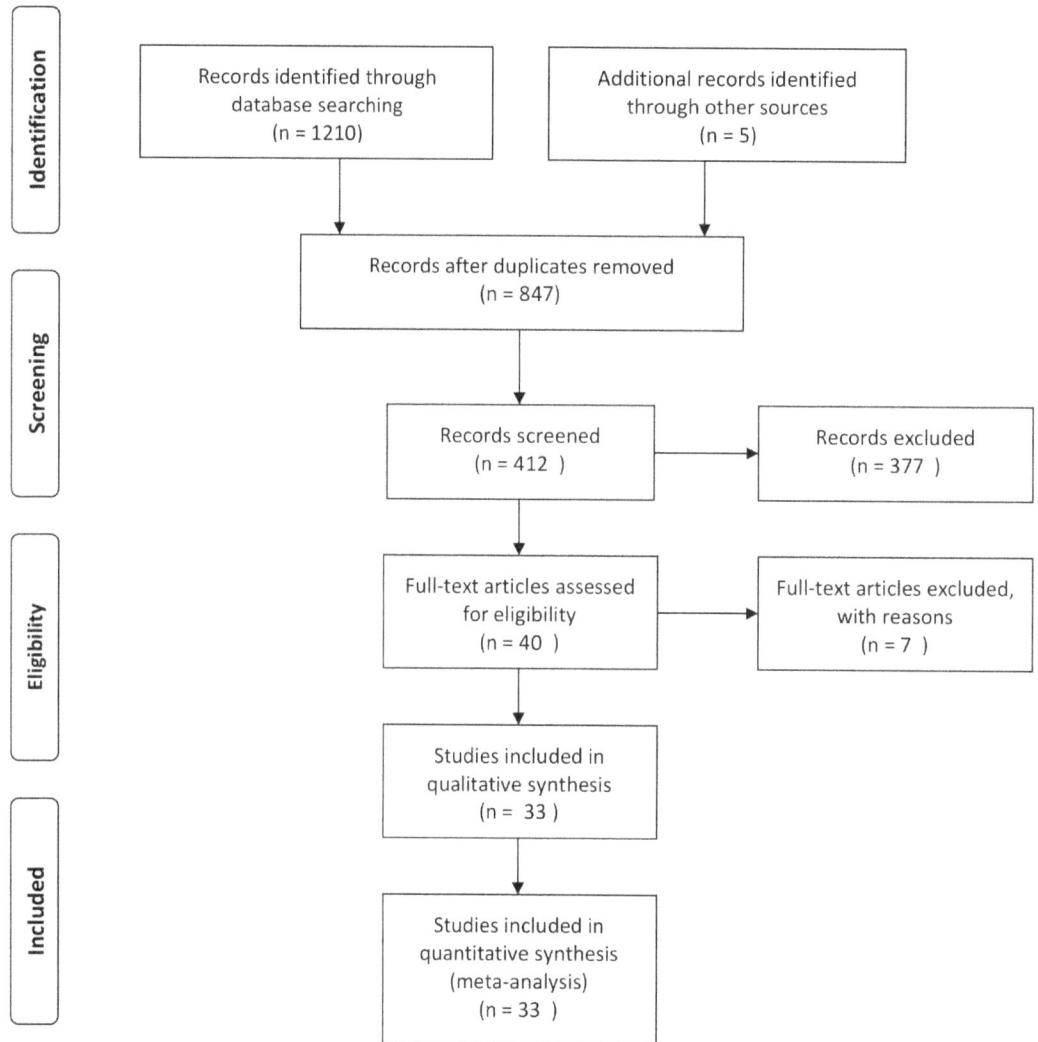

Figure 1. Flow of article selection.

conducted meta-analysis was ten or less [9–14]. In 2010–2014, many new studies were performed to evaluate the effectiveness of probiotics supplementation increasing the evidence base. 2) There are various genera of probiotics used in clinical practice and most meta-analyses [9,11,12,13] solely concentrated on one specific strain. 3) Opposite conclusions were drawn regarding side effects: Sachdeva [10], Zou [9], and Zheng [13] report no significant reduction of overall side effects in probiotics groups. However, Tong [8], Szajewska [11], Wang [12], and Li [14] arrive at opposite conclusions. Moreover, no sensitivity analysis regarding side effects has been conducted so far. 4) While all meta-analysis conducted so far reported increased eradication rates due to probiotics supplementation, no sensitivity analysis with regard to the effectiveness of the antibiotic therapy has been conducted, i.e. it is conceivable that probiotics supplementation will solely increase effectiveness in relatively ineffective antibiotic regimens.

This meta-analysis is thus designed to evaluate the current evidence regarding effects of probiotics supplementation compared with eradication triple therapy only on H. pylori eradication rates and side effects. Moreover, we aim to compare different strains of probiotics as well as differentially effective antibiotic therapies and to evaluate the quality of the trials conducted so far.

Methods

Search Strategy

We performed a literature search in Pubmed and Web of Science, covering papers published up to May 2014. A combination of the following keywords was used: (probiotics OR prebiotics OR *Bifidobacterium* OR *Lactobacillus* OR *Saccharomyces*) AND (*Helicobacter pylori* OR *H. pylori*). The reference lists of the selected papers and the seven previous meta-analyses were also screened for other eligible articles that may have been missed in the initial search. Next we scanned the titles and abstracts of the trials identified in the computerized search to exclude studies that were obviously irrelevant and scrutinized the full-texts of the remaining studies.

Inclusion and Exclusion Criteria

The following criteria were used for the selection of relevant articles: 1) studies should be randomized controlled trials (RCTs); 2) study populations should never have been treated for *H. pylori* infection before; 3) studies should include at least two branches of treatment consisting of (a) studies should have patients in a control group who received antibiotic therapy, (b) there should be patients

Table 1. Characteristica of studies included in the meta-analysis.

Authors (references)	Year	Area	Case Number (treatment/control)	Patients	Diagnositic Methods	Eradication Therapy	Probiotics Supplementation	% Eradication of Treatment (no. of patients)	% Rradication of Control (no. of patients)	% Side Effects (no. of patients) (treatment/control)
Canducci [15]	2000	Italy	120 (60/60)	symptomatic adults	UBT +Histology/UBT-Histology (6 weeks later)	rabeprazole 20 mg bid, clarithromycin 250 mg tid, amoxicillin 500 mg tid, 7 days	Lacteol Fort (Bruschettini s.r.l., Genoa, Italy), containing 5×10^9 heat-killed L. acidophilus per capsule, tid, 10 days	ITT 86.7% (52/60) PP 89.7% (52/58)	ITT 70.0% (42/60) PP 71.2% (42/59)	10.0% (6/60) 10.0% (6/60)
Armuzzi a[16]	2001	Italy	60 (30/30)	asymptomatic adults	UBT +ELISA/UBT (6 weeks later)	rabeprazole 20 mg bid, clarithromycin 500 mg bid, tinidazole 500 mg bid, 7 days	a probiotic preparation containing 6×10^9 Lactobacillus GG (Giflorex, Errekappa Euroterapici S.p.A, Milan, Italy), bid, 14 days	ITT 83.3% (25/30)	ITT 80.0% (24/30)	40.0% (12/30) 33.3% (10/30)
Armuzzi b 17]	2001	Italy	120 (60/60)	symptomatic adults	UBT +ELISA/UBT (6 weeks later)	pantoprazole 40 mg bid, clarithromycin 500 mg bid, tinidazole 500 mg bid, 7 days	a probiotic preparation containing 6×10^9 Lactobacillus GG (Giflorex, Errekappa Euroterapici S.p.A, Milan, Italy), bid, 14 days	ITT 80.0% (48/60) PP 80.0% (48/60)	ITT 70.0% (42/60) PP 80.7% (46/57)	43.3% (26/60) 61.7% (37/60)
Cremonini LB[18]	2002	Italy	42 (21/21)	asymptomatic adults	UBT/UBT (5–7 weeks later)	rabeprazole 20 mg bid, clarithromycin 500 mg bid, tinidazole 500 mg bid, 7 days	a probiotic preparation containing 6×10^9 Lactobacillus GG (Giflorex, Errekappa Euroterapici, Milan, Italy), bid, 14 days	ITT 76.2% (16/21) PP 76.2% (16/21)	ITT 76.2% (16/21) PP 80.0% (16/20)	14.3% (3/21) 57.1% (12/20)
Cremonini SM[18]	2002	Italy	42 (22/21)	asymptomatic adults	UBT/UBT (5–7 weeks later)	rabeprazole 20 mg bid, clarithromycin 500 mg bid, tinidazole 500 mg bid, 7 days	a probiotic preparation containing 5×10^9 Saccharomyces boulardii (Codex, SmithKline Beecham, Milan, Italy), bid, 14 days	ITT 77.3% (17/22) PP 85.0% (17/20)	ITT 76.2% (16/21) PP 80.0% (16/20)	13.6% (3/22) (57.1% (12/20)
Cremonini MS[18]	2002	Italy	42 (21/21)	asymptomatic adults	UBT/UBT (5–7 weeks later)	rabeprazole 20 mg bid, clarithromycin 500 mg bid, tinidazole 500 mg bid, 7 days	a combination of 5×10^9 live Lactobacillus acidophilus and Bifidobacterium lactis (Ferzym, Specchiasol, Milan, Italy), bid, 14 days	ITT 85.7% (18/21) PP 85.7% (18/21)	ITT 76.2% (16/21) PP 80.0% (16/20)	23.8% (5/21) 57.1% (12/20)
Sheu [19]	2002	China (Taiwan)	160 (80/80)	symptomatic adults	Histology+RUT/UBT (8 weeks later)	lansoprazole 30 mg bid, clarithromycin 500 mg bid, amoxicillin 1 g tid, 7 days	200 mL AB-Yogurt containing 5×10^9 Lactobacillus and Bifidobacterium (President Corp, Tainan, Taiwan), bid, 5 weeks	ITT 91.3% (73/80) PP 94.8% (73/77)	ITT 78.8% (63/80) PP 87.5% (63/72)	/
Nista [20]	2004	Italy	106 (54/52)	asymptomatic adults	UBT/UBT (6 weeks later)	rabeprazole 20 mg bid, clarithromycin 500 mg bid, amoxicillin 1 g bid, 7 days	a probiotic preparation containing 2×10^9 B. clausii (Enterogermina, Sanofi-Synthelabo OTC, Milan, Italy), tid, 14 days	ITT 72.2% (39/54) PP 78.0% (39/50)	ITT 71.2% (37/52) PP 74.0% (37/50)	74.0% (40/54) 82.7% (43/50)
Myllyuom [21]	2005	Finland	47 (23/24)	asymptomatic adults	UBT +ELISA/UBT+ ELISA (6 weeks later)	lansoprazole 30 mg bid, clarithromycin 500 mg bid, amoxicillin 1 g tid, 7 days	65 mL milk-based fruit drink containing 1×10^9 cfu/mL of Lactobacillus GG, L. rhamnosus LC705, Bifidobacterium breve Bb99 and Propionibacterium freudenreichii ssp. shermanii JS (Valio Ltd, Helsinki, Finland), bid during the eradication treatment and qd for the following 3 weeks	ITT 91.3% (21/23) PP 91.3% (21/23)	ITT 79.2% (19/24) PP 79.2% (19/24)	86.9% (20/23) 91.7% (22/24)
Sykora[22]	2005	UK	86 (39/47)	symptomatic children	Histology +RUT+ HpSA/UBT + HpSA (4 weeks later)	omeprazole 10 mg (15–30 kg) or 20 mg (30 kg) bid, clarithromycin 7.5 mg/kg bid, amoxicillin 25 mg/kg bid, 7 days	100 mL of fermented milk containing 10^10 cfu/mL L. casei DN-114001, qd, 14 days	ITT 84.6% (33/39) PP 91.7% (33/36)	ITT 57.4% (27/47) PP 79.2% (27/44)	23.1% (9/39) 21.3% (10/47)

Table 1. Cont.

Authors (references)	Year	Area	Case Number (treatment/control)	Patients	Diagnositic Methods	Eradication Therapy	Probiotics Supplementation	% Eradication of Treatment (no. of patients)	% Rradication of Control (no. of patients)	% Side Effects (no. of patients) (treatment/control)
Goldman[23]	2006	Argentina	65 (33/32)	symptomatic children	UBT/UBT (4 weeks later)	omeprazole 1 mg/kg qd, clarithromycin 15 mg/kg qd, amoxicillin 50 mg/kg qd, 7 days	250 mL commercial yogurt containing 10^7 cfu/mL *Bifidobacterium animalis* and *Lactobacillus casei*	ITT 45.5% (15/33) PP 45.5% (15/33)	ITT 37.5% (12/32) PP 37.5% (12/32)	/
Lionetti[24]	2006	Italy	40 (20/20)	symptomatic children	Histology +RUT+ UBT/UBT (20 days later)	omeprazole 1 mg/kg/d plus amoxicillin 50 mg/kg/d for 5 days; omeprazole 1 mg/kg/d, clarithromycin 15 mg/kg/d and tinidazole 20 mg/kg/d for the next 5 days	each pill containing 10^8 cfu of *L. reuteri* ATCC 55730 (SD2112) (BioGaia, Sweden), qd, 20 days	ITT 85.0% (17/20) PP 89.5% (17/19)	ITT 80.0% (16/20) PP 84.2% (16/19)	/
Zieminak[25]	2006	Poland	245 (53/192)	symptomatic adults	UBT/UBT (6 weeks later)	pantoprazole 40 mg bid, clarithromycin 500 mg bid, amoxicillin 1 g bid, 10 days	Lacidofil containing *Lactobacillus acidophilus* and *Lactobacillus rhamnosus*, bid, 10 days	ITT 96.2% (51/53) PP 96.2% (51/53)	ITT 85.9% (165/192) PP 85.9% (165/192)	/
Cindoruk[26]	2007	Turkey	124 (62/62)	symptomatic adults	Histology/UBT (6 weeks later)	lansoprazole 30 mg bid, clarithromycin 500 mg bid, amoxicillin 1 g bid, 14 days	250 mg sachets of *Saccharomyces boulardii* (Reflor, Sanofi-Synthelabo Ilac A.S., Istanbul, Turkey), bid, 14 days	ITT 71.0% (44/62) PP 71.0% (44/62)	ITT 59.7% (37/62) PP 59.7% (37/62)	22.6% (14/62) 59.7% (37/62)
De Bortoli N [27]	2007	Italy	206 (105/101)	symptomatic adults	HpSA+ UBT/UBT (8 weeks later)	esomeprazole 20 mg bid, clarithromycin 500 mg bid, amoxicillin 1 g bid, 7 days	blf 20 mg bid; Pb containing 5×10^9 *Lactobacillus plantarum*, 2×10^9 *L. reuterii*, 2×10^9 *L. caseisubsp.rhamnosus*, 2×10^9 *Bifidobacterium infantis* and *B.longum*, 1×10^9 *L. salivarius*, 1×10^9 *L. acidophilus*, 5×10^9 *Streptococcus termophilus*, and 1×10^9 *L. sporogenes*, bid, 7 days	ITT 88.6% (93/105) PP 92.1% (93/101)	ITT 72.3% (73/101) PP 76.0% (73/96)	9.5% (10/105) 40.6% (41/101)
Francavill [28]	2008	Italy	40 (20/20)	symptomatic adults	Histology+HpSA+ UBT/HpSA+ UBT (4 weeks later)	rabeprazole 20 mg bid plus amoxicillin 1 g, bid for 5 days; rabeprazole 20 mg bid, clarithromycin 500 mg bid and tinidazole 500 mg bid for the next 5 days	each tablet containing 10^8 *L. reuteri* ATCC 55730 (Reuterin, Nòos), qd, 4 weeks	ITT 85.0% (17/20) PP 85.0% (17/20)	ITT 80.0% (16/20) PP 80.0% (16/20)	/
Kim[29]	2008	Korea	347 (168/179)	symptomatic adults	Histology+UBT/UBT (4 weeks later)	standard PPI bid, clarithromycin 500 mg bid, amoxicillin 1 g bid, 7 days	150 mL Will yogurt containing *L. acidophilus* HY 2177 (> 10^5 cfu/mL), *L. casei* HY 2743 (> 10^5 cfu/mL), *B. longum* HY 8001 (> 10^6 cfu/mL), and *S. thermophilus* B-1 (> 10^8 cfu/mL) (Korea Yakult Company Limited; Chunan-Si, Chungnam, S. Korea), qd, 3 weeks	ITT 79.2% (133/168) PP 87.5% (133/152)	ITT 72.1% (129/179) PP 78.6% (129/164)	41.1% (69/168) 26.3% (47/179)
Scaccianoce LB[30]	2008	Italy	33 (17/16)	symptomatic adults	Histology/UBT (4–6 weeks later)	lansoprazole 30 mg bid, clarithromycin 500 mg bid, amoxicillin 1 g bid, 7 days	each tablet containing 10^8 *L. reuteri* ATCC 55730 (Reuterin, Nòos), bid, 7 days	ITT 52.9% (9/17) PP 52.9% (9/17)	ITT 62.5% (10/16) PP 66.7% (10/15)	5.9% (1/17) 25.0% (4/16)

Table 1. Cont.

Authors (references)	Year	Area	Case Number (treatment/control)	Patients	Diagnositic Methods	Eradication Therapy	Probiotics Supplementation	% Eradication of Treatment (no. of patients)	% Rradication of Control (no. of patients)	% Side Effects (no. of patients) (treatment/control)
Scaccianoce MS[30]	2008	Italy	31 (15/16)	symptomatic adults	Histology/UBT (4–6 weeks later)	lansoprazole 30 mg bid, clarithromycin 500 mg bid, amoxicillin 1 g bid, 7 days	a probiotic mixture with *Lactobacillus plantarum* (5×10^9), *L. reuteri* (2×10^9), *Lactobacillus casei subsp. Rhamnosus* (2×10^9), *Bifidobacterium infantis* (2×10^9), *Bifidobacterium longum* (2×10^9), *Lactobacillus salivarius* (1×10^9), *Lactobacillus acidophilus* (1×10^9), *Streptococcus termophilus* (5×10^9), and *Lactobacillus sporogenes* (1×10^9); 5 g/dose, bid, 7 days	ITT 53.3% (8/15) PP 53.3% (8/15)	ITT 62.5% (10/16) PP 66.7% (10/15)	20.0% (3/15) 25.0% (4/16)
Hurduc[31]	2009	Romania	90 (48/42)	symptomatic children	Histology+RUT/ Histology+RUT (4–6 weeks later)	omeprozole or esomeprazole 0.5 mg/kg bid, clarithromycin 7.5 mg/kg bid, amoxicillin 25 mg/kg bid, 7–10 days	*Saccharomyces boulardii* (Enterol; Biocodex, Gentilly Cedex, France), 250 mg/day, bid, 4 weeks	ITT 93.8% (45/48) PP 100% (45/45)	ITT 80.9% (34/42) PP 75.6% (34/42)	8.3% (4/48) 30.9% (13/42)
Szajewsk[32]	2009	Poland	66 (34/32)	asymptomatic children	Histology+ RUT+UBT/ UBT (4–6 weeks later)	omeprozole 0.5 mg/kg bid, clarithromycin 10 mg/kg bid, amoxicillin 25 mg/kg bid, 7 days	10^9 cfu/mL *Lactobacillus GG*, bid, 7 days	ITT 67.6% (23/34) PP 67.6% (23/34)	ITT 68.8% (22/32) PP 68.8% (22/32)	52.9% (18/34) 40.6% (13/32)
Song[33]	2010	Korea	661 (330/331)	symptomatic adults	UBT/UBT (4 weeks later)	omeprazole 20 mg bid, clarithromycin 500 mg bid, amoxicillin 1 g bid, 7 days	3×10^9 cfu/g *Saccharomyces boulardii* (Bioflor250, Kuhnil Pharmacy, Seoul, Korea), tid, 4 weeks	ITT 80.0% (264/330) PP 85.4% (264/309)	ITT 71.6% (237/331) PP 80.0% (237/296)	/
Yasar[34]	2010	Turkey	76 (38/38)	symptomatic children	Histology/UBT (4 weeks later)	pantoprazole 40 mg bid, clarithromycin 500 mg bid, amoxicillin 1 g bid, 14 days	125 ml yogurt containing 10^10 cfu/g *Bifidobacterium DN-173 010*, qd, 14 d	ITT 65.8% (25/38) PP 65.8% (25/38)	ITT 52.6% (20/38) PP 52.6% (20/38)	/
Bekar[35]	2011	Turkey	82 (46/36)	symptomatic adults	UBT/UBT (45 days later)	lansoprazole 30 mg bid, clarithromycin 500 mg bid, amoxicillin 1 g bid, 14 days	250 mL kefir, bid, 141d	ITT 78.3% (36/46) PP 78.3% (36/46)	ITT 50.0% (18/36) PP 50.0% (18/36)	/
Madeiros[36]	2011	France	62 (31/31)	asymptomatic adults	Histology+ culture/ UBT (≥6 weeks later)	esomeprazole 20 mg bid, clarithromycin 500 mg bid, amoxicillin 1 g bid, 8 days	Lacteol (BioSaúde laboratories, Portugal), capsule containing 5×10^9 *L. acidophilus*, bid, 8 days	ITT 83.9% (26/31) PP 83.9% (26/31)	ITT 77.4% (24/31) PP 77.4% (24/31)	/
Ozdil[37]	2011	Turkey	193 (98/95)	symptomatic adults	Histology/HpSA (5 weeks later)	lansoprazole 30 mg bid, clarithromycin 500 mg bid, amoxicillin 1 g bid, 14 days	250 mg capsules containing *saccharomyces boulardii*, qd, 14 days	ITT 72.4% (71/98) PP 77.2% (71/92)	ITT 86.3% (82/95) PP 89.1% (82/92)	/
Deguchi[38]	2012	Japan	229 (115/114)	symptomatic adults	Histology-RUT-culture/ UBT+HpSA (8 weeks later)	rabeprazole 10 mg bid, clarithromycin 200 mg bid, amoxicillin 750 mg bid, 7 days	yogurt containing 112 g *L. gasseri OLL2716* (>10 cfu/g), bid, 4 weeks	ITT 82.6% (95/115) PP 89.6% (95/106)	ITT 69.3% (79/114) PP 71.2% (79/111)	5.2% (6/115) 3.5% (4/114)

Table 1. Cont.

Authors (references)	Year	Area	Case Number (treatment/control)	Patients	Diagnositic Methods	Eradication Therapy	Probiotics Supplementation	% Eradication of Treatment (no. of patients)	% Rradication of Control (no. of patients)	% Side Effects (no. of patients) (treatment/control)
Du YQ[39]	2012	China	157 (78/79)	symptomatic adults	Histology+RUT+UBT/UBT (4 weeks later)	omeprazole 20 mg bid, clarithromycin 500 mg bid, amoxicillin 1 g bid, 7 days	probiotics, containing 3×10^7 Lactobacillus acidophilus, qd, 14 d	ITT 79.5% (62/78) PP 81.6% (62/76)	ITT 60.7% (48/79) PP 61.5% (48/78)	1.3% (1/78) 0.0% (0/79)
Manfredi[40]	2012	Italy	149 (73/76)	symptomatic adults	Histology+ HpSA/HpSA (8–10 weeks later)	esomeprazole 20 mg bid plus amoxicillin 1 g, bid for 5 days; esomeprazole 20 mg bid, clarithromycin 500 mg bid and tinidazole 500 mg bid for the next 5 days	Lactogermine plus (Humana Italia s.p.a., Milano, Italy), containing 10^9 Lactobccillus acidophilus, 5×10^8 Bifidobacterium bifidum, 10^9 Streptococcus thermophilus, 10^9 Lactobacillus bulgaricus	ITT 91.8% (67/73) PP 94.4% (67/71)	ITT 85.5% (65/76) PP 92.8% (65/70)	39.7% (29/73) 65.8% (50/76)
Mirzaee[41]	2012	Iran	68 (34/34)	symptomatic adults	UBT/UBT (4 weeks later)	pantoprazole 40 mg qd, clarithromycin 500 mg bid, amoxicillin 1 g bid, 7 days	150 mg probiotic yogurt, bid, 7 days	ITT 55.9% (19/34) PP 61.3% (19/31)	ITT 58.8% (20/34) PP 64.5% (20/31)	64.7% (22/34) 67.6% (23/34)
Tolone[42]	2012	Italy	68 (34/34)	symptomatic children	Histology+ UBT/UBT (4 weeks later)	omeprazole 1 mg/kg qd, amoxicillin 50 mg/kg bid, clarithromycin 15 mg/kg bid, 7 days	the PB supplement containing 5×10^9 Lactobacillus plantarum, 2×10^9 L. reuterii, 2×10^9 L.casei subsp. rhamnosus, 2×10^9 Bifidobacterium infantis and B. longum, 1×10^9 L. salivarius, 1×10^9 L. acidophilus, 5×10^9 Streptococcus termophilus, and 1×10^9 L. sporogenes, qd, 7 days	ITT 88.2% (30/34) PP 88.2% (30/34)	ITT 76.5% (26/34) PP 76.5% (26/34)	/
Ahmad[43]	2013	Iran	66 (33/33)	symptomatic children	Histology+RUT/HpSA (4–8 weeks later)	omeprazole 0.5 mg/kg bid, amoxicillin 25 mg/kg bid, furazolidone 3 mg/kg bid, 7 days	probiotic combination (restore, 10^9 cfu/ each sachet, Protexin Co, UK) containing Lactobacillus acidophilus, Lactobacillus rhamnosus, Lactobacillus bulgaricus, Lactobacillus casei, Streptococcus thermophilus, Bifidobacterium infantis and Bifidobacterium breve, qd, 7 days	ITT 90.9% (30/33) PP 90.9% (30/33)	ITT 69.7% (23/33) PP 69.7% (23/33)	/
Navarro[44]	2013	Brazil	107 (55/52)	symptomatic adults	Histology+RUT+UBT/UBT (4 weeks later)	lansoprazole 30 mg bid, tetracycline 500 mg bid, furazolidone 200 mg bid, 7 days	the probiotic compound containing 1.25×10^9 CFUs Lactobacillus acidophilus, 1.25×10^9 CFUs Lactobacillus rhamnosus, 1.25×10^9 CFUs Bifidobacterium bifidum and 1.25×10^9 CFUs Streptococcus faecium, bid, 30 days	ITT 89.1% (49/55) PP 96.1% (49/51)	ITT 84.6% (44/52) PP 89.8% (44/49)	58.2% (32/55) 71.2% (37/52)
Shavakhi[45]	2013	Iran	180 (90/90)	symptomatic adults	Histology+ RUT/UBT (4 weeks later)	omeprazole 20 mg bid, bismuth subcitrate 240 mg bid, amoxicillin 1000 mg bid, clarithromycin 500 mg bid, 14 days	Balance capsule containing Lactobacillus strains (L. casei, L. rhamnosus, L. acidophilus, and L. bulgaricus), Bifidobacterium strains (B. breve and B. longum), and Streptococcus thermophiles, total viable count is 10^8 CFU/per capsule, bid, 14 days	ITT 76.7% (69/90) PP 82.1% (69/84)	ITT 81.1% (73/90) PP 84.9% (73/86)	18.9% (17/90) 16.7% (15/90)
Dajani[46]	2013	United Arab Emirates	206 (100/106)	symptomatic adults	Histology+RUT+UBT+HpSA/UBT (6–8 weeks later)	omeprazole 20 mg bid, clarithromycin 500 mg bid, amoxicillin 1 g bid, 7 days	3×10^9 B. infantis 2036, bid, 7 days	ITT 83% (83/100) PP 83% (83/100)	ITT 68.9% (73/106) PP 68.9% (73/106)	/

Table 1. Cont.

Authors (references)	Year	Area	Case Number (treatment/ control)	Patients	Diagnositic Methods	Eradication Therapy	Probiotics Supplementation	% Eradication of Treatment (no. of patients)	% Rradication of Control (no. of patients)	% Side Effects (no. of patients) (treatment/ control)
Wang YH[47]	2013	China	100 (49/51)	symptomatic children	UBT/UBT (6 weeks later)	PPI 0.6–0.8 mg/kg bid, clarithromycin 10–15 mg/kg bid, amoxicillin 30–50 mg/kg bid, and metronidazole 15–20 mg/kg bid for penicillin allergic children, 14 days	each packs of probiotics containing 2 g $L.$ $acidophilus$-5 (4.7×10^9 cfu/100 g) and 2 g $B.$ $bifidum$-12 (4.3×10^8 cfu/100 g), <5 years old, qd; >5 years old, bid, 14 days	ITT 73.5% (36/49) PP 83.7% (26/43)	ITT 56.9% (29/51) PP 64.4% (29/45)	10.2% (5/49) 23.5% (12/51)

in an intervention group who received probiotics plus the identical antibiotic therapy; 4) there should be data on successful eradication rates available; 5) for the analysis of side effects data on occurrence of side effects during treatment were required as well. Exclusion criteria were as follows: 1) the design and the definition of the trials were obviously different from those specified above; 2) essential information was not provided; 3) papers were letters, commentaries, editorials, reviews and duplicate publications; 4) articles were written in a language other than English.

Data Extraction

Two authors extracted data independently from all eligible papers. For conflicting evaluations, another author was consulted to solve the dispute and a final decision was made by the majority of the votes. The data extracted included authors, year of publication, base characteristics of the patients, details of the *H. pylori* eradication therapy, details related to interventions, primary and secondary outcomes, and confirmation methods of *H. pylori* infection. Extracted information was entered into a database.

Assessment of Study Quality

We used the Cochrane Tool of Bias [15] to assess study quality. Two authors independently evaluated all studies. Results were then compared and discussed to form consensus. If consensus could not be reached, another author was consulted and a decision made by the majority of the votes.

Statistical Analysis

Statistical heterogeneity was analyzed with Chi-squared distributed Chochran's Q and the I-squared statistics ($I^2 = 100\% \times$ (Q-df)/Q). I-square indicates the percentage of variation across studies due to hetereogeneity as opposed to chance. We assumed sufficient homogeneity to apply a fixed-effects model (Mantel-Haenszel) for the meta-analysis if I-square was under 40% and/or Q was not significant at p<0.05. Otherwise, we opted for a random-effects model (DerSimonian and Laird). The jacknife was used to assess the influence of individual studies, i.e. estimation of the overall effect was repeated while omitting one study at each time. Funnel plots and Harbord's modified test for small study effects were used to assess publication bias [16]. The influence of the probiotic strain applied (multi-strain interventions were excluded from this analysis), blinded vs. non-blinded trials, and in the case of eradication rates, the effectiveness of antibiotic therapy in the control group (eradication rate >80% vs. <80%) was assessed with sub-group analysis. Concerning eradication rates, meta-regression was used to determine the influence of studies in pediatric vs. adult as well as symptomatic and asymptomatic populations. All statistical analysis was performed with STATA 12.0 (Stata Corporation, Texas, USA).

Results

Study Characteristics

The bibliographic search yielded 1114 articles, 40 of which were reviewed in full text (Figure 1) [17–56]. Of these studies, 33 RCTs [17–49] met the inclusion criteria for the analysis of eradication rates and 20 were eligible for analyzing side effects. These trials randomized a total of 4459 patients, 4261 of which were followed up. Nine studies enrolled only children [24–26,33,34,36,44,45,49] and 24 were undertaken exclusively with adults [17–23,27–32,35,37–43,46–48]. From all the included trials, 29 used PPI-triple therapy [17 25,27 29,31 41,43–46,48,49], three used sequential therapy [26,30,42] and one used bismuth-quadruple therapy [47]. These studies were undertaken

A

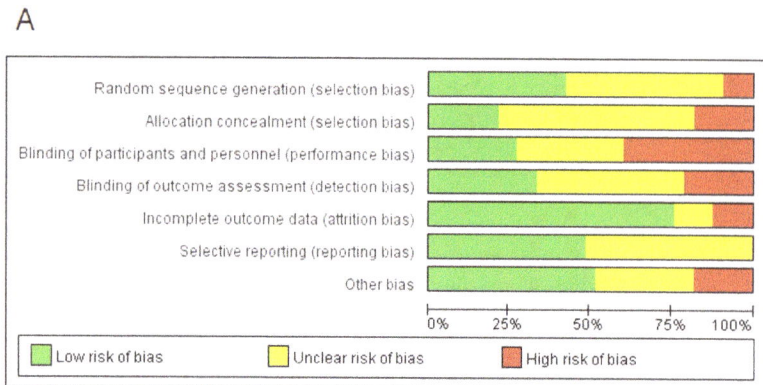

| | Low risk of bias | | Unclear risk of bias | | High risk of bias |

B

	Random sequence generation (selection bias)	Allocation concealment (selection bias)	Blinding of participants and personnel (performance bias)	Blinding of outcome assessment (detection bias)	Incomplete outcome data (attrition bias)	Selective reporting (reporting bias)	Other bias
2000 Canducci	+	?	−	−	+	?	?
2001 Armuzzi a	?	+	+	+	+	+	?
2001 Armuzzi b	?	+	+	−	+	+	+
2002 Cremonini	+	?	+	+	+	+	?
2002 Sheu	?	−	−	−	?	?	?
2004 Nista	+	?	+	?	+	+	?
2005 Myllyuom	+	?	+	+	+	+	+
2005 Sykora	+	?	?	+	+	+	?
2006 Goldman	?	+	?	?	+	+	+
2006 Lionetti	?	+	+	+	−	?	+
2006 Zieminak	?	−	−	?	+	+	−
2007 Cindoruk	+	?	?	+	+	+	+
2007 de Bortoli N	?	?	?	+	+	+	+
2008 Francavill	+	+	+	+	+	?	+
2008 Kim	−	−	−	−	+	+	?
2008 Scaccianoce	?	?	?	?	+	?	?
2009 Hurduc	+	?	?	+	+	+	+
2009 Szajewsk	+	+	+	+	+	?	+
2010 Song	?	−	−	?	+	+	−
2010 Yasar	?	−	−	?	+	?	+
2011 Bekar	?	?	?	?	−	?	−
2011 Madeiros	?	?	−	−	−	?	?
2011 Ozdil	?	?	?	?	+	?	+
2012 Deguchi	+	?	?	?	+	?	+
2012 Du YQ	?	?	?	?	+	?	+
2012 Manfredi	+	?	−	−	?	+	−
2012 Mirzaee	+	?	−	?	?	+	?
2012 Tolone	?	−	−	?	+	?	+
2013 Dajani	−	?	−	−	+	?	−
2013 Khodada	?	?	+	+	+	?	+
2013 Navarro	+	?	?	+	+	+	+
2013 Shavakhi	+	+	+	+	+	?	+
2013 Wang	−	?	−	?	?	?	−

Figure 2. Risk of bias graph (A): review authors' judgements about each risk of bias item presented as percentages across all included studies. (B) Risk of bias summary: review authors' judgements about each risk of bias item for each included study. (+) = low risk of bias, (?) = unclear, (-) = high risk of bias.

in Argentina [25], Brazil [46], mainland China [41,49], China (Taiwan) [21], Finland [23], France [38], Iran [43,45,47], Italy [17–20,22,26,29,30,32,42,44], Japan [40], Korea [31,35], Poland [27,34], Romania [33], Turkey [28,36,37,39], United Kingdom [24] and the United Arabian Emirates [48].(Table 1)

Trial Quality Assessment

Figure 2A shows authors' judgements for each Cochrane risk of bias item and Figure 2B presents the Cochrane risk of bias score for each citation included.

Eradication Rates

Eradication risk ratios (RRs) were available for 4459 patients (2189 in the probiotics supplementation group and 2270 in the control group). Heterogeneity was found to be low for the overall incidence of eradication rates in ITT analysis ($\chi^2 = 42.97$, p = 0.167, $I^2 = 18.5\%$) but higher for PP analysis ($\chi^2 = 58.55$, p = 0.006, $I^2 = 41.9\%$). Therefore, a fixed effects model based on Mantel-Haenszel's estimation method was used in the case of ITT and a random effects model in the case of PP analysis. Nonetheless, the pooled RRs from intention-to-treat (ITT) and from pre-protocol (PP) analysis for the probiotics supplementation group over controls were very similar with 1.122 (95% CI 1.086–1.159) and 1.114 (95% CI 1.070–1.159) respectively (Figure 3A&B).

Omitting individual studies from the meta-analysis did not change occurrence of RRs being significantly higher than 1. Funnel-plot and Harbord's modified test showed no evidence for publication bias (p = 0.64). Neither the age of the population (pediatric vs. adult, p = 0.127) nor symptom status (p = 0.314) played a role according to the meta-regression.

Studies used different strains of probiotics with 13 applying *Lactobacillus* [17–20,24,26,27,30,32,34,38,40,41], two using *Bifidobacterium* [36,48], one using *Bacillus clausii* [22], five using *Saccharomyces* [20,28,33,35,39] and 15 using multistrain [20,21,23,26,29,31,32,37,42–47,49]. Among those studies applying individual probiotic supplementation, four specific species were effective: *Lactobacillus acidophilus* (pooled RR = 1.235, 95% CI 1.090–1.400), *Lactobacillus casei* DN-114001 (pooled RR = 1.473, 95% CI 1.13–1.949), *Lactobacillus gasseri* (pooled RR = 1.192, 95% CI 1.028–1.382), *Bifidobacterium infantis* 2036 (pooled RR = 1.205, 95% CI 1.031–1.408). While double-blinded (pooled RR = 1.118, 1.045–1.196) and non-blinded trials (pooled RR = 1.120, 95% CI 1.080–1.162) basically arrived at the same results, sub-group analysis by effectiveness of the antibiotic therapy in the control group revealed that supplementation with probiotics solely increased eradication rates in relatively ineffective therapies.

Another sub-group analysis was done according to the effectiveness of eradication regimens. The less effective antibiotic therapies were, the more useful probiotic supplementation was: when eradication rate <60%, pooled RR = 1.28, 1.12–1.45; when eradication rate within 60%–69%, pooled RR = 1.18, 1.10–1.27; when eradication rate within 70%–79%, pooled RR = 1.11, 1.06–1.17; while if eradication rate over 80%, the supplementation was useless (pooled RR = 1.01, 0.96–1.77). (Figure 4)

Side Effects

There were 20 trials (2487 patients, 1269 in the probiotics supplementation group and 1218 in the control group) which provided data on the overall incidence of side effects. Significant heterogeneity was found for the overall occurrence of side effects ($I^2 = 72.2\%$, P<0.001). Therefore, the random effects model was used. The pooled RR in the probiotics supplementation over control was 0.735 (95% CI 0.598–0.902) (Figure 5). While Harbord's modified test for publication bias was insignificant (p = 0.17), the Funnel Plot did reveal some asymmetry (Figure 6): Small studies showing a considerable reduction in side-effects occur more often than those only showing a small reduction. When stratified by probiotics, only the pooled RR for *Saccharomyces boulardii* indicates a significant reduction in side effects (0.335, 95% CI 0.220–0.510; LB: 0.892, 95% CI 0.632–1.259; MS: 0.760, 95% CI 0.568–1.017; BB: 0.895: 0.737–1.087).

Only eight studies reporting the overall incidence of side effects were blinded and solely results from non-blinded trials provided evidence for the reduction of side-effects (pooled RR = 0.589, 95% CI: 0.412–0.842), while double blinded studies did not (pooled RR = 0.889, 95% CI: 0.728–1.085) (Figure 7).

Discussion

This is the largest meta-analysis on the effects of probiotics on *H. pylori* eradication and side effects conducted to date. The quality of the trials was medium to low and no study had a low risk of bias across all Cochrane criteria. Overall, this analysis of 33 RCTs suggests that supplementation of antibiotic therapy with probiotics increases eradication rates compared to a placebo or no intervention. However, regarding individual probiotic strains, this could only be confirmed for several strains of *Lactobacillus* and one strain of *Bifidobacterium*. Moreover, probiotics only demonstrated additional effects if the eradication rates in the control groups were relatively low, i.e. cases where the antibiotic therapy was relatively ineffective. Although, we found an overall decrease of total side effects, this held true only for *Saccharomyces boulardii* and non-blinded RCTs.

In comparison with previously published meta-analysis [8,9,10,11,12,13,14], our updated study represents the most comprehensive analysis. Various potential influence factors were taken into consideration such as age, symptom status, eradication therapy regime, eradication rate, bacterial strain and blinding method. Moreover, from the previous studies, six [8,10,11,12,13,14] excluded trials using Quadruple therapy as the co-intervention and none performed subgroup analysis with regard to the efficacy of antibiotic therapy. Moreover, sensitivity analysis for side effects had not been conducted so far. Our results confirm previous meta-analyses only to some extent. At a first glance, probiotics seem effective in increasing eradication rates and decreasing side effects. At a closer look, however, evidence only supports these claims for specific probiotic strains, ineffective antibiotic therapies and low-quality trials (i.e. non-blinded studies) as far as side effects are concerned.

Before Marshall found *H. pylori* in 1984 [57], the stomach was considered to be a sterile organ due to its low pH level. In the aftermath, 'H. pylori has been intensively studied and recent sequencing analysis of other gastric microbiota shows that *H. pylori* is not alone' [58]. Although there are only 10^2-4 cfu/g in the gastric mucosa [59], these commensal bacteria can play an important role in maintaining human health. Probiotic supple-

A

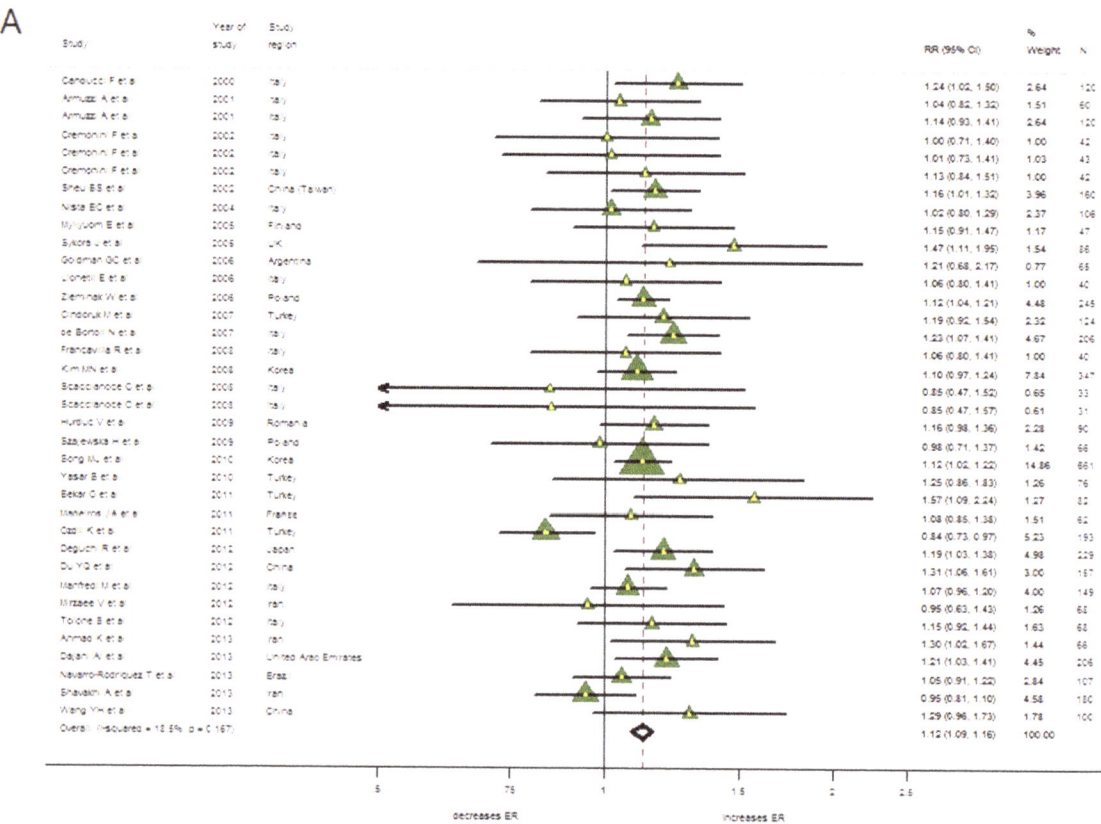

B

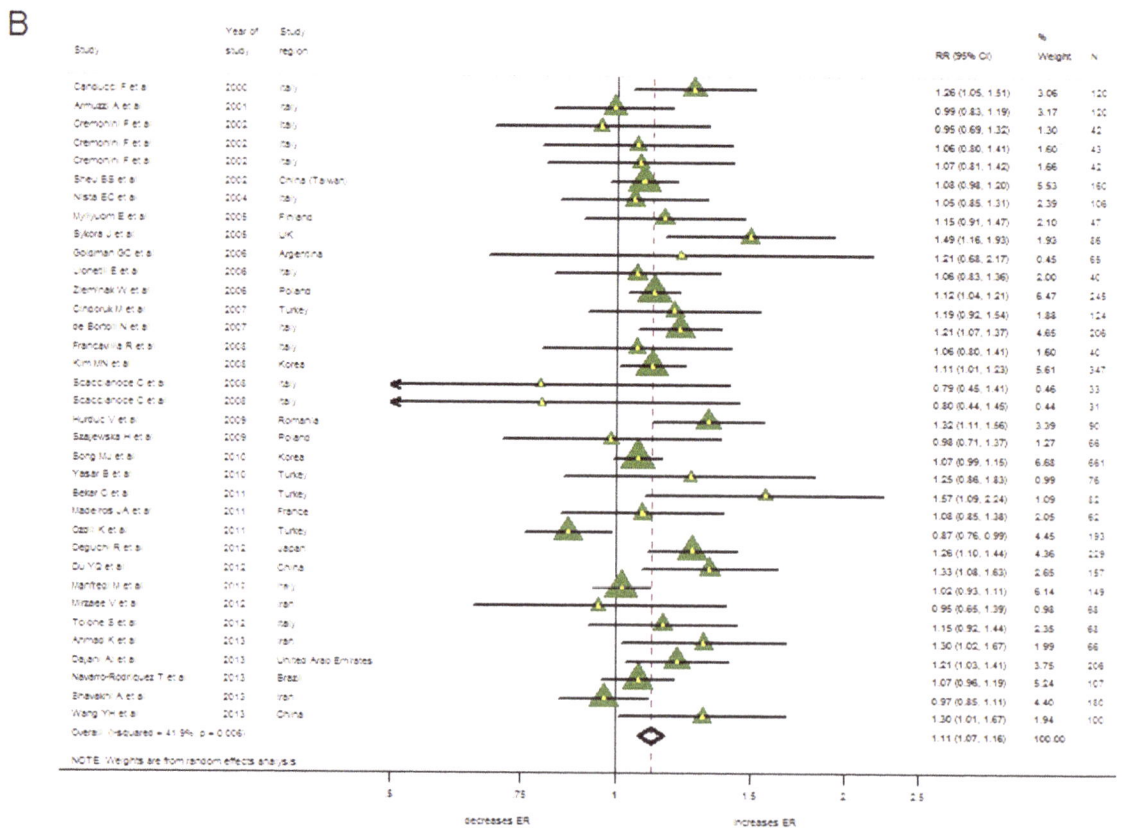

Figure 3. The effect of probiotics supplementation vs. without probiotics on eradication rates by intention-to-treat analysis in (A). RR, risk ratio; CI, confidence interval. The triangles represent individual studies and the size of the triangle represents the weight given to each study in the meta-analysis. The diamond represents the combined results. (B) The effect of probiotics supplementation vs. without probiotics on eradication rates by pre-protocol analysis. RR, risk ratio; CI, confidence interval. The triangles represent individual studies and the size of the triangle represents the weight given to each study in the meta-analysis. The diamond represents the combined results.

mentation as an approach to manipulate gastrointestinal flora has been intensively studied.

While all previously conducted meta-analysis concluded that supplementation of antibiotic therapy with probiotics is effective in increasing eradication rates [8,9,10,11,12], our study identified only four effective individual strain: *Lactobacillus acidophilus* (Pubmed ID: 190198), *Lactobacillus casei* DN-114001 (Pubmed ID: 35628), *Lactobacillus gasseri* (Pubmed ID: 35528), *Bifidobac-*

terium infantis 2036 (Pubmed ID: 41468). However, more than one trial was conducted only for *Lactobacillus acidophilus* so far, limiting the generalizability of the findings regarding the other three strains. While *Lactobacillus GG* had previously been reported to be an effective strain [8], we do reach the same conclusion here.

Moreover, our results suggest that probiotic supplementation is only useful in less effective (eradication rate <80%) antibiotic

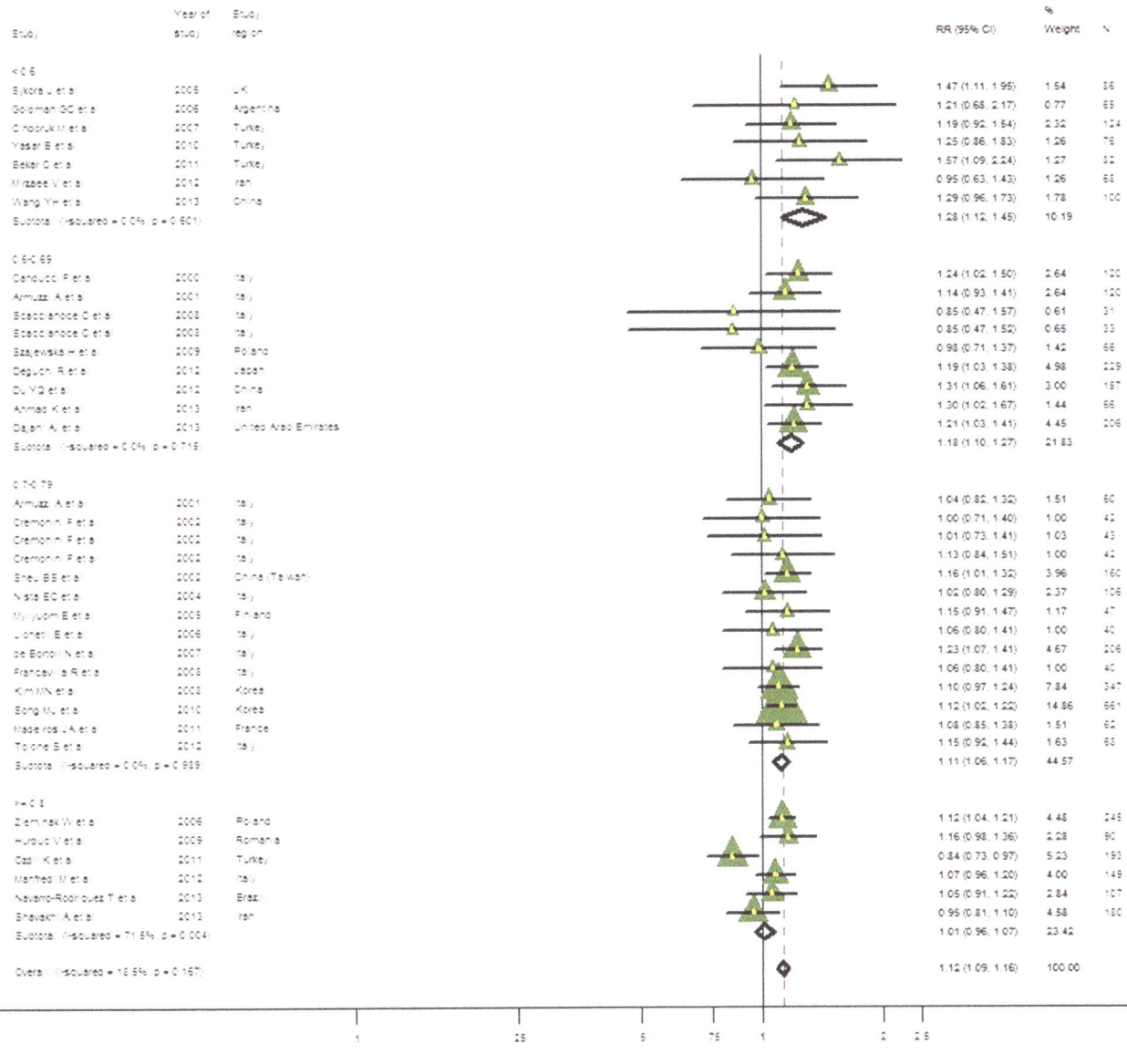

Figure 4. Meta-analysis of eradication rates by eradication rates due to antibiotic therapy only (control group). RR, risk ratio; CI, confidence interval. The triangles represent individual studies and the size of the triangle represents the weight given to each study in the meta-analysis. The diamond represents the combined results.

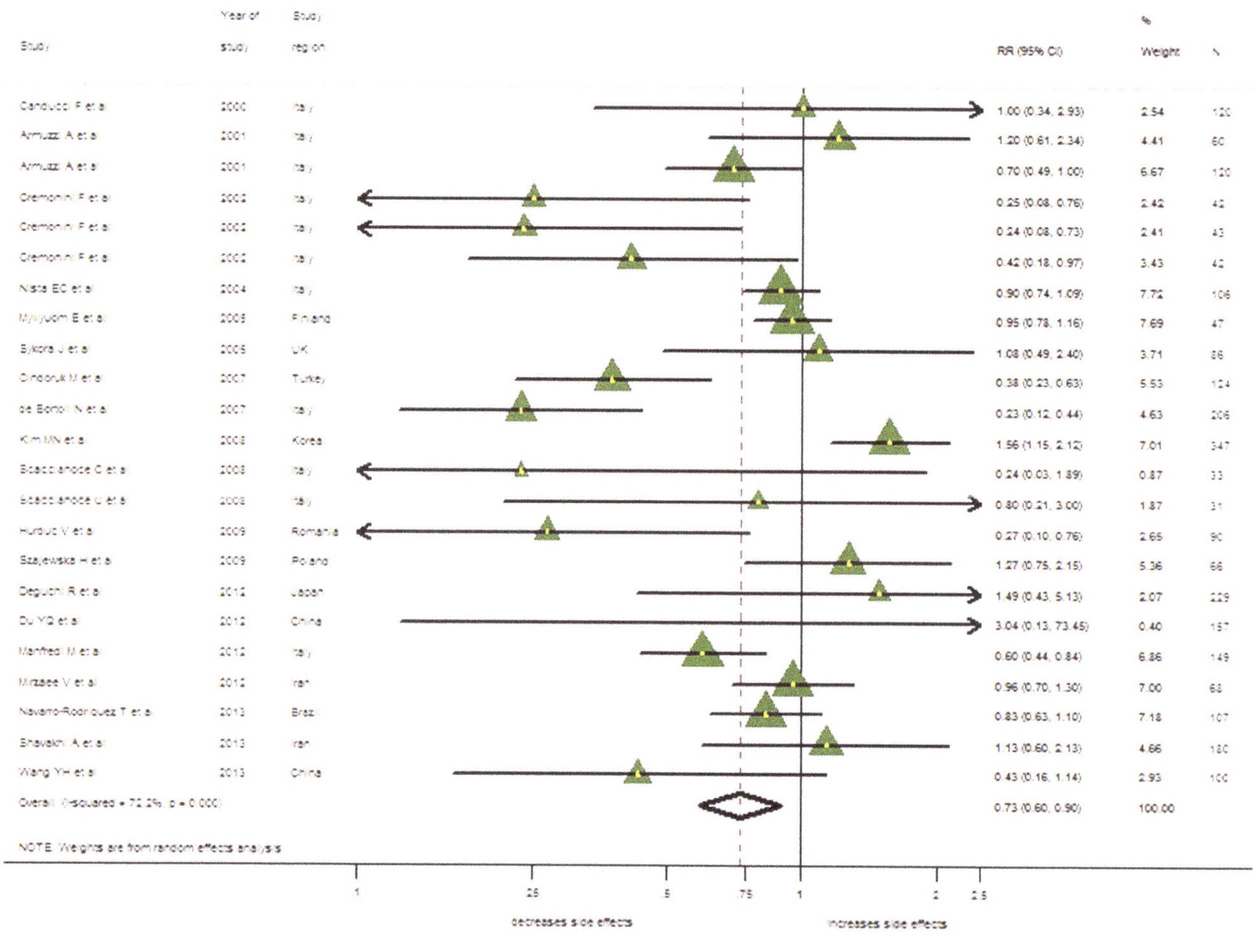

Study	Year of study	Study region	RR (95% CI)	% Weight	N
Canducci F et al	2000	Italy	1.00 (0.34, 2.93)	2.54	120
Armuzzi A et al	2001	Italy	1.20 (0.61, 2.34)	4.41	60
Armuzzi A et al	2001	Italy	0.70 (0.49, 1.00)	6.67	120
Cremonini F et al	2002	Italy	0.25 (0.08, 0.76)	2.42	42
Cremonini F et al	2002	Italy	0.24 (0.08, 0.73)	2.41	43
Cremonini F et al	2002	Italy	0.42 (0.18, 0.97)	3.43	42
Nista EC et al	2004	Italy	0.90 (0.74, 1.09)	7.72	106
Myllyuoma E et al	2005	Finland	0.95 (0.78, 1.16)	7.69	47
Sykora J et al	2005	UK	1.08 (0.49, 2.40)	3.71	86
Cindoruk M et al	2007	Turkey	0.38 (0.23, 0.63)	5.53	124
de Bortoli N et al	2007	Italy	0.23 (0.12, 0.44)	4.63	206
Kim MN et al	2008	Korea	1.56 (1.15, 2.12)	7.01	347
Scaccianoce G et al	2008	Italy	0.24 (0.03, 1.89)	0.87	33
Scaccianoce G et al	2008	Italy	0.80 (0.21, 3.00)	1.87	31
Hurduc V et al	2009	Romania	0.27 (0.10, 0.76)	2.65	90
Szajewska H et al	2009	Poland	1.27 (0.75, 2.15)	5.36	66
Deguchi R et al	2012	Japan	1.49 (0.43, 5.13)	2.07	229
Du YQ et al	2012	China	3.04 (0.13, 73.45)	0.40	157
Manfredi M et al	2012	Italy	0.60 (0.44, 0.84)	6.86	149
Mirzaee V et al	2012	Iran	0.96 (0.70, 1.30)	7.00	68
Navarro-Rodriguez T et al	2013	Brazil	0.83 (0.63, 1.10)	7.18	107
Shavakhi A et al	2013	Iran	1.13 (0.60, 2.13)	4.66	150
Wang YH et al	2013	China	0.43 (0.16, 1.14)	2.93	100
Overall (I-squared = 72.2% p = 0.000)			0.73 (0.60, 0.90)	100.00	

NOTE: Weights are from random effects analysis

decreases side effects | increases side effects

Figure 5. The effect of probiotics supplementation vs. without probiotics on total side effects. RR, risk ratio; CI, confidence interval. The triangles represent individual studies and the size of the triangle represents the weight given to each study in the meta-analysis. The diamond represents the combined results.

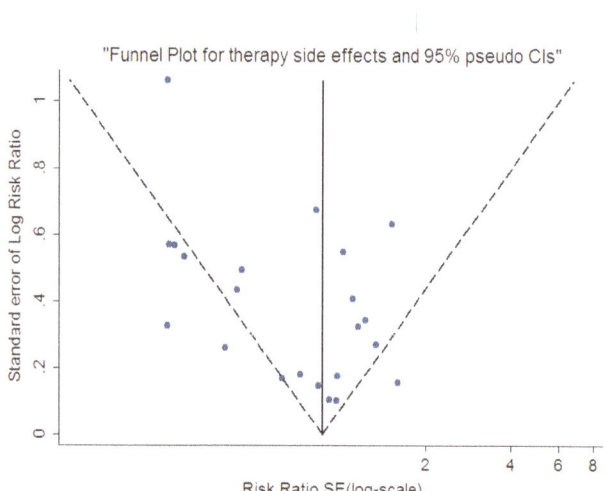

Figure 6. Funnel plot analysis of publication bias for side effects.

therapies. Acceptable success rates have often been defined as 80% or more on an ITT basis [2]. In those effective regimens, probiotic supplementation may not be needed. On the other hand, however, increasing resistance to antibiotics, for example clarithromycin, has been noted [2]. Probiotics supplementation may be a potential approach for eliminating resistant strains. However, no study included in this meta-analysis reported the detection of antibiotic resistance. Further research is warranted to clarify this issue.

The occurrence of adverse effects is one of the major drawbacks of antibiotic treatment. Although antibiotics may modify the composition of intestinal bacteria, broad spectrum antibiotics also often lead to gastrointestinal side effects [20,60]. Results regarding the effectiveness of probiotics in reducing side effects from previous meta-analyses have been inconclusive [8,9,10,11,12]. While we found that probiotics reduce the overall occurrence of side effects in the pooled data, this result must be taken with caution. Only one specific strain significantly reduced side effects and this overall result was only confirmed for non-blinded trials. While eradication rates can be determined with objective measures, assessment of side effects must rely on patients' subjective reporting. When patients are not blinded, their reports are likely to be biased due to awareness of a respective intervention.

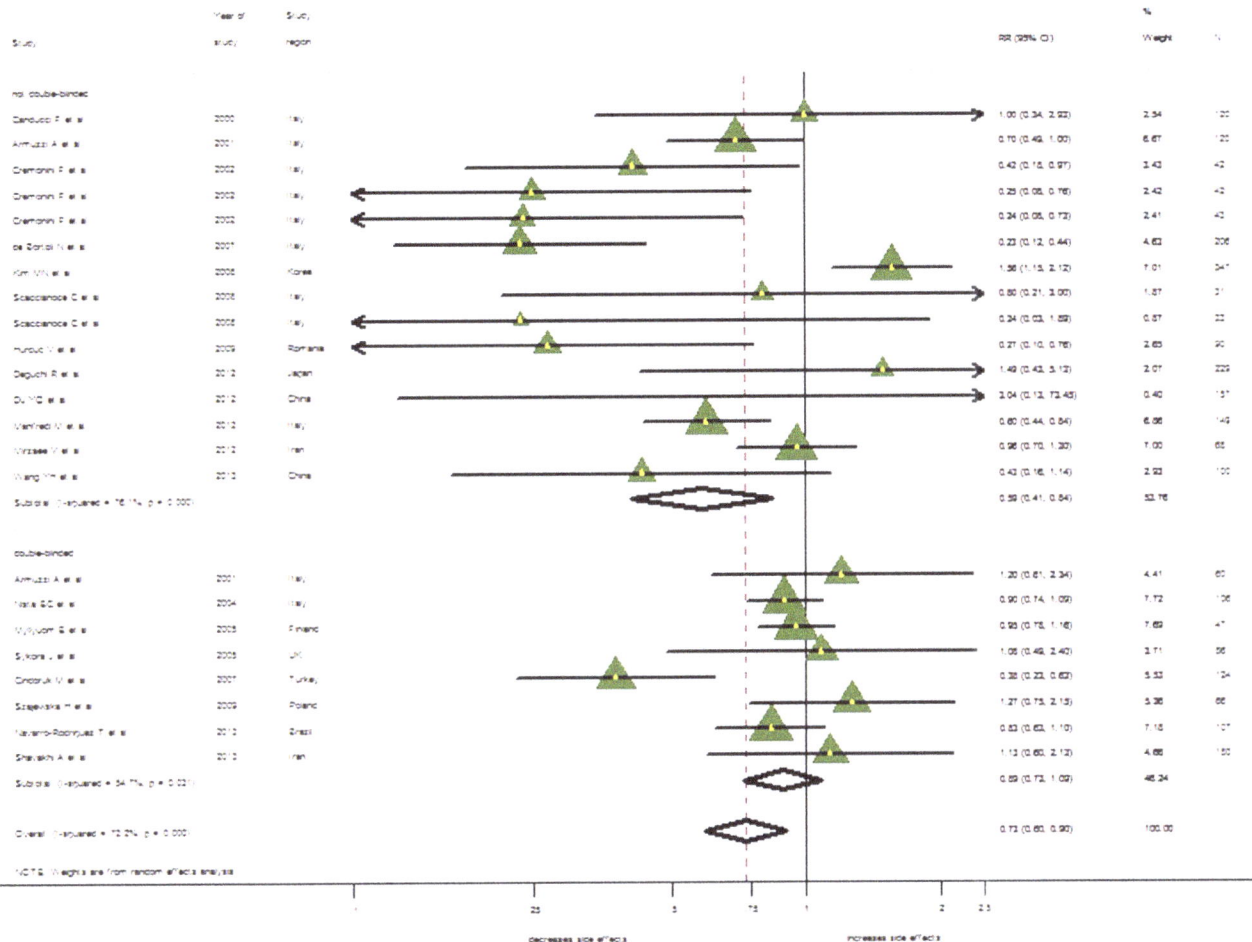

Figure 7. Meta-analysis of side effects by double-blinded and nondouble-blinded trials. RR, risk ratio; CI, confidence interval. The triangles represent individual studies and the size of the triangle represents the weight given to each study in the meta-analysis. The diamond represents the combined results.

Limitations of this meta-analysis include a limited number of high quality trials that could be analyzed, particularly regarding individual probiotic strains and side effects. Only one controlled trial was conducted for the following specific species *Bacillus clausii*, *Bifidobacterium* DN-173 010, *Lactobacillus casei* DN-114001, *Lactobacillus gasseri*, *Bifidobacterium infantis* 2036. Accordingly, results regarding those strains need to be interpreted with caution. Thirteen of the analyzed trials did not provide data on overall occurrence of side effects. Moreover, tools used for measuring side effects and reporting individual side effects largely differed across studies. From 20 studies reporting on the overall incidence of side effects, only eight were blinded.

Conclusions

The pooled data suggests that supplementation with specific strains of probiotics compared with eradication therapy may be considered as an option for increasing eradication rates of *H.*

pylori, particularly when antibiotic therapies are relatively ineffective.

Acknowledgments

We acknowledge the help of Dr. Andrew Pennycott in correcting the manuscript for grammar and style.

Author Contributions

Conceived and designed the experiments: YD GZ. Performed the experiments: YD XZ. Analyzed the data: YD JDR. Contributed reagents/materials/analysis tools: XZ. Wrote the paper: YD JDR GZ.

References

1. Go MF (2002) Review article: natural history and epidemiology of *Helicobacter pylori* infection. Aliment Pharmacol Ther 16: 3–15.
2. Malfertheiner P, Meqraud F, O'Morain CA, Atherton J, Axon AT, et al. (2012) Management of *Helicobacter pylori* infection–the Masstricht IV/Florence Consensus Report. Gut 61: 646–64.
3. Meqraud F (2004) Basis of the management of drug-resistant *Helicobacter pylori* infection. Drugs 64: 1893–904.
4. Yoqueswaran K, Chen G, Cohen L, Cooper MA, Yong E, et al. (2011) How well is *Helicobacter pylori* treated in usual practice? Can J Gastroenterol 25: 543–6.

5. Lionetii E, Indrio F, Pavone L, Borrello G, Cavallo L, et al. (2010) Role of probiotics in pediatric patients with *Helicobacter pylori* infection: a comprehensive review of the literature. Helicobacter 15: 79–87.

6. FAO/WHO (2001) Report on Joint FAO/WHO Expert Consultation on Evaluation of Health and Nutritional Properties of Probiotics in Food Including Powder Milk with Live Lactic Acid Bacteria. Cordoba, FAO/WHO, 2001.

7. Saxelin M, Tynkkynen S, Mattila-Sandholm T, de Vos WM (2005) Probiotic and other functional microbes: from markets to mechanisms. Curr Opin Biotechnol 16: 204–11.

8. Tong JL, Ran ZH, Shen J, Zhang CX, Xiao SD (2007) Meta-analysis: the effect of supplementation with probiotics on eradication rates and adverse events during *Helicobacter pylori* eradication therapy. Aliment Pharmacol Ther 25: 155–68.

9. Zou J, Dong J, Yu X (2009) Meta-analysis: Lactobacillus containing quadruple therapy versus standard triple first-line therapy for *Helicobacter pylori* eradication. Helicobacter 14: 449–459.

10. Sachdeva A, Naqpal J (2009) Effect of fermented milk-based probiotic preparations on *Helicobacter pylori* eradication: a systematic review and meta-analysis of randomized-controlled trials. Eur J Gastroenterol Hepatol 21: 45–53.

11. Szajewska H, Horvath A, Piwowarczyk A (2010) Meta-analysis: the effects of Saccharomyces boulardii supplementation on *Helicobacter pylori* eradication rates and side effects during treatment. Aliment Pharmacol Ther 32: 1069–79.

12. Wang ZH, Gao QY, Fang JY (2013) Meta-analysis of the efficacy and safety of Lactobacillus-containing and Bifidobacterium-containning probiotic compound preparation in *Helicobacter pylori* eradication therapy J Clin Gastroenterol 47: 25–32.

13. Zheng X, Lyu L, Mei Z (2013) Lactobacillus-containingprobiotic supplementation increases Helicobacter plyori eradication rate: evidence from a meta-analysis. Rev Esp Enferm Dig 105: 445–53.

14. Li S, Huang XL, Sui JZ, Chen SY, Xie YT, et al. (2014) Meta-analysis of randomized controlled trails on the efficacy of probiotics in Helicobacter pylori eradication therapy in children. Eur J Pediatr 173: 153–61.

15. Higgins JP, Altman DG, Gotzsche PC, Juni P, Moher D, et al. (2011) The Cochrane Collaboration's tool for assessing risk of bias in randomized trials. BMJ 343: d5928.

16. Roger MH, Ross JH, Jonathan AC (2009) Updated tests for small-study effects in meta-analyses. The Stata Journal 9: 197–210.

17. Canducci F, Armuzzi A, Cremonini F, Cammarota G, Bartolozzi F, et al. (2000) A lyophilized and inactivated culture of Lactobacillus acidophilus increases *Helicobacter pylori* eradication rates. Aliment Pharmacol Ther 14: 1625–9.

18. Armuzzi A, Cremonini F, Bartolozzi F, Canducci F, Candelli M, et al. (2001) The effect of oral administration of Lactobacillus GG on antibiotic-associated gastrointestinal side-effects during *Helicobacter pylori* eradication therapy. Aliment Pharmacol Ther 15: 163–9.

19. Armuzzi A, Cremonini F, Ojetti V, Bartolozzi F, Canducci F, et al. (2001) Effect of Lactobacillus GG supplementation on antibiotic-associated gastrointestinal side effects during *Helicobacter pylori* eradication therapy: a pilot study. Digestion 63: 1–7.

20. Cremonini F, Di Caro S, Covino M, Armuzzi A, Gabrielli M, et al. (2002) Effect of different probiotic preparations on anti-*Helicobacter pylori* therapy-related side effects: a parallel proup, triple blind, placebo-controlled study. Am J Gastroenterol 97: 2744–9.

21. Sheu BS, Wu JJ, Lo CY, Wu HW, Chen JH, et al. (2002) Impact of supplement with Lactobacillus- and Bifidobacterium-containing yogurt on triple therapy for *Helicobacter pylori* eradication. Aliment Pharmacol Ther 16: 1669–75.

22. Nista EC, Candelli M, Cremonini F, Cazzato IA, Zocco MA, et al. (2004) Bacillus clausii therapy to reduce side-effects of anti-*Helicobacter pylori* treatment: randomized, double-blind, placebo controlled trial. Aliment Pharmacol Ther 20: 1181–8.

23. Myllyluoma E, Veijola L, Ahlroos T, Tynkkynen S, Kankuri E, et al. (2005) Probiotic supplementation improves tolerance to *Helicobacter pylori* eradication therapy-a placebo-controlled, double-blind randomized pilot study. Aliment Pharmacol Ther 21: 1263–72.

24. Sykora J, Valeckova K, Amlerova J, Siala K, Dedek P, et al. (2005) Effects of a specially designed fermented milk product containing probiotic Lactobacillus casei DN-114 001 and the eradication of *H. pylori* in children: a prospective randomized double-blind study. J Clin Gastroenterol 39: 692–8.

25. Goldman CG, Barrado DA, Balcarce N, Rua EC, Oshiro M, et al. (2006) Effect of a probiotic food as an adjuvant to triple therapy for eradication of *Helicobacter pylori* infection in children. Nutrition 22: 984–8.

26. Lionetti E, Miniello VL, Castellaneta SP, Maqista AM, de Canio A, et al. (2006) Lactobacillus reuteri therapy to reduce side-effects during anti-*Helicobacter pylori* treatment in children: a randomized placebo controlled trial. Aliment Pharmacol Ther 24: 1461–8.

27. Ziemniak W (2006) Efficacy of *Helicobacter pylori* eradication taking into account its resistance to antibiotics. J Physiol Pharmacol 57: 123–41.

28. Cindoruk M, Erkan G, Karakan T, Dursun A, Unal S (2007) Efficacy and safety of Saccharomyces boulardii in the 14-day triple anti-*Helicobacter pylori* therapy: a prospective randomized placebo-contrilled double-blind study. Helicobacter 12: 309–16.

29. de Bortoli N, Leonardi G, Ciancia E, Merlo A, Bellini M, et al. (2007) *Helicobacter pylori* eradication: a randomized prospective study of triple therapy versus triple therapy plus lactoferrin and probiotics. Am J Gastroenterol 102: 951–6.

30. Francavilla R, Lionetii E, Casrellaneta SP, Maqusta AM, Mauroqivanni G, et al. (2008) Inhibition of *Helicobacter pylori* infection in humans by Lactobacillus reuteri ATCC 55730 and effect on eradication therapy: a pilot study. Helicobacter 13: 127–34.

31. Kim MN, Kim N, Lee SH, Park YS, Hwang JH, et al. (2008) The effects of probiotics on PPI-triple therapy for *Helicobacter pylori* eradication. Helicobacter 13: 261–8.

32. Scaccianoce G, Zullo A, Hassan C, Gentili F, Cristofari F, et al. (2008) Triple therapies plus different probiotics for *Helicobacter pylori* eradication. Eur Rev Med Pharmacol Sci. 12: 251–6.

33. Hurduc V, Plesca D, Dragomir D, Sajin M, Vandenplas Y (2009) A randomized, open trial evaluating the effect of Saccharomyces boulardii on the eradication rate of *Helicobacter pylori* infection children. Acta Paediatr 98: 127–31.

34. Szajewska H, Albrecht P, Topczewska-Cabanek A (2009) Randomized, double-blind, placebo-controlled trial: effect of lactobacillus GG supplementation on *Helicobacter pylori* eradication rates and side effects during treatment in children. J Pediatr Gastroenterol Nutr 48: 431–6.

35. Song MJ, Park DI, Park JH, Kim HJ, Cho YK, et al. (2010) The effect of probiotics and mucoprotective agents on PPI-based triple therapy for eradication of *Helicobacter pylori*. Helicobacter 15: 206–13.

36. Yasar B, Abut E, Kayadibi H, Toros B, Sezikli M, et al. (2010) Efficacy of probiotics in *Helicobacter pylori* eradication therapy. Turk J Gastroenterol 21: 212–7.

37. Bekar O, Yilmaz Y, Gulten M (2011) Kefir improves the efficacy and tolerability of triple therapy in eradicating *Helicobacter pylori* 14: 344–7.

38. Medeiros JA, Goncalves TM, Boyanova L, Pereira MI, de Carvalho JN, et al. (2011) Evaluation of *Helicobacter pylori* eradication by triple therapy plus Lactobacillus acidophilus compared to triple therapy alone. Eur J Clin Microbiol Infect Dis 30: 555–9.

39. Ozdil K, Calhan T, Sahin A, Senates E, Kahraman R, et al. (2011) Levofloxacin based sequential and triple therapy compared with standard plus probiotic combination for *Helicobacter pylori* eradication. Hepatogastroenterology 58: 1148–52.

40. Deguchi R, Nakaminami H, Rimbara E, Noquchi N, Sasatsu M, et al. (2012) Effect of pretreatment with Lactobacillus gasseri OLL2716 on first-line *Helicobacter pylori* eradication therapy. J Gastroenterol Hepatol 27: 888–92.

41. Du YQ, Su T, Fan JG, Lu YX, Zheng P, et al. (2012) Adjuvant probiotics improve the eradication effect of triple therapy for *Helicobacter pylori* infection. World J Gastroenterol 18: 6302–7.

42. Manfredi M, Bizzarri B, Sacchero RI, Maccari S, Calabrese L, et al. (2012) *Helicobacter pylori* infection in clinical practice: probiotics and a combination of probiotics + lactoferrin improve compliance, but not eradication, in sequential therapy. Helicobacter 17: 254–63.

43. Mizaee V, Rezahosseini O (2012) Randomized control trial: Comparison of Triple Therapy plus Probiotic Yogurt vs. Standard Triple Therapy on *Helicobacter pylori* Eradication. Iran Red Crescent Med J 14: 657–66.

44. Tolone S, Pellino V, Vitaliti G, Lanzafame A, Tolone C (2012) Evaluation of *Helicobacter pylori* eradication in pediatric patients by triple therapy plus lactoferrin and probiotics compared to triple therapy alone. Ital J Pediatr 38: 63.

45. Ahmad K, Fatemeh F, Mehri N, Maryam S (2013) Probiotics for the treatment of pediatric *Helicobacter pylori* infection: a randomized double blind clinical trial. Iran H Pediatr 23: 79–84.

46. Navarro-Rodriquez T, Silva FM, Barbuti RC, Mattar R, Moraes-Filho JP, et al. (2013) Association of a probiotic to a *Helicobacter pylori* eradication regimen does not increase efficacy or decreases the adverse effects of the treatment: a prospective, randomized, double-blind, placebo-controlled study. BMC Gastroenterol 13: 56.

47. Shavakhi A, Tabesh E, Yaqhoutkar A, Hashemi H, Tabesh F, et al. (2013) The effects of multistrain probiotics compound on bismuth-containing quadruple therapy for *Helicobacter pylori* infection: A randomized placebo-controlled triple-blind study. Helicobacter 18: 280–4.

48. Dajani AI, Abu Hammour AM, Yang DH, Chung PC, Nounou MA, et al. (2013) Do probiotics improve eradication response to *Helicobacter pylori* on standard triple or sequential therapy? Saudi J Gastroenterol 19: 113–120.

49. Wang YH, Huang Y (2013) Effect of Lactobacillus acidophilus and Bifidobacterium bifidum supplementation to standard triple therapy on Helicobacter pylori eradication and dynamic changes in intestinal flora. World J Microbiol Biotechnol 30: 847–53.

50. Sheu BS, Cheng HC, Kao AW, Wang ST, Yang YJ, et al. (2006) Pretreatment with Lactobacillus- and Bifidobacterium-containng yogurt can improve the efficacy of quadruple therapy in eradicating residual *Helicobacter pylori* infection after failed triple therapy. Am J Clin Nutr 83: 864–9.

51. Boonyaritichaikij S, Kuwabara K, Nagano J, Naqano J, Kobayashi K, et al. (2009) Long-term administration of probiotics to asymptomatic pre-school children for either the eradication or the prevention of *Helicobacter pylori* infection. Helicobacter 14: 202–7.

52. Yoon H, Kim N, Kim JY, Park SY, Park JH, et al. (2011) Effects of multistrain probiotic containing yogurt on second-line triple therapy for Hp infection. J Gastroenterol Hepatol 26: 44–8.

53. Selinger CP, Bell A, Cairns A, Lockett M, Sebastian S, et al. (2013) Probiotics VSL#3 prevents antibiotic-associated diarrhea in a double-blind, randomized, placebo-controlled clinical trial. J Hosp Infect 84: 159–65.

54. Mehling H, Busjahn A (2013) Non-viable Lactobacillus reuteri DSMZ 17648 (Pylopass™) as a new approach to Helicobacter pylori control in humans. Nutrients 5: 3062–73.

55. Dore MP, Cuccu M, Pes GM, Manca A, Graham DY (2014) Lactobacillus reuteri in the treatment of *Helicobacter pylori* infection. Intern Emerg Med 9: 649–54.

56. Francavilla R, Polimeno L, Demichina A, Mauroqiovanni G, Principi B, et al. (2014) Lactobacillus reuteri strain combination in *Helicobacter pylori* infection: a randomized, double-blind, placebo-controlled study. J Clin Gastroenterol 48: 407–13.

57. Marshall BJ, Warren JR (1984) Unidentified curved bacilli in the stomach of patients with gastritis and peptic ulceration. Lancet 1: 1311–5.

58. Walker MM, Talley NJ (2014) Review article: bacteria and pathogenesis of disease in the upper gastrointestinal tract—beyond the era of Helicobacter pylori. Aliment Pharmacol Ther 39: 767–79.

59. Delqado S, Cabrera-Rubio R, Mira A, Suarez A, Mayo B (2013) Microbiological survey of the human gastric ecosystem using culturing and pyrosequencing methods. Microb Ecol 65: 763–72.

60. Patel A, Shah N, Prajapati JB (2014) Clinical appliance of probiotics in the treatment of *Helicobacter pylori* infection-A brief review. J Microbiol Immunol Infect 47: 429–37.

21

Lactobacillus gasseri SBT2055 Induces TGF-β Expression in Dendritic Cells and Activates TLR2 Signal to Produce IgA in the Small Intestine

Fumihiko Sakai[1⑨], Tomohiro Hosoya[1⑨], Aiko Ono-Ohmachi[1], Ken Ukibe[1], Akihiro Ogawa[1], Tomohiro Moriya[1], Yukio Kadooka[1], Takuya Shiozaki[2], Hisako Nakagawa[2], Yosuke Nakayama[2], Tadaaki Miyazaki[2*]

1 Milk Science Research Institute, Megmilk Snow Brand Co. Ltd., Minamidai, Kawagoe, Saitama, Japan, 2 Department of Probiotics Immunology, Institute for Genetic Medicine, Hokkaido University, Kita-ku, Sapporo, Japan

Abstract

Probiotic bacteria provide benefits in enhancing host immune responses and protecting against infection. Induction of IgA production by oral administration of probiotic bacteria in the intestine has been considered to be one reason for this beneficial effect, but the mechanisms of the effect are poorly understood. *Lactobacillus gasseri* SBT2055 (LG2055) is a probiotic bacterium with properties such as bile tolerance, ability to improve the intestinal environment, and it has preventive effects related to abdominal adiposity. In this study, we have found that oral administration of LG2055 induced IgA production and increased the rate of IgA$^+$ cell population in Peyer's patch and in the lamina propria of the mouse small intestine. The LG2055 markedly increased the amount of IgA in a co-culture of B cells and bone marrow derived dendritic cells (BMDC), and TLR2 signal is critical for it. In addition, it is demonstrated that LG2055 stimulates BMDC to promote the production of TGF-β, BAFF, IL-6, and IL-10, all critical for IgA production from B cells. Combined stimulation of B cells with BAFF and LG2055 enhanced the induction of IgA production. Further, TGF-β signal was shown to be critical for LG2055-induced IgA production in the B cell and BMDC co-culture system, but TGF-β did not induce IgA production in a culture of only B cells stimulated with LG2055. Furthermore, TGF-β was critical for the production of BAFF, IL-6, IL-10, and TGF-β itself from LG2055-stimulated BMDC. These results demonstrate that TGF-β was produced by BMDC stimulated with LG2055 and it has an autocrine/paracrine function essential for BMDC to induce the production of BAFF, IL-6, and IL-10.

Editor: Emiko Mizoguchi, Massachusetts General Hospital, United States of America

Funding: This study was funded by Megmilk Snow Brand Co., Ltd. The funder provided support in the form of salaries for authors (Fumihiko Sakai, Tomohiro Hosoya, Aiko Ono-Ohmachi, Ken Ukibe, Akihiro Ogawa, Tomohiro Moriya, and Yukio Kadooka), but did not have any additional role in the study design, data collection and analysis, decision to publish, or preparation of the manuscript. The specific roles of these authors are articulated in the 'author contributions' section.

Competing Interests: Fumihiko Sakai, Tomohiro Hosoya, Aiko Ono-Ohmachi, Ken Ukibe, Akihiro Ogawa, Tomohiro Moriya, and Yukio Kadooka are employees of Megmilk Snow Brand Co., Ltd. There are no other patents, products in development or marketed products to declare.

* Email: miyazaki@pop.med.hokudai.ac.jp

⑨ These authors contributed equally to this work.

Introduction

Probiotics are live microorganisms which when they are administered in adequate amounts confer health benefits to the host [1]. Probiotic bacteria, mainly belonging to the class of lactic acid bacteria (LAB), are well known to induce beneficial effects in human and animal health. In particular, lactobacilli are characterized by the production of lactic acid and are commonly applied to many vegetable, meat, and dairy fermentations. These bacteria can influence the composition and activity of the gut microbiota. Currently, there is a general consensus that orally administered probiotic bacteria contribute to immune homeostasis by altering the microbial balance or by interacting with the host immune system [2–4]. In particular, the interplay between the mucosa-associated immune system and microbiota certainly plays a pivotal role in mucosal tissue homeostasis as well as in protection against infectious and inflammatory diseases occurring at mucosal sites [5].

In the intestinal tract, IgA is the most abundant immunoglobulin isotype, with up to 3 g of secretory IgA secreted into the human intestinal lumen per day [6,7]. The IgA plays an important role in the host defense against mucosally transmitted pathogens, preventing commensal bacteria from binding to epithelial cells, and neutralizing their toxins to maintain homeostasis at the mucosal surfaces [8]. These functions are beneficial for the host as they reduce the risk of infection and maintain an intestinal environment accommodating to the appropriate commensal population. In humans, individuals with IgA deficiency have increased rates of respiratory and gastrointestinal infectious diseases, and lympho-proliferative disorders of the small intestine [9]. It has been reported that intestinal commensal bacteria induce IgA production by developing gut associated lymphoid tissue

(GALT) in the small and large intestine [10–13]. Within the network of intestinal immunity, dendritic cells (DCs) play a critical role in the switching between stimulating immune regulation or activating immune responses of commensal microbiota [14]. It has been reported that administration of some strains of lactobacilli or bifidobateria increase the mucosal IgA production [15–19]. However, the mechanism of the induction of IgA production by probiotic bacteria has not been established in detail.

The *L. gasseri* strain SBT2055 (LG2055) is a human intestine-originating probiotic bacterium with properties including bile tolerance [20], the ability to become established in the intestine, and lowered both faecal bacterial population of *Staphylococcus* and faecal concentration of p-cresol. [21,22], having a cholesterol lowering effect in humans with mild hepercholesterolemia [23], and preventing abdominal adiposity in rats [24,25] and humans [26], among others. A further recent finding regarding LG2055 has reported that LG2055-fed mouse dams reduced rotavirus infections in pups and elevated RV-specific IgA levels in breast milk originating from the stomach [27]. This finding raises the possibility that administration of LG2055 may induce IgA production in the intestinal tract, where IgA is most abundantly produced in the tissue.

In the present study, we examined whether the administration of LG2055 to mice augmented IgA levels in the intestine, and also the precise molecular mechanisms for the IgA induction by an *in vitro* culture system using bone marrow derived dendritic cells.

Materials and Methods

Mice

Male SPF BALB/c mice were purchased from Charles River Japan (Yokohama, Japan) or SLC Inc. (Shizuoka, Japan) and maintained at the animal experimental facilities of Megmilk Snow brand Co., Ltd. or Hokkaido University. The mice were given free access to food and distilled water. All procedures for animal care and use complied with the animal experimentation regulations of the Milk Science Research Institute of Megmilk Snow Brand Co., which is based on the guidelines proposed by the Science Council of Japan, and the guidelines of the Bioscience Committee of Hokkaido University. All animal experiments were approved by the Animal Care and Use Committee of Milk Science Research Institute of Megmilk Snow Brand Co. and the Animal Care and Use Committee of Hokkaido University.

Preparation and growth condition for *Lactobacillus gasseri* SBT2055 (LG2055)

LG2055 is a bacterial strain derived from a fecal specimen of a healthy adult, which had been isolated by Fujiwara et al. [21] and deposited in the International Patent Organism Depository, National Institute of Advanced Industrial Science and Technology (Tsukuba, Ibaraki 305-8566, Japan). LG2055 was cultivated at 37°C in 2 liters of MRS broth (Difco, Detroit, Mich.) for 18 hours and harvested by centrifugation at 5,000×g for 10 minutes at 4°C. To prepare the LG2055 cells powder for *in vivo* studies, the harvested cells were washed twice with sterile distilled water and suspended in 500 mL of 10% (w/v) lactose solution. Suspension of LG2055 was lyophilized to make powder. The lyophilized powder consist of $1.2×10^{11}$ cfu/g viable LG2055 and 65% lactose. LG2055-lactose powder was stored at −80°C for experiments. To prepare the LG2055 cells for *in vitro* studies, cells washed with distilled water were suspended in distilled water, and lyophilized to make powder. LG2055 powder was resuspended in PBS and heated at 80°C for 30 min. The ability of other strains of *Lactobacillus* spp., which were *L. gasseri* JCM1131^{T} (LG1131T),

L. helveticus SBT2171 (LH2171), *L. acidophilus* SBT2062 (LA2062), to induce the IgA and cytokines production in an *in vitro* culture system were examined for comparison with that of LG2055. LH2171 is used for manufacturing of natural cheeses [28] and LA2062 is a probiotic bacterium utilized for manufacturing of fermented milk products in past times [29]. The cells of these three strains were prepared by the same methods used to prepare the LG2055 cells for *in vitro* studies.

Orally administration of LG2055 to mice

Six-week-old male Balb/c mice were used in this study. After a one-week adaptation period, the mice were separated into two groups: the control group (n = 10), and the LG2055-treated group (n = 10). Each group mouse had a similar mean body weight. Mice were given an experimental diet prepared according to AIN93 formation supplemented with 1% LG2055-lactose powder for LG2055 group or AIN93 supplemented with 0.65% lactose for control group to equalize lactose contents in both diets. Experimental diets were replaced on a daily basis by fresh diets. Stocks of experimental diets were stored at −20°C and the viable cell number of LG2055 in the diet was confirmed as more than $1.0×10^{9}$ cfu/g throughout the experiments. All mice were fed ad libitum experimental diets and distilled water for 5 weeks.

Extraction of intestinal tissues

After the mice were sacrificed under isoflurane anesthesia, the small intestine and colon were removed carefully from mice of control or LG2055-treated group. Small intestinal lavage fluid was collected by washing out with 10 ml of ice-cold PBS for determination of IgA level. Peyer's patches were excised from the small intestine for FACS analysis. The colon was isolated, open longitudinally to remove the intestinal contents, and then washed with PBS. The tissue samples from jejunum, ileum, and colon were weighted and added a 40-fold volume of PBS containing protease inhibitor cocktail (Roche, Schweiz) in 50 mM Tris-HCl (pH 6.8) buffer. These tissues were homogenized on ice using a ULTRA-TURRAX T25 (IKA-Werke GmbH & CO., Germany). The suspension was centrifuged at 10,000×g for 15 min, and the supernatant was used for the detection of intestinal IgA levels by mouse IgA ELISA (Bethyl Laboratories, Inc., TX) and total protein levels by BCA Protein Assay kits (PIERCE, IL).

Fecal sample preparation

The fecal samples collected were lyophilized and suspended in PBS containing 50 mM EDTA and 0.1 mg/ml soybean trypsin inhibitor. Fecal suspension was vigorously vortexed and centrifuged at 15,000×g for 10 min at 4°C. The supernatants were diluted appropriately for mouse IgA ELISA analysis.

Preparation of Peyer's patch cells and lamina propria cells from mice after oral administration of LG2055

Peyer's patches (PP) cells and lamina propria (LP) cells from small intestine were prepared for analysis of IgA^{+} cells by FACS. PP dissected from the small intestine were mechanically disrupted in the cell culture medium (RPMI-1640 containing 10% FBS, 10 mM HEPES buffer, 2 mM L-glutamine, 100 U/mL of penicillin, 100 μg/mL streptomycin, and 0.05 mM 2-mercaptoethanol). PP cells were washed with PBS and resuspended in the cell culture medium. For isolation of LP cells, the small intestines were removed after excluding PP, and then cut into 5–7 pieces. The small intestines were washed three times with 40 mL of Hanks' balanced salt solution (HBSS) (sigma, MO) supplemented with 5% FBS and 5 mM EDTA in a 50 mL tube and incubated at

37°C for 30 min with shaking at 150 rpm. The tissues were cut into smaller pieces, and were incubated with RPMI1640 supplemented with 1 mg/ml collagenase (Sigma) and 10 µU/ml DNase I (Roche) at 37°C for 60 min with stirring. Collected cells were placed on the boundary between 44% and 70% concentration of Percoll solution (GE Healthcare, UK), and were centrifuged at 1500×g for 20 min. These cells were washed and used as LP cells.

Flow cytometric analysis

PP cells and LP cells were pretreated with FcR blocking reagent (Miltenyi Biotec) and then stained with the following mAbs: FITC-labeled anti-mouse IgA (C10-3, BD Bioscience, NJ), PE-labeled anti-mouse B220 (RA3-6B2, eBioscience, CA). Live cells were gated based on 7AAD exclusion during acquisition on a FACS CantII (BD Bioscience). Flow cytometric analysis was performed using FACS Diva software (BD Bioscience).

Depletion or isolation of CD11c$^+$ cells from PP cells, and naïve B cell isolation from spleen cells

Six to ten-week-old male Balb/c mice were euthanized by overdose of inhalant anesthetic and sacrificed. PPs collected from small intestines were shaken for 45 min at 37°C in PBS containing 2 mM EDTA to remove epithelial cells and subsequently treated with 2 mg/ml collagenase D (Roche) and 10 µg/ml DNaseI (Roche) in complete RPMI1640 at 37°C for 30 min. Digested PPs were harvested and passed through a 70 µm cell strainer (BD biosciences). Single cells were labeled with microbeads coated with anti-CD11c antibody (Miltenyi Biotec) according to the manufacturer's instruction. For depletion of CD11c$^+$ cells, labeled PP cells were applied to LD column (Miltenyi Biotec) and passing through unlabeled cells were collected, used as CD11c depleted cells. For isolation of CD11c$^+$ cells, labeled PP cells were applied to LS column (Miltenyi Biotec) and positively selected cells were collected. To obtain the CD11c$^+$ cells further purified to greater than 95%, magnetic-isolated cells were labeled with APC-conjugated anti-mouse CD11c (N418, BioLegend), and PE-conjugated anti-mouse CD19 (BD biosciences) antibodies for depletion of the contaminating B cells. CD11c$^+$CD19$^-$ cells were collected by sorting using a FACSAria (BD biosciences), and used as CD11c$^+$ cells derived from PP cells (purity >95%, CD11c$^+$). To prepare B cells, spleen cells from naïve BALB/c mice were separated by anti-mouse IgM coated MicroBeads (Miltenyi Biotec). Positively selected IgM$^+$ cells were used as B cells (purity >95%, IgM$^+$).

Generation of bone marrow derived dendritic cell (BMDC)

Six to ten-week-old male Balb/c mice were euthanized by overdose of inhalant anesthetic and sacrificed. The femurs and tibias were removed, cleaned, and sterilized. The bone marrow was flushed from bones by use of a syringe containing culture medium. For BMDC isolation, the bone marrow cells were washed and cultured at $4.0×10^6$ cells/dish (10 cm culture dish) in culture medium supplemented with 40 ng/ml GM-CSF (Wako, Japan) for 8 days. Harvested cells were blocked with anti-mouse Fc receptor (BioLegend) and then stained with FITC-labeled anti-CD11c (N418, BioLegend) and PE-labeld anti-mouse MHC-II (I-A/I-E; M5, BioLegend) antibodies. CD11c$^+$ MHC-II$^+$ cells were sorted by BD FACSAria II (BD Biosciences, CA) and used as BMDC (purity >95%, CD11c$^+$ MHC-II$^+$ cells). For analysis of the function of LG2055 against BMDC, BMDC ($5.0×10^5$ cells/mL) were cultured with or without LG2055 (20 µg/mL) in presence or

absence of SB505124 (5 µM, Sigma) in 12 well plate (BD Bioscience) for 48 h. After incubation, culture supernatants were collected for measurement of cytokines and total RNA were prepared from BMDC for real time-PCR analysis.

Assay for IgA amounts by *in vitro* cell culture system

To evaluate the mechanism for IgA production *in vitro*, B cells ($5×10^5$) were either directly stimulated with LG2055 (10 µg/ml), BAFF (500 ng/ml, Sigma), TGF-β1 (0.001–1 ng/ml, R&D Systems), and LPS (10 µg/mL, InvivoGen) or co-cultured with CD11C$^+$ cells derived from PP cells ($0.5×10^5$) or BMDC ($1×10^5$) in the presence or absence of LG2055 (10 µg/ml), LE135 (500 nM), SB505124 (5 µM), anti-human/mouse TLR2 (5 µg/ml), Pam3CSK4 (1 µg/mL, InvivoGen, CA), and FSL-1 (1 µg/mL, InvivoGen) in 96 well-round bottom plate (BD Bioscience) for 7 days. To block cell-to-cell contact, B cells ($2×10^6$) and BMDC ($4×10^5$) were cultured in 24 well culture plate (BD Bioscience) with or without a transwell device (BD Bioscience) in the presence or absence of LG2055 (10 µg/ml) for 7 days. The level of IgA in the culture supernatants was measured by mouse IgA ELISA.

RNA extraction and quantitative real time-PCR analysis

Total RNA from BMDC was extracted with TRIzol reagent (Molecular Research Center, Inc.) and reverse transcribed by the TaqMan Reverse Transcription Reagent kit (Applied Biosystems) according to the manufacturer's instructions. The following TaqMan Gene Expression Assay were purchased (Applied Biosystems): BAFF (Assay ID Mm00446347_m1), APRIL (Assay ID Mm00840215_g1), RALDH2 (Assay ID Mm00501306_m1), β-actin (Assay ID Mm00607939_s1). Amplifications were carried out in a total volume of 20 µL containing 1x TaqMan Universal PCR Master Mix (Applied Biosystems). The cycling parameters were initiated by 20 s at 95°C, followed by 40 cycles of 3 s at 95°C and 30 s at 60°C using the ABI Prism 7000 (Applied Biosystems). The amplifications were normalized by the expression of β-actin encoding gene.

Quantification of antibody isotypes and cytokine by ELISA

Tissue extract, intestinal gavage, serum, and culture supernatants were analyzed for the amount of total IgA or IgG antibody by mouse IgA or IgG ELISA (Bethyl Laboratories, Inc., TX). Mouse IL-5, IL-6, IL-10, latent TGF-β (BioLegend), and BAFF (R&D systems Inc.) in cell culture supernatants were analyzed using commercial available ELISA kits according to the manufacturer's instructions.

Statistical analysis

Data were expressed as means ± standard deviations. Level of significance was determined by one-way ANOVA and Tukey-Kramer post test in experiment for comparison of IgA induction of non-treated LG2055 with that of heat-treated LG2055, one-way ANOVA and Dunnett's post test in examinations of IgA induction by four strains of *Lactobacillus* spp., and Student t test in the other experiments and p values<0.05 were considered to be statistically significant.

Results

LG2055 induces the production of IgA in the small intestine

To identify the *Lactobacillus* strain to induce IgA production, we assessed its induction in PP cells cultured with or without four strains of *Lactobacillus* species. LG2055 exhibited a marked induction of IgA production in comparison to the other strains (Figure 1A). Induction of IgA production by LG2055 increased in a time- and dose-dependent manner (Figure 1B, C). Next, we evaluated the IgA induction in the intestine of mice by the oral-administration of LG2055. Mice were given the control diet or the diet supplemented with LG2055 for 5 weeks. Throughout the experiments, the food intake and body weight of the control and LG2055 diet groups did not differ (data not shown). The amount of total IgA in the intestinal tissue extracts of the jejunum and ileum was significantly higher in the LG2055 fed group compared with the control group (Figure 1D). In the intestinal lavage fluid, the amount of IgA in the LG2055 group was significantly higher than that in the control group, but the amount of IgG in the serum was not different for the two groups (Figure S1A, C). The amount of fecal IgA in the LG2055 fed group was significantly higher after 5 weeks than that in the control group (Figure S1B). The populations of IgA$^+$ lymphocytes (IgA$^+$B220$^+$ cells) in PP cells and IgA$^+$ plasma cells (IgA$^+$B220$^-$ cells) in LP cells were higher in the LG2055 group than that in the control group (Figure 1E, F). When mice were fed the control diet or the diet supplemented with LG2055 for 10 days, the amount of total IgA in the intestine of the LG2055 fed group were slightly higher in the jejunum and significantly higher in the colon compared with the control group (Figure S2), suggesting that oral administration of LG2055 for at least 10 days is required to induce the IgA production in the intestine. We also ascertained that administration of heat-treated LG2055 significantly increased the total IgA level in the jejunum extracts, but that the effect was not as strong as with the non-heat-treated LG2055 (Figure S3).

Dendritic cells are crucial for the induction of IgA production by LG2055

To investigate the mechanisms for the induction of IgA production by LG2055, we determined and compared the amounts of total IgA in the supernatants of PP cells and PP cells depleted of CD11c$^+$ cells cultured with or without LG2055 for 7 days. In the control PP cells, LG2055 markedly induced the IgA production, while in PP cells depleted of CD11c$^+$ cells, the induction level of IgA by LG2055 was significantly lower than that in the control PP cells (Figure 2A). To determine the function of DC that is significant in the induction of IgA production by LG2055, B cells were cultured with or without CD11c$^+$ cells isolated from PP cells or BMDC in the presence or absence of LG2055. The LG2055 increased IgA production when B cells alone were cultured, however the amount of IgA production when B cells were cultured with CD11c$^+$ cells from PP cells or BMDC was more strongly increased (Figure 2B, C). Though LG2055 also increased the amount of IgG production when B cells were cultured with BMDC, the level of IgG production was very little (about 10 ng/ml) (Figure S4), compared with that of IgA (about 2 µg/ml) (Figure 2C).The induction level of IgA production by LG2055 in the B cell and BMDC co-culture system was highest among the tested four strains (Figure S3). To elucidate whether diffusible factors or interaction between B cells and BMDC was critical for the IgA production by LG2055, BMDC and B cells were cultured by a transwell arrangement where the BMDC were unable to directly contact the B cells, thereby restricting the interaction to only diffusible factors, and measured the IgA amounts generated. The LG2055-induced IgA production levels with the transwell arrangement were similar to those in the conventional culture (Figure 2D). This result suggests that diffusible factor(s) from the BMDC induced by LG2055 is/are

necessary for the LG2055-induced IgA production in this culture system.

Identification of BMDC-derived factors involved in IgA production by LG2055 stimulation

To identify diffusible factors from BMDC induced by LG2055, BMDC were stimulated with LG2055 for 48 h and the gene expression level of IgA-inducing factors in BMDC and the amount of IgA-inducing cytokines in the culture supernatants were measured. The gene expression of BAFF and RALDH2 increased significantly in the LG2055-treated cells compared with the control cells, but the expression of APRIL did not increase (Figure 3). The amounts of TGF-β, IL-6, and IL-10 increased significantly in the supernatants of LG2055-treated cells when compared with those of the control cells, but the amount of IL-5 did not increase (Figure 3). The induction level of these cytokines production from BMDC stimulated with LG2055 was significantly higher than that with LG1131T or LA2062 (Figure S4). These results suggest that there are multiple factors that are critical for IgA production and are induced in the BMDC stimulated with LG2055, and that these factors promote the secretion of IgA from B cells.

To investigate the functional significance of TGF-β and RALDH2 in LG2055-induced IgA production, a TGF-β receptor I inhibitor (SB505124) or an RAR antagonist (LE135) blocking TGF-β or RA signaling was added in the B cell and BMDC co-culture. Blocking of TGF-β signaling inhibited LG2055-induced IgA production, but the RAR antagonist did not (Figure 4A), indicating that TGF-β signaling plays a critical role in LG2055-induced IgA production. As TGF-β is known as critical for IgA-class switching of naïve B cells, it was further examined whether TGF-β induced the IgA production from B cells in the presence or absence of LG2055. The TGF-β induced IgA production from B cells stimulated with LPS, as previously reported [30,31], while with or without LG2055 stimulation it did not induce IgA production from B cells (Figure 4B). Next the effects of BAFF on IgA production in the presence or absence of LG2055 and LA2062 were evaluated. Each of BAFF and LG2055 induced IgA production, and combined stimulation with BAFF and LG2055 enhanced the induction of IgA production more markedly (Figure 4C), but combined stimulation with BAFF and LA2062 did not (Figure 4D).

TGF-β signaling is crucial for the production of IgA-inducing factors from BMDC stimulated by LG2055

We investigated whether TGF-β signaling was critical for the production of IgA-inducing factors from BMDC stimulated by LG2055. TGF-β type I receptor inhibitor (SB505124) completely inhibited BAFF and TGF-β induction and partially inhibited IL-10 and IL-6 induction in BMDC stimulated by LG2055 (Figure 5). These results demonstrate that endogenous TGF-β contributes to the production of BAFF, TGF-β, IL-10, and IL-6 from BMDC stimulated by LG2055.

LG2055 induces IgA production via TLRs

A previous report has shown that LG2055 stimulated total IgA production in splenocyte cultures from TLR4-knockout mice, but not in splenocyte cultures from TLR2-knockouts [27], and here we examined whether LG2055-induced IgA production in the B cell and BMDC co-culture was dependent on the function of TLR2. Addition of anti-TLR2 antibodies significantly inhibited LG2055-induced IgA production in both B cells alone and B cell and BMDC co-cultures (Figure 6A). Furthermore, Pam3CSK4 and

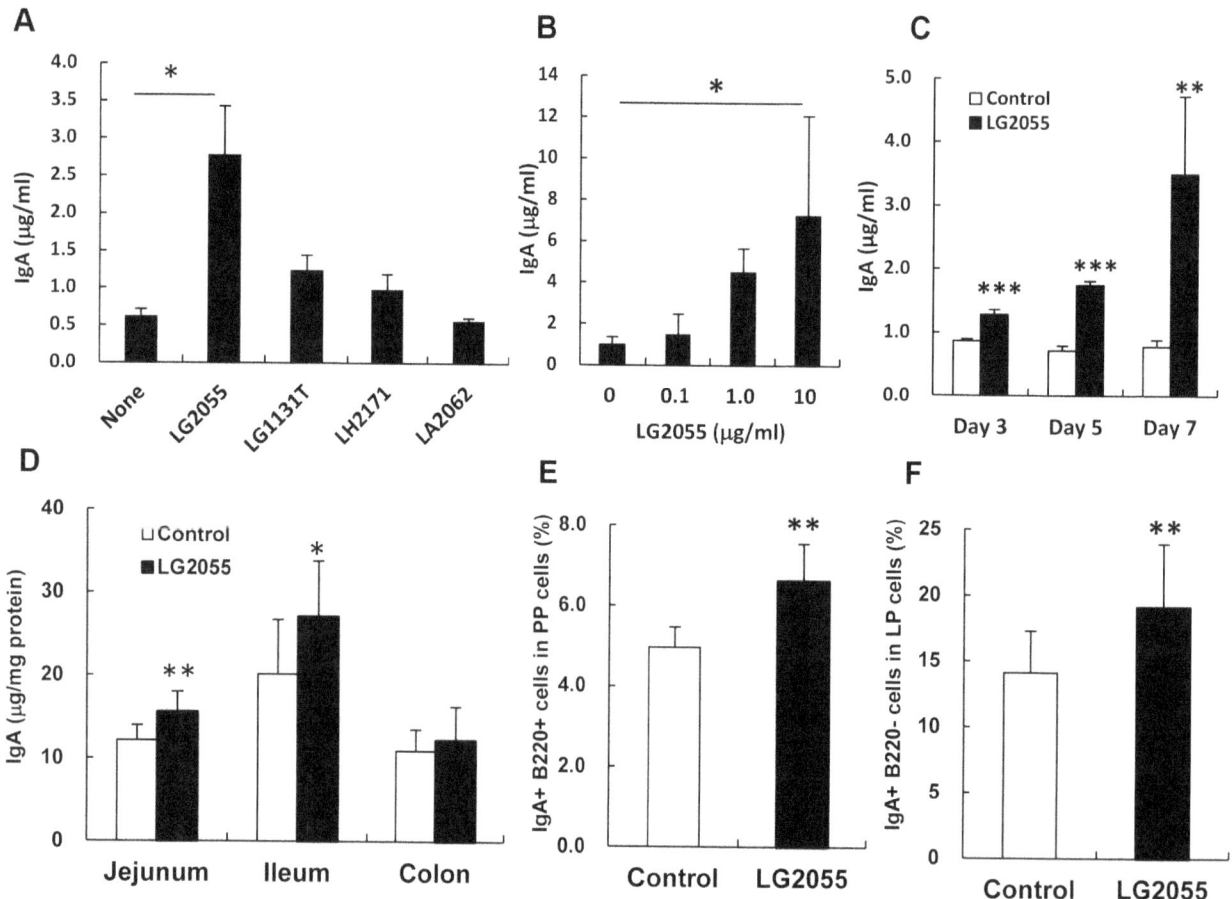

Figure 1. Augmentation of IgA production by LG2055 *in vitro* and *in vivo*. Whole PP cells were cultured with or without each of the four *Lactobacillus* strains (LG2055, *L. gasseri* JCM1131^t (LG1131T), *L. helveticus* SBT2171 (LH2171), *L. acidophilus* SBT2062 (LA2062), 10 µg/ml) for 7 days (A). Whole PP cells were cultured with 0, 0.1, 1.0, and 10 µg/ml of LG2055 for 7 days (B). Whole PP cells were cultured with or without LG2055 (10 µg/ml) for 3, 5, and 7 days (C) The amounts of IgA in culture supernatants were determined by ELISA. Each experiment was done with tripricate cultures; data are shown as the mean ± SD. The values for cells cultured with lactic acid bacteria are compared with that of without the bacteria by one-way ANOVA, Dunnett's post test (A and B) and the *t*-test (C). Significant differences are indicated by * P<0.05, ** P<0.01, *** P<0.001. LG2055 was orally administrated to BALB/c mice for 5 weeks. Amounts of total IgA in intestinal tissue extracts (D) were determined by ELISA. The population of IgA$^+$ B220$^+$ cells in PP cells (E) and IgA$^+$ B220$^-$ cells in LP cells (F) was analyzed by FACS. Representative data from two independent experiments are shown. Data are shown as the mean ± SD (number of mice n = 10). Significant difference from control group at *P<0.05, **P<0.01 was shown by the *t*-test (D, E, and F).

FSL-1, which are ligands of TLR1/2 and TLR2/6, respectively, induced the IgA production in B cell and BMDC co-cultures (Figure 6B). These results demonstrate that LG2055 induces IgA production via TLR2, in B cells only, as well as in B cell and BMDC co-cultures.

Discussion

L. gasseri is a predominant *Lactobacillus* species in the human small intestine [32] and the *L. gasseri* strain, LG2055 has been isolated from human feces. We have previously reported that LG2055-fed mouse dams reduced rotavirus (RV) infections in pups and showed elevated RV-specific IgA levels in the breast milk originating from the stomach [27]. In the present study, we demonstrate that administration of LG2055 induces IgA production in the mouse small intestine. Induction of IgA production by LG2055 is considered to prevent the invasion of harmful microorganisms and toxins through the epithelial cells in the small intestine, the initial site of adherence and infection of pathogens in the gut. There are many reports that some probiotic

bacteria strains induce IgA production in the intestine. However, not all probiotic strains have the ability to induce production of IgA *in vivo*, even when the induction of IgA production is confirmed by *in vitro* studies [13,33,34]. It is known that different probiotic strains have different properties and that it is not possible to extrapolate the effects of one probiotic strain to others or the effect of one strain with a specific pathogen to other pathogens [35]. In this investigation it is established that LG2055 exhibits the induction of IgA production both *in vivo* and *in vitro*. Furthermore, the effect of induction of IgA production by LG2055 is clearly stronger than that of the *L. gasseri* type strain or of other *Lactobacillus* species *in vitro* (Figure 1A and S3). Thus, LG2055 might be a strain that has the high potential to induce IgA production among *Lactobacillus* strains. Recent findings have shown that secretion of IgA is critical in the regulation of the composition of the microbial community in the gut [36–38]. Oral administration of LG2055 in humans is able to lower both faecal bacterial population of *Staphylococcus* and faecal concentration of p-cresol. [21,22]. Induction of IgA production by LG2055 should play an important role in the control of the intestinal microflora.

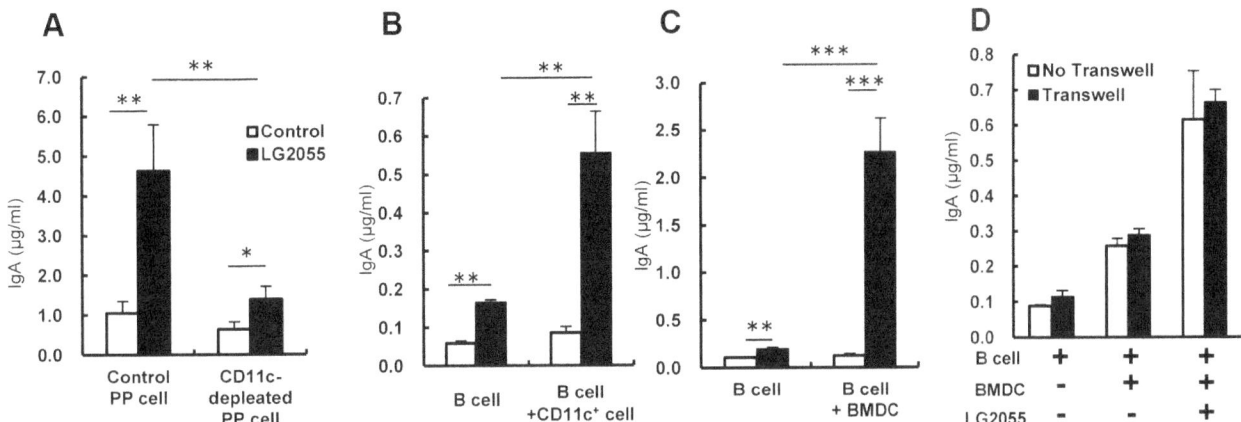

Figure 2. Significance of dendritic cell for the induction of IgA production by LG2055. (A) Whole PP cells or CD11c⁺ cell-depleted PP cells were cultured with or without heat treated LG2055 (10 μg/ml) for 7 days. (B) B cells from the spleen were co-cultured with or without CD11c⁺ cells derived from PP cells (left) or BMDC (right) in the presence or absence of the LG2055 for 7 days. (C) B cells were co-cultured with or without BMDC in the presence or absence of LG2055 in Transwell system for 7 days. The amounts of IgA in culture supernatants were determined by ELISA. Representative data from three for PP cells or four for BMDC independent experiments are shown. Each experiment was done with triplicate cultures; data are shown as the mean ± SD. * P<0.05, ** P<0.01, *** P<0.001 was shown by t-test.

The DCs play a critical role in the induction of IgA production in the gastrointestinal tract [39,40], and a large number of soluble factors produced by the DCs in the gut are involved in the induction of IgA production (e.g., TGF-β, IL-6, IL-10, APRIL, BAFF, and retinoic acid) [41]. It has been reported that interaction between probiotic bacteria and DCs modulates cytokine produc-

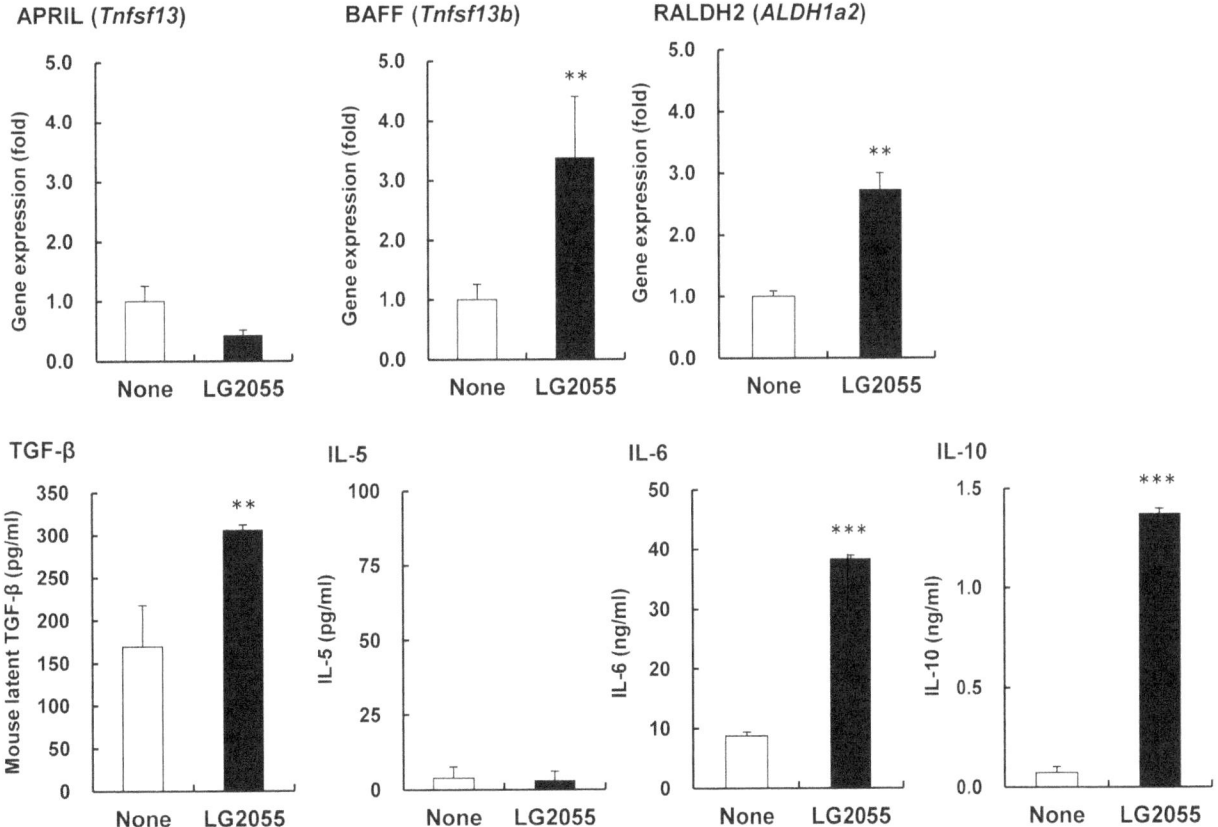

Figure 3. Gene expression and cytokine production of BMDC stimulated by LG2055. BMDC was cultured with or without LG2055 for 48 hours. Gene expression of APRIL (tnfsp13), BAFF (tnfsp13b), RALDH2 (aldh1a2) in BMDC was determined by quantitative PCR. Amounts of TGF-β, IL-5, IL-6, and IL-10 in the culture supernatants were determined by ELISA. Representative data from three independent experiments are shown. Each experiment was done with triplicate cultures; data are shown as the mean ± SD. ** P<0.01, *** P<0.001 was shown by t-test.

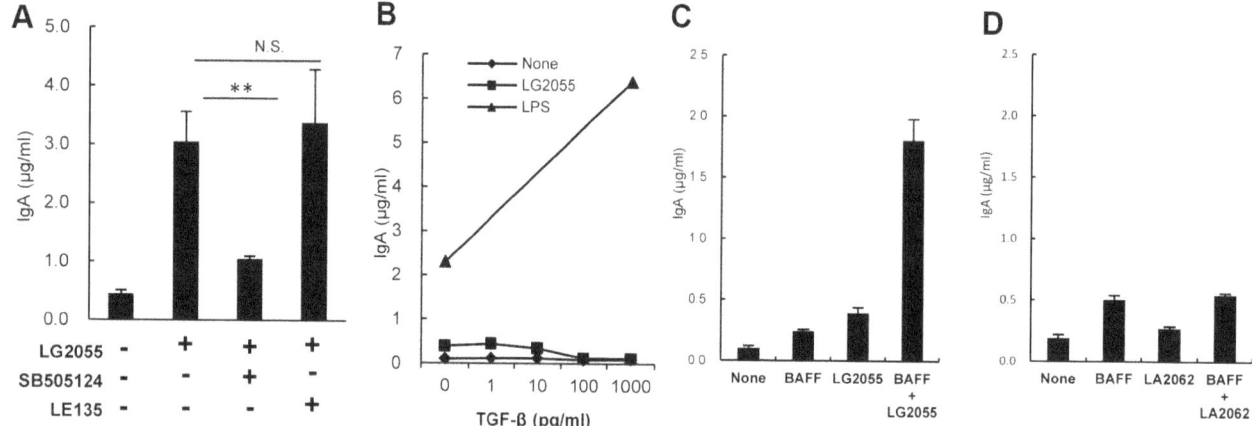

Figure 4. Critical factors for induction of IgA production by LG2055. (A) B cells and BMDC were co-cultured with or without LG2055 in the presence or absence of the TGF-β type I receptor inhibitor SB505124 or RAR antagonist LE135 for 7 days. (B) B cells were cultured with or without LG2055 or LPS (10 μg/ml) in the presence or absence of TGF-β for 7 days. B cells were cultured with or without LG2055 (C) or LA2062 (D), in the presence or absence of BAFF (500 ng/ml) for 7 days. IgA amounts in the supernatants were determined by ELISA. Representative data from three independent experiments are shown. Each experiment was done with triplicate cultures; data are shown as the mean ± SD. ** P<0.01, was shown by t-test.

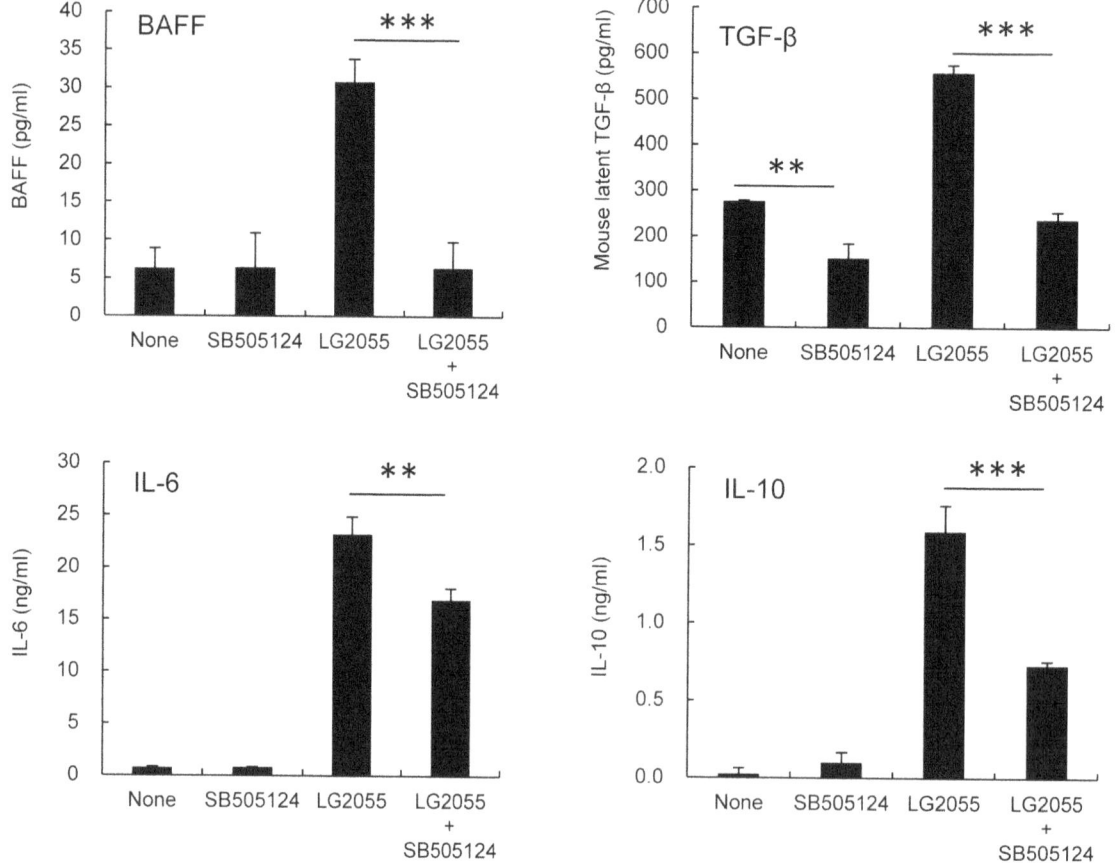

Figure 5. Importance of TGF-β type I signaling for the cytokine production from BMDC by LG2055. BMDC was cultured in the presence or absence of LG2055 and TGF-β type I receptor inhibitor SB505124 for 48 hours. Amounts of TGF-β, IL-6, and IL-10 in the culture supernatants were determined by ELISA. Representative data from two independent experiments are shown. Each experiment was done with triplicate cultures; data are shown as the mean ± SD. ** P<0.01, *** P<0.001 was shown by t-test.

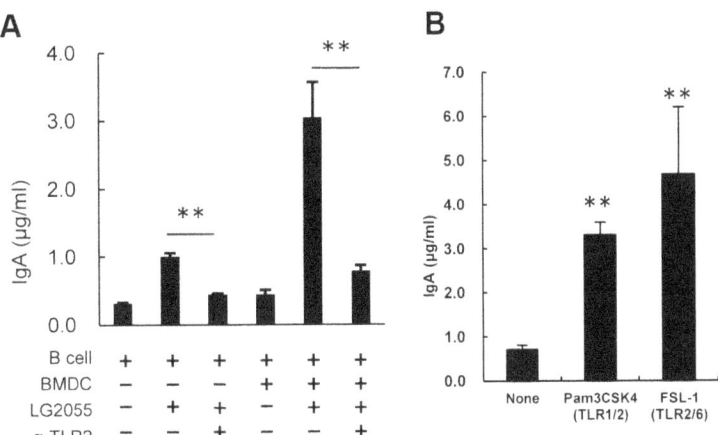

Figure 6. Significance of TLR2 in the induction of IgA production by LG2055. B cells or both B cells and BMDC were cultured with or without LG2055 in the presence or absence of the anti TLR2 antibody for 7 days. (B) Pam3CSK4 (TLR1/2 ligand) or FSL-1 (TLR2/6 ligand) was added to the B cell and BMDC co-culture system, and cultured for 7 days. The amounts of IgA in culture supernatants were determined by ELISA. Representative data from three independent experiments are shown. Each experiment was done with triplicate cultures; data are shown as the mean ± SD. ** P<0.01 was shown by t-test.

tion as well as the function of DCs [42,43]. These reports prompted us to investigate the role of DCs in the induction of IgA production by LG2055. The data here using an *in vitro* co-culture of B cells with either CD11C+ cells derived from PP cells or BMDC showed that the DCs play an important role in the augmentation of IgA production by LG2055 (Figure 2). We also observed that the gene expression and protein production of IgA-inducing mediators, BAFF, RALDH2, TGF-β, IL-6, and IL-10, were up-regulated in LG2055-stimulated BMDC (Figure 3). As being reported, TGF-β and BAFF induce an IgA class-switch recombination, and IL-6 and IL-10 induce differentiation of IgA-producing plasma cells in IgA+ B cells [30]. It is possible that the interaction between LG2055 and DCs may contribute to both IgA class-switching and plasma cell differentiation. This is supported by the result that oral administration of LG2055 in mice increased the rate of IgA+ lymphocytes in PP cells and IgA+ plasma cells in LP cells (Figure 1E, F). The induction level of BAFF, TGF-β, IL-6, and IL-10 production from BMDC stimulated with LG2055 was significantly higher than that stimulated with LG1131T or LA2062 which had low potential to induce the IgA production (Figure S4). The high induction of these cytokines by LG2055 may be one of the reasons for the higher induction level of IgA production by LG2055 compared with that by the other *Lactobacillus* strains.

Recently, it has been reported that DCs-derived retinoic acid promotes IgA secretion [44] and retinoic acid-treated BMDC induces IgA-secreting B cells, through promotion of TGF-β production from BMDCs [45]. Though the gene expression of RALDH2 (*ALDH1a2*), which is an enzyme responsible for the conversion of retinal into retinoic acid and the major isoform expressed in the gut-associated DCs, was up-regulated in LG2055-stimulated BMDC, LG2055 induced IgA production in the presence of the RAR antagonist, LE135 (Figure 4A). At the same time, with the TGF-β type I receptor inhibitor SB505124 there was no LG2055-induced IgA production (Figure 4A), indicating that the TGF-β signal transduction is critical for IgA production by LG2055 in the B cell-BMDC co-culture, while RAR signal transduction is not essential. Recent evidence has indicated that gut-associated DCs induce the expression of gut-homing receptors

on B cells, via a mechanism that depends on retinoic acid. This fact implies the possibility that LG2055 contributes to the induction of IgA production via a promotion of the gut-homing of B cells, as the gene expression of RALDH2 was increased in the small intestine of the LG2055-administrated mice (data not shown).

Our results confirm previous observations that TGF-β induced IgA production in LPS-stimulated B cells [30], but not in LG2055-stimulated B cells (Figure 4B). It is noteworthy that the amount of IgA produced from B cells stimulated by LG2055 markedly increased in the presence of BAFF (Figure 4C). The BAFF and APRIL are shown to enhance immunoglobulin class switching to IgA in B cells [46,47]. In addition, it is reported that TACI expression is up-regulated in B cells stimulated by oligodeoxynucleotides (CpG ODN), and the up-regulation of TACI, working as BAFF and APRIL receptors, on B cells render these to induce IgA secretion [48]. Further, it is demonstrated that the TLR ligands cooperate with the TACI ligands to induce antibody secretion [49]. The results here show that LG2055-induced IgA production was strongly inhibited by the treatment of anti TLR2 antibody both in the B cell only culture and in the B cell-BMDC co-culture (Figure 6A). From this it may be surmised that LG2055, cooperatively with BAFF stimulates B cells to induce IgA production, mainly through the TLR2 signaling pathway. Our results indicated that LG2055 solely induced IgA-production from B cells in the absence of BMDC or BAFF, but the reasons for this practice are unclear. As we also showed, LPS-stimulated B cells induce IgA-production to some extend in the absence of TGF-β, though they massively induce IgA-production in the presence of TGF-β. The induction level of IgA-production from B cells stimulated with LG2055 and/or BAFF appears to be similar to that stimulated with LPS and/or TGF-β. LPS and LG2055 may be involved in proliferation or survival of B cells, post-switched, but not yet expressed IgA. Further investigation is needed to elucidate how LPS or LG2055 solely induces IgA-production from B cells. LA2062 did not stimulate B cells with BAFF to induce IgA production (Figure 4D). The biochemical mechanism for the difference between LG2055 and LA2062 effect remains unexplained. Amounts of peptidoglycan and lipoteichoic acid present in the cell walls of *Lactobacillus* species widely differ with respect to the each strain [50]. The difference of the capability of IgA-induction between LG2055 and LA2062 may depend on the amounts of TLR2 ligand present in the each strain.

The results in Figure 4A demonstrate that LG2055-induced IgA production in B cells and the BMDC co-culture was inhibited by the TGF-β type I receptor inhibitor, even when the addition of exogenous TGF-β1 did not induce IgA production by LG2055-stimulated B cells (Figure 4B). These results suggest that the target cell of TGF-β may be BMDC, and not the B cells. The results here also show that the induction of IgA-inducing factors in LG2055-stimulated BMDC is inhibited by TGF-β type I receptor inhibitors, among which BAFF and TGF-β are completely inhibited, IL-10 and IL-6 partially (Figure 5). These results demonstrate that the TGF-β signal contributes to the production of IL-6, IL-10, BAFF, and TGF-β itself by LG2055-stimulated BMDC. It has been reported that exogenous TGF-β1 stimulates the induction of BAFF expression on mouse macrophages [51]. We confirmed that the addition of exogenous TGF-β1 enhanced the gene expression of BAFF in BMDC (data not shown). The TGF-β1 acts directly on skin-resident DCs in an autocrine/paracrine manner to develop Langerhans cells and to inhibit inflammation-induced migration [52,53]. Further, it is revealed that autocrine TGF-β sustains the default tolerogenic function of CD8+ DCs [54]. Yet, no details are known of TGF-β capability in an autocrine/paracrine manner on the regulation of IgA production by DCs. Here our results imply that endogenous TGF-β from LG2055-stimulated BMDC act on BMDC itself in an autocrine/paracrine manner and contributes to the production of IgA-inducing factors, at least in this *in vitro* culture system. On the other hand, given that TGF-β1 cooperates with the CD40 ligand (CD40L) to generate antigen-specific IgA+ B cells in PPs [55,56], this allows the assumption that TGF-β produced by LG2055-stimulated DCs could also act directly on B cells with CD40L and contribute to the induction of IgA production *in vivo*.

In conclusion, this study demonstrated the augmentation of IgA production in the mouse small intestine by oral administration of LG2055 and elucidated the detailed molecular mechanisms for the induction of IgA production by using a B cell and BMDC co-culture system *in vitro*. Specifically, the results suggest that LG2055 activates both DCs and B cells to induce the IgA production, and TLR2 signal is critical for its production. Further, we show that TGF-β produced by LG2055-stimulated BMDC acts on BMDC in an autocrine/paracrine manner and induces the production of IL-6, IL-10, BAFF, and TGF-β itself from BMDC to induce the subsequent IgA production (Figure S5).

Supporting Information

Figure S1 Effect of oral administration of LG2055 on amounts of IgA in the intestinal lavage fluid and feces, and IgG in the serum. LG2055 was orally administrated to BALB/c mice for 5 weeks. Amounts of total IgA in intestinal lavage fluid (A), feces on 0 and 5 weeks after administration (B), and IgG in the serum (C) were determined by ELISA. Representative data from two independent experiments are shown. Data are shown as the mean ± SD (number of mice n = 10). Significant difference from control group at *P<0.05, **P<0.01 was shown by the *t*-test.

Figure S2 Effect of oral administration of LG2055 for 10 days on the production of IgA in the intestine. LG2055 was orally administrated to BALB/c mice for 10 days. Amounts of total IgA in intestinal tissue extracts were determined by ELISA.

Data are shown as the mean ± SD (number of mice n = 10). Significant difference from control group at *P<0.05, **P<0.01 was shown by the *t*-test.

Figure S3 Comparison of IgA induction of heat-treated LG2055 with that of non-treated LG2055 in the mouse small intestine. Heat-treated LG2055 (heated at 80°C for 30 min) or non-treated LG2055 was orally administrated to BALB/c mice for 5 weeks. Amounts of total IgA in small intestinal tissue extracts were determined by ELISA. Data are shown as the mean ± SD (number of mice n = 10). Significant difference among groups at *P<0.05, **P<0.01 was shown by one-way ANOVA and Tukey-Kramer post test.

Figure S4 Effect of LG2055 treatment on IgG production by B cell co-cultured with or without BMDC. B cells from the spleen were co-cultured with or without BMDC in the presence or absence of the heat treated LG2055 (10 μg/ml) for 7 days. The amounts of IgG in culture supernatants were determined by ELISA. Each experiment was done with triplicate cultures; data are shown as the mean ± SD. * P<0.05 was shown by *t*-test.

Figure S5 Comparison of IgA induction among four strains of *Lactobacillus* species. Each of the four *Lactobacillus* strains (LG2055, *L. gasseri* JCM1131^t (LG1131T), *L. helveticus* SBT2171 (LH2171), *L. acidophilus* SBT2062 (LA2062)) was added to the B cell and BMDC co-culture system, and cultured for 7 days. The amounts of IgA in culture supernatants were determined by ELISA. Each experiment was done with triplicate cultures; data are shown as the mean ± SD. Values for stimulated cells are compared with value for non-stimulated cells by one-way ANOVA and Dunnett's post test. Significant differences are indicated by ** P<0.01.

Figure S6 Comparison of cytokine production of BMDC among three strains of *Lactobacillus* species. BMDC was cultured with or without LG2055, LG1131T, LA2062 for 48 hours. Amounts of BAFF, TGF-β, IL-6, and IL-10 in the culture supernatants were determined by ELISA. Each experiment was done with triplicate cultures; data are shown as the mean ± SD. Values not sharing a common letter are significantly different by Tukey-Kraner multiple comparison test at p<0.05.

Figure S7 Schematic illustration of hypothetical model for enhancement of IgA production by LG2055. LG2055 activates both DC and B cell. TGF-β produced by LG2055-stimulated BMDC acts on BMDC in an autocrine/paracrine manner and induces the production of IL-6, IL-10, BAFF, and TGF-β itself from BMDC to induce subsequent IgA production. TLR2 signal is critical for the induction of IgA by LG2055, at least for B cell stimulation by LG2055.

Acknowledgments

We express sincere appreciation to Ms. Noriko Ikeda (Hokkaido University) for excellent technical assistance, and Dr. Koyu Ito and Dr. Junko Morimoto (Hokkaido University) for providing valuable suggestion.

Author Contributions

Conceived and designed the experiments: FS TH T. Miyazaki. Performed the experiments: FS TH A. Ono-Ohmachi KU A. Ogawa TS. Analyzed the data: HN YN. Contributed reagents/materials/analysis tools: T. Moriya YK. Wrote the paper: FS TH T. Miyazaki.

References

1. group RoajFWw (2002) Guidelines for evaluation of probiotics in food. London, Ontario, Canada: FAO/WHO.
2. Corthesy B, Gaskins HR, Mercenier A (2007) Cross-talk between probiotic bacteria and the host immune system. J Nutr 137: 781S–790S.
3. Matsuzaki T, Takagi A, Ikemura H, Matsuguchi T, Yokokura T (2007) Intestinal microflora: probiotics and autoimmunity. J Nutr 137: 798S–802S.
4. Ouwehand AC (2007) Antiallergic effects of probiotics. J Nutr 137: 794S–797S.
5. Dongarra ML, Rizzello V, Muccio L, Fries W, Cascio A, et al. (2013) Mucosal immunology and probiotics. Curr Allergy Asthma Rep 13: 19–26.
6. Conley ME, Delacroix DL (1987) Intravascular and mucosal immunoglobulin A: two separate but related systems of immune defense? Ann Intern Med 106: 892–899.
7. Mestecky J, Russell MW, Jackson S, Brown TA (1986) The human IgA system: a reassessment. Clin Immunol Immunopathol 40: 105–114.
8. Fagarasan S (2008) Evolution, development, mechanism and function of IgA in the gut. Curr Opin Immunol 20: 170–177.
9. Burt RW JR (1999) Polyposis syndromes. In: T Yamada DA, L . Laine, C . Owyang, D . Powell, editor. In textbook of Gastroenterology. Philadelphia: Lippincott William & Wilkins pp. 1995–2022.
10. Umesaki Y, Setoyama H, Matsumoto S, Imaoka A, Itoh K (1999) Differential roles of segmented filamentous bacteria and clostridia in development of the intestinal immune system. Infect Immun 67: 3504–3511.
11. Talham GL, Jiang HQ, Bos NA, Cebra JJ (1999) Segmented filamentous bacteria are potent stimuli of a physiologically normal state of the murine gut mucosal immune system. Infect Immun 67: 1992–2000.
12. Shroff KE, Meslin K, Cebra JJ (1995) Commensal enteric bacteria engender a self-limiting humoral mucosal immune response while permanently colonizing the gut. Infect Immun 63: 3904–3913.
13. Yanagibashi T, Hosono A, Oyama A, Tsuda M, Hachimura S, et al. (2009) Bacteroides induce higher IgA production than Lactobacillus by increasing activation-induced cytidine deaminase expression in B cells in murine Peyer's patches. Biosci Biotechnol Biochem 73: 372–377.
14. Coombes JL, Powrie F (2008) Dendritic cells in intestinal immune regulation. Nat Rev Immunol 8: 435–446.
15. Park JH, Um JI, Lee BJ, Goh JS, Park SY, et al. (2002) Encapsulated Bifidobacterium bifidum potentiates intestinal IgA production. Cell Immunol 219: 22–27.
16. Park JH, Lee Y, Moon E, Seok SH, Cho SA, et al. (2005) Immunoenhancing effects of a new probiotic strain, Lactobacillus fermentum PL9005. J Food Prot 68: 571–576.
17. Vinderola G, Perdigon G, Duarte J, Farnworth E, Matar C (2006) Effects of the oral administration of the exopolysaccharide produced by Lactobacillus kefiranofaciens on the gut mucosal immunity. Cytokine 36: 254–260.
18. Galdeano CM, Perdigon G (2006) The probiotic bacterium Lactobacillus casei induces activation of the gut mucosal immune system through innate immunity. Clin Vaccine Immunol 13: 219–226.
19. Galdeano CM, de Leblanc Ade M, Carmuega E, Weill R, Perdigon G (2009) Mechanisms involved in the immunostimulation by probiotic fermented milk. J Dairy Res 76: 446–454.
20. Usman, Hosono A (1999) Bile tolerance, taurocholate deconjugation, and binding of cholesterol by Lactobacillus gasseri strains. J Dairy Sci 82: 243–248.
21. Fujiwara S, Seto Y, Kimura A, Hashiba H (2001) Establishment of orally-administered Lactobacillus gasseri SBT2055SR in the gastrointestinal tract of humans and its influence on intestinal microflora and metabolism. J Appl Microbiol 90: 343–352.
22. Takahashi H, Fujita T, Suzuki Y, Benno Y (2006) Monitoring and survival of Lactobacillus gasseri SBT2055 in the human intestinal tract. Microbiol Immunol 50: 867–870.
23. Kajimoto O HH, Aoe S, Takahoshi T, Suzuki Y, Tanaka H (2002) Fermented milk containing Lactobacillus gasseri (SP strain) decreases serum cholesterol concentration in men with boundary and mild hypercholesterolemia. Jpn J Lactic acid Bacteria 13: 114–124.
24. Sato M, Uzu K, Yoshida T, Hamad EM, Kawakami H, et al. (2008) Effects of milk fermented by Lactobacillus gasseri SBT2055 on adipocyte size in rats. Br J Nutr 99: 1013–1017.
25. Hamad EM, Sato M, Uzu K, Yoshida T, Higashi S, et al. (2009) Milk fermented by Lactobacillus gasseri SBT2055 influences adipocyte size via inhibition of dietary fat absorption in Zucker rats. Br J Nutr 101: 716–724.
26. Kadooka Y, Sato M, Imaizumi K, Ogawa A, Ikuyama K, et al. (2010) Regulation of abdominal adiposity by probiotics (Lactobacillus gasseri SBT2055) in adults with obese tendencies in a randomized controlled trial. Eur J Clin Nutr 64: 636–643.
27. Kadooka Y, Tominari K, Sakai F, Yasui H (2012) Prevention of rotavirus-induced diarrhea by preferential secretion of IgA in breast milk via maternal administration of Lactobacillus gasseri SBT2055. J Pediatr Gastroenterol Nutr 55: 66–71.

28. Sasaki M, Bosman BW, Tan PS (1996) A new, broad-substrate-specificity aminopeptidase from the dairy organism Lactobacillus helveticus SBT 2171. Microbiology 142 (Pt 4): 799–808.
29. Takiguchi R, Miyamoto M, Mochizuki E, Suzuki Y, Kageyama R, et al. (1998) Effect of fermented-milk administration of defecation and fecal microflora of healthy volunteers. Journal of Intestinal Microbiology 11: 117–122.
30. Coffman RL, Lebman DA, Shrader B (1989) Transforming growth factor beta specifically enhances IgA production by lipopolysaccharide-stimulated murine B lymphocytes. J Exp Med 170: 1039–1044.
31. Sonoda E, Matsumoto R, Hitoshi Y, Ishii T, Sugimoto M, et al. (1989) Transforming growth factor beta induces IgA production and acts additively with interleukin 5 for IgA production. J Exp Med 170: 1415–1420.
32. Reuter G (2001) The Lactobacillus and Bifidobacterium microflora of the human intestine: composition and succession. Curr Issues Intest Microbiol 2: 43–53.
33. Nakanishi Y, Hosono A, Hiramatsu Y, Kimura T, Nakamura R, et al. (2005) Characteristic Immune Response in Peyer's Patch Cells Induced by Oral Administration of Bifidobacterium Components. Cytotechnology 47: 69–77.
34. Harata G, He F, Kawase M, Hosono A, Takahashi K, et al. (2009) Differentiated implication of Lactobacillus GG and L. gasseri TMC0356 to immune responses of murine Peyer's patch. Microbiol Immunol 53: 475–480.
35. Perdigon G, Fuller R, Raya R (2001) Lactic acid bacteria and their effect on the immune system. Curr Issues Intest Microbiol 2: 27–42.
36. Fagarasan S, Muramatsu M, Suzuki K, Nagaoka H, Hiai H, et al. (2002) Critical roles of activation-induced cytidine deaminase in the homeostasis of gut flora. Science 298: 1424–1427.
37. Kawamoto S, Tran TH, Maruya M, Suzuki K, Doi Y, et al. (2012) The inhibitory receptor PD-1 regulates IgA selection and bacterial composition in the gut. Science 336: 485–489.
38. Suzuki K, Meek B, Doi Y, Muramatsu M, Chiba T, et al. (2004) Aberrant expansion of segmented filamentous bacteria in IgA-deficient gut. Proc Natl Acad Sci U S A 101: 1981–1986.
39. Fayette J, Durand I, Bridon JM, Arpin C, Dubois B, et al. (1998) Dendritic cells enhance the differentiation of naive B cells into plasma cells in vitro. Scand J Immunol 48: 563–570.
40. Massacand JC, Kaiser P, Ernst B, Tardivel A, Burki K, et al. (2008) Intestinal bacteria condition dendritic cells to promote IgA production. PLoS One 3: e2588.
41. Fagarasan S, Kawamoto S, Kanagawa O, Suzuki K (2010) Adaptive immune regulation in the gut: T cell-dependent and T cell-independent IgA synthesis. Annu Rev Immunol 28: 243–273.
42. Evrard B, Coudeyras S, Dosgilbert A, Charbonnel N, Alame J, et al. (2011) Dose-dependent immunomodulation of human dendritic cells by the probiotic Lactobacillus rhamnosus Lcr35. PLoS One 6: e18735.
43. Foligne B, Zoumpopoulou G, Dewulf J, Ben Younes A, Chareyre F, et al. (2007) A key role of dendritic cells in probiotic functionality. PLoS One 2: e313.
44. Mora JR, Iwata M, Eksteen B, Song SY, Junt T, et al. (2006) Generation of gut-homing IgA-secreting B cells by intestinal dendritic cells. Science 314: 1157–1160.
45. Feng T, Cong Y, Qin H, Benveniste EN, Elson CO (2010) Generation of mucosal dendritic cells from bone marrow reveals a critical role of retinoic acid. J Immunol 185: 5915–5925.
46. Castigli E, Wilson SA, Scott S, Dedeoglu F, Xu S, et al. (2005) TACI and BAFF-R mediate isotype switching in B cells. J Exp Med 201: 35–39.
47. Moore PA, Belvedere O, Orr A, Pieri K, LaFleur DW, et al. (1999) BLyS: member of the tumor necrosis factor family and B lymphocyte stimulator. Science 285: 260–263.
48. Kanswal S, Katsenelson N, Selvapandiyan A, Bram RJ, Akkoyunlu M (2008) Deficient TACI expression on B lymphocytes of newborn mice leads to defective Ig secretion in response to BAFF or APRIL. J Immunol 181: 976–990.
49. He B, Santamaria R, Xu W, Cols M, Chen K, et al. (2010) The transmembrane activator TACI triggers immunoglobulin class switching by activating B cells through the adaptor MyD88. Nat Immunol 11: 836–845.
50. Sashihara T, Sueki N, Ikegami S (2006) An analysis of the effectiveness of heat-killed lactic acid bacteria in alleviating allergic diseases. J Dairy Sci 89: 2846–2855.
51. Kim HA, Jeon SH, Seo GY, Park JB, Kim PH (2008) TGF-beta1 and IFN-gamma stimulate mouse macrophages to express BAFF via different signaling pathways. J Leukoc Biol 83: 1431–1439.
52. Bobr A, Igyarto BZ, Haley KM, Li MO, Flavell RA, et al. (2012) Autocrine/paracrine TGF-beta1 inhibits Langerhans cell migration. Proc Natl Acad Sci U S A 109: 10492–10497.
53. Kaplan DH, Li MO, Jenison MC, Shlomchik WD, Flavell RA, et al. (2007) Autocrine/paracrine TGFbeta1 is required for the development of epidermal Langerhans cells. J Exp Med 204: 2545–2552.

54. Belladonna ML, Volpi C, Bianchi R, Vacca C, Orabona C, et al. (2008) Cutting edge: Autocrine TGF-beta sustains default tolerogenesis by IDO-competent dendritic cells. J Immunol 181: 5194–5198.

55. Zan H, Cerutti A, Dramitinos P, Schaffer A, Casali P (1998) CD40 engagement triggers switching to IgA1 and IgA2 in human B cells through induction of endogenous TGF-beta: evidence for TGF-beta but not IL-10-dependent direct S mu-->S alpha and sequential S mu-->S gamma, S gamma-->S alpha DNA recombination. J Immunol 161: 5217–5225.

56. Fayette J, Dubois B, Vandenabeele S, Bridon JM, Vanbervliet B, et al. (1997) Human dendritic cells skew isotype switching of CD40-activated naive B cells towards IgA1 and IgA2. J Exp Med 185: 1909–1918.

Permissions

All chapters in this book were first published in PLOS ONE, by The Public Library of Science; hereby published with permission under the Creative Commons Attribution License or equivalent. Every chapter published in this book has been scrutinized by our experts. Their significance has been extensively debated. The topics covered herein carry significant findings which will fuel the growth of the discipline. They may even be implemented as practical applications or may be referred to as a beginning point for another development.

The contributors of this book come from diverse backgrounds, making this book a truly international effort. This book will bring forth new frontiers with its revolutionizing research information and detailed analysis of the nascent developments around the world.

We would like to thank all the contributing authors for lending their expertise to make the book truly unique. They have played a crucial role in the development of this book. Without their invaluable contributions this book wouldn't have been possible. They have made vital efforts to compile up to date information on the varied aspects of this subject to make this book a valuable addition to the collection of many professionals and students.

This book was conceptualized with the vision of imparting up-to-date information and advanced data in this field. To ensure the same, a matchless editorial board was set up. Every individual on the board went through rigorous rounds of assessment to prove their worth. After which they invested a large part of their time researching and compiling the most relevant data for our readers.

The editorial board has been involved in producing this book since its inception. They have spent rigorous hours researching and exploring the diverse topics which have resulted in the successful publishing of this book. They have passed on their knowledge of decades through this book. To expedite this challenging task, the publisher supported the team at every step. A small team of assistant editors was also appointed to further simplify the editing procedure and attain best results for the readers.

Apart from the editorial board, the designing team has also invested a significant amount of their time in understanding the subject and creating the most relevant covers. They scrutinized every image to scout for the most suitable representation of the subject and create an appropriate cover for the book.

The publishing team has been an ardent support to the editorial, designing and production team. Their endless efforts to recruit the best for this project, has resulted in the accomplishment of this book. They are a veteran in the field of academics and their pool of knowledge is as vast as their experience in printing. Their expertise and guidance has proved useful at every step. Their uncompromising quality standards have made this book an exceptional effort. Their encouragement from time to time has been an inspiration for everyone.

The publisher and the editorial board hope that this book will prove to be a valuable piece of knowledge for researchers, students, practitioners and scholars across the globe.

List of Contributors

Miriam Bermudez-Brito, Sergio Muñoz-Quezada, Angel Gil and Carolina Gomez-Llorente
Department of Biochemistry and Molecular Biology II, Institute of Nutrition and Food Technology "José Mataix", Biomedical Research Center, University of Granada, Granada, Spain

Esther Matencio, Maria J. Bernal and Fernando Romero
Global Centre for Child Nutrition Technology, Hero Group, Alcantarilla, Murcia, Spain

Elie A. Akl
Department of Internal Medicine, American University of Beirut, Beirut, Lebanon
Department of Medicine, State University of New York at Buffalo, Buffalo, New York, United States of America
Department of Clinical Epidemiology and Biostatistics, McMaster University, Hamilton, Ontario, Canada

Bradley C. Johnston
Department of Clinical Epidemiology and Biostatistics, McMaster University, Hamilton, Ontario, Canada
Department of Anesthesia and Pain Medicine, The Hospital for Sick Children, Toronto, Ontario, Canada
Child Health Evaluative Sciences, SickKids Research Institute, Toronto, Ontario, Canada

Pablo Alonso-Coello
Iberoamerican Cochrane Centre, CIBERESP-IIB Sant Pau, Barcelona, Spain

Ignacio Neumann and Shanil Ebrahim
Department of Clinical Epidemiology and Biostatistics, McMaster University, Hamilton, Ontario, Canada

Matthias Briel
Department of Clinical Epidemiology and Biostatistics, McMaster University, Hamilton, Ontario, Canada
Basel Institute for Clinical Epidemiology and Biostatistics, University Hospital Basel, Basel, Switzerland

Deborah J. Cook and Gordon H. Guyatt
Department of Clinical Epidemiology and Biostatistics, McMaster University, Hamilton, Ontario, Canada
Department of Medicine, McMaster University, Hamilton, Ontario, Canada

Verity Ann Sattler and Viviana Klose
University of Natural Resources and Applied Life Sciences, Department for Agrobiotechnology, Tulln, Austria

Michaela Mohnl
BIOMIN Research Center, Tulln, Austria

Young-Wook Kim, Young-Ju Jeong, Hyun-Hee Son, Jong-Am Lee, Cheong-Hwan Jung and Chae-Hyun Kim
KBNP Technology Institute, KBNP Inc., Yesan, Korea

Ah-Young Kim
School of Biological Sciences and Technology, Chonnam National University, Gwangju, Korea

Jaeman Kim
Department of Biology, Mokpo National University, Muan, Korea

Huan Wang, Jing Gong, Yu Fu, Wei Qian and Xiaohua Hou
Division of Gastroenterology, Union Hospital of Tongji Medical College, Huazhong University of Science and Technology, Wuhan, Hubei, China

Wenfeng Wang
Division of Gastroenterology, Hospital of Dongfeng Motor Company of Hubei University of Medicine, Shiyan, Hubei, China

Yanqin Long
Division of Gastroenterology, Sir Run Run Shaw Hospital, School of Medicine, Zhejiang University, Hangzhou, Zhejiang, China

Xiaochao Fu
Division of Culture Collection, Hubei Center of Industrial Culture Collection and Research, Wuhan, Hubei, China

Alberto Finamore, Marianna Roselli, Ambra Imbinto and Elena Mengheri
Consiglio per la Ricerca e la Sperimentazione in Agricoltura (CRA), Centro di Ricerca per gli Alimenti e la Nutrizione (Research Center on Food and Nutrition, CRA-NUT), Rome, Italy

Julie Seeboth and Isabelle P. Oswald
INRA, UMR 1331 Toxalim, Research Center in Food Toxicology, Toulouse, France
University of Toulouse, National Polytechnic Institute of Toulouse (INP), UMR 1331 Toxalim, Toulouse, France

Cristian Botta and Luca Cocolin
Department of Forestry, Agriculture and Food Sciences, University of Torino, Torino, Italy

Tomaz Langerholc and Avrelija Cencič
Department of Microbiology, Biochemistry, Molecular Biology and Biotechnology, Faculty of Agriculture and Life Sciences, University of Maribor, Maribor, Slovenia

Elaine Galicia-Connolly and Denise Adams
CARE Program, Department of Pediatrics, University of Alberta, Edmonton, Alberta, Canada

Justin Bateman
Faculty of Medicine & Dentistry, University of Alberta, Edmonton, Alberta, Canada

Simon Dagenais
Palladian Health, West Seneca, New York, United States of America

Tammy Clifford
Departments of Pediatrics and of Epidemiology & Community Medicine, University of
Ottawa, Ottawa, Ontario, Canada; Canadian Agency for Drugs and Technologies in Health, Ottawa, Ontario, Canada
Department of Pediatrics, Faculty of Medicine and Dentistry, University of Alberta, Edmonton, Alberta, Canada

Lola Baydala
Department of Pediatrics, Faculty of Medicine and Dentistry, University of Alberta, Edmonton, Alberta, Canada

W. James King
Division of Pediatric Medicine, Department of Pediatrics, University of Ottawa, Ottawa, Ontario, Canada; Children's Hospital of Eastern Ontario, Ottawa, Ontario, Canada

Sunita Vohra
CARE Program, PedCAM Network, Department of Pediatrics, Faculty of Medicine & Dentistry and School of Public Health, University of Alberta, Edmonton, Alberta, Canada

Claes Ohlsson, Sara H. Windahl, Helen H. Farman, Sofia Movérare-Skrtic and Klara Sjögren
Centre for Bone and Arthritis Research, Institute of Medicine, Sahlgrenska Academy at University of Gothenburg, Gothenburg, Sweden

Cecilia Engdahl, Ulrika Islander and Annica Andersson
Centre for Bone and Arthritis Research, Institute of Medicine, Sahlgrenska Academy at University of Gothenburg, Gothenburg, Sweden
Department of Rheumatology and Inflammation Research, Institute of Medicine, Sahlgrenska Academy at University of Gothenburg, Gothenburg, Sweden

Frida Fåk
Applied Nutrition and Food Chemistry, Department of Food Technology, Engineering and Nutrition, Lund University, Lund, Sweden

Bin Zheng, Mary E. Morgan, Aletta D. Kraneveld, Saskia Overbeek, Hendrik J. G. van de Kant and Gert Folkerts
Division of Pharmacology, Utrecht Institute for Pharmaceutical Science, Faculty of Science, Utrecht University, Utrecht, The Netherlands

Jeroen van Bergenhenegouwen and Johan Garssen
Division of Pharmacology, Utrecht Institute for Pharmaceutical Science, Faculty of Science, Utrecht University, Utrecht, The Netherlands
Nutricia Research, Utrecht, The Netherlands

Paul Vos
Nutricia Research, Utrecht, The Netherlands

Giacomo Ross i and Angela Palumbo Piccionello
School of Veterinary Medical Sciences, University of Camerino, Camerino, Italy

Graziano Pengo
Clinic "St. Antonio", Cremona, Italy

Marco Caldin
San Marco Laboratories, Padova, Italy

Jörg M. Steiner and Jan S. Suchodolski
Gastrointestinal Laboratory, Department of Small Animal Clinical Sciences, College of Veterinary Medicine and Biomedical Sciences, Texas A&M University, College Station, Texas, United States of America

Noah D. Cohen
Department of Large Animal Clinical Sciences, College of Veterinary Medicine and Biomedical Sciences, Texas A&M University, College Station, Texas, United States of America

Albert E. Jergens
Department of Veterinary Clinical Sciences, College of Veterinary Medicine, Iowa State University, Ames, Iowa, United States of America

Mangala A. Nadkarni and Neil Hunter
Institute of Dental Research, Westmead Centre for Oral Health and Westmead Millennium Institute, Westmead, New South Wales, Australia
Faculty of Dentistry, The University of Sydney, Sydney, New South Wales, Australia

Zhiliang Chen and Marc R. Wilkins
Systems Biology Initiative, School of Biotechnology and Biomolecular Sciences, The University of New South Wales, Sydney, New South Wales, Australia

Jordan E. Bisanz and Amy McMillan
Canadian Centre for Human Microbiome and Probiotic Research, Lawson Health Research Institute, London, Canada
Microbiology and Immunology, The University of Western Ontario, London, Canada

Shannon Seney
Canadian Centre for Human Microbiome and Probiotic Research, Lawson Health Research Institute, London, Canada

Rebecca Vongsa, David Koenig, LungFai Wong and Barbara Dvoracek
Kimberly Clark Corporation, Corporate Research and Engineering-Microbial Control, Neenah, Wisconsin, United States of America

Gregory B. Gloor
Canadian Centre for Human Microbiome and Probiotic Research, Lawson Health Research Institute, London, Canada
Biochemistry, The University of Western Ontario, London, Canada

Mark Sumarah
Agriculture and Agri-Food Canada, London, Canada

Brenda Ford6, Dorli Herman
Springbank Medical Clinic, London, Canada

Jeremy P. Burton
Canadian Centre for Human Microbiome and Probiotic Research, Lawson Health Research Institute, London, Canada
Microbiology and Immunology, The University of Western Ontario, London, Canada
Surgery, The University of Western Ontario, London, Canada
Division of Urology, The University of Western Ontario, London, Canada

Gregor Reid
Canadian Centre for Human Microbiome and Probiotic Research, Lawson Health Research Institute, London, Canada
Microbiology and Immunology, The University of Western Ontario, London, Canada
Surgery, The University of Western Ontario, London, Canada

Douglas C. Woodhams
Institute of Evolutionary Biology and Environmental Studies, University of Zurich, Zurich, Switzerland
Department of Ecology and Evolutionary Biology, University of Colorado, Boulder, Colorado, United States of America

Hannelore Brandt, Simone Baumgartner, Christian Bel, Sandro Hodel and Leyla R. Davis
Institute of Evolutionary Biology and Environmental Studies, University of Zurich, Zurich, Switzerland

Eliane Küpfer
Institute of Evolutionary Biology and Environmental Studies, University of Zurich, Zurich, Switzerland
Department of Evolutionary Biology, Technical University of Braunschweig, Braunschweig, Germany

Ursina Tobler and Benedikt R. Schmidt
Institute of Evolutionary Biology and Environmental Studies, University of Zurich, Zurich, Switzerland KARCH, Neuchâtel, Switzerland

Rob Knight
Howard Hughes Medical Institute and Department of Chemistry and Biochemistry, BioFrontiers Institute, University of Colorado, Boulder, Colorado, United States of America

Valerie McKenzie
Department of Ecology and Evolutionary Biology, University of Colorado, Boulder, Colorado, United States of America

Jos Kielgast
Section for Freshwater Biology, Department of Biology, University of Copenhagen, Copenhagen, Denmark

M. Tanweer Khan, Jan Maarten van Dijl and Hermie J. M. Harmsen
Department of Medical Microbiology, University of Groningen, University Medical Center Groningen, Groningen, The Netherlands

Ida M. Smith
Health & Nutrition Division Discovery, Chr. Hansen A/S, Hørsholm, Denmark
Department of Food Science, University of Copenhagen, Frederiksberg, Denmark

Jeffrey E. Christensen
Health & Nutrition Division Discovery, Chr. Hansen A/S, Hørsholm, Denmark

Nils Arneborg and Lene Jespersen
Department of Food Science, University of Copenhagen, Frederiksberg, Denmark

Eleonora Distrutti
S.C. di Gastroenterologia ed Epatologia, Azienda Ospedaliera di Perugia, Perugia, Italy

Sabrina Cipriani, Andrea Mencarelli, Barbara Renga and Stefano Fiorucci
Dipartimento di Medicina Clinica e Sperimentale, Università degli Studi di Perugia, Perugia, Italy

Lu Mei
Department of Microecology, School of Basic Medical Science, Dalian Medical University, Dalian, Liaoning, China

Department of Gastroenterology, the Second Affiliated Hospital of Zhengzhou University, Zhengzhou, Henan, China

Ming Li, Dianbin Yang, Ao Xie and Jieli Yuan
Department of Microecology, School of Basic Medical Science, Dalian Medical University, Dalian, Liaoning, China

Lin Yuan
Faculty of Agricultural, Life and Environmental Sciences, University of Alberta, Edmonton, Alberta, Canada

Eleonora Distrutti
S.C. di Gastroenterologia ed Epatologia, Azienda Ospedaliera di Perugia, Perugia, Italy

Julie-Ann O'Reilly, Claire McDonald and Marina A. Lynch
Trinity College Institute for Neuroscience, Department of Physiology, Trinity College, Dublin, Ireland

Sabrina Cipriani
Dipartimento di Medicina, Università degli Studi di Perugia, Perugia, Italy

Barbara Renga and Stefano Fiorucci
Dipartimento di Scienze Chirurgiche e Biomediche, Università degli Studi di Perugia, Perugia, Italy

Keita Nishiyama, Yuji Yamamoto and Takao Mukai
Department of Animal Science, School of Veterinary Medicine, Kitasato University, Towada, Aomori, Japan

Yasuyuki Seto
Milk Science Research Institute, Megmilk Snow Brand Co. Ltd., Kawagoe, Saitama, Japan

Kazuki Yoshioka, Tsutomu Kakuda and Shinji Takai
Faculty of Veterinary Medicine, School of Veterinary Medicine, Kitasato University, Towada, Aomori, Japan

Yini Dang, Xiaoying Zhou and Guoxin Zhang
Department of Gastroenterology, The First Affiliated Hospital of Nanjing Medical University, Nanjing, China, and First Clinical Medical College of Nanjing Medical University, Nanjing, China

Jan D. Reinhardt
Institute for Disaster Management and Reconstruction, Sichuan University, Chengdu, China, and Hong Kong Polytechnical University, Hung Hom, Hong Kong, Chin
Department of Health Sciences, University of Lucerne, Lucerne, Switzerland
Swiss Paraplegic Research, Nottwil, Switzerland

Fumihiko Sakai, Tomohiro Hosoya, Aiko Ono-Ohmachi, Ken Ukibe, Akihiro Ogawa Tomohiro Moriya and Yukio Kadooka
Milk Science Research Institute, Megmilk Snow Brand Co. Ltd., Minamidai, Kawagoe, Saitama, Japan

Takuya Shiozaki, Hisako Nakagawa, Yosuke Nakayama and Tadaaki Miyazaki
Department of Probiotics Immunology, Institute for Genetic Medicine, Hokkaido University, Kita-ku, Sapporo, Japan

Index